Holomorphic Functions in the Plane and n-dimensional Space

Klaus Gürlebeck
Klaus Habetha
Wolfgang Sprößig

Birkhäuser
Basel · Boston · Berlin

Klaus Gürlebeck
Institut für Angewandte Mathematik
Bauhaus-Universität Weimar
Coudraystr. 13
99423 Weimar
Germany
e-mail: klaus.guerlebeck@bauing.uni-weimar.de

Klaus Habetha
Lehrstuhl II für Mathematik
RWTH Aachen
Templergraben 55
52062 Aachen
Germany
e-mail: habetha@math2.rwth-aachen.de

Wolfgang Sprößig
Institut für Angewandte Analysis
Fakultät für Mathematik
TU Bergakademie Freiberg
Prüferstr. 9
09596 Freiberg
Germany
e-mail: sproessig@math.tu-freiberg.de

2000 Mathematical Subject Classification: 30-01, 30G30

Library of Congress Control Number: 2007933912

Bibliographic information published by Die Deutsche Bibliothek.
Die Deutsche Bibliothek lists this publication in the Deutsche Nationalbibliografie;
detailed bibliographic data is available in the Internet at http://dnb.ddb.de

ISBN 978-3-7643-8271-1 Birkhäuser Verlag AG, Basel - Boston - Berlin

© 2008 Birkhäuser Verlag AG
Basel · Boston · Berlin
P.O. Box 133, CH-4010 Basel, Switzerland
Part of Springer Science+Business Media
Printed on acid-free paper produced from chlorine-free pulp. TCF ∞

ISBN 978-3-7643-8271-1

ISBN 978-3-7643-8272-8 (eBook)

9 8 7 6 5 4 3 2 1

www.birkhauser.ch

Contents

Preface to the German Edition

Complex analysis has produced a large number of deep recumbent and aesthetic results in its more than 200 year-old history. In the classical context complex analysis is the theory of complex differentiable functions of a complex variable, or also the theory of *holomorphic functions*. These are the solutions of a (2×2)-system of partial differential equations, which usually are called *Cauchy–Riemann differential equations (CRD)*.

Indeed, the algebra of the real quaternions of W.R. Hamilton has been available since 1843 and the real Clifford algebra of W.K. Clifford since 1878, but until the 1930s the prevailing view was that complex analysis is a purely two-dimensional theory. Only the group around the Swiss mathematician R. Fueter and the Romanian mathematicians G.C. Moisil and N. Teodorescu around 1930 started to develop a hypercomplex analysis in the algebra of real quaternions and in real Clifford algebras. In the late 1960s a group of Belgian mathematicians around R. Delanghe in Ghent created a rich higher dimensional analogy to complex analysis, called Clifford analysis. Since 1990 the number of relevant articles and books has increased significantly. Today intensive research is going on in Clifford analysis to which the more than 9000 entries in our database on the relevant literature testify. The database can be found on the CD attached to our book.

The purpose of the present textbook is to collect the essentials of classical complex analysis and to present its higher dimensional generalizations at a level suitable for university instruction. The typical users we have in mind are, first of all, students of mathematics and physics, but also students of any discipline requiring some sophisticated mathematics from the second year onward. The material to be covered is extensive, and we have attempted to make it as self-sufficient as possible within the limits of a modest size book. We have covered not only analytical and geometrical aspects but numerical procedures as well. Historical references outline the development of the field and present some of the personal characteristics of the most important personalities who have contributed to that history.

In the first chapter complex numbers, quaternions, and the Clifford numbers are introduced. We have emphasized the parallelism of our presentation. Quaternions and Clifford numbers take up more space than complex numbers. Besides the algebraic and geometrical properties we treat in particular also rotations and representations.

In Section 4 we illustrate the topological and analytical basic facts for the treatment of functions up to Riemann spheres. This section is deliberately kept short in view of its relationship to classical analysis. Section 5 treats some of the possible definitions of holomorphic functions. We keep this name also in higher dimensions, because the definitions are almost independent of dimension. The standard literature uses here mostly the concept of Weierstrass monogenic functions. However, it seems to us at least debatable whether this best describes the meaning of the definition (cf. end of Section 5). Also, the notion of holomorphic functions fits conceptually better that of meromorphic functions. Since the articles by H. Malonek the concept of holomorphic functions can be introduced also via local approximation by suitable linear functions, so even in that context the analogy holds in all dimensions. Section 6 is devoted to "simple" functions, namely powers and Möbius transforms. The polynomials named after R. Fueter are suitable as power functions in higher dimensions since they have many nice qualities. Unfortunately, the reduction of Fueter polynomials to the planar case leads to powers $(-iz)^n$ and not to z^n; however, the parallelism is still given. In particular L.V. Ahlfors has studied Möbius transforms in higher dimensions. Here too the clear comparability of all dimensions can be recognized.

We have put together the necessary aids for integration in Appendix 2, and a short introduction to the theory of differential forms in Appendix 1. We believe that this can be helpful, because in lectures to beginners these areas are often treated only very briefly, if at all. Indeed, we do not include the proof of Stokes' integral theorem as this would lead too far away from the subject. Then in Chapter III Cauchy' s integral theorem and the Borel–Pompeiu formula are easy consequences of Stokes' theorem. However, we also consider the boundary value formulae of Plemelj–Sokhotzki. Conclusions on Cauchy's integral formula follow. Moreover, the Teodorescu transform is examined and the Hodge decomposition of quaternionic Hilbert space is treated. The needed functional spaces are briefly introduced in Appendix 3.

Chapter IV is devoted to different areas of hypercomplex analysis. We firstly treat Taylor and Laurent series. The effort is clearly larger in higher dimensions than in the plane, but the similarity helps. Unfortunately, the Taylor series in dimensions greater than 2 are not orthogonal expansions. For quaternions orthogonal expansions are introduced, which are especially adequate for numerical purposes.

The elementary functions in the plane have no special difficulties. They are given more or less canonically. For all generalizations to higher dimensions, a royal way does not exist symptomatically. Different generalizations of the exponential function are pointed out, and one generalization given by the method of separation of variables is developed. This exponential function has some nice properties and is an appropriate kernel for the Fourier transform of quaternion-valued functions.

Section 12, which explores the local structure of holomorphic functions, shows that in higher dimensions this is still an active field of research. The pleasant qualities of zeros and isolated singularities in the plane at first sight get lost in space. There is still no suitable structure in which to understand all the relevant phenomena. At least the residue theorem can be transferred, and also first attempts for an argument principle were found.

Section 13 deals with special functions. The Gamma function and the Riemann Zeta function are treated, followed by considerations about automorphic functions and forms in $C\ell(n)$ which offer an insight into the latest research in this field.

Problems at the end of every section should help the reader toward a better understanding of the corresponding area. The use of the skewfield structure of the real quaternions allows one to formulate some statements more precisely and in a more readable form than in general Clifford algebras. Since applications in $\mathbb{R}^3$ and $\mathbb{R}^4$ are of special interest, we have sometimes waived the more general case of real Clifford algebras $C\ell(n)$.

Results of many colleagues working in the area of Clifford analysis are used in the presentation of higher dimensional results. We thank especially our colleagues Professor Krüger (Kaiserslautern/Germany), Professor Malonek (Aveiro/Portugal), and Professor Kraußhar (Leuven/Belgium), who helped us to write some of the sections. We discussed details with Professor Sommen (Ghent/Belgium) and Professor Shapiro (Mexico-City/Mexico). We thank Professor Jank (Aachen/Germany) and the editor Dr. Hempfling (Birkhäuser) for improving the typescript. Critical remarks by the referees of a first version of the book were of great value for us. M. Sprößig and T. Lahmer helped us by very carefully
long work for this book.

Weimar, Aachen and Freiberg, August 2005

Klaus Gürlebeck, Klaus Habetha and Wolfgang Sprößig

Preface to the English Edition

We thank the publisher for the opportunity to translate this book into English. Of course we have corrected all mistakes we found in the German edition, other changes have been made only rarely.

We thank in particular Professor E. Venturino (Torino/Italy) who translated one chapter of the book, but also improved the translation of the rest of the book. Some sections have been translated by M. Schneider and A. Schlichting who are students in Freiberg/Germany.

Weimar, Aachen, Freiberg, May 2007

Klaus Gürlebeck, Klaus Habetha, Wolfgang Sprößig

Chapter I

Numbers

1 Complex numbers

1.1 The History of Their Discovery

The European Renaissance as rebirth of humanistic thought began in the middle of the fourteenth century in Italy and had its maximal development in the sixteenth century, in which with some delay the ideas of renewal of natural sciences and mathematics were accepted. In particular, algebra experienced a push forward, but also important results came from mechanics, astronomy, trigonometry and geometry. A few learned men began to build on the mathematical results of the ancient scientists. Thus the works of ARCHIMEDES, APOLLONIUS, EUCLID and HERON OF ALEXANDRIA became of remarkable interest. It was discovered that in the year 50 A.D. already HERON in his book *Stereometria* had described roots of negative numbers.

The main influence on the algebraically oriented mathematics of this time came from the Arabic world. Rules for finding roots of negative numbers were described for the first time in 499 in the principal work *Āryabhaṭyīa* of ĀRYABHAṬA THE OLDER. Through the method of "completion of the square", the Babylonians were able to solve quadratic equations with positive coefficients.

Girolamo Cardano

The algebraists of the Renaissance had two principal tasks: to extend the number system in order to understand the meaning of negative and complex numbers, and to develop an efficient mathematical symbolism.

With the exception of DIOPHANTUS OF ALEXANDRIA, the ancient Greek and Arabic mathematicians were accustomed to describing mathematical situations by rhetorical means. With his principal work, printed in 1545 in Nürnberg *Artis magnae sive de regulis algebraicis* or simply *Ars magna*, GIROLAMO CARDANO (1501–1576) established the foundations of modern mathematics. In his book he not only showed how to solve cubic equations, but using roots of negative numbers he gave also the formulas for solving algebraic equations of fourth order. CARDANO himself is seen as personification of the Renaissance. In a singular way he was able to bridge the Middle Age and modern perspectives. On the one hand he was an adept and active practitioner of occultism and of natural magic. He authored general treatises on palmistry and the meaning of dreams, and he wrote about spirits, angels, and deamons. On the other hand, his researches are completely free from mystical and supernatural influxes. The solution methods that he made public were based on firmly established results of his predecessors.

These methods date back to SCIPIONE DAL FERRO (1465–1526) and NICCOLÒ FONTANA ("TARTAGLIA") (1500–1557). Although he swore to TARTAGLIA, who gave him the "secret knowledge" for solving algebraic equations of third and fourth order, not to publish these new methods without TARTAGLIA's consent, CARDANO considered it his duty to make them known to the mathematical world of that time, indicating their discoverer. The story of the *Ars magna* is indeed a fascinating historical episode.

Niccolò Fontana

A new stage began with RENÉ DESCARTES (1596–1650). He introduced in 1637 the terms *real* and *imaginary* and with his works in general tried to popularize these new concepts. ALBERT GIRARD (1595–1632) gave geometrical interpretations of roots of negative numbers. Also references to imaginary numbers are found in the writings of the German universal scientist GOTTFRIED WILHELM LEIBNIZ (1646–1716). He writes in 1675 that an imaginary number is "a wonderful creature of an ideal world, of amphibic nature between the things which are and those which are not." Later, general contributions were made by JOHN WALLIS (1616–1703), ROGER COTES (1682–1716) and above all by LEONHARD EULER (1707–1783), who in 1777 introduced also the symbol $i = \sqrt{-1}$, shortly after having discovered the relation

$$e^{i\varphi} = \cos \varphi + i \sin \varphi$$

and from it the astonishing result:

$$e^{i\pi} + 1 = 0.$$

The concept *complex number* was introduced only in 1832 by CARL FRIEDRICH GAUSS. Since then, for a complex number z the notation $z = x + iy$ became standard. The introduction of complex numbers as pairs of real numbers dates back to WILLIAM ROWAN HAMILTON (1837).

1.2 Definition and Properties

The complex numbers can be defined in several different ways: we try to remain consistent through the following chapters and begin by defining an extended field of the real numbers $\mathbb{R}$. We have then to establish a set and corresponding relationships.

Definition 1.1 (Field of the complex numbers). Let the set $\mathbb{C} := \{z : z = (x,y),\ x, y \in \mathbb{R}\}$ be the set of ordered pairs of real numbers. The numbers x and y are called the *coordinates* of z.

Componentwise *Addition* is defined as

$$z_1 + z_2 := (x_1, y_1) + (x_2, y_2) := (x_1 + x_2, y_1 + y_2)$$

for

$$z_i = (x_i, y_i) \in \mathbb{C},$$

and multiplication with a real number a as

$$a(x, y) := (ax, ay).$$

Multiplication is defined as the linear continuation of the multiplication of the basis elements $\mathbf{1}$ and i, where $\mathbf{1} := (1, 0)$ represents the multiplicative unit element and $i := (0, 1)$ satisfies the rule $i^2 = -1$.

Some remarks: multiplication with real numbers allows the usual representation for the *complex numbers*:

$$z = (x, y) = 1x + iy = x + iy.$$

Thus in place of $x + i0$ we can simply write x, and in place of $0 + iy$ simply iy, in particular in place of $\mathbf{1}$ only 1 (or nothing at all). The continuation of multiplication of the basis elements is carried out through formal expansion of the product

$$z_1 z_2 = (x_1 + iy_1)(x_2 + iy_2) = (x_1 x_2 - y_1 y_2) + i(x_1 y_2 + y_1 x_2),$$

where we apply $i^2 = -1$.

The pair $(0, 0)$ is the neutral element of addition. To prove that addition and multiplication in $\mathbb{C}$ are associative and distributive is a somewhat tiresome exercise, while commutativity follows immediately from the definition. All nonzero elements have multiplicative inverses. Thus $\mathbb{C}$ is a field, the *field of complex numbers*.

If we identify the complex numbers $x + i0$ with the real numbers, then addition and multiplication in $\mathbb{C}$ correspond to those in $\mathbb{R}$, so that $\mathbb{C}$ is an extension field of $\mathbb{R}$. Also the above defined multiplication by a real number, which leads to a vector space structure, is encompassed by the multiplication in $\mathbb{C}$.

The complex numbers $x + iy$ clearly correspond in a one-to-one manner to the vectors $\binom{x}{y}$ of the vector space $\mathbb{R}^2$, however vectors cannot be multiplied as complex numbers, so that the structure of $\mathbb{C}$ is richer.

In higher dimensions the basis elements are frequently denoted by e_0, e_1 and so on: in the plane this is not yet necessary, as here the use of summation signs does not introduce any simplification.

Definition 1.2. $x =: \operatorname{Re} z$ is called the *real part* and $y := \operatorname{Im} z$ the *imaginary part* of the complex number $z = (x, y)$. The number $\bar{z} := x - iy$ is called the *complex conjugate* of z. The expression $|z| := \sqrt{z\bar{z}} = \sqrt{x^2 + y^2}$ will be denoted as *modulus* or *absolute value* of z.

1. Complex numbers

The modulus of a complex number is obviously equal to the Euclidean norm of the corresponding two-dimensional vector.

Proposition 1.3. *Let $z, z_1, z_2 \in \mathbb{C}$ be given. Then the following relations hold:*

(i) $\operatorname{Re} z = \frac{z + \bar{z}}{2}$,

(ii) $\operatorname{Im} z = \frac{z - \bar{z}}{2i}$,

(iii) $\frac{1}{z} = \frac{\bar{z}}{|z|^2}$, $z \neq 0$,

(iv) $\overline{z_1 + z_2} = \overline{z_1} + \overline{z_2}$,

(v) $\overline{z_1 z_2} = \overline{z_1}\, \overline{z_2}$,

(vi) $\overline{\overline{z}} = z$,

(vii) $|z_1 z_2| = |z_1||z_2|$, *in particular for all $n \in \mathbb{Z}$*

(viii) $|z^n| = |z|^n$,

(ix) $|\bar{z}| = |-z| = |z|$.

Proof. The relations (i)–(vi) and (ix) require only simple calculations. For (vii) write $|z_1 z_2|^2 = \overline{z_1 z_2} z_1 z_2 = \overline{z_1} z_1 \overline{z_2} z_2 = |z_1|^2 |z_2|^2$. $\qquad\square$

This proposition shows that conjugation is exchangeable with addition and multiplication in $\mathbb{C}$: (iv), (v) and (vi) mean that conjugation is an *involution*. Thus in physics and operator theory frequently the notation z^* is used in place of $\bar{z}$. The reader should convince him/herself that conjugation is the only automorphism in the algebra $\mathbb{C}$ apart from the identity. An example follows:

Example 1.4. If we want to split a fraction of complex numbers in its real and imaginary parts, we have to expand it with the complex conjugate part of the denominator as follows:

$$\frac{i+3}{2i-4} = \left(\frac{i+3}{2i-4}\right)\left(\frac{-2i-4}{-2i-4}\right) = \frac{-10-10i}{20} = -\frac{1}{2} - \frac{i}{2}.$$

Remark 1.5. Proposition 1.3 contains the following statements: a sum of two squares can be written as the product of linear expressions, obviously using the complex unit i:

$$x^2 + y^2 = (x + iy)(x - iy).$$

Moreover, the *two-squares-theorem* holds:

$$(x_1^2 + y_1^2)(x_2^2 + y_2^2) = (x_1 x_2 - y_1 y_2)^2 + (x_1 y_2 + y_1 x_2)^2,$$

which says that a product of two sums of two squares can be written again as the sum of squares. In higher dimensions an extensive theory has been constructed around this theorem (see [66]).

Now a few inequalities concerning the modulus of a complex number follow:

Proposition 1.6. *For the modulus of a complex number we have*

(i) $|\operatorname{Re} z| \leq |z|$, $\quad |\operatorname{Im} z| \leq |z|$,

(ii) $|z_1 + z_2| \leq |z_1| + |z_2|$ *(triangular inequality)*,

(iii) $||z_1| - |z_2|| \leq |z_1 - z_2|$.

With the so-called *Euclidean distance* $|z_1 - z_2|$ of two complex numbers, $\mathbb{C}$ becomes a metric space. In Section 4.1 we will investigate this more closely.

Proof. We show only the triangular inequality and (iii) with the help of Proposition 1.3 and (i):

$$
\begin{aligned}
|z_1 + z_2|^2 &= (z_1 + z_2)(\overline{z_1 + z_2}) = |z_1|^2 + |z_2|^2 + 2\mathrm{Re}(z_2\overline{z_1}) \\
&\leq |z_1|^2 + |z_2|^2 + 2|z_2||\overline{z_1}| = (|z_1| + |z_2|)^2.
\end{aligned}
$$

We obtain (iii) from (ii), since $|z_1| = |z_2 + z_1 - z_2| \leq |z_2| + |z_1 - z_2|$, also $|z_1| - |z_2| \leq |z_1 - z_2|$. Now we can interchange z_1 and z_2, one of the two left sides equals $||z_1| - |z_2||$, so that (iii) follows.

$\square$

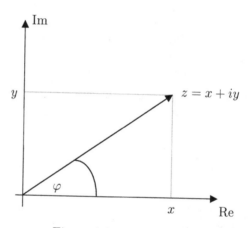

Figure 1.1

Complex numbers can be represented as points in the so-called GAUSS or ARGAND plane, also Gaussian plane, with rectangular coordinates or with polar coordinates (see Figure 1.1). Every complex number $z = x + iy$ with $r := |z|$ can be described as follows:

$$
\begin{aligned}
z &= r\,\frac{z}{r} = r\left(\frac{x}{r} + i\frac{y}{r}\right) = r\left(\frac{x}{\sqrt{x^2 + y^2}} + i\frac{y}{\sqrt{x^2 + y^2}}\right) \\
&= r(\cos\varphi + i\sin\varphi).
\end{aligned}
$$

The last representation is called the *trigonometric form* or *polar form* of the complex number z. As in $\mathbb{R}^2$, r and φ are called the *polar coordinates* of z. Thus r gives the distance of the point z from the origin of the Gaussian plane and φ describes the angle between the positive real axis and the segment joining 0 and z. The coordinate φ is called the *argument* of z and is denoted by arg z. For $-\pi < \varphi \leq \pi$, we speak of the *principal value* of the argument. In general we use in calculations the principal value. But fundamentally the argument is determined up to integer multiples of 2π.

The formulae relating Euclidean and polar coordinates are, for $x \neq 0$:

$$x = r \cos \varphi, \ y = r \sin \varphi, \ r = \sqrt{x^2 + y^2}, \ \varphi = \arctan \frac{y}{x}.$$

If φ lies in the second quadrant, one has to add π, while if it lies in the third quadrant, one has to subtract π to obtain the principal value of the argument. We assume that the arctangent has values between $-\pi/2$ and $\pi/2$ as its argument moves from $-\infty$ to ∞.

In 1799 CASPAR WESSEL represented complex numbers by geometric figures in the plane. However little notice of his work was taken. Independently, the accountant JEAN ROBERT ARGAND found in 1806 a geometric interpretation of complex numbers. Finally it was C. F. GAUSS who represented complex numbers by means of arrows in the plane, a notation that remained in constant use thereafter.

Due to simplicity of use, complex numbers in polar coordinates are particularly suited for multiplication and division (Figure 1.2).

Proposition 1.7. *Complex numbers are multiplied by multiplying their moduli and adding their arguments. They are divided by dividing the moduli and subtracting the arguments.*

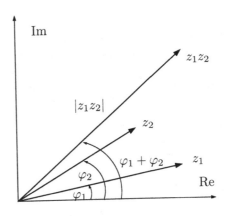

Figure 1.2

Proof. Let two complex numbers be given in trigonometric form $z_1 = r_1(\cos\varphi_1 + i\sin\varphi_1)$ and $z_2 = r_2(\cos\varphi_2 + i\sin\varphi_2)$. By expanding the multiplication we have

$$
\begin{aligned}
z_1 z_2 &= r_1 r_2 [(\cos\varphi_1 \cos\varphi_2 - \sin\varphi_1 \sin\varphi_2) + i(\cos\varphi_1 \sin\varphi_2 + \sin\varphi_1 \cos\varphi_2)] \\
&= r_1 r_2 [\cos(\varphi_1 + \varphi_2) + i\sin(\varphi_1 + \varphi_2)].
\end{aligned}
$$

For the division we obtain

$$
\frac{z_1}{z_2} = \frac{r_1}{r_2}\left(\cos(\varphi_1 - \varphi_2) + i\sin(\varphi_1 - \varphi_2)\right). \qquad \square
$$

An important consequence is the formula of year 1707 named after ABRAHAM DE MOIVRE (1667–1754)

Corollary 1.8 (De Moivre formula). *For all $n \in \mathbb{Z}$ with $z = r(\cos\varphi + i\sin\varphi)$ we have*

$$
z^n = |z|^n(\cos n\varphi + i\sin n\varphi),
$$

where any argument φ of z can be used.

Proof. The proof for positive $n \in \mathbb{N}$ follows by a simple mathematical induction. For $n = 0$ on both sides of the formula we have 1 and for negative n one has to change the sign of the formula for positive n, paying attention to the fact that $(\cos n\varphi + i\sin n\varphi) \cdot (\cos n\varphi - i\sin n\varphi) = 1$.
$\square$

The de Moivre formula allows the root extraction of every complex number:

Proposition 1.9 (Rootfinding). *Let the complex number $a \neq 0$ have the trigonometric representation*

$$
a = |a|(\cos\psi + i\sin\psi).
$$

The number a possesses for each natural n exactly n different n-th roots in $\mathbb{C}$, i.e., the solutions of the equation $z^n = a$. We can calculate them by the formula

$$
z_k = \sqrt[n]{|a|}\left(\cos\frac{\psi + 2k\pi}{n} + i\sin\frac{\psi + 2k\pi}{n}\right)
$$

for $k = 0, \ldots, n-1$.

From this formula it follows that all z_k are points lying on the circle of radius $\sqrt[n]{|a|}$, and neighboring points are distinguished by the difference of $\gamma = 2\pi/n$ in their argument. By imposing $k = 0, 1, \ldots, n-1$ in the above formula, we obtain all these points from the periodicity of trigonometric functions.

Example 1.10. Let $z^3 = 1$. The right-hand side has then the representation $1 = 1(\cos 0 + i\sin 0)$. It follows (Figure 1.3)

$$
\begin{aligned}
z_1 &= \cos\frac{0}{3} + i\sin\frac{0}{3} = 1, \\
z_2 &= \cos\frac{2\pi}{3} + i\sin\frac{2\pi}{3} \quad \text{(in the second quadrant)}, \\
z_3 &= \cos\frac{4\pi}{3} + i\sin\frac{4\pi}{3} \quad \text{(in the third quadrant)}.
\end{aligned}
$$

Remark 1.11. The de Moivre formula shows that the partitioning of the circle, i.e., inscribing a regular n-polygon in the unit circle $\{|z| = 1\}$ can be solved by solving the equation

$$z^n = 1,$$

i.e., by

$$z_k = \cos\frac{2\pi i k}{n} + i\sin\frac{2\pi i k}{n} \quad (0 \le k \le n-1).$$

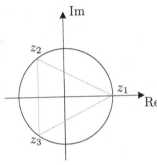

Figure 1.3

Partitioning of the circle can be viewed as a special case of the solution of the general algebraic equation of n-th degree:

$$P_n(z) = a_n z^n + a_{n-1} z^{n-1} + \cdots + a_1 z + a_0 = 0, \; a_n \ne 0.$$

For this we have the *fundamental theorem of algebra*:

Theorem 1.12 (Fundamental theorem of algebra). *The n-th degree polynomial $P_n(z)$ possesses exactly n zeros, provided we count each one according to its multiplicity, i.e., $P_n(z)$ can be written in the form*

$$P_n(z) = a_n (z - z_1)^{n_1} (z - z_2)^{n_2} \cdots (z - z_p)^{n_p}$$

with $n_1 + n_2 + \cdots + n_p = n$ and $a_n \ne 0$. If the complex numbers z_j are pairwise distinct, then n_j indicates the multiplicity of the zero z_j.

We will prove this theorem later, when the necessary preliminary results and tools will be ready for use.

The Flemish mathematician ALBERT GIRARD in 1629 expressed for the first time the fact that every algebraic equation of n-th order has n roots, which are to be sought in a larger domain than $\mathbb{R}$. The first serious attempt to prove this statement was undertaken in 1746 by JEAN-BAPTISTE LE ROND D'ALEMBERT.

For polynomials with real coefficients there was in 1749 a proof by L. EULER, based on an idea of "Ars Magna". Under the assumption that the solutions exist, in 1795 PIERRE SIMON LAPLACE was able to provide an elegant proof. In C. F. GAUSS' doctoral thesis the first complete proof of the fundamental theorem was given. However probably the simplest proof, based on the idea of D'ALEMBERT, goes back to J. R. ARGAND and was published in 1814. A constructive proof was finally given only in 1940 by HELLMUTH KNESER.

1.3 Representations and geometric aspects

The multiplication of a complex number with $\cos\varphi + i\sin\varphi$ describes a *rotation* of the Gaussian plane of an angle φ, since the argument of every complex number z would be augmented by the angle φ. All these rotations form a group, the *special orthogonal group* $SO(2)$, where the number 2 denotes the dimension of the underlying $\mathbb{R}$-vector space $\mathbb{C}$. Clearly 1 is the neutral element of this group, the inverse element of $\cos\varphi + i\sin\varphi$ being given by $\cos\varphi - i\sin\varphi$.

Since the rotations in the plane can be represented also by multiplication of the vector $\binom{x}{y}$ by a matrix, one may ask what is the relationship of complex numbers with this multiplication by a matrix. In fact we have

Proposition 1.13. *The mapping* $M : z \to M(z)$ *of* $z = x + iy \in \mathbb{C}$ *on the real* (2×2)-*matrices of the special form*

$$M(z) = \begin{pmatrix} x & -y \\ y & x \end{pmatrix}$$

is a ring isomorphism from $\mathbb{C}$ *onto the subring of* $\mathbb{R}^{2\times2}$-*matrices of the above form.*

Proof. The equations $M(z + z') = M(z) + M(z')$ and $M(zz') = M(z)M(z')$ for two complex numbers z and z' are easily checked. $\qquad\square$

We then have found a so-called *representation* of complex numbers by matrices, thus an isomorphic mapping from $\mathbb{C}$ into the linear mappings of $\mathbb{R}^2$ onto itself. In this way $M^\top$ corresponds to $\bar{z}$, as well as $\det M$ to $|z|^2$ and the trace M to $2\mathrm{Re}\,z$. The rotations considered above correspond to the known planar rotation matrices, i.e.,

$$R(\varphi) := M(\cos\varphi + i\sin\varphi) = \begin{pmatrix} \cos\varphi & -\sin\varphi \\ \sin\varphi & \cos\varphi \end{pmatrix},$$

with $\varphi \in \mathbb{R}$. It is easily shown that

$$R(\varphi + \psi) = R(\varphi)R(\psi) \qquad (\varphi, \psi \in \mathbb{R}).$$

The mapping $R : \varphi \to R(\varphi)$ represents a homomorphism of the additive group $\mathbb{R}$ onto the multiplicative group of the given rotation matrices. Each such matrix, in view of our association, corresponds to a point on the *unit circle* $S^1 = \{z : |z|=1\}$, the boundary of the *unit disk* $\mathbb{D} := \{z : |z| < 1\}$. In general here the multiplicity of the argument plays no role, since the addition of $2k\pi$, $k \in \mathbb{Z}$, to φ does not change the value of $\cos\varphi$ and $\sin\varphi$. In view of $R(\pi) = R(-\pi)$ the elements of the matrix of $R(\varphi)$ remain continuous, if we let the value of φ jump from π on the negative real axis to $-\pi$.

Remark 1.14. In conclusion we hint at the relationship between *rotations* and *reflections*. By conjugation every complex number is reflected about the real axis. In this way we can describe the reflection about an arbitrary straight line through the origin as follows: if the line makes the angle α with the positive real axis, it will be turned by $(\cos\alpha - i\sin\alpha)z$ into the real axis; then reflection is given by conjugation. Finally one has to rotate the line in the opposite direction, in a way that the reflection point z' of z is expressed through

$$z' = (\cos\alpha + i\sin\alpha)^2\overline{z} = (\cos 2\alpha + i\sin 2\alpha)\overline{z}.$$

If we perform another reflection, then we obtain

$$z'' = \big(\cos(2\beta + 2\alpha) + i\sin(2\beta + 2\alpha)\big)z.$$

But this is a rotation about the origin. We can state then:

Two reflections of $\mathbb{C}$ on straight lines through the origin give a rotation of $\mathbb{C}$ about the origin. Conversely we can split any rotation into two reflections.

We are now in the position to state a few facts of planar analytic geometry through complex numbers. A straight line in the plane is described by an equation of the form $ax + by + c = 0$ with real a, b, c, and $a^2 + b^2 > 0$. On using the formulae of Proposition 1.3 we obtain

$$(a - ib)z + (a + ib)\overline{z} + 2c = 0$$

or

$$\overline{N}\,z + N\,\overline{z} + 2c = 0$$

with the complex number $N := a + ib \neq 0$ and a real c. This can also be written in the form $\mathrm{Re}\,(\overline{N}\,z) + c = 0$.

For two vectors $\binom{x_1}{y_1}$ and $\binom{x_2}{y_2}$ of $\mathbb{R}^2$ the scalar product is defined by

$$\begin{pmatrix} x_1 \\ y_1 \end{pmatrix} \cdot \begin{pmatrix} x_2 \\ y_2 \end{pmatrix} := x_1 x_2 + y_1 y_2.$$

For the corresponding complex numbers $z_1 = x_1 + iy_1$ and $z_2 = x_2 + iy_2$ the expression

$$\mathrm{Re}(\overline{z_1} z_2) = x_1 x_2 + y_1 y_2 = \mathrm{Re}(z_1 \overline{z_2})$$

is called the *scalar product*. For the *linear equation*

$$\mathrm{Re}(\overline{N}z) + c = 0,$$

which means that for two points z_1 and z_2 on the straight line we have

$$\mathrm{Re}\left(\overline{N}(z_1 - z_2)\right) = 0.$$

This says that the vector corresponding to N lies perpendicular to the line. We have then introduced the *orthogonality* of complex numbers, which can be illustrated in the Gaussian plane. We can observe easily that the point on the line closest to the origin is given by

$$z_0 = \frac{-cN}{|N|^2}.$$

Finally, we recall that a straight line can also be described by a *parametric representation*. If A denotes a complex number, corresponding to a vector in the direction of the line — thus perpendicular to N — and z_0 denotes a point on the line, then the parametric representation is

$$z(t) = At + z_0, \quad -\infty < t < \infty.$$

One can easily go from one representation of the equation of the straight line to the other ones.

In a way similar to the scalar product, we can also introduce in $\mathbb{C}$ a *vector product* (or *cross product*) by

$$[z_1, z_2] := x_1 y_2 - x_2 y_1 = \mathrm{Im}(\overline{z_1} z_2).$$

It corresponds to the operation $\begin{pmatrix} x_1 \\ y_1 \end{pmatrix} \cdot J \begin{pmatrix} x_2 \\ y_2 \end{pmatrix}$ in $\mathbb{R}^2$, where

$$J = \begin{pmatrix} 0 & 1 \\ -1 & 0 \end{pmatrix}$$

represents the rotation matrix by the angle $-\pi/2$. This gives the formula $[z_1, z_2] = \mathrm{Re}(\overline{z_1}(-i)z_2) = \mathrm{Im}\,\overline{z_1} z_2$ for the vector product. The matrix J shows up also in the curvature theory of planar curves where it is called a CARTAN *matrix*. If $[z_1, z_2] = 0$, then z_1 and z_2 lie on a straight line through the origin, they are then *collinear*. Moreover the relationship holds

$$\overline{z_1} z_2 = z_1 \cdot z_2 + i[z_1, z_2].$$

A further elementary geometrical figure in the plane is the circle, which is given by its *center* z_0 and *radius* R. It contains all points at distance R from z_0,

$$S_R(z_0) := \{z \ : \ |z - z_0| = R\}.$$

1. Complex numbers

The *equation of the circle* can be written in the from

$$(z - z_0)(\overline{z} - \overline{z_0}) = z\overline{z} - 2\text{Re}\, z\overline{z_0} + z_0\overline{z_0} = R^2.$$

The *tangent* to such a circle at one of its points $z_1 \in S_R(z_0)$ is the straight line through z_1 orthogonal to the difference $z_1 - z_0$ (Figure 1.4). Its equation is thus

$$\text{Re}\,(\overline{z_1 - z_0})(z - z_1) = 0 \quad \text{or} \quad \text{Re}\,(\overline{z_1 - z_0})z = \text{Re}\,(\overline{z_1 - z_0})z_1.$$

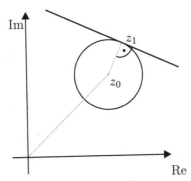

Figure 1.4

A circle also can be described by a *parametric representation*. It is simply

$$z(t) = z_0 + R(\cos t + i \sin t), \quad -\pi < t \leq \pi.$$

Both line and circle equations can be summarized by the formula

$$Az\overline{z} + \text{Re}(\overline{B}z) + C = 0,$$

here A and C represent real numbers, B is an arbitrary complex number with $|B|^2 - AC > 0$. For $A \neq 0$ we obtain a circumference, for $A = 0$ a straight line. $\text{Re}(\overline{B}z)$ is the scalar product of the vectors corresponding to B and z.

1.4 Exercises

1. For arbitrary complex numbers z_1 and z_2 prove the equality *(Apollonius identity)*

$$|z_1 + z_2|^2 + |z_1 - z_2|^2 = 2(|z_1|^2 + |z_2|^2).$$

2. Determine the geometric locus of all points of the complex plane for which the following relationship holds:

$$\text{a)} \;\; \text{Im}\frac{z-1}{z+i} \leq 0, \qquad \text{b)} \;\; |z - 2| - |z + 2| < 2.$$

3. Prove de Moivre's formula and investigate whether it can be extended to rational exponents.

4. Show that the sum of all roots of the equation $z^n = 1$ vanishes. Prove that these roots are the corners of a regular n-polygon.

5. Calculate the roots of the equation $32z^5 = (z + 1)^5$.

6. With the help of complex numbers show

$$\text{a) } \frac{1}{2} + \cos\theta + \cos 2\theta + \cdots + \cos n\theta = \frac{\sin(n + 1/2)\theta}{2\sin\theta/2},$$

$$\text{b) } \sin\theta + \sin 2\theta + \cdots + \sin n\theta = \frac{\sin(n + 1)\theta/2}{\sin\theta/2}\sin(n\theta)/2).$$

7. Prove that the pairwise distinct points z_1, z_2, z_3 all lie on a same straight line, if and only if the expression $(z_3 - z_1)/(z_2 - z_1)$ is real.

8. The mapping $z' = f(z)$ is called *reflection about the unit circle*, when both z and z' lie on the same ray emanating from the origin and $|z||z'| = 1$. Find $f(z)$ explicitly and determine z' geometrically.

2 Quaternions

2.1 The history of their discovery

It is often the case that mathematical results appear at first as the work of several people in obscurity and then in a very short time are suddenly formulated by several scientists. It is difficult to determine who precisely had the decisive idea and should therefore be named the discoverer. Often the opinions on that clearly diverge. Thus W. Blaschke, celebrating the 250th birth anniversary of Euler in his conference talk entitled "Euler und die Geometrie (Euler and the geometry)", (Berlin, 23.03.1957), stated that Euler had been the first one to define quaternions (see [12]). In a letter to Christian Goldbach of May 4th, 1748 in the framework of his researches on parametric representations of movements in Euclidean space, Euler defined quaternions, without explicitly naming them.

It seems however that Euler did not recognize the fundamental character of this structure. He employed only "vectorial quaternions" in his kinematic researches, without studying in depth this new type of numbers. This idea was forgotten for a long time. Almost 100 years later the French mathematician and philosopher OLINDE RODRIGUES began to use numbers similar to quaternions to describe rotations in 3-dimensional space. Also GAUSS worked already with quaternionic formulae, that he used in an 1819 note which then was not published.

Leonhard Euler (1707–1783)

The discoverer of quaternions is one of the most fascinating scientists of the nineteenth century, SIR WILLIAM ROWAN HAMILTON. Already in the early thirties he was involved in investigating algebraic questions. Thus in 1833 he was able to show that complex numbers build an algebra, in case the unities 1 and i are used with $1^2 = 1$ and $i^2 = -1$. All the elements of his algebra have then the form $x + iy$, where x, y denote real numbers.

He tried for more than ten years to extend this result to the so-called *triples*, i.e., the real unit 1 would be studied together with two other "imaginary" units i and j. In later papers he named these triples of numbers *vectors*. He was able to figure out how these vectors are to be added and multiplied, but he did not succeed in finding a suitable division — he was very unhappy about that. In view of his so far brilliant scientific performance the inability to solve this problem was for him a situation to which he was really unaccustomed. Only after introducing a further imaginary unit and dropping commutativity was he able to divide vectors. About the discovery of quaternions the following anecdote is told:

It was Monday, October 16th, 1843 in the morning, Hamilton had to chair a meeting of the Royal Irish Academy. His wife walked with him along the Royal Canals in Dublin. At that moment he got the decisive idea for the solution of his ten years old problem. With his pocketknife he wrote in the stones of the Broome bridge the famous formula

$$\mathbf{i}^2 = \mathbf{j}^2 = \mathbf{k}^2 = \mathbf{ijk} = -1.$$

In a letter to his eldest son just shortly before his death, he indicated the above bridge erroneously as the Brougham bridge, as it is named also today.

From then on Hamilton called the new numbers

$$q = a + b\mathbf{i} + c\mathbf{j} + d\mathbf{k} \quad (a, b, c, d \in \mathbb{R})$$

quaternions. The first paper about quaternions appeared on November 14th, 1843, in the Council Books of the Royal Academy at the "First General Meeting of the Session" (see [6]). Concerning the naming, P.G. TAIT, the only pupil of Hamilton, expresses himself as follows:

Quaternion in Latin means "set of four", the Greek translation of this word is "Tetractys" . Hamilton who knew the Greek language in depth and revered the pythagorical school, shows to have built a bridge between his structure and the pythagorical Tetractys, which was considered as the source of all things.

Further interesting interpretations of the origin of the notation "Quaternion", can be found in S.L. ALTMAN's book [6].

2.2 Definition and properties

We want now to begin with a systematic exposition of the real quaternions. As in $\mathbb{C}$ we need to define a set with corresponding connections:

Definition 2.1. Let $\mathbb{H} := \{x : x = (x_0, x_1, x_2, x_3), x_k \in \mathbb{R}, k = 0, 1, 2, 3\}$ be the set of the ordered quadruples of real numbers. The numbers x_0, x_1, x_2, x_3 are called the *coordinates of x*. Two quadruples $x = (x_0, x_1, x_2, x_3)$ and $y = (y_0, y_1, y_2, y_3)$ are then equal, when the single *coordinates* are equal: $x_k = y_k$, $k = 0, 1, 2, 3$. *Addition* is defined coordinatewise:

$$x + y := (x_0 + y_0, x_1 + y_1, x_2 + y_2, x_3 + y_3),$$

and similarly *multiplication with a real number* λ:

$$\lambda x := (\lambda x_0, \lambda x_1, \lambda x_2, \lambda x_3).$$

Multiplication is introduced as linear continuation of the multiplication of the basis elements of the *standard basis* of $\mathbb{R}^4$, namely $e_0 := (1, 0, 0, 0)$, $e_1 := (0, 1, 0, 0)$,

$e_2 := (0,0,1,0)$, $e_3 := (0,0,0,1)$. e_0 should be the unit element of multiplication, the other three basis elements satisfy the relations

$$e_i e_j + e_j e_i = -2\delta_{ij}, \quad i,j = 1,2,3,$$

with the Kronecker symbol δ_{ij} and the relation

$$e_1 e_2 = e_3.$$

The properties that we have learnt in $\mathbb{C}$ and $\mathbb{H}$ have the following general structure which is called a *vector space*. $\mathbb{R}^n$ for $n = 1,2,3,4$ becomes in our situation a special case of this structure.

Definition 2.2. A set V with the following properties is called an $\mathbb{R}$-*vector space* or *real vector space* if:

  (i) on V an addition is defined and V with respect to this addition is a commutative group,

  (ii) a multiplication of numbers from $\mathbb{R}$ and vectors of V is defined, which satisfies the following rules: for $a, b \in \mathbb{R}$, $\mathbf{v}, \mathbf{w} \in V$ and 1 denoting the unit element of $\mathbb{R}$ (thus the real number 1) we have

$$a(\mathbf{v} + \mathbf{w}) = a\mathbf{v} + a\mathbf{w}, \quad (a + b)\mathbf{v} = a\mathbf{v} + b\mathbf{v} \quad \textit{(distributivity)},$$
$$(ab)\mathbf{v} = a(b\mathbf{v}) \quad \textit{(associativity)},$$
$$1\mathbf{v} = \mathbf{v} \quad \textit{(existence of the unit element)}.$$

The vector space is called real, since for multiplication only real numbers are allowed, but of course other noncommutative fields could be employed, as for instance $\mathbb{C}$.

Vector spaces will be necessary in many places. As observed earlier, in this way the usual vector space structure can be introduced in $\mathbb{R}^4$ and the canonical representation of the *real quaternions* is made possible

$$x = x_0 e_0 + x_1 e_1 + x_2 e_2 + x_3 e_3 = \sum_{k=0}^{3} x_k e_k.$$

Often e_0 as unit element is not even written.

For clarity's sake the above multiplication rule can also be written as

$$e_1^2 = e_2^2 = e_3^2 = -1; \quad e_i e_j = -e_j e_i, \quad i \neq j = 1,2,3.$$

The last relation can be written cyclically:

$$e_{i+1} e_{i+2} = e_{i+3}, \quad i = 0,1,2 \pmod 3.$$

In general HAMILTON has denoted the basis elements by $i := e_1$, $j := e_2$, $k := e_3$, so that the relationship with $\mathbb{C}$, or better the extension of complex numbers, is even clearer. Since $i^2 = j^2 = k^2 = -1$ the basis elements can also be denoted as imaginary units. We can immediately see that multiplication is not commutative, thus quaternions do not build a field, rather a noncommutative or skew field. The continuation of multiplication on general quaternions is obtained through formal expansion:

$$
\begin{aligned}
xy &= (x_0 + x_1 e_1 + x_2 e_2 + x_3 e_3)(y_0 + y_1 e_1 + y_2 e_2 + y_3 e_3) \\
&= (x_0 y_0 - x_1 y_1 - x_2 y_2 - x_3 y_3) \\
&\quad + (x_0 y_1 + x_1 y_0 + x_2 y_3 - x_3 y_2) e_1 \\
&\quad + (x_0 y_2 - x_1 y_3 + x_2 y_0 + x_3 y_1) e_2 \\
&\quad + (x_0 y_3 + x_1 y_2 - x_2 y_1 + x_3 y_0) e_3.
\end{aligned}
$$

ARTHUR CAYLEY, an English mathematician (1821–1895), who wrote the first papers with quaternions after HAMILTON , developed a simple scheme to describe this multiplication, which today is known as the *Cayley table*:

	1	e_1	e_2	e_3
1	1	e_1	e_2	e_3
e_1	e_1	-1	e_3	$-e_2$
e_2	e_2	$-e_3$	-1	e_1
e_3	e_3	e_2	$-e_1$	-1

Also the following diagram is often very useful:

The Irish political figure and patriot EAMON DE VALERA (1882–1975), who was three times prime minister and from 1959 to 1973 was president of the Irish republic, should have been executed in Dublin on Monday April 24th, 1916, for high treason against the British crown. Because of lucky circumstances his death sentence was commuted to a long jail detention. Valera, who had been a mathematics teacher, full of national pride engraved the defining equations of quaternions on the walls of his cell.

The quadruple $(0,0,0,0)$ is clearly the neutral element of addition and the proof that addition and multiplication are associative and distributive is an even more tiring task to show than in $\mathbb{C}$: the addition is naturally commutative, the multiplication is not as just remarked. The inverse element of x with respect to addition is clearly $-x$, the one with respect to multiplication will shortly be simply described, so that the real quaternions build a noncommutative field, the *noncommutative field of the real quaternions* $\mathbb{H}$.

2. Quaternions

If we identify the quaternions of the form $x_0 + e_1 x_1$ with the complex numbers $x_0 + i x_1$, then we can convince ourselves easily that multiplication in $\mathbb{H}$ is mapped into complex multiplication, hence $\mathbb{C}$ is a subfield of $\mathbb{H}$.

Clearly the quaternions x can be uniquely associated to vectors $(x_0, x_1, x_2, x_3)^\top$ in $\mathbb{R}^4$, (transposition indicates that vectors should in general be written as column vectors). But as in $\mathbb{C}$ the structure of quaternions is richer than the one of vector space.

Definition 2.3. For a quaternion $x = x_0 + x_1 e_1 + x_2 e_2 + x_3 e_3$ the real number x_0 is called the *scalar part* $\mathrm{Sc}(x)$ of the quaternion. The quaternion $\mathbf{x} := x_1 e_1 + x_2 e_2 + x_3 e_3$ is called the *vector part* $\mathrm{Vec}(x)$ of the quaternion, so that we can write $x = x_0 + \mathbf{x}$. The quaternion $\overline{x} := x_0 - \mathbf{x}$ associated with x is called the *conjugate quaternion* of x. The expression $|x| := \sqrt{x_0^2 + x_1^2 + x_2^2 + x_3^2}$ is then indicated as the *modulus* or *absolute value* of the quaternion. For the *set of all vectors* we write $\mathrm{Vec}\,\mathbb{H}$, for the *set of all scalars* $\mathrm{Sc}\,\mathbb{H}$.

It is then clear that the vector part of a quaternion contains the basis elements and therefore is not a real number, as it happens with the imaginary part of complex numbers. The modulus of a quaternion is also here equal to the modulus of the corresponding vector in $\mathbb{R}^4$. $\mathrm{Vec}\,\mathbb{H}$ and $\mathrm{Sc}\,\mathbb{H}$ are real linear subspaces of $\mathbb{H}$, but unfortunately $\mathrm{Vec}\,\mathbb{H}$ is not closed with respect to quaternion multiplication. $\mathrm{Sc}\,\mathbb{H}$ can clearly be identified with $\mathbb{R}$ and $\mathrm{Vec}\,\mathbb{H}$ with $\mathbb{R}^3$.

Since a physical meaning could always be assigned to the vector part of a quaternion, W.R. HAMILTON introduced in 1846 the concept of *vector*. Even nowadays in engineering lectures Hamilton's notation $\mathbf{i}$, $\mathbf{j}$ and $\mathbf{k}$ for the basis elements in $\mathbb{R}^3$ is often used. Thus a vector in the sense of Hamilton is representable in the form

$$\mathbf{x} = x_1 \mathbf{i} + x_2 \mathbf{j} + x_3 \mathbf{k}.$$

Corresponding rules of complex numbers hold:

Proposition 2.4. *Let us take* $x, y \in \mathbb{H}$, *then the following relations are verified.*

(i) $\mathrm{Sc}(x) = \frac{(x + \overline{x})}{2}$, (ii) $\mathrm{Vec}(x) = \frac{(x - \overline{x})}{2}$,

(iii) $x\overline{x} = \overline{x}x = |x|^2$, (iv) $x^{-1} = \frac{\overline{x}}{|x|^2}$, $x \neq 0$,

(v) $\overline{x + y} = \overline{x} + \overline{y}$, (vi) $\overline{xy} = \overline{y}\,\overline{x}$,

(vii) $\overline{\overline{x}} = x$, (viii) $|xy| = |x||y|$,

(ix) $|\overline{x}| = |-x| = |x|$, (x) $(xy)^{-1} = y^{-1}x^{-1}$, $xy \neq 0$.

Property (iv) indicates that $\mathbb{H}$ is in fact a noncommutative field, since here the inverse element for multiplication is given; (x) follows from the noncommutativity and must be considered with particular attention since it is unusual.

Proof. The relations (i)–(vii) and (ix) require only simple calculations. For (viii) we write $|xy|^2 = xy\overline{xy} = xy\overline{y}\,\overline{x} = |y|^2|x|^2$. (x) The associativity gives

$$(xy)(y^{-1}x^{-1}) = x(yy^{-1})x^{-1} = xx^{-1} = 1. \qquad \square$$

Remark 2.5. It should be noted that in a similar way as for $\mathbb{C}$ in $\mathbb{H}$ the fourfold quadratic sum

$$x_0^2 + x_1^2 + x_2^2 + x_3^2 = (x_0 + x_1e_1 + x_2e_2 + x_3e_3)(x_0 - x_1e_1 - x_2e_2 - x_3e_3)$$

can be represented as a product of two linear expressions, as in $\mathbb{C}$ this is not evident. Also the *four squares theorem* holds

$$
\begin{aligned}
&(x_0^2 + x_1^2 + x_2^2 + x_3^2)(y_0^2 + y_1^2 + y_2^2 + y_3^2) \\
&= (x_0y_0 - x_1y_1 - x_2y_2 - x_3y_3)^2 + (x_0y_1 + x_1y_0 + x_2y_3 - x_3y_2)^2 \\
&\quad + (x_0y_2 - x_1y_3 + x_2y_0 + x_3y_1)^2 + (x_0y_3 + x_1y_2 - x_2y_1 + x_3y_0)^2,
\end{aligned}
$$

which says that a product of two fourfold quadratic sums can be written again as a quadratic sum (see [66]).

Propositions on the modulus of quaternions correspond to the complex case:

Proposition 2.6. *Let x and y be quaternions, then for the modulus we have*

(i) $|\mathrm{Sc}(x)| \le |x|$, $\quad |\mathrm{Vec}(x)| \le |x|$,

(ii) $|x + y| \le |x| + |y|$ \qquad *(triangular inequality)*,

(iii) $\big||x| - |y|\big| \le |x - y|$.

Proof. The proof mimicks the one for $\mathbb{C}$. $\qquad \square$

$\mathbb{H}$ becomes a metric space with the introduction of the *Euclidean distance* $|x - y|$. These propositions correspond to those for complex numbers, which will be treated in Section 4.1. We now investigate specific properties of quaternions: The multiplication of two quaternions gives

$$xy = (x_0 + \mathbf{x})(y_0 + \mathbf{y}) = x_0y_0 + x_0\mathbf{y} + y_0\mathbf{x} + \mathbf{xy}.$$

We want to consider more closely the last product $\mathbf{xy}$:

$$
\begin{aligned}
\mathbf{xy} = \ &-(x_1y_1 + x_2y_2 + x_3y_3) \\
&+ (x_2y_3 - x_3y_2)e_1 + (x_3y_1 - x_1y_3)e_2 + (x_1y_2 - x_2y_1)e_3 \\
= \ &-\mathbf{x} \cdot \mathbf{y} + \mathbf{x} \times \mathbf{y},
\end{aligned}
$$

with the *scalar product* $\mathbf{x} \cdot \mathbf{y}$ and the *vector* or *cross product* $\mathbf{x} \times \mathbf{y}$ of the two vectors $\mathbf{x}$ and $\mathbf{y}$. Historically these products were introduced in this way and then later were "emancipated" from the theory of quaternions. But observe that

$$\mathbf{x} \cdot \mathbf{y} := x_1y_1 + x_2y_2 + x_3y_3 \quad \text{and} \quad \mathbf{x} \times \mathbf{y} := \begin{vmatrix} \mathbf{e}_1 & \mathbf{e}_2 & \mathbf{e}_3 \\ x_1 & x_2 & x_3 \\ y_1 & y_2 & y_3 \end{vmatrix}.$$

We remark also that
$$|x|^2 = x_0^2 - \mathbf{x}^2 = x_0^2 + \mathbf{x} \cdot \mathbf{x}.$$

A scalar product does not exist only for vectors, but also for quaternions. Only the analogous operation to the vector product is not explained so easily:

Definition 2.7 (Scalar product). Let $x, y \in \mathbb{H}$. The product

$$\begin{aligned}
x \cdot y &= \frac{1}{2}(x\bar{y} + y\bar{x}) = \mathrm{Sc}(x\bar{y}) = \mathrm{Sc}(y\bar{x}) \\
&= x_0 y_0 + x_1 y_1 + x_2 y_2 + x_3 y_3
\end{aligned}$$

is called the *scalar product* of the quaternions x and y. For $x \cdot y = 0$, x and y are *orthogonal* to each other.

Proposition 2.8. *Let x, y, z be arbitrary elements in $\mathbb{H}$. Then*

(i) $\mathrm{Sc}\,(xyz) = \mathrm{Sc}(yzx) = \mathrm{Sc}(zxy)$,

(ii) $\mathrm{Sc}(xyz) = \bar{x} \cdot (yz)$ *is a real number, which is also called a scalar mixed (or triple) product of the quaternions x, y, z (in this order).*

Proof. The proof is left to the reader (see Exercise 2.6.1). □

Remark 2.9. So far we have worked with $\mathbb{R}$, $\mathbb{C}$ and $\mathbb{H}$, sets that have both a vector space structure and a field structure. They are called *algebras*. Since all elements different from zero have a multiplicative inverse, we speak of *division algebras*. The next theorem shows that no other division algebra over the field of real numbers exists:

Theorem 2.10 (Frobenius theorem). *The only finite-dimensional associative division algebras over the field of real numbers $\mathbb{R}$ are $\mathbb{R}$, $\mathbb{C}$ and $\mathbb{H}$.*

Proof. We refer to the book [72]. □

The solution behavior of quadratic equations manifest themselves in a completely different way than for complex numbers. We formulate this in the next proposition:

Proposition 2.11. (i) *The reciprocally conjugate quaternions $x = x_0 + \mathbf{x}$ and $\bar{x} = x_0 - \mathbf{x}$ satisfy the quadratic equation with real coefficients*

$$\xi^2 - 2x_0 \xi + |x|^2 = 0.$$

(ii) *A quaternion is a vector different from zero if and only if x^2 is real and negative.*

(iii) *A quaternion different from zero is real if and only if x^2 is real and positive.*

(iv) *The solution set of a quadratic equation in $\mathbb{H}$ with real coefficients consists either of one element, two elements or a two-dimensional sphere.*

Proof. (i) Substitution into the given equation delivers

$$(x_0 \pm \mathbf{x})^2 - 2x_0(x_0 \pm \mathbf{x}) + x_0^2 - \mathbf{x}^2 = 0$$

and therefore the desired result. (ii) For an arbitrary $x \in \mathbb{H}$ it follows that

$$x^2 = (x_0 + \mathbf{x})(x_0 + \mathbf{x}) = x_0^2 - |\mathbf{x}|^2 + 2x_0\mathbf{x}.$$

At first let us consider $x = \mathbf{x} \neq \mathbf{0}$, then $x^2 = -|\mathbf{x}|^2 < 0$. Conversely if we assume $\mathbb{R} \ni x^2 < 0$ we then find $2x_0\mathbf{x} = x^2 - x_0^2 + |\mathbf{x}|^2 \in \mathbb{R}$. It thus follows that $x_0 = 0$ or $\mathbf{x} = 0$. In case $x_0 = 0$, we are done. For $\mathbf{x} = 0$, we would then have $x^2 = x_0^2 > 0$, a contradiction with the assumption. Thus the second case cannot arise. (iii) The proof is completely analogous to the proof of (ii).
(iv) With completion of the square from the equation

$$x^2 + 2ax + b = 0, \quad (a, b \in \mathbb{R})$$

we find as usual

$$(x + a)^2 = a^2 - b.$$

If the right-hand side is non-negative, then by means of (iii) we find the well-known real roots of the quadratic equation. If the right-hand side is negative, then from (ii) $x + a$ must be a vector, whose modulus is $\sqrt{b - a^2}$. But this is a sphere in $\mathbb{R}^3$ with precisely this radius. $\qquad\square$

In the following proposition we want to list a number of important properties:

Proposition 2.12. *Let x and y be two quaternions, correspondingly $\mathbf{x}$ and $\mathbf{y}$ two vectors. Then the relations hold:*

(i) $\mathbf{x} \cdot \mathbf{y} = -\mathrm{Sc}(\mathbf{xy}) = -\frac{1}{2}(\mathbf{xy} + \mathbf{yx})$.

(ii) $\mathbf{x} \times \mathbf{y} = \mathrm{Vec}(\mathbf{xy}) = \frac{1}{2}(\mathbf{xy} - \mathbf{yx})$.

(iii) *From $x^2 = y^2$ it does not necessarily follow that $x = \pm y$.*

(iv) *A quaternion x is real if and only if for every other quaternion y we have $yx = xy$.*

Proof. The properties (i) and (ii) follow immediately from the definition. For (iii) we should observe that, from the previous proposition for $x^2 = y^2$ real and negative, it follows only that x, correspondingly y, lie on a sphere in $\mathbb{R}^3$ and therefore in no way must be equal apart from the sign. (iv) Real numbers commute naturally with an arbitrary quaternion. Moreover $yx = xy$ for all $y \in \mathbb{H}$. Then in particular for $y = e_1$ it follows that

$$xe_1 = -x_1 + x_0e_1 - x_2e_3 + x_3e_2 = -x_1 + x_0e_1 + x_2e_3 - x_3e_2 = e_1x,$$

from which $0 = x_2e_3 - x_3e_2$. Thus $x_2 = x_3 = 0$ must be true. In the same way we also obtain that $x_1 = 0$. $\qquad\square$

A proposition on the relationship between quaternions and vectors follows:

Proposition 2.13. (i) *Let x be a quaternion. Then there exists a vector $\mathbf{a} \neq 0$, so that $x\mathbf{a}$ is again a vector.*

(ii) *An arbitrary quaternion x is representable as product of two vectors.*

(iii) *From each quaternion a we can calculate at least a root in $\mathbb{H}$, i.e., there is an $x \in \mathbb{H}$ with $x^2 = a$.*

(iv) *Every quaternion e with $|e| = 1$ can be represented in the form*

$$e = \mathbf{xyx}^{-1}\mathbf{y}^{-1}.$$

Proof. (i) For $x = x_0 + \mathbf{x}$ and $\mathbf{x} = 0$ we can choose $\mathbf{a}$ arbitrarily. For $\mathbf{x} \neq 0$ let $\mathbf{a} \neq 0$ be a vector orthogonal to $\mathbf{x}$. Then in view of $\mathbf{x} \cdot \mathbf{a} = 0$ we have

$$x\,\mathbf{a} = x_0\mathbf{a} + \mathbf{x} \times \mathbf{a} \in \text{Vec } \mathbb{H}.$$

(ii) Let $\mathbf{a}$ be chosen as in (i), then we have

$$x\,\mathbf{a} = \mathbf{b}, \text{ thus } x = \mathbf{ba}^{-1},$$

and the inverse of a vector is again a vector. The proof of the points (iii) and (iv) is left to the reader (compare Exercises 2.6.2, 2.6.3). □

The real quaternions show remarkable analogies with complex numbers besides these elementary properties, provided naturally that commutativity is ignored.

Theorem 2.14. *Every quaternion $x \in \mathbb{H}$ with $\mathbf{x} \neq 0$ satisfies the trigonometric representation*

$$x = |x|(\cos \varphi + \boldsymbol{\omega}(\mathbf{x}) \sin \varphi).$$

Since $x_0^2 + |\mathbf{x}|^2 = |x|^2$ we find the relationships

$$\cos \varphi = \frac{x_0}{|x|}, \quad \sin \varphi = \frac{|\mathbf{x}|}{|x|}, \quad \boldsymbol{\omega}(\mathbf{x}) = \frac{\mathbf{x}}{|\mathbf{x}|} \in S^2 \quad (-\pi \leq \varphi \leq \pi),$$

where S^2 denotes the unit sphere in $\mathbb{R}^3$.

Proof. We have clearly

$$x = x_0 + \mathbf{x} = |x| \left(\frac{x_0}{|x|} + \frac{\mathbf{x}}{|\mathbf{x}|} \frac{|\mathbf{x}|}{|x|} \right) = |x|(\cos \varphi + \boldsymbol{\omega}(\mathbf{x}) \sin \varphi). \qquad \square$$

Example 2.15. Let $x = 1 + 2e_1 + 2e_2 + e_3$, it follows then that $\varphi = \arccos(1/\sqrt{10}) = \arcsin(3/\sqrt{10})$. With this we obtain the representation

$$x = \sqrt{10} \left[\frac{1}{\sqrt{10}} + \frac{(2e_1 + 2e_2 + e_3)}{3} \frac{3}{\sqrt{10}} \right]$$

and see also that a calculation does not represent a problem.

Corollary 2.16 (de Moivre's formula). *Let $x = x_0 + \mathbf{x} \in \mathbb{H}$, $x \neq 0$, $n \in \mathbb{Z}$ be given; we then have*

$$(\cos \varphi + \boldsymbol{\omega}(\mathbf{x}) \sin \varphi)^n = \cos n\varphi + \boldsymbol{\omega}(\mathbf{x}) \sin n\varphi.$$

Proof. We observe at first that contrary to the complex case we must replace here the imaginary unit i by an element of the two-dimensional unit sphere. But we nevertheless have

$$\omega^2(\mathbf{x}) = \left(\frac{\mathbf{x}}{|\mathbf{x}|}\right)^2 = -1.$$

For positive n the proof follows then by mathematical induction. The case $n = 0$ gives 1 on both sides, and for negative n we have

$$(\cos\varphi + \omega(\mathbf{x})\sin\varphi)^{-n} = (\cos\varphi + \overline{\omega}(\mathbf{x})\sin\varphi)^n = \cos n\varphi - \omega(\mathbf{x})\sin n\varphi. \qquad \square$$

2.3 Mappings and representations

Up to now we have considered mainly the structural side of the real quaternions. We will now pay attention to the study of algebraic automorphisms, correspondingly antiautomorphisms of $\mathbb{H}$, which are specific mappings within the quaternionic algebra closely connected with rotations and reflections in $\mathbb{R}^4$ as well as in $\mathbb{R}^3$. In this context the automorphism h has to satisfy among other properties the multiplicative property $h(xy) = h(x)h(y)$, $x, y \in \mathbb{H}$, while an antiautomorphism k should verify the relation $k(xy) = k(y)k(x)$, $x, y \in \mathbb{H}$. Our presentation relies in part on results which are contained in the monographs [119] and [6]. Finally we will investigate representations of the quaternionic algebra in the matrix ring $\mathbb{R}^{4\times4}$.

2.3.1 Basic maps

We observe at first that we do not need to distinguish between the Euclidean space $\mathbb{R}^3$ and the $\mathbb{R}$-linear subspace $\text{Vec}\,\mathbb{H}$ of $\mathbb{H}$. As real vector spaces $\mathbb{H}$ and $\mathbb{R}^4$ are also naturally isomorphic.

In comparison with $\mathbb{C}$ in $\mathbb{H}$ there are more possibilities to define an involution. We thus distinguish different involutions:

Definition 2.17. (i) Let $x \in \mathbb{H}$ be given. The mapping $x \to \overline{x}$ with

$$\overline{x} = \text{Sc}\,(x) - \text{Vec}\,(x) \in \mathbb{H}$$

is called *conjugation* in $\mathbb{H}$. As we already know, the corresponding element $\overline{x}$ is then the *conjugate quaternion* of x. Moreover $\overline{xy} = \overline{y}\,\overline{x}$, so that we have an antiautomorphism.

(ii) The mapping $x \to \hat{x}$,

$$\hat{x} := e_2 x e_2^{-1} \in \mathbb{H},$$

is called the *principal involution* in $\mathbb{H}$. The element $\hat{x}$ is called the *involute* of the quaternion x.

(iii) Finally the composition of conjugation and principal involution

$$\tilde{x} := \hat{\bar{x}} = \bar{\hat{x}}$$

is called the *reversion* in $\mathbb{H}$. The element $\tilde{x}$ is the *reverse* of the quaternion x.

Remark 2.18. We can consider geometrically $x \to \hat{x}$ as a reflection about the plane $\{\lambda + \mu e_2 : \lambda, \mu \in \mathbb{R}\}$. Quaternions are then reflected about the plane spanned by e_0 and e_2, the vectors in $\mathbb{R}^3$ are reflected about the $\mathbf{e_2}$-axis and moreover $\widehat{xy} = \hat{x}\,\hat{y}$. The principal involution is thus an automorphism.

We also see that by using reversion the order of the factors in the quaternionic product is reversed, thus it is in fact an antiautomorphism. It is exactly this last property that justifies the name reversion.

We would like now to characterize the automorphisms of $\mathbb{H}$:

Theorem 2.19 (Rodrigues, Porteous). *An arbitrary automorphism or antiautomorphism m of the algebra $\mathbb{H}$ has always the representation*

$$m(x) := \mathrm{Sc}\ (x) + h\big(\mathrm{Vec}\ (x)\big), \quad x \in \mathbb{H},$$

with an orthogonal automorphism h of $\mathbb{R}^3$.

Proof. Let $m(1) = y_0 + \mathbf{y}$ with $y_0 \in \mathbb{R}$, $\mathbf{y} \in \mathbb{R}^3$. Then since $m(1) = m(1^2) = m^2(1)$ we have

$$y_0^2 - |\mathbf{y}|^2 + 2y_0\mathbf{y} = y_0 + \mathbf{y}.$$

For $\mathbf{y} \neq \mathbf{0}$ we would have $2y_0 - 1 = 0$, which cannot hold in view of $y_0 - y_0^2 = 1/4 = -|\mathbf{y}|^2$. Thus it must be that $\mathbf{y} = \mathbf{0}, y_0 = 1$ and $m(x_0) = x_0$ for a real x_0. For arbitrary $x = x_0 + \mathbf{x}$ it then follows that

$$m(x) = x_0 + m(\mathbf{x}).$$

If it were now $m(\mathbf{x}) = y_0 + \mathbf{y}$, then we would have

$$m(\mathbf{x}^2) = -m(|\mathbf{x}|^2) = -|\mathbf{x}|^2 = y_0^2 - |\mathbf{y}|^2 + 2y_0\mathbf{y}.$$

Here we must have $y_0\mathbf{y} = \mathbf{0}$, since $\mathbf{y} = \mathbf{0}$ is excluded in view of $-|\mathbf{x}|^2 = y_0^2$, then we find only $y_0 = 0$ and $m(\mathbf{x}) \in \mathbb{R}^3$. Then the imbedding $m|_{\mathbb{R}^3} =: h$ is an automorphism of $\mathbb{R}^3$, and in view of $m(|\mathbf{x}|^2) = |\mathbf{x}|^2$ it is norm preserving and therefore orthogonal. $\square$

A very important automorphism of $\mathbb{H}$ is defined by

$$\rho_y(x) := yxy^{-1}, \quad y \in \mathbb{H},$$

clearly with $y \neq 0$. The mapping thus introduced will now be investigated in view of its algebraic and topological properties.

Theorem 2.20. *For $x, x' \in \mathbb{H}$ and $\lambda, \lambda' \in \mathbb{R}$ the mapping ρ_y possesses the following properties:*

(i) $\rho_y(\lambda x + \lambda' x') = \lambda \rho_y(x) + \lambda' \rho_y(x')$ (*$\mathbb{R}$-linearity*).

(ii) $\rho_y(xx') = \rho_y(x)\rho_y(x')$ *(multiplicativity).*

(iii) ρ_y *is an isometric automorphism on* $\mathbb{H}$.

(iv) *The canonical scalar product in* $\mathbb{R}^4$ *is invariant under the mapping* ρ_y, *i.e.,*

$$\rho_y(x) \cdot \rho_y(x') = x \cdot x'.$$

(v) *We have* $\rho_y\rho_{y'} = \rho_{yy'}$.

Proof. Since the relations (i) and (ii) are very easy to show, we only prove the last three properties. We first show that $\rho_y^{-1} : x \in \mathbb{H} \to y^{-1}xy$ is the inverse of ρ_y, then $\rho_y^{-1} = \rho_{y^{-1}}$. Thus ρ_y is an automorphism. Moreover

$$|\rho_y(x)| = |yxy^{-1}| = |x|,$$

so that (iii) is proven. In order to obtain (iv), we calculate as follows:

$$
\begin{aligned}
\rho_y(x) \cdot \rho_y(x') &= \frac{1}{2}\left[\rho_y(x')\overline{\rho_y(x)} + \rho_y(x)\overline{\rho_y(x')}\right] \\
&= \frac{1}{2}\left[yx'y^{-1}\overline{yxy^{-1}} + yxy^{-1}\overline{yx'y^{-1}}\right] \\
&= \frac{1}{2}\left[yx'y^{-1}\overline{y^{-1}}\,\overline{x}\,\overline{y} + yxy^{-1}\overline{y^{-1}}\,\overline{x'}\,\overline{y}\right] \\
&= \frac{1}{2}y(x'\overline{x} + x\overline{x'})\overline{y}\frac{1}{|y|^2} = x \cdot x'.
\end{aligned}
$$

Finally the last relation follows from the associativity:

$$\rho_y\rho_{y'}(x) = y(y'xy'^{-1})y^{-1} = yy'x(yy')^{-1} = \rho_{yy'}(x). \qquad \square$$

Since ρ_y is independent of $|y|$ we can consider only those y with $|y| = 1$, i.e., the elements of the unit sphere S^3 in $\mathbb{R}^4$, so that

$$S^3 := \{y \in \mathbb{H} : |y| = 1\}.$$

Let $x, y \in S^3$ be given, then since $|xy| = |x||y|$ it follows that also $xy \in S^3$. Further we find $1 \in S^3$ and $\overline{y} = y^{-1} \in S^3$. Thus we have shown the

Proposition 2.21. S^3 *is a subgroup of* $\mathbb{H}$ *and the mapping* $y \to \rho_y$ *is a homomorphism of* $\mathbb{H}$ *onto* S^3.

We want now to study the properties of the automorphism ρ_y in $\mathbb{R}^3$ from which we will gather insights into the behavior in $\mathbb{H}$.

2.3.2 Rotations in $\mathbb{R}^3$

We begin by considering ρ_y in $\mathbb{R}^3$. In this case as well as in higher dimensions a *rotation* is a mapping

$$\mathbf{x}' = A\mathbf{x}$$

with an orthogonal matrix A, i.e., $A^{-1} = A^\top$ and det $A = 1$. These matrices build the group $SO(3)$. Here det $A = 1$ means that the orientation is maintained, for det $A = -1$ we obtain instead a *reflection*. At first we can state that in view of the theorem of Rodrigues–Porteous 2.19 $\rho_y(\mathbf{x})$ is again a vector, so that a scalar part of $\mathbf{x}$ does not exist. Thus ρ_y (we use the same notation also for the imbedding onto $\mathbb{R}^3$) is also an automorphism of $\mathbb{R}^3$ with the properties listed in Theorem 2.20. Moreover ρ_y in $\mathbb{R}^3$ is exchangeable with the vector product, so that

$$
\begin{aligned}
\rho_y(\mathbf{x}) \times \rho_y(\mathbf{x}') &= \frac{1}{2}\left[\rho_y(\mathbf{x})\,\rho_y(\mathbf{x}') - \rho_y(\mathbf{x}')\,\rho_y(\mathbf{x})\right] \\
&= \frac{1}{2}\left[yxy^{-1}\,yx'y^{-1} - yx'y^{-1}\,yxy^{-1}\right] \\
&= \frac{1}{2}y\left[xx' - x'x\right]y^{-1} = \rho_y(\mathbf{x} \times \mathbf{x}').
\end{aligned}
$$

We can thus summarize as follows:

The mapping ρ_y is an automorphism of $\mathbb{R}^3$ which leaves the canonical scalar product invariant. It is homomorphic with respect to the vector product, as it follows immediately from the above calculation.

In view of the last property, ρ_y leaves also the orientation invariant, thus it is a rotation, a result we will see also in the next proposition.

The consideration of rotations was originally not connected with the use of quaternions. Already LEONHARD EULER in 1775 tried to describe the problem of the composition of two affine transformations. After elimination of three parameters the remaining ones appeared as rotation axes and corresponding rotation angles. He reduced the statement of the task to a pure algebraic problem. But he could not find closed form algebraic expressions. BENJAMIN OLINDE RODRIGUES (1794–1851), the son of a banker in Bordeaux, was instead more successful. He considered rotations as general movements on a sphere and in 1840 solved Euler's problem of the composition of two rotations in a constructive way.

With the help of quaternion theory we compute in a simpler way the mapping $\mathbf{x}'$ of a vector $\mathbf{x}$ under the mapping ρ_y:

Let $y \in S^3$ be given. As we already know this can be put in the form $y = y_0 + \mathbf{y} = y_0 + \omega|\mathbf{y}|$ with $y_0^2 + |\mathbf{y}|^2 = 1$ and $\omega^2 = -1$ by means of Theorem 2.14. As usual we set $y_0 =: \cos\varphi$ and $|\mathbf{y}| =: \sin\varphi$. At first we have

$$
\rho_y(\omega) = (y_0 + \omega|\mathbf{y}|)\omega(y_0 - \omega|\mathbf{y}|) = (y_0^2 + |\mathbf{y}|^2)\omega = \omega.
$$

We have thus shown that the vector ω is invariant under the map ρ_y. It follows that

$$
\begin{aligned}
\rho_y(\mathbf{x}) &= y_0 x y_0 - y_0 x|\mathbf{y}|\omega + |\mathbf{y}|\omega x y_0 - |\mathbf{y}|^2 \omega x \omega \\
&= y_0^2 \mathbf{x} + y_0|\mathbf{y}|(\omega \mathbf{x} - \mathbf{x}\omega) - |\mathbf{y}|^2 \omega \mathbf{x} \omega \\
&= \mathbf{x}\cos^2\varphi + (\omega \times \mathbf{x})\sin 2\varphi - \omega \mathbf{x} \omega \sin^2\varphi.
\end{aligned}
$$

In view of

$$\omega \mathbf{x} \omega = \omega \mathbf{x} \omega + \mathbf{x} \omega \omega - \mathbf{x} \omega \omega = \mathbf{x} + (\omega \mathbf{x} + \mathbf{x} \omega) \omega = \mathbf{x} - 2(\omega \cdot \mathbf{x}) \omega$$

we obtain the following important result:

Proposition 2.22. a) **Euler–Rodrigues formula.** *Let the point $\mathbf{x} \in \mathbb{R}^3$ be mapped by means of ρ_y into a point $\mathbf{x}'$. Then with $y = y_0 + \mathbf{y} = y_0 + \omega |\mathbf{y}|$ and $\omega^2 = -1$, $y_0 = \cos \varphi$, $|\mathbf{y}| = \sin \varphi$ we have*

$$\mathbf{x}' := \rho_y(\mathbf{x}) = \mathbf{x} \cos 2\varphi + (\omega \times \mathbf{x}) \sin 2\varphi + (1 - \cos 2\varphi)(\omega \cdot \mathbf{x}) \omega.$$

b) *Each map ρ_y is a rotation about the axis ω and the angle 2φ. Conversely each rotation in $\mathbb{R}^3$ can be represented by an automorphism of the form $\rho_y = yxy^{-1}$ with $y \in \mathbb{H}$.*

Remark 2.23. The rotations in $\mathbb{R}^3$ can be more advantageously described with the help of quaternionic exponential functions, which will be introduced later in Section 11.2. The rotation axis of consecutive rotations can then be calculated in an elegant way, see also [55].

Proof. Part a) was proven before the formulation of the proposition, we have only to show part b). Since in fact we have a rotation about the axis ω, we can best proceed by decomposing the vectors $\mathbf{x}$ and $\mathbf{x}'$ in the components parallel to ω and those perpendicular to it. We can easily convince ourselves that this decomposition can be written as

$$\mathbf{x} =: \mathbf{z} + (\omega \cdot \mathbf{x}) \omega, \quad \mathbf{x}' =: \mathbf{z}' + (\omega \cdot \mathbf{x}') \omega.$$

In view of the invariance of the scalar product according to Theorem 2.20 (iv) and $\rho_y(\omega) = \omega$ it follows that $\omega \cdot \mathbf{x} = \omega \cdot \mathbf{x}'$. Substitution into the Euler–Rodrigues formula 2.22 a) yields

$$
\begin{aligned}
\mathbf{z}' &= \mathbf{x}' - (\omega \cdot \mathbf{x}') \omega \\
&= \mathbf{x} \cos 2\varphi + (\omega \times \mathbf{x}) \sin 2\varphi - (\omega \cdot \mathbf{x}) \omega \cos 2\varphi \\
&= \mathbf{z} \cos 2\varphi + (\omega \times \mathbf{z}) \sin 2\varphi.
\end{aligned}
$$

The last equation says that $\mathbf{z}$ gets rotated by the angle 2φ in the plane through the origin orthogonal to ω since in this plane $\mathbf{z}$ and $\omega \times \mathbf{z}$ build an orthogonal coordinate system. The component of $\mathbf{x}$ in the direction ω, i.e., the distance to the plane, remains unchanged. This then describes the rotation of $\mathbb{R}^3$ about the axis ω of the angle 2φ. We here have described a rotation; that this also agrees with the definition given at the beginning of this subsection follows from what we mentioned above.

We have still to show the converse, that each rotation can be represented by a ρ_y: in order to do that the rotation axis ω and the rotation angle 2φ must be given; from these y_0 and $|\mathbf{y}|$ can be determined immediately, and therefore also $\mathbf{y}$ and y. □

If we consider ρ_y with a vector $\mathbf{y}$, then $y_0 = 0$ and therefore $\varphi = \pi/2$. This means a rotation of an angle π about the $\mathbf{y}$-axis. This can also be interpreted as a reflection about the $\mathbf{y}$-axis. Since according to Proposition 2.13 (ii) every quaternion can be represented as the product of two vectors, we can state the

Corollary 2.24. *Every rotation in $\mathbb{R}^3$ can be represented as the product of two reflections about straight lines through the origin with associated vectors* **a** *and* **b**,

$$\rho_y = \rho_{\mathbf{a}}(\rho_{\mathbf{b}}) = \rho_{\mathbf{ab}}.$$

We stress however that these are reflections about straight lines in $\mathbb{R}^3$ and not about planes.

Reflections about a plane (through the origin) with the normal **n**, $|\mathbf{n}| = 1$, can easily be described by splitting **x** in a component parallel to the plane and another one in the direction **n**,

$$\mathbf{x} = \mathbf{z} + (\mathbf{n} \cdot \mathbf{x})\mathbf{n},$$

and the reflection means that the part in the direction **n** gets the minus sign, so that for the reflection point $\mathbf{x}'$ we have

$$\mathbf{x}' = \mathbf{z} - (\mathbf{n} \cdot \mathbf{x})\mathbf{n} = \mathbf{x} - 2(\mathbf{n} \cdot \mathbf{x})\mathbf{n}.$$

Now the successive reflections about two mutually perpendicular planes give exactly the reflection about the straight line which is the intersection of the two planes, and this can also be calculated. Since $\mathbf{n} \cdot \mathbf{n}' = 0$ on the one hand we have

$$\begin{aligned}
\mathbf{x}'' &= \mathbf{x}' - 2(\mathbf{n}' \cdot \mathbf{x}')\mathbf{n}' \\
&= \mathbf{x} - 2(\mathbf{n} \cdot \mathbf{x})\mathbf{n} - 2(\mathbf{n}' \cdot \mathbf{x})\mathbf{n}',
\end{aligned}$$

while on the other hand the Euler–Rodrigues formula with $\boldsymbol{\omega} := \mathbf{n} \times \mathbf{n}'$ yields

$$\mathbf{x}'' = -\mathbf{x} + 2(\boldsymbol{\omega} \cdot \mathbf{x})\boldsymbol{\omega}.$$

Equating the two formulae we finally find

$$\mathbf{x} = (\mathbf{n} \cdot \mathbf{x})\mathbf{n} + (\mathbf{n}' \cdot \mathbf{x})\mathbf{n}' + (\boldsymbol{\omega} \cdot \mathbf{x})\boldsymbol{\omega},$$

which is true since it is the representation of **x** in the coordinate system given by **n**, **n**' and $\boldsymbol{\omega}$. We have thus shown the

Corollary 2.25. *Every rotation in $\mathbb{R}^3$ can at most be composed by four reflections about planes.*

Finally we still show a proposition for the more precise determination of the relationship between S^3 and $SO(3)$.

Proposition 2.26 (Porteous). *The map*

$$\rho : S^3 \to SO(3) \quad \text{with} \quad \rho(y) = \rho_{\mathbf{y}}$$

is a surjective group homomorphism with $\ker \rho = \{-1, 1\}$.

Proof. The last propositions say, among other things, that every rotation in $\mathbb{R}^3$, and thus every element of $SO(3)$ appears as an image, so that we have a surjective map. From Theorem 2.20 (v) we conclude that yy' is mapped into $\rho_{yy'}$ so that we have a homomorphism. If $y \in \ker \rho$, then ρ_y must be the identity rotation, which leaves all vectors invariant. We have then

$$\rho_y(\mathbf{x}) = y\mathbf{x}y^{-1} = \mathbf{x} \quad (\mathbf{x} \in \mathbb{R}^3),$$

from which it follows that $y\mathbf{x} = \mathbf{x}y$ for all $\mathbf{x}$. Clearly we can add to $\mathbf{x}$ an arbitrary scalar part, so that also $yx = xy$ holds for all quaternions x. From Proposition 2.12 (iv) y must then be real. Since however $|y| = 1$ it follows that $y = \pm 1$. $\qquad\square$

2.3.3   Rotations of $\mathbb{R}^4$

Finally we consider also the rotations in $\mathbb{R}^4$. As we remarked above, these are introduced through orthogonal matrices, which leave distances invariant and whose determinant have the value 1. This matrix group is denoted by $SO(4)$. Since the mappings ρ_y leave always invariant a straight line in the direction ω, and moreover the real axis in $\mathbb{H}$, a 2-dimensional plane E_1 in $\mathbb{H}$ remains invariant under their action. In the two-dimensional plane E_2, which has only the origin in common with E_1 (in $\mathbb{R}^4$ our imagination ability fails!) it will then be rotated about an angle of 2φ. A vector $\mathbf{y}$ will be reflected through $\rho_{\mathbf{y}}$ about the plane spanned by $\mathbf{y}$ and e_0. Hence we cannot describe by means of ρ_y all the movements of $\mathbb{H}$.

An orthogonal matrix of fourth order can in general have as first case 4 real eigenvalues, which must all equal 1 in view of the length invariance; thus we obtain the identity. As second case two real eigenvalues can arise, and two complex conjugate ones. The former must once again equal 1, the corresponding eigenvectors determine an invariant plane, and those corresponding to the complex eigenvalues determine also a plane, in which one turns by a suitable angle. In general this is the case previously examined. As third and last possibility, the rotation matrix can have two pairs of complex conjugate eigenvalues; in the mutually orthogonal planes spanned by the corresponding eigenvectors, one rotates through a suitable angle. This case has not been previously discussed. Since every rotation can be decomposed as a product of reflections, this property continues to hold also for rotations in $\mathbb{R}^4$.

The fundamental theorem on rotations in $\mathbb{R}^4$ is however the following

Theorem 2.27 (Cayley's theorem). *The rotations of $\mathbb{H}$ are exactly those mappings*

$$x \to x' = axb$$

with $|a| = |b| = 1$ and $a, b \in \mathbb{H}$.

Proof. At first we observe that it is an orthogonal mapping, since

$$
\begin{aligned}
x' \cdot y' &= \frac{1}{2}(x'\overline{y'} + y'\overline{x'}) \\
&= \frac{1}{2}a(x b \overline{b} \overline{y} + y b \overline{b} \overline{x})\overline{a} = x \cdot y.
\end{aligned}
$$

2. Quaternions

In order to show that ax, correspondingly xb, define rotations, we have to find the value 1 for the determinant of the corresponding matrix. In view of the multiplication rule for ax the matrix reads

$$A = \begin{pmatrix} a_0 & -a_1 & -a_2 & -a_3 \\ a_1 & a_0 & -a_3 & a_2 \\ a_2 & a_3 & a_0 & -a_1 \\ a_3 & -a_2 & a_1 & a_0 \end{pmatrix}$$

with $\det A = |a|^4 = 1$; in a similar way we obtain the result for xb. Since $SO(4)$ is a group, the composition of two rotations gives again a rotation, so that $x' = axb$ defines indeed a rotation.

Conversely, if the rotation T is given, then let $T(e_0) =: a$. It follows that $a^{-1}T$ is also a rotation T_1 with $T_1(e_0) = e_0$. In this way the real numbers remain invariant under T_1, T_1 is therefore a rotation in $\mathbb{R}^3$, which in view of the previous subsection has the form $T_1(x) = bxb^{-1}$. Thus $T(x) = abxb^{-1}$, and this completes the second part of the claim. $\qquad\square$

2.3.4 Representations

As known, all real correspondingly complex $(n \times n)$-matrices constitute a ring, the so-called *complete matrix ring* $\mathbb{R}^{n \times n}$, correspondingly $\mathbb{C}^{n \times n}$. Often it is useful to look at how algebraic structures to be studied represent themselves as isomorphic images in the matrix ring, therefore as automorphisms of $\mathbb{R}^4$ or also of $\mathbb{C}^2$. Our goal is to find suitable subfields of the skew field of the real quaternions in $\mathbb{R}^{4 \times 4}$ as well as in $\mathbb{C}^{2 \times 2}$, which appear as isomorphic images.

Let two arbitrary quaternions be given with $x = x_0 + x_1 e_1 + x_2 e_2 + x_3 e_3$ and $y = y_0 + y_1 e_1 + y_2 e_2 + y_3 e_3$. By long multiplication of the quaternions xy we obtain

$$\begin{aligned} xy = \quad & (x_0 y_0 - x_1 y_1 - x_2 y_2 - x_3 y_3) \\ & + (x_0 y_1 + x_1 y_0 + x_2 y_3 - x_3 y_2) e_1 \\ & + (x_0 y_2 - x_1 y_3 + x_2 y_0 + x_3 y_1) e_2 \\ & + (x_0 y_3 + x_1 y_2 - x_2 y_1 + x_3 y_0) e_3. \end{aligned}$$

By means of the usual isomorphy this quaternion will be associated with the $\mathbb{R}^4$-vector

$$\begin{pmatrix} x_0 y_0 - x_1 y_1 - x_2 y_2 - x_3 y_3 \\ x_1 y_0 + x_0 y_1 - x_3 y_2 + x_2 y_3 \\ x_2 y_0 + x_3 y_1 + x_0 y_2 - x_1 y_3 \\ x_3 y_0 - x_2 y_1 + x_1 y_2 + x_0 y_3 \end{pmatrix}.$$

This vector is nothing else than the result of a left multiplication by the matrix

$$L_x := \begin{pmatrix} x_0 & -x_1 & -x_2 & -x_3 \\ x_1 & x_0 & -x_3 & x_2 \\ x_2 & x_3 & x_0 & -x_1 \\ x_3 & -x_2 & x_1 & x_0 \end{pmatrix}$$

on the $\mathbb{R}^4$-vector $y = (y_0, y_1, y_2, y_3)^\top$, i.e., we have

$$\begin{pmatrix} x_0 y_0 - x_1 y_1 - x_2 y_2 - x_3 y_3 \\ x_1 y_0 + x_0 y_1 - x_3 y_2 + x_2 y_3 \\ x_2 y_0 + x_3 y_1 + x_0 y_2 - x_1 y_3 \\ x_3 y_0 - x_2 y_1 + x_1 y_2 + x_0 y_3 \end{pmatrix} = L_x y.$$

In this way the matrix L_x will be associated in a natural way to the quaternion x,

$$x \to L_x \quad \text{with} \quad xy = L_x y$$

for all $y \in \mathbb{H}$, so that L_x will be called a *left-representation* of the quaternion x in $\mathbb{R}^{4 \times 4}$. It is not difficult to prove the properties

$$\begin{aligned} \text{(i)} \qquad & L_1 = E, \\ \text{(ii)} \qquad & L_{\overline{x}} = L_x^\top, \\ \text{(iii)} \qquad & L_{x\tilde{x}} = L_x L_{\tilde{x}}, \\ \text{(iv)} \qquad & \det L_x = |x|^4, \end{aligned}$$

where E denotes the unit matrix in $\mathbb{R}^{4 \times 4}$. We finally consider the decomposition

$$L_x = x_0 E + X$$

with $X^\top = -X$.

In a completely analogous way also a *right-representation*

$$R_x = \begin{pmatrix} x_0 & -x_1 & -x_2 & -x_3 \\ x_1 & x_0 & x_3 & -x_2 \\ x_2 & -x_3 & x_0 & x_1 \\ x_3 & x_2 & -x_1 & x_0 \end{pmatrix}$$

of the quaternion x in $\mathbb{R}^{4 \times 4}$ can be obtained, where $yx = R_x y$. The properties (i), (ii) and (iv) continue to hold, while (iii) gets replaced by

$$R_{x\tilde{x}} = R_{\tilde{x}} R_x.$$

Other matrix representations in $\mathbb{R}^{4 \times 4}$ are possible, but will not be discussed here. Following the presentation of [95], we can show that

$$L_{e_1} L_{e_2} L_{e_3} = -E, \quad R_{e_1} R_{e_2} R_{e_3} = E.$$

Both sets $\{L_x \in \mathbb{R}^{4 \times 4} : x \in \mathbb{H}\}$ and $\{R_x \in \mathbb{R}^{4 \times 4} : x \in \mathbb{H}\}$ build for $\mathbb{H}$ isomorphic subalgebras of $\mathbb{R}^{4 \times 4}$.

Remark 2.28. Quaternions can also be represented by means of matrices in $\mathbb{C}^{2\times 2}$. An association close to physics is the following one:

$$x_0 e_0 + e_1 x_1 + e_2 x_2 + e_3 x_3 \rightarrow \begin{pmatrix} x_0 - ix_3 & -ix_1 - x_2 \\ -ix_1 + x_2 & x_0 + ix_3 \end{pmatrix}.$$

This association appears naturally, if the orthogonal unit vectors e_0, e_1, e_2, e_3 are mapped on variants of the so-called *Pauli matrices* , i.e., one sets as Pauli matrix

$$\sigma_0 := \begin{pmatrix} 1 & 0 \\ 0 & 1 \end{pmatrix}, \sigma_1 := \begin{pmatrix} 0 & 1 \\ 1 & 0 \end{pmatrix}, \sigma_2 := \begin{pmatrix} 0 & -i \\ i & 0 \end{pmatrix}, \sigma_3 := \begin{pmatrix} 1 & 0 \\ 0 & -1 \end{pmatrix}$$

and identifies the e_j successively with $\sigma_0, -i\sigma_1, -i\sigma_2, -i\sigma_3 = \sigma_1 \sigma_2$. We thus obtain a subalgebra of $\mathbb{C}^{2\times 2}$.

Remark 2.29. In analogy to the non-commutative quaternions CLYDE DAVENPORT published a monograph with the title *"A commutative hypercomplex calculus with applications to special relativity"* [30]. In place of the Pauli matrices he used the so-called *Davenport numbers*, which are

$$e_1 := \begin{pmatrix} 0 & -i \\ i & 0 \end{pmatrix}, \quad e_2 := \begin{pmatrix} 0 & -1 \\ 1 & 0 \end{pmatrix}, \quad e_3 := \begin{pmatrix} i & 0 \\ 0 & i \end{pmatrix}.$$

We can show that after introduction of the usual addition and of a suitable multiplication the totality of all Davenport numbers becomes a commutative algebra, which is also called *D-space algebra*. Such algebras have been closely investigated in particular by B. PEIRCE (1881) [113] and E. STUDY (1889) [149]. Similar to the Study numbers in the plane the mathematical abundance of the Davenport numbers is not given, and therefore these are introduced only for very special tasks. Further information on these structures can be found in [30].

2.4 Vectors and geometrical aspects

In July 1846 a paper on quaternions by W.R. Hamilton appeared in *Philosophical Magazine* of the Royal Irish Academy, in which among other things for the first time he introduced the words *vector* and *scalar* as parts of a quaternion. He considered a quaternion with the notation

$$q = w + \mathbf{i}x + \mathbf{j}y + \mathbf{k}z.$$

He considers ([29], S. 31):

The algebraically real part may receive ... all values contained on the one scale of progression of number from negative to positive infinity; we shall call it therefore the scalar part, or simply the scalar of the quaternion,... On the other hand the algebraically imaginary part, being geometrically constructed by a straight line or radius vector, which has in general for each determined quaternion, a determined length and determined direction in space, may be called vector part, or simply the vector of a quaternion.

He introduced for the scalar part of the quaternion q the notation "S.q" resp.
"Scal.q". The vector part of the upper quaternion was abbreviated to "V.q" corresp.
"Vect.q". W.R. Hamilton created these symbols only in order to quickly separate
the real and imaginary parts of a quaternion. By considering the two vectors

$$v = \mathbf{i}x + \mathbf{j}y + \mathbf{k}z \quad \text{and} \quad v' = \mathbf{i}x' + \mathbf{j}y' + \mathbf{k}z'$$

and calculating their quaternion product, Hamilton obtained the suitable definitions of scalar and cross product. The latter appeared as

$$\text{S.}vv' = -(xx' + yy' + zz'),$$
$$\text{V.}vv' = \mathbf{i}(yz' - zy') + \mathbf{j}(zx' - xz') + \mathbf{k}(xy' - yx').$$

Nowadays the scalar product is usually introduced as $-\text{S.}vv'$, while the definition for the cross product remains unchanged. Thus Hamilton can absolutely be
regarded as the founder of vector calculus, the later competitor of quaternion calculus. The vector calculus in particular developed by the American JOSIAH W.
GIBBS (1839–1903) and the English OLIVER HEAVISIDE (1850–1925) has emancipated as easier to learn from the mother, the quaternions and has become an
independent field.

However a large number of physical quantities like mass, charge, time, temperature
are defined just by giving a single number, a scalar. But in several situations also
the direction of the physical effects and of the flows is of interest. Quantities of this
sort are for instance path, force, velocity, and electrical field and are described by
means of vectors. The character of the quantities of this sort can be very different,
depending on the type of applications. We want to provide a few examples.

Example 2.30 (Locally applied vectors). Particles of a flowing fluid medium possess
individual velocities, which are different in module and direction. We describe the
velocities of the single particle by vectors, which are linked to the actual position;
we speak of *locally applied vectors*. The description of the whole current is obtained
through a *velocity field*. Such *vector fields* arise frequently in describing natural
phenomena.

Example 2.31 (Line vector). Force vectors can be displaced along their line of
action, without affecting the physical situation and are therefore called *line vectors*.

Since line vectors are meaningful only in special physical applications, in the following we want only to consider locally applied vectors and free vectors in $\mathbb{R}^3$. To
this end we endow the surrounding space with an orthonormal coordinate system,
so that every point can be described by its three coordinates: $A = (a_1, a_2, a_3)$.
To each pair of points A, B we can attach the locally applied vector $\overrightarrow{AB}$ in
A, to which a vector in $\text{Vec}\,\mathbb{H}$ or in $\mathbb{R}^3$ can be assigned with no problems by
$B - A = (b_1 - a_1, b_2 - a_2, b_3 - a_3)$.

The vectors locally applied into the origin are called in a special way, the *position
vectors*, since they describe the positions in space. The set of all applied vectors

at point A is called *tangential space* $T_A(\mathbb{R}^3)$ of $\mathbb{R}^3$ at the point A. The totality of the position vectors corresponds then to the tangential space $T_0(\mathbb{R}^3)$. We describe now the so-called *free vectors*.

Example 2.32. In the set of all locally applied vectors $\overrightarrow{AB}$ we introduce an equivalence relation

$$\overrightarrow{AB} \sim \overrightarrow{CD} \text{ if and only if } B - A = D - C.$$

We can easily convince ourselves that the latter is really an equivalence relation, since it is symmetric, transitive and reflexive; the corresponding equivalence classes $[\overrightarrow{AB}]$ are called *free vectors*. These classes contain all the locally applied vectors, which possess the same coordinate differences, and are then assigned to the same element of $\mathbb{R}^3$. Each class contains as representative exactly one position vector. In this way we can carry properties of the vectors in $\mathbb{R}^3$ over to the geometry of our surrounding space, an action that we will demonstrate in what follows.

Since the quantities just considered depend only on the coordinate differences of two points A and B, they are well defined for the free vectors and represent corresponding geometrical objects. Among these is the *Euclidean distance* $d(A, A')$ of two points A and A', given by the norm of the associated vector $|\overrightarrow{AA'}|$. For $|\overrightarrow{AA'}| = 1$, the applied vector in A as well as the free vector $[\overrightarrow{AA'}]$ are called *unit vectors*. In general we abbreviate the unit vectors as a lowercase boldface $\mathbf{e}$, as well as position vectors $\overrightarrow{OX}$ with a lower case boldface $\mathbf{x}$.

We can reconsider the geometrical meaning of the already known vector operations: the addition of two position vectors $\overrightarrow{OA}$ and $\overrightarrow{OB}$ can be illustrated by the parallelogram that both vectors span, the sum being represented by the diagonal of the parallelogram originating in O. If we add two free vectors $[\overrightarrow{OA}]$ and $[\overrightarrow{OB}]$, we can select also from the second class a representative which is locally applied at A whose endpoint is then the sum of both vectors. Clearly the commutativity and associativity of addition are easily proven. The *null vector* $\mathbf{0}$ corresponds to a vector of zero length, $[\overrightarrow{AA}]$, which must absolutely be distinguished from the origin O. Since this addition corresponds also for instance to addition of forces in physics, vectors are an important tool for describing forces.

If we multiply a vector with a real number $r > 0$, we do not change its direction, but only its length by a factor r. In case the factor is $r < 0$, in particular $r = -1$, then $-[\overrightarrow{AB}]$ is the *opposite vector* $[\overrightarrow{BA}]$, the addition of $[\overrightarrow{AB}]$ and $[\overrightarrow{BA}]$ gives then the null vector. All the remaining computational rules of the vector space are transferred naturally to position vectors or to free vectors. In the future we will not particularly mention the geometrical interpretation but will assume the concept is understood.

We want to mention another important concept of vector space theory:

Definition 2.33 (Linear dependence). The vectors $\mathbf{a}_1, \mathbf{a}_2, \ldots, \mathbf{a}_n$ are called *linearly dependent*, if there are real numbers $r_1, r_2, \ldots, r_n$ for which $r_1^2 + r_2^2 + \cdots + r_n^2 \neq 0$, so that $r_1 \mathbf{a}_1 + \cdots + r_n \mathbf{a}_n = \mathbf{0}$. In case such numbers cannot be found, the vectors $\mathbf{a}_1, \mathbf{a}_2, \ldots, \mathbf{a}_n$ are called *linearly independent*.

Remark 2.34. For $n = 2$ the concept of linear dependence is indicated as *collinearity*. In case of linear dependence for three vectors we say that they are *coplanars*.

The following remark is fundamental for this concept. If two vectors are collinear, then one is a real multiple of the other one. We could describe this with the vector product introduced in Section 2.2, since the latter vanishes if and only if both vectors are parallel or collinear.

For the description of the coplanarity we need a threefold product, which we will learn only later.

We add a remark from mechanics:

Example 2.35 (Center of mass). The center of gravity (or of mass) $\mathbf{s}$ of several point masses m_k located at the points $\mathbf{x}_k$ is calculated by the formula

$$\mathbf{s} = \frac{\sum m_k \mathbf{x}_k}{\sum m_k}.$$

For any two distinct points the center of gravity bisects the segment joining the points in the proportion $m_1 : m_2$, since

$$\mathbf{s}_{12} - \mathbf{x}_1 = \frac{m_2}{m_1 + m_2}(\mathbf{x}_2 - \mathbf{x}_1), \quad \mathbf{s}_{12} - \mathbf{x}_2 = \frac{m_1}{m_1 + m_2}(\mathbf{x}_1 - \mathbf{x}_2).$$

The segment lengths stand in the proportion $m_1 : m_2$.

If we add another point mass at $\mathbf{x}_3$ with mass m_3, for the center of mass it follows that

$$\mathbf{s}_{123} = \frac{m_1 \mathbf{x}_1 + m_2 \mathbf{x}_2 + m_3 \mathbf{x}_3}{m_1 + m_2 + m_3} = \mathbf{s}_{12} + \frac{m_3}{m_1 + m_2 + m_3}(\mathbf{x}_3 - \mathbf{s}_{12}).$$

From $(m_1 + m_2 + m_3)\mathbf{s}_{123} = m_3 \mathbf{x}_3 + (m_1 + m_2)\mathbf{s}_{12}$ it follows that the segment $\overline{X_3 S_{12}}$ is divided in the proportion $(m_1 + m_2) : m_3$ by S_{123}. A corresponding fact holds also for the lines through the other vertices. It then follows that the center of gravity or center of mass of three point masses concentrated at the corners of a triangle is the intersection of the line joining the vertex with the center of mass or of gravity of the opposite side. Here S_{ij} cuts the segment $\overline{X_i X_j}$ in the proportion $m_i : m_j$ for $i, j = 1, 2, 3$, $i \neq j$. It follows immediately that

$$\frac{m_2}{m_1} \cdot \frac{m_3}{m_2} \cdot \frac{m_1}{m_3} = 1 .$$

Thus we have shown the well-known theorem of Giovanni Ceva (1648 to 1734):

Theorem 2.36 (Ceva's Theorem). *If the three lines through the three vertices of a triangle cut each other all at the same point, then the product of the proportions of their side subdivisions is equal to one.*

Corollary 2.37. *If* $m_1 = m_2 = m_3 = 1$ *holds, then* s_{123} *cuts any one of the three medians, i.e., the segments joining a vertex with the center of the opposite side, in the proportion* $2 : 1$.

2.4.1 Bilinear products

In Section 2.2 we encountered the vector products introduced by Hamilton, which later have become independent with vector calculus. We want to prove properties of these products as well as investigate more closely their geometrical meaning. We will always move in the space $\mathrm{Vec}\,\mathbb{H}$ or also in $\mathbb{H}$, and dealing with geometrical questions we interchange freely point and free vectors. The products allow us to comprehend more situations in the applications.

We recall that for the two vectors $\mathbf{x} = x_1\mathbf{e_1} + x_2\mathbf{e_2} + x_3\mathbf{e_3}$ and $\mathbf{y} = y_1\mathbf{e_1} + y_2\mathbf{e_2} + y_3\mathbf{e_3}$ in Section 2.2 we have defined both the following products,

the *scalar product*

$$\mathbf{x} \cdot \mathbf{y} = -\mathrm{Sc}\,\mathbf{xy} = -\frac{1}{2}(\mathbf{xy} + \mathbf{yx}) = x_1y_1 + x_2y_2 + x_3y_3,$$

and the *vector product*

$$\mathbf{x} \times \mathbf{y} = \mathrm{Vec}\,\mathbf{xy} = \frac{1}{2}(\mathbf{xy} - \mathbf{yx}) = \begin{vmatrix} \mathbf{e_1} & \mathbf{e_2} & \mathbf{e_3} \\ x_1 & x_2 & x_3 \\ y_1 & y_2 & y_3 \end{vmatrix}.$$

Here the determinant is formally defined and is to be developed along the first row. We determine at first the algebraic properties of both products and interpret their geometrical meaning. The vector product was adopted in the vector calculus in a work by Gibbs in 1901.

Proposition 2.38. *Let* $\mathbf{x}, \mathbf{y} \in \mathrm{Vec}\,\mathbb{H}$.

(i) *The scalar and vector products are both homogeneous, i.e., for a real number* r *we have*

$$r(\mathbf{x} \cdot \mathbf{y}) = (r\mathbf{x}) \cdot \mathbf{y} = \mathbf{x} \cdot (r\mathbf{y}) \text{ as well as } r(\mathbf{x} \times \mathbf{y}) = (r\mathbf{x}) \times \mathbf{y} = \mathbf{x} \times (r\mathbf{y}).$$

(ii) *The scalar and vector products are both left and right distributive, i.e.,*

$$\mathbf{x} \cdot (\mathbf{y} + \mathbf{z}) = \mathbf{x} \cdot \mathbf{y} + \mathbf{x} \cdot \mathbf{z} \text{ and } (\mathbf{x} + \mathbf{y}) \cdot \mathbf{z} = \mathbf{x} \cdot \mathbf{z} + \mathbf{y} \cdot \mathbf{z},$$
$$\mathbf{x} \times (\mathbf{y} + \mathbf{z}) = \mathbf{x} \times \mathbf{y} + \mathbf{x} \times \mathbf{z} \text{ and } (\mathbf{x} + \mathbf{y}) \times \mathbf{z} = \mathbf{x} \times \mathbf{z} + \mathbf{y} \times \mathbf{z}.$$

(iii) *The scalar product is commutative, the vector product is anticommutative, i.e.,*

$$\mathbf{x} \cdot \mathbf{y} = \mathbf{y} \cdot \mathbf{x} \text{ und } \mathbf{x} \times \mathbf{y} = -\mathbf{y} \times \mathbf{x}.$$

Proof. (i) The proofs for scalar and vector product are completely parallel. Since real numbers can be exchanged with quaternions, from

$$r(\mathbf{x} \cdot \mathbf{y}) = -\frac{r}{2}(\mathbf{xy} + \mathbf{yx})$$

it follows immediately that r can be pulled inside the product near $\mathbf{x}$ as well as near $\mathbf{y}$.
(ii) Again the proof can be done in parallel for the scalar and vector products. Let us choose one of the four equations. From the distributivity of the quaternion multiplication it follows that

$$\mathbf{x} \times (\mathbf{y} + \mathbf{z}) = \frac{1}{2}(\mathbf{x}(\mathbf{y} + \mathbf{z}) - (\mathbf{y} + \mathbf{z})\mathbf{x})$$
$$= \frac{1}{2}(\mathbf{xy} + \mathbf{xz} - \mathbf{yx} - \mathbf{zx}) = \mathbf{x} \times \mathbf{y} + \mathbf{x} \times \mathbf{z}.$$

(iii) The commutavity of the scalar product and the anticommutativity of the vector product are an easy consequence of the definition. □

We examine now the geometrical meaning of the products, at first for the scalar product. In view of $\overline{\mathbf{x}} = -\mathbf{x}$ we have

$$\mathbf{x} \cdot \mathbf{x} = -\mathbf{x}^2 = \mathbf{x}\overline{\mathbf{x}} = |\mathbf{x}|^2,$$

the scalar product of a vector by itself is then the square of its length. Thus

$$|\mathbf{x} - \mathbf{y}|^2 = \mathbf{x} \cdot \mathbf{x} + \mathbf{y} \cdot \mathbf{y} - 2\mathbf{x} \cdot \mathbf{y},$$

and from the cosine theorem of planar geometry we have finally

$$\mathbf{x} \cdot \mathbf{y} = |\mathbf{x}||\mathbf{y}| \cos \alpha,$$

where α denotes the angle of both vectors $\mathbf{x}$ and $\mathbf{y}$.

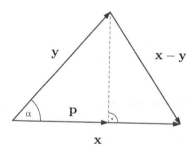

Figure 2.1

From the picture we infer that $|\mathbf{y}| \cos \alpha$ is the length of the projection $\mathbf{p}$ of $\mathbf{y}$ upon $\mathbf{x}$, which corresponds to a theorem in geometry. Moreover we state

The scalar product of two vectors vanishes if and only if the two vectors are orthogonal to each other, $\mathbf{x} \perp \mathbf{y}$.

In addition we have shown:

Corollary 2.39. *Since* $|\cos(\mathbf{x}, \mathbf{y})| \leq 1$ *we have*

$$|\mathbf{x} \cdot \mathbf{y}| \leq |\mathbf{x}||\mathbf{y}|;$$

which is the well-known Schwarz's inequality

HERMANN A. SCHWARZ (1843–1921), German mathematician, active in Halle, Zürich, Göttingen and Berlin. He worked in analysis and published important papers in function theory.

We now turn to the simple geometric figures in $\mathbb{R}^3$, straight lines, planes and spheres. If we solve an equation of the form $\mathbf{n} \cdot \mathbf{x} = d$ for $\mathbf{x}$ we find

Proposition 2.40 (Equation of a plane). *Let* $\mathbf{n} \neq \mathbf{0}$ *be a given vector and* d *a real number. Then the general solution of the equation*

$$\mathbf{n} \cdot \mathbf{x} = d$$

is given by

$$\mathbf{x} = \frac{\mathbf{n}}{|\mathbf{n}|^2} d + \mathbf{y} \tag{2.1}$$

where $\mathbf{y}$ *represents an arbitrary vector; orthogonal to* $\mathbf{n}$*. The equation* (2.1) *defines the plane through the point* $\mathbf{n}d/|\mathbf{n}|^2$ *orthogonal to* $\mathbf{n}$*. Thus* $\mathbf{n} \cdot \mathbf{x} = d$ *is an equation of a plane in* $\mathbb{R}^3$*, and if the vector* $\mathbf{n}$ *is a unit vector, we speak of the Hesse normal form of the plane equation,* $\mathbf{n}$ *represents then the normal of the plane.*

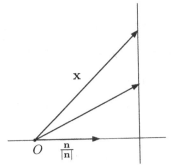

Figure 2.2

Proof. Clearly $\mathbf{x}_0 := \mathbf{n}d/|\mathbf{n}|^2$ is a solution of the equation. To it we can add all the vectors $\mathbf{y}$, for which $\mathbf{n} \cdot \mathbf{y} = 0$, thus obtaining the above expression for the general solution. Geometrically a vector $\mathbf{n}$ applied at a point $\mathbf{x}_0$ in $\mathbb{R}^3$ determines a plane orthogonal to $\mathbf{n}$, which is indeed specified by our equation. $\qquad\square$

Thus we have described one of the first geometrical figures, the plane.

Example 2.41 (Sphere). a) Let X be a point on the *sphere* with *center* M and *radius* r. Clearly we have $|\overline{XM}|^2 = r^2$, so that

$$(\mathbf{x} - \mathbf{m}) \cdot (\mathbf{x} - \mathbf{m}) = |\mathbf{x} - \mathbf{m}|^2 = r^2.$$

For $\mathbf{m} = \mathbf{0}$ it follows that $|\mathbf{x}| = r$. In a similar way as in $\mathbb{C}$ the *tangent plane* at the point $\mathbf{x}_0$ is the plane which is orthogonal to the radius vector $\mathbf{x}_0 - \mathbf{m}$ at the point $\mathbf{x}_0$. Its equation is then

$$\mathbf{x} \cdot (\mathbf{x}_0 - \mathbf{m}) = \mathbf{x}_0 \cdot (\mathbf{x}_0 - \mathbf{m}).$$

b) As a further example let us consider the *parametric representation of a plane*: if $\mathbf{a}$ and $\mathbf{b}$ denote two vectors in the considered plane through the point $\mathbf{x}_0$ which are not collinear, then one can reach every point of the plane by adding to $\mathbf{x}_0$ arbitrary real multiples of $\mathbf{a}$ and $\mathbf{b}$ giving the equation

$$\mathbf{x} = \mathbf{x}_0 + s\mathbf{a} + t\mathbf{b}, \quad s, t \in \mathbb{R}.$$

Before we make geometrical remarks on the vector product, we want to prove a proposition due to Lagrange:

JOSEPH L. LAGRANGE (1736–1813) was born of French parents in Turin in Italy and educated there. He worked in Turin, was president of the Prussian Academy in Berlin, and later was made president of the committee for reforming the standards of weights and measures in Paris. He was one of the greatest mathematicians of his century, his fields were mainly mechanics and analysis.

Proposition 2.42 (Lagrange identity). *Let* $\mathbf{x}, \mathbf{y}$ *be given vectors. Then*

$$|\mathbf{x}|^2 |\mathbf{y}|^2 = |\mathbf{x} \cdot \mathbf{y}|^2 + |\mathbf{x} \times \mathbf{y}|^2.$$

Proof. Following Hamilton we had the relationship

$$\mathbf{x}\mathbf{y} = -\mathbf{x} \cdot \mathbf{y} + \mathbf{x} \times \mathbf{y},$$

from which taking moduli it follows that

$$|\mathbf{x}\mathbf{y}|^2 = |\mathbf{x} \cdot \mathbf{y}|^2 + |\mathbf{x} \times \mathbf{y}|^2,$$

i.e., the statement. $\qquad\qquad\qquad\qquad\qquad\qquad\qquad\qquad\qquad\qquad\qquad\qquad\quad$ □

We now consider geometrical properties of the vector product. The equation

$$4\mathbf{x} \cdot (\mathbf{x} \times \mathbf{y}) = -\mathbf{x}(\mathbf{x}\mathbf{y} - \mathbf{y}\mathbf{x}) - (\mathbf{x}\mathbf{y} - \mathbf{y}\mathbf{x})\mathbf{x} = -\mathbf{x}^2\mathbf{y} + \mathbf{x}\mathbf{y}\mathbf{x} - \mathbf{x}\mathbf{y}\mathbf{x} + \mathbf{y}\mathbf{x}^2 = 0$$

(in view of the real $\mathbf{x}^2$) gives immediately that $\mathbf{x} \times \mathbf{y}$ is orthogonal to $\mathbf{x}$. In a similar way this holds also for $\mathbf{y}$, so that in summary we can state that

the vector $\mathbf{x} \times \mathbf{y}$ *is orthogonal to* $\mathbf{x}$ *and* $\mathbf{y}$.

From the Lagrange identity it follows that

$$|\mathbf{x} \times \mathbf{y}|^2 = |\mathbf{x}|^2|\mathbf{y}|^2 - |\mathbf{x} \cdot \mathbf{y}|^2 = |\mathbf{x}|^2|\mathbf{y}|^2 \sin^2 \alpha$$

with the angle $\alpha = \angle(\mathbf{x}, \mathbf{y})$. Thus the modulus of the vector product represents the surface area of the parallelogram spanned by $\mathbf{x}$ and $\mathbf{y}$. Also the vector $\mathbf{x} \times \mathbf{y}$ admits the representation

$$\mathbf{x} \times \mathbf{y} = |\mathbf{x}||\mathbf{y}|(\sin \alpha)\, \mathbf{e}_{\mathbf{x} \times \mathbf{y}},$$

which is also called *oriented surface area*, since $\mathbf{e}_{\mathbf{x} \times \mathbf{y}}$ is a unit vector orthogonal to $\mathbf{x}$ and $\mathbf{y}$. This follows from the special case $\mathbf{e}_1, \mathbf{e}_2, \mathbf{e}_3$ and

$$\mathbf{e}_1 \times \mathbf{e}_2 = \frac{1}{2}(\mathbf{e}_1\mathbf{e}_2 - \mathbf{e}_2\mathbf{e}_1) = \mathbf{e}_3$$

which satisfy the *right-hand rule*, i.e., if the thumb of the right hand points to the direction of $\mathbf{x}$ and the index finger to the direction of $\mathbf{y}$, then the middle finger indicates the direction $\mathbf{x} \times \mathbf{y}$, in the natural assumption that our coordinate frame of reference also satisfies this right-hand rule. We clearly have

$$\mathbf{e}_{(-\mathbf{x}) \times \mathbf{y}} = \mathbf{e}_{\mathbf{x} \times (-\mathbf{y})} = -\mathbf{e}_{\mathbf{x} \times \mathbf{y}}.$$

In a way similar to that for the scalar product we can solve the vector equation $\mathbf{a} \times \mathbf{x} = \mathbf{b}$ with the following result:

Proposition 2.43 (Plücker's equation of the line). *Let* $\mathbf{a}$ *and* $\mathbf{b}$ *be given vectors with* $\mathbf{a} \neq 0$ *and* $\mathbf{a} \perp \mathbf{b}$. *Then the general solution of the equation*

$$\mathbf{a} \times \mathbf{x} = \mathbf{b}$$

is given by

$$\mathbf{x} = \frac{\mathbf{b} \times \mathbf{a}}{|\mathbf{a}|^2} + t\mathbf{a},$$

where t *represents an arbitrary real number. This is a straight line through the point* $\mathbf{b} \times \mathbf{a}/|\mathbf{a}|^2$ *in the direction* $\mathbf{a}$; *the last equation is also called Plücker's equation of the line.*

The German mathematician JULIUS PLÜCKER (1801–1868) worked in several universities and finally for over 30 years as professor of mathematics and physics at the University of Bonn. His important mathematical works concern analytic geometry.

Proof. The necessary solvability condition $\mathbf{b} \perp \mathbf{a}$ follows from the direction of the cross product orthogonal to $\mathbf{a}$. The solution itself can only be sought in the direction orthogonal to $\mathbf{b}$, i.e., we can make the attempt with $\mathbf{x}_0 = \mathbf{b} \times \mathbf{a}$; in view of $\mathbf{a} \cdot (\mathbf{b} \times \mathbf{a}) = 0$ we have

$$
\begin{aligned}
\mathbf{a} \times (\mathbf{b} \times \mathbf{a}) &= \mathbf{a}(\mathbf{b} \times \mathbf{a}) + \mathbf{a} \cdot (\mathbf{b} \times \mathbf{a}) = \frac{1}{2}\mathbf{a}(\mathbf{b}\mathbf{a} - \mathbf{a}\mathbf{b}) \\
&= \frac{1}{2}(\mathbf{a} \cdot \mathbf{a})\mathbf{b} - \frac{1}{2}(\mathbf{a} \cdot \mathbf{b})\mathbf{a} + \frac{1}{2}(\mathbf{a} \times \mathbf{b}) \times \mathbf{a}
\end{aligned}
$$

and therefore
$$\mathbf{a} \times (\mathbf{b} \times \mathbf{a}) = (\mathbf{a} \cdot \mathbf{a})\mathbf{b} - (\mathbf{a} \cdot \mathbf{b})\mathbf{a}.$$
Thus $\mathbf{x}_0 = (\mathbf{b} \times \mathbf{a})/|\mathbf{a}|^2$ is indeed a solution of our equation to which all vectors collinear with $\mathbf{a}$ can be added, since their vector product with $\mathbf{a}$ vanishes. This is the given solution. Clearly the latter describes a straight line in view of the free parameter. Our starting equation goes back to Plücker who has provided in such a way a compact description of a straight line in $\mathbb{R}^3$, resp. in Vec $\mathbb{H}$. □

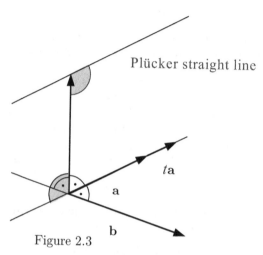

Plücker straight line

$t\mathbf{a}$

$\mathbf{a}$

$\mathbf{b}$

Figure 2.3

To characterize the geometrical position by algebraic means we can use special properties of the quaternion product of two vectors $\mathbf{x}$ and $\mathbf{y}$:

Proposition 2.44. *Let* $\mathbf{x}, \mathbf{y}$ *be vectors. We then have*

(i) $\mathbf{x}\mathbf{y} = \mathbf{y}\mathbf{x}$ *if and only if* $\mathbf{x}$ *and* $\mathbf{y}$ *are collinear and*

(ii) $\mathbf{x}\mathbf{y} = -\mathbf{y}\mathbf{x}$ *if and only if* $\mathbf{x}$ *is orthogonal to* $\mathbf{y}$.

Proof. (i) and (ii) follow immediately from the relations
$$0 = \mathbf{x}\mathbf{y} - \mathbf{y}\mathbf{x} = 2(\mathbf{x} \times \mathbf{y}) \quad \text{and} \quad 0 = \mathbf{x}\mathbf{y} + \mathbf{y}\mathbf{x} = -2\mathbf{x} \cdot \mathbf{y}.$$ □

2.4.2 Multilinear products

A multiplication combination of more than two vectors in scalar or vector product form entails some detours. While the vector product is nonassociative, i.e., in general the vectors $(\mathbf{x} \times \mathbf{y}) \times \mathbf{z}$ and $\mathbf{x} \times (\mathbf{y} \times \mathbf{z})$ are different from each other, the scalar product is mainly defined only for two vectors. An alternating application of both products is possible only with limitations, since in any case the vector product has to be calculated first. Here we can remark that multiple products are always to be expected with very specific properties. Instead the use of quaternion multiplication is completely without problems. The latter is computable without problems and in addition associative, a fact that in the investigation of multiple products will play an important role.

Let $\mathbf{x}, \mathbf{y}, \mathbf{z}$ be vectors. In view of quaternion calculus we have $(\mathbf{x}\,\mathbf{y})\mathbf{z} = \mathbf{x}(\mathbf{y}\,\mathbf{z})$. However we cannot guarantee anymore that the quaternion product of two vectors is again a vector. We already know that

$$\mathbf{x}\,\mathbf{y} = -\mathbf{x} \cdot \mathbf{y} + \mathbf{x} \times \mathbf{y},$$

so that we have

$$(\mathbf{x}\,\mathbf{y})\mathbf{z} = -(\mathbf{x} \cdot \mathbf{y})\mathbf{z} - (\mathbf{x} \times \mathbf{y}) \cdot \mathbf{z} + (\mathbf{x} \times \mathbf{y}) \times \mathbf{z} \ .$$

On the other hand we have the identity

$$\mathbf{x}(\mathbf{y}\,\mathbf{z}) = -\mathbf{x}(\mathbf{y} \cdot \mathbf{z}) - \mathbf{x} \cdot (\mathbf{y} \times \mathbf{z}) + \mathbf{x} \times (\mathbf{y} \times \mathbf{z}) \ .$$

Two quaternions are equal when their scalar and vector parts coincide. By comparison of the scalar and vector parts we have

(i) $\qquad \mathbf{x} \cdot (\mathbf{y} \times \mathbf{z}) = (\mathbf{x} \times \mathbf{y}) \cdot \mathbf{z},$

(ii) $\qquad \mathbf{x} \times (\mathbf{y} \times \mathbf{z}) + \mathbf{z}(\mathbf{x} \cdot \mathbf{y}) = (\mathbf{x} \times \mathbf{y}) \times \mathbf{z} + \mathbf{x}(\mathbf{y} \cdot \mathbf{z}) \ .$

The interpretation of (i) gives that the signs "$\cdot$" and "$\times$" in the double product are interchangeable. From this we can obtain a new notation which no longer contains explicitly the multiplication sign. We define:

Definition 2.45 (Mixed product). For three vectors $\mathbf{x}, \mathbf{y}, \mathbf{z}$ the product

$$(\mathbf{x}, \mathbf{y}, \mathbf{z}) := \mathbf{x} \cdot (\mathbf{y} \times \mathbf{z})$$

is called their *mixed product*.

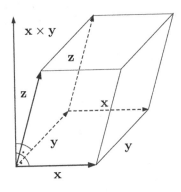

Figure 2.4

We summarize a few properties of the mixed product:

Proposition 2.46. (i) *The mixed product defines a trilinear vector form, i.e., in each component it is $\mathbb{R}$-homogeneous and distributive, for instance*

$$(\mathbf{x}, s\mathbf{y} + t\mathbf{y}', \mathbf{z}) = s(\mathbf{x}, \mathbf{y}, \mathbf{z}) + t(\mathbf{x}, \mathbf{y}', \mathbf{z}), \quad s, t \in \mathbb{R}.$$

(ii) *By cyclically interchanging the vectors in the mixed product the sign does not change, while by an anticyclical interchange it does change, so that*

$$(\mathbf{x}, \mathbf{y}, \mathbf{z}) = (\mathbf{y}, \mathbf{z}, \mathbf{x}) = (\mathbf{z}, \mathbf{x}, \mathbf{y}) = -(\mathbf{y}, \mathbf{x}, \mathbf{z}) = -(\mathbf{x}, \mathbf{z}, \mathbf{y}) = -(\mathbf{z}, \mathbf{y}, \mathbf{x}).$$

(iii) *The mixed product corresponds to the oriented volume of the parallelepiped spanned by the three vectors. Using the definitions of scalar and vector product we have then*

$$(\mathbf{x}, \mathbf{y}, \mathbf{z}) = |\mathbf{x}||\mathbf{y}||\mathbf{z}| \sin \angle(\mathbf{x}, \mathbf{y}) \cos \angle(\mathbf{x} \times \mathbf{y}, \mathbf{z}).$$

Then the mixed product vanishes if and only if the three vectors are coplanar.

(iv) *We have the determinant representation*

$$(\mathbf{x}, \mathbf{y}, \mathbf{z}) = \begin{vmatrix} x_1 & x_2 & x_3 \\ y_1 & y_2 & y_3 \\ z_1 & z_2 & z_3 \end{vmatrix}.$$

Proof. (i) This property holds simply because it is satisfied by the scalar and vector products.

(ii) The statement follows from the interchangeability of "·" and "×", as shown above, and from the commutativity of the scalar product as well as the anticommutativity of the vector product.

(iii) The proof follows immediately from the geometrical meaning of scalar and vector product. Only when the volume of such a solid figure vanishes are the three vectors coplanar. We thus have obtained a formula for the computation of coplanarity.

(iv) We obtain immediately

$$
\begin{aligned}
(\mathbf{x}, \mathbf{y}, \mathbf{z}) &= (\mathbf{x} \times \mathbf{y}) \cdot \mathbf{z} \\
&= \left(\begin{vmatrix} x_2 x_3 \\ y_2 y_3 \end{vmatrix} \mathbf{e}_1 + \begin{vmatrix} x_3 x_1 \\ y_3 y_1 \end{vmatrix} \mathbf{e}_2 + \begin{vmatrix} x_1 x_2 \\ y_1 y_2 \end{vmatrix} \mathbf{e}_3 \right) \cdot \left(z_1 \mathbf{e}_1 + z_2 \mathbf{e}_2 + z_3 \mathbf{e}_3 \right) \\
&= z_1 \begin{vmatrix} x_2 x_3 \\ y_2 y_3 \end{vmatrix} + z_2 \begin{vmatrix} x_3 x_1 \\ y_3 y_1 \end{vmatrix} + z_3 \begin{vmatrix} x_1 x_2 \\ y_1 y_2 \end{vmatrix}
\end{aligned}
$$

and from this the statement. □

The following discussion is devoted to the double vector product.

Proposition 2.47 (Development formula). *Let* x, y, z *be vectors. We then have*

$$x \times (y \times z) = (x \cdot z)y - (x \cdot y)z.$$

Proof. Using $2(x \cdot z)y = (x \cdot z)y + y(x \cdot z)$ for each term from the definition of scalar and vector product it follows that

$$4[(x \cdot z)y - (x \cdot y)z]$$
$$= -(xz + zx)y - y(xz + zx) + (xy + yx)z + z(xy + yx)$$
$$= x(yz - zy) - (yz - zy)x = 4x \times (y \times z).$$
□

As immediate consequence we find

Corollary 2.48 (Identity sum of the double vector product). *Let* x, y, z *be vectors. Then*

$$x \times (y \times z) + y \times (z \times x) + z \times (x \times y) = 0.$$

Proof. If we apply the development theorem subsequently to each of the three products, we obtain

$$x \times (y \times z) + y \times (z \times x) + z \times (x \times y)$$
$$= y(x \cdot z) - z(x \cdot y) + z(y \cdot x) - x(y \cdot z) + x(z \cdot y) - y(z \cdot x) = 0.$$
□

The development theorem for the double vector product allows the inclusion of other vectors, which leads to a vector form of the Lagrange identity:

Proposition 2.49 (Lagrange identity). *Let* w *be a further vector. For the scalar product of two vector products we have*

$$(x \times y) \cdot (z \times w) = \begin{vmatrix} x \cdot z & x \cdot w \\ y \cdot z & y \cdot w \end{vmatrix}.$$

Proof. We have

$$(x \times y) \cdot (z \times w) = [(x \times y) \times z] \cdot w$$
$$= [y(x \cdot z) - x(y \cdot z)] \cdot w = (y \cdot w)(x \cdot z) - (x \cdot w)(y \cdot z).$$

From this the determinant representation follows immediately.
□

If we now consider the factor $z \times w$ substituting z in the double vector product, the fourfold vector product can be calculated as follows:

$$(x \times y) \times (z \times w) = y(x \cdot (z \times w)) - x(y \cdot (z \times w)) = y(x, z, w) - x(y, z, w).$$

In particular the following relation is proven by substituting z by y and w by z:

Proposition 2.50 (Double factor rule). *Let* x, y, z *be given vectors. Then*

$$(x \times y) \times (y \times z) = y(x, y, z).$$

2.5 Applications

2.5.1 Visualization of the sphere S^3

To an arbitrary quaternion $x = x_0 + x_1\mathbf{i} + x_2\mathbf{j} + x_3\mathbf{k}$ let us associate the complex numbers $z_1 := x_0 + \mathbf{i}x_1$ and $z_2 := x_2 + \mathbf{i}x_3$. These allow the representation $x = z_1 + z_2\mathbf{j}$. A quaternion $x \in S^3$ corresponds then to a pair of complex numbers $(z_1, z_2) \in \mathbb{C} \times \mathbb{C} = \mathbb{C}^2$ with $|x|^2 = |z_1|^2 + |z_2|^2 = 1$. In a natural way we find now a mapping m, which associates to x a unitary 2×2 matrix with complex entries. We define:

$$m(x) := \begin{pmatrix} z_1 & z_2 \\ -\overline{z}_2 & \overline{z}_1 \end{pmatrix} \in \mathbb{C}^{2 \times 2}.$$

All unitary matrices with det $m(x) = 1$ make the *unitary group* $SU(2)$. Moreover we have $\overline{m^T} = m^{-1}$.

Proposition 2.51. *The mapping* $m : S^3 \rightarrow SU(2)$ *with* $x \rightarrow m(x)$ *is an isomorphism.*

Proof. The identity $m(xx') = m(x)m(x')$ needs to be shown. In fact let $x' = z_1' + z_2'\mathbf{j}$ be a further quaternion with $z_1', z_2' \in \mathbb{C}$. We then have

$$(z_1 + z_2\mathbf{j})(z_1' + z_2'\mathbf{j}) = (z_1 z_1' - z_2\overline{z_2'}) + (z_1 z_2' + z_2\overline{z_1'})\mathbf{j}.$$

The corresponding matrix multiplication reads

$$\begin{pmatrix} z_1 & z_2 \\ -\overline{z}_2 & \overline{z}_1 \end{pmatrix} \begin{pmatrix} z_1' & z_2' \\ -\overline{z}_2' & \overline{z}_1' \end{pmatrix} = \begin{pmatrix} z_1 z_1' - z_2\overline{z_2'} & z_1 z_2' + z_2\overline{z_1'} \\ -(z_1 z_2' + z_2\overline{z_1'}) & \overline{z_1 z_1' - z_2\overline{z_2'}} \end{pmatrix},$$

which then proves our relationship. $\qquad\square$

The identification of S^3 and $SU(2)$ allows us to describe with a formal algebraic calculation poles, meridians and parallels. To this end, we consider first the characteristic polynomial of the matrix $m(x) = m(z_1, z_2) \in SU(2)$, which is given by the equation

$$\det(m(z_1, z_2) - \lambda E) = (z_1 - \lambda)(\overline{z}_1 - \lambda) + |z_2|^2 = \lambda^2 - (z_1 + \overline{z}_1)\lambda + 1.$$

The value $x_0 = \operatorname{Re} z_1 \in [-1, 1]$ is exactly half the trace of the matrix $m(x)$ or the scalar part of the quaternion x. This can be used to describe parallels on S^3 in a completely analogous way as for $\mathbb{R}^3$. Unfortunately we only have an analogy to express this similarity. A "parallel" on S^3 can be described by the formula

$$x_1^2 + x_2^2 + x_3^2 = 1 - x_0^2.$$

These are two-dimensional spheres, for $x_0 = 0$ we get the "equator". For $x_0 = \pm 1$ it follows that $x_1 = x_2 = x_3 = 0$, and these would then be the poles. In an analogous way the parallels in $\mathbb{R}^3$ would be described by $x_1^2 + x_2^2 = 1 - x_0^2$. The parameter value $x_0 = 0$ corresponds to the equator and $x_0 = \pm 1$ with $x_1 = x_2 = 0$ to the poles.

Remark 2.52. The parallels are associated to a whole class $\mathcal{M}_m$ of unitary matrices, since all matrices

$$\mathcal{M}_m = \{m'mm'^{-1} : m' \in SU(2)\}$$

possess the same trace x_0. Conversely each such similarity class is uniquely tied with a parallel on S^3.

A more refined geometrical visualization of the sphere S^3 can be illustrated by means of the function $\psi : S^3 \to [-1,1]$ with $\psi(z_1, z_2) = |z_1|^2 - |z_2|^2$. We consider the following levels:

$$N_{\psi_0} := \{(z_1, z_2) \in S^3 : \psi(z_1, z_2) = \psi_0, \ \psi_0 \in [-1,1]\}.$$

Since $|z_1|^2 + |z_2|^2 = 1$ we have

$$|z_1|^2 = \frac{1 + \psi_0}{2} \ \text{and} \ |z_2|^2 = \frac{1 - \psi_0}{2}.$$

It is immediately clear that N_1 and N_{-1} correspond indeed to the unit circles in the first, resp. second factor of $\mathbb{C}^2$. Finally we obtain that N_{ψ_0} with $-1 < \psi_0 < 1$ is the characteristic product of both circles, i.e., through

$$N_{\psi_0} := \left\{(z_1, z_2) \in \mathbb{C}^2 : |z_1|^2 = \frac{1 + \psi_0}{2} ; |z_2|^2 = \frac{1 - \psi_0}{2}\right\}$$

a torus is obtained. We also say that the tori (with $\psi_0 \in (-1,1)$), which are called *Clifford-Tori*, are slices of the sphere S^3.

2.5.2 Elements of spherical trigonometry

Multiple products can be applied in a particularly comfortable way to found the elementary relationships of spherical trigonometry. Spherical trigonometry is important in particular directions of the engineering sciences, in particular mining and topography. It is not our goal to completely treat spherical trigonometry, but we present only some chosen examples to show the usefulness of the quaternionic calculus.

A spherical triangle is obtained when we cut out of the unit sphere a tetrahedron with a vertex in the center of the sphere, which serves as origin and therefore is denoted by O. The vertices of the spherical triangle A, B, C correspond to the position vectors $\mathbf{a}, \mathbf{b}, \mathbf{c}$. The angles between the edge vectors will be consecutively denoted by $\angle(\mathbf{a}, \mathbf{b}) = \gamma$, $\angle(\mathbf{b}, \mathbf{c}) = \alpha$, $\angle(\mathbf{c}, \mathbf{a}) = \beta$. The angles between the side surfaces of the tetrahedron are seen as angles in the spherical triangle. These angles in the points A, B, C are denoted by α', β', γ'. From the picture it follows that for $\alpha'' = \angle(\mathbf{c} \times \mathbf{a}, \mathbf{a} \times \mathbf{b})$ we have $\alpha'' = \pi - \alpha'$. We should observe that the twice appearing vectors in this angle description stand inwards. It then follows trivially that $\cos \alpha'' = -\cos \alpha'$, $\sin \alpha'' = \sin \alpha'$. In an analogous way a similar relation holds also for the remaining angles.

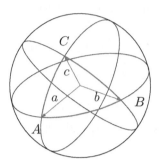

Figure 2.5

Proposition 2.53 (Spherical cosine theorem). *We have*

$$\cos \beta = \cos \gamma \cos \alpha + \sin \gamma \sin \alpha \cos \beta'.$$

Proof. Let $\mathbf{a}, \mathbf{b}, \mathbf{c}$ be vectors with $|\mathbf{a}| = |\mathbf{b}| = |\mathbf{c}| = 1$. The Lagrange identity delivers

$$(\mathbf{a} \times \mathbf{b}) \cdot (\mathbf{b} \times \mathbf{c}) = \begin{vmatrix} \mathbf{a} \cdot \mathbf{b} & \mathbf{a} \cdot \mathbf{c} \\ \mathbf{b} \cdot \mathbf{b} & \mathbf{b} \cdot \mathbf{c} \end{vmatrix} = (\mathbf{a} \cdot \mathbf{b})(\mathbf{b} \cdot \mathbf{c}) - \mathbf{a} \cdot \mathbf{c} = \cos \gamma \cos \alpha - \cos \beta .$$

For the left side we obtain

$$(\mathbf{a} \times \mathbf{b}) \cdot (\mathbf{b} \times \mathbf{c}) = \sin \gamma \sin \alpha \cos \beta'' = -\sin \gamma \sin \alpha \cos \beta' ,$$

from which the desired relation follows. $\square$

Proposition 2.54 (Spherical sine-cosine theorem). *With the already introduced notation we have the following relationship:*

$$\sin \alpha \cos \gamma' = \cos \gamma \sin \beta - \cos \beta \sin \gamma \cos \alpha'.$$

Proof. We start from the obvious identity

$$\begin{vmatrix} \mathbf{a} \cdot \mathbf{a} & \mathbf{a} \cdot \mathbf{b} & \mathbf{a} \cdot \mathbf{c} \\ \mathbf{a} \cdot \mathbf{a} & \mathbf{a} \cdot \mathbf{b} & \mathbf{a} \cdot \mathbf{c} \\ \mathbf{c} \cdot \mathbf{a} & \mathbf{c} \cdot \mathbf{b} & \mathbf{c} \cdot \mathbf{c} \end{vmatrix} = 0 \qquad \text{(first row = second row!)}$$

and develop along the first row, to get

$$(\mathbf{a} \cdot \mathbf{a}) \begin{vmatrix} \mathbf{a} \cdot \mathbf{b} & \mathbf{a} \cdot \mathbf{c} \\ \mathbf{c} \cdot \mathbf{b} & \mathbf{c} \cdot \mathbf{c} \end{vmatrix} + (\mathbf{a} \cdot \mathbf{b}) \begin{vmatrix} \mathbf{a} \cdot \mathbf{c} & \mathbf{a} \cdot \mathbf{a} \\ \mathbf{c} \cdot \mathbf{c} & \mathbf{c} \cdot \mathbf{a} \end{vmatrix} + (\mathbf{a} \cdot \mathbf{c}) \begin{vmatrix} \mathbf{a} \cdot \mathbf{a} & \mathbf{a} \cdot \mathbf{b} \\ \mathbf{c} \cdot \mathbf{a} & \mathbf{c} \cdot \mathbf{b} \end{vmatrix} = 0.$$

The application of the Lagrange identity gives

$$(\mathbf{a} \cdot \mathbf{a})[(\mathbf{a} \times \mathbf{c}) \cdot (\mathbf{b} \times \mathbf{c})] + (\mathbf{a} \cdot \mathbf{b})[(\mathbf{a} \times \mathbf{c}) \cdot (\mathbf{c} \times \mathbf{a})] + (\mathbf{a} \cdot \mathbf{c})[(\mathbf{a} \times \mathbf{c}) \cdot (\mathbf{a} \times \mathbf{b})] = 0,$$

so that

$$(\mathbf{a} \times \mathbf{c}) \cdot (\mathbf{b} \times \mathbf{c}) - (\mathbf{a} \cdot \mathbf{b})|\mathbf{a} \times \mathbf{c}|^2 - (\mathbf{a} \cdot \mathbf{c})[(\mathbf{c} \times \mathbf{a}) \cdot (\mathbf{a} \times \mathbf{b})] = 0$$

and therefore

$$- \sin \beta \sin \alpha \cos \gamma'' - \cos \gamma \sin^2 \beta - \cos \beta \sin \beta \sin \gamma \cos \alpha'' = 0 \ .$$

The division by $\sin \beta$ and the equation $\cos \gamma'' = - \cos \gamma'$ resp. $\cos \alpha'' = - \cos \alpha'$ prove
our assertion. $\qquad\square$

Proposition 2.55 (Spherical sine theorem). *We have*

$$\frac{\sin \beta'}{\sin \beta} = \frac{\sin \gamma'}{\sin \gamma} = \frac{\sin \alpha'}{\sin \alpha}.$$

Proof. From the double factor rule on the edge vector **b** it follows that

$$|\mathbf{a} \times \mathbf{b}||\mathbf{b} \times \mathbf{c}| \sin(\mathbf{a} \times \mathbf{b}, \mathbf{b} \times \mathbf{c}) = V|\mathbf{b}| \ , \quad \text{where } V := |(\mathbf{a}, \mathbf{b}, \mathbf{c})| \ .$$

Since $|\mathbf{a}| = |\mathbf{b}| = |\mathbf{c}| = 1$ we obtain

$$\sin \gamma \sin \alpha \sin \beta' = V \ .$$

Since a corresponding formula holds also on the edge vectors **a** and **c** we find

$$V = \sin \alpha \sin \beta \sin \gamma' = \sin \beta \sin \gamma \sin \alpha' = \sin \gamma \sin \alpha \sin \beta'$$

and therefore the assertion. $\qquad\square$

2.6 Exercises

1. Let x, y, z be arbitrary elements in $\mathbb{H}$. Prove Proposition 2.8:

   (a) $\text{Sc}(xyz) = \text{Sc}(yzx) = \text{Sc}(zxy)$.

   (b) $\text{Sc}(xyz) = \overline{x} \cdot (yz)$ is a real number.

2. Show that from every quaternion a in $\mathbb{H}$ at least one root can be calculated, i.e., there is at least one $x \in \mathbb{H}$ with $x^2 = a$.

3. Prove Proposition 2.13 (iv): every quaternion e with $|e| = 1$ can be represented in the form

   $$e = \mathbf{x}\mathbf{y}\mathbf{x}^{-1}\mathbf{y}^{-1}.$$

4. We consider a tetrahedron spanned by $\mathbf{x}, \mathbf{y}, \mathbf{z}$; the remaining edges are then suitable differences of these three vectors. Show that the sum of the so-called oriented surface area of all side surfaces (all surfaces at the same time oriented toward the interior or toward the exterior !) equals zero. Can this result be extended to an arbitrary polyhedron?

3   Clifford numbers

3.1   History of the discovery

While observation is a general foundation of mathematical knowledge up to dimension three, in higher dimensional spaces we have to free ourselves from any spatial imagination. Thus H. GRASSMANN writes in 1844 in his nowadays famous book *Die lineale Ausdehnungslehre (The linear extension theory)*:

Since long time it has become clear to me that geometry is to be regarded as a branch of mathematics in no way similar to arithmetic or combinatorics, but rather geometry refers to something already given by nature (indeed the space) and that therefore there should be a branch of mathematics which in a purely abstract way produces similar laws as they appear linked to space in geometry.

Hermann G. Grassmann (1809–1877)

This point of view set forth by H. Graßmann, led D. HILBERT in his paper of 1899 to completely strike out from geometry the concept of visualization. Graßmann's work remained in his time widely unnoticed and poorly understood, probably because the most part of his colleagues thought in the framework of three-dimensional space.

Only 18 years later in 1862 appeared the methodically strongly improved second edition of his earlier book. The latter, particularly popularized by H. HANKEL's book (1867) *Theory of the complex number systems, in particular of the common imaginary numbers and of the Hamiltonian quaternions with their geometrical representation* lastingly influenced the development of several fundamental theories like tensor calculus, vector analysis or also Clifford analysis. The algebraic fundamental terms of the extension theory were denoted as *extensive quantities* or *elementary quantities*, which can be combined through two product constructions, the *inner* and the *outer product*. The latter leads to k-vectors and finally to the antisymmetric tensors. In modern notation Graßmann required his elementary quantities $e_1, \ldots, e_n$ to satisfy the following algebraic properties:

(i) $e_i e_j + e_j e_i = 0, \quad e_i^2 = 0 \quad (i, j = 1, 2, \ldots, n),$

(ii) $e_i(e_j + e_k) = e_i e_j + e_i e_k.$

The basic relations of the algebra that bears his name, *Graßmann algebra* are thus defined.

HERMANN GÜNTER GRASSMANN (1809–1877) was born in 1809 as the third of 12 children. In 1827 he began to study theology and philology at the Berlin University. After his return to Stettin he continued to study mathematics and physics on his own while teaching. In 1834 he took a position at the Berlin Industry Institute (later Berlin Institute of Technology), but returned already in 1835 to Stettin. Around 1840 he wrote a book on the theory of tides and flows as dissertation. He applied several times for a university position, sadly without success. So he remained for his whole life as professor at the Stettin high school (Gymnasium). He married in 1849 and got 11 children in the following years. He was not only an excellent mathematician, but also a worldwide known linguist. He created in 1875 the *dictionary to Rig-Veda*. The Veda is the step before Sanscrit, the Rig-Veda is the foundation of Hinduismus. Graßmann was interested in particular in special affinities between the Latin and the Greek languages. In addition he busied himself with Gothic, old Prussian, Russian, old Persian, Lithuanian and slavic languages. In linguistics there is even the "Grassmannian aspiration law". In 1876 he got the honorary doctoral degree of the University of Tübingen. He passed away a year later of a kidney disease.

The genial combination of Graßmann's extension theory and Hamilton's quaternions led W.K. CLIFFORD in 1876–1878 to a structure of a *geometrical algebra*, as Clifford called it. In his famous work, published in 1878, *Applications of Grassmann's Extensive Algebra* he built a new algebra, made of scalars, vectors and in general of k-vectors ($1 \leq k \leq n$), the elements of which are today called *Clifford numbers*. The k-vectors were built by means of Graßmann's outer product. Every real linear combination of k-vectors and scalars is thus a Clifford number.

William K. Clifford (1845–1879)

The English philosopher and geometer WILLIAM KINGDON CLIFFORD (1845–1879) was appointed professor of applied mathematics at the London University College. He shortly thereafter became Fellow of the Royal Society. He was only 35 years old when he died of tuberculosis in Madeira.

Complex numbers and real quaternions are simple examples of Clifford numbers. In M. CHISHOLM's book *Such Silver Currents*, Clifford's relation with J. C. MAXWELL, who was one of the referees of his call, is described in this way:

After 1874 Clifford and Maxwell often met as Fellows of the Royal Society. They shared a mutual belief in the importance of Hamilton's "quaternion methods", and used them in their teachings and writings. Later religious differences would complicate this relationship and their attitudes to fundamental research.

Another significant mathematician of his time was the German RUDOLF LIP-SCHITZ (1832–1903), who discovered again the *geometric algebra* in 1880 [93] while studying sums of squares. He was also the first one to formulate geometric applications in higher dimensional spaces. K. THEODOR VAHLEN (1869–1945) introduced in 1902 [156] a multiplication rule between two basis elements of a Clifford algebra. Only in 1986 was this rule generalized in a paper by P.E. HAGMARK and P. LOUNESTO by using the Walsh functions. A significant progress was achieved in 1908 by E. Cartan, who discovered relationships between general Clifford algebras and matrix algebras and showed the periodicity theorem.

3.2 Definition and properties

3.2.1 Definition of the Clifford algebra

W.K. CLIFFORD has extended Hamilton's idea of quaternions to $\mathbb{R}^n$; in order to do that he had to define the multiplication in $\mathbb{R}^n$ which he did in a way completely analogous to that of Hamilton:

Definition 3.1. Let the space $\mathbb{R}^{n+1}$ be given, with the basis $\{e_0, e_1, \ldots, e_n\}$. For the multiplication let the following rules hold: Let us consider $p \in \{0, \ldots, n\}$, $q := n - p$; we define

$$e_0 e_i = e_i e_0 = e_0, \quad i = 1, \ldots, n,$$
$$e_i e_j = -e_j e_i, \quad i \neq j, \ i, j = 1, \ldots n,$$
$$e_0^2 = e_1^2 = \cdots = e_p^2 = 1, \ e_{p+1}^2 = \cdots = e_{p+q}^2 = -1.$$

Thus we obtain a basis of an algebra $\mathcal{A} =: Cl_{p,q}$:

$$e_0; e_1, \ldots, e_n; e_1 e_2, \ldots, e_{n-1} e_n; e_1 e_2 e_3, \ldots; e_1 e_2 \ldots e_n$$

with e_0 as unit element. The addition and the multiplication with a real number are defined coordinatewise. Further, let the condition hold

$$e_1 e_2 \ldots e_n \neq \pm 1 \ \text{if} \ p - q \equiv 1 \,(\text{mod}\, 4).$$

The algebra found in this way is called *(universal) Clifford algebra* $Cl_{p,q}$.

A list of remarks follows:

Remark 3.2. a) It is easy to see that $Cl_{p,q}$ is a real vector space and it is also easy to show that there are exactly 2^n given basis elements, since squares of the e_i are reduced in view of the assumptions. The further rules of an algebra are satisfied without problems; commutativity clearly does not hold (in case $n > 1$), associativity is contained in the definition, since we give the basis elements without brackets. Distributivity is easily calculated, if in the bracket there are two basis elements of the same degree, otherwise it is defined by means of linear continuation. There are other generalizations of this type to $\mathbb{R}^{n+1}$, among which some of the basis elements have the square 0.

b) The basis elements can also be given in the more easily understandable form

$$e_{i_1 i_2 \ldots i_p} := e_{i_1} e_{i_2} \ldots e_{i_p}.$$

The latter can be abbreviated even more, since as indices of the basis elements we can use the elements A of the set $\mathcal{P}_n$ containing the subsets of $\{1, \ldots, n\}$ where in these subsets the numbers are naturally ordered according to their size. Here the empty set corresponds to the index 0; thus we describe the elements of $C\ell_{p,q}$ in the form

$$x = \sum_{A \in \mathcal{P}_n} x_A e_A.$$

Such a number is called a *Clifford number*. We generally use $|A|$ to denote the cardinality of A. There are also other notations for $C\ell_{p,q}$, for instance the notation $\mathbb{R}_{p,q}$ introduced by F. Sommen, which has its own advantages.

c) The signature (p, q) remains invariant by a change of basis, according to the theorem of Sylvester in linear algebra, so that the above definition makes sense. We will mainly consider the algebra $C\ell_{0,n}$ and for it we introduce a special notation:

$$C\ell_{0,n} =: C\ell(n).$$

d) The Clifford algebra of an arbitrary (real) vector space V can also be introduced by means of a quadratic form $Q(x)$ over V. Then the product is reduced by the condition

$$x^2 = Q(x);$$

in our case above we would choose

$$Q(x) = x_1^2 + x_2^2 + \cdots + x_p^2 - x_{p+1}^2 - \cdots - x_{p+q}^2 \quad (p + q = n).$$

For $i = 1, \ldots, p$ after substitution of e_i,

$$e_i^2 = Q(e_i) = 1,$$

and correspondingly for the remaining e_i we obtain the equation $e_i^2 = -1$. If we insert the vector $e_i + e_j$ into $x^2 = Q(x)$ we obtain the exchange rule $e_i e_j + e_j e_i = 0$. Such a pair (V, Q) thus actually defines a Clifford algebra.

Example 3.3. a) *(Algebra of the real numbers.)* The index $n = 0$ is allowed. In this case we obtain the real numbers $\mathbb{R}$ as a Clifford algebra.

b) *(Algebra of the complex numbers.)* Let $n = 1$ and $p = 0$ be given. Then $C\ell_{0,1}$ is the vector space spanned by 1 and e_1 with $e_1^2 = -1$. Thus it is the set of complex numbers $\mathbb{C}$, where naturally 1 and $e_1 = i$ are to be taken as basis vectors in our sense.

c) *(Algebra of the dual numbers.)* Let $n = 1$ and $p = 1$ be given. Then $C\ell_{1,0}$ is the algebra of the *dual numbers*, spanned by the elements 1 and e_1 with $e_1^2 = 1$. The big disadvantage against the complex numbers is the appearance of zero divisors, since

$$(1 + e_1)(1 - e_1) = 0,$$

although the factors are not zero.

d) *(Algebra of the real quaternions.)* Here we take $n = 2$ and $p = 0$. Then we have as basis the elements $1, e_1, e_2, e_1e_2$, which correspond exactly to the basis elements of $\mathbb{H}$, also with respect to the calculation rules. The isomorphy $C\ell_{0,2} \cong \mathbb{H}$ follows, where as in $\mathbb{H}$ the basis elements are to be understood as vectors in $\mathbb{R}^4$.

e) *(Algebra $C\ell_{0,3}$.)* Let us now take $n = 3$ and $p = 0$. Then $C\ell_{0,3}$ becomes a Clifford algebra, which is generated by the anti-euclidean space with $e_1^2 = e_2^2 = e_3^2 = -1$. The basis consists of the vectors e_1, e_2, e_3, the bivectors (or 2-vectors) e_{12}, e_{23}, e_{13} and the 3-vector e_{123}. On the basis of the algebraic relation

$$e_{123}^2 = 1$$

we speak of e_{123} as a *pseudoscalar*. It is easy to show that

$$e_{123}e_i = e_i e_{123}$$

holds, i.e., e_{123} also belongs to the center of the algebra. We can also connect this algebra with the Pauli matrices introduced in Remark 2.28. This will be done below in Example 3.21 (7). Here an element

$$x = x_0e_0 + x_1e_1 + x_2e_2 + x_3e_3 + x_4e_1e_2 + x_5e_2e_3 + x_6e_3e_1 + x_7e_1e_2e_3$$

of the algebra is rewritten in the following way:

$$x = (x_0 + x_1e_1) + (x_4 - x_2e_1)e_1e_2 + (x_5 + x_7e_1)e_2e_3 + (x_6 + x_3e_1)e_3e_1.$$

f) *(Spacetime algebra)*. Here we choose $n = 4$ and $p = 1$. Then the algebra $C\ell_{1,3}$ is spanned by $1, e_1, e_2, e_3, e_4, e_{12}, \ldots, e_{1234}$; the generating vector space has the signature $(1, 3)$ and is called the *Minkowski space*. Let us remark that $\mathbb{H}$ as 4-dimensional space has also the signature $(1, 3)$, and we can imbed the Minkowski space into an algebra in different ways.

3. *Clifford numbers*

3.2.2 Structures and automorphisms

The $\mathbb{R}$-linear hull

$$\mathrm{span}\{e_A : |A| = k\}$$

generates the $\mathbb{R}$-linear subspace of the so-called *k-vectors*, which we will from now on denote by $C\ell_{p,q}^k$. Clearly in this space there are $\binom{n}{k}$ basis elements. It follows then that the vector space dimension of the (universal) algebra $C\ell_{p,q}$ can be calculated by

$$\dim_{\mathbb{R}} C\ell_{p,q} = \binom{n}{0} + \binom{n}{1} + \cdots + \binom{n}{n} = 2^n.$$

Let $[\cdot]_k : C\ell_{p,q} \to C\ell_{p,q}^k$ be a linear projection with

$$[x]_k := \sum_{|A|=k} x_A e_A.$$

Thus an arbitrary element $x \in C\ell_{p,q}$ can be written in the form

$$x = [x]_0 + [x]_1 + \cdots + [x]_n.$$

The elements of the form $x = [x]_0$ are called *scalars*, the elements of the form $x = [x]_1$ are called *vectors*, and finally the elements of the form $x = [x]_0 + [x]_1$ are called *paravectors*; they are the sum of a scalar and a vector. Moreover let

$$C\ell_{p,q}^+ = \bigoplus_{2 \le 2\ell \le n} C\ell_{p,q}^{2\ell} \quad \text{and} \quad C\ell_{p,q}^- = \bigoplus_{1 \le 2\ell+1 \le n} C\ell_{p,q}^{2\ell+1}.$$

The dimensions of these subspaces are both exactly 2^{n-1}, as it is easy to see. The subspace with the even degrees is even a subalgebra, since an even number of factors is again obtained by multiplication of the corresponding basis elements.

We want to state an algebraic property of $C\ell(n)$ where the *center* consists of the elements of the algebra, which are exchangeable with all other elements. The proof is skipped here, (cf. Exercise 3.5.3).

Proposition 3.4. *The center of the algebra $C\ell(n)$ for even n consists of the real numbers $\mathbb{R}$; for odd n it is generated from e_0 and the* pseudoscalar $e_1 e_2 \cdots e_n$.

We are now in place to describe important automorphisms with certain known invariant properties. We define next the principal involution (Inv M), the conjugation (Inv C) and the reversion (Inv R):

Definition 3.5. In case the identities

(i) $\mathrm{Inv}\, M(xy) = \mathrm{Inv}\, M(x) \mathrm{Inv}\, M(y)$,

(ii) $\mathrm{Inv}\, M(e_i) = -e_i \quad (i = 1, \ldots, n)$

are satisfied for arbitrary elements $x, y \in C\ell_{p,q}$, Inv M is called the *principal involution* or *inversion*. We write generally $\mathrm{Inv}\, M(x) =: \tilde{x}$.

The following proposition shows that really such an involution exists:

Proposition 3.6. *Let $x \in C\ell_{p,q}$ be given, then*

$$\operatorname{Inv} M(x) = \tilde{x} = [x]_0 - [x]_1 + [x]_2 - [x]_3 + \cdots ,$$

i.e., $C\ell_{p,q}^+$ and $C\ell_{p,q}^-$ are eigenspaces of the operator $\operatorname{Inv} M$.

Proof. The proof is an easy consequence of the definition. $\qquad\square$

In general for a k-vector x we have $\tilde{x} := (-1)^k x$. For arbitrary Clifford numbers we have $\widetilde{xy} = \tilde{x}\,\tilde{y}$.

Definition 3.7. In case the arbitrary elements $x, y \in C\ell_{p,q}$ satisfy the relations

(i) $\quad \operatorname{Inv} C(xy) = \operatorname{Inv} C(y) \operatorname{Inv} C(x),$

(ii) $\quad \operatorname{Inv} C(e_i) = -e_i \quad (i = 1, 2, \ldots, n),$

then $\operatorname{Inv} C$ is called the *Clifford conjugation*, for which we write

$$\operatorname{Inv} C(x) =: \overline{x}.$$

In case of no confusion we can speak of the *conjugation* which naturally corresponds to conjugations in $\mathbb{H}$ and $\mathbb{C}$. Here also we need to discuss the existence of such an involution:

Proposition 3.8. *For an arbitrary $x \in C\ell_{p,q}$ we have*

$$\operatorname{Inv} C(x) = \overline{x} = [x]_0 - [x]_1 - [x]_2 + [x]_3 + [x]_4 - \cdots ,$$

i.e., for $x \in C\ell_{p,q}^k$ we have

$$\overline{x} = x, \quad for \quad k \equiv 0, 3 \,(\mathrm{mod}\,4),$$
$$\overline{x} = -x, \quad for \quad k \equiv 1, 2 \,(\mathrm{mod}\,4).$$

Proof. The proof follows again immediately from the definition. $\qquad\square$

For a k-vector x we have the formula

$$\operatorname{Inv} C(x) = \overline{x} = (-1)^{\frac{k(k+1)}{2}} x,$$

and for arbitrary x, y it follows that $\overline{xy} = \overline{y}\,\overline{x}$, we have thus an antiautomorphism. Finally we want to define the reversion, which is the subsequent composition of conjugation and principal involution.

Definition 3.9. In case for arbitrary $x, y \in C\ell_{p,q}$ the relations

(i) $\quad \operatorname{Inv} R(xy) = \operatorname{Inv} R(y) \operatorname{Inv} R(x),$

(ii) $\quad \operatorname{Inv} R(e_i) = e_i \quad (i = 1, 2, \ldots n)$

are satisfied, then $\operatorname{Inv} R$ is called *reversion*. We write then $\operatorname{Inv} R(x) =: \hat{x}.$

3. *Clifford numbers*

As before we provide an existence proposition for such an involution:

Proposition 3.10. *For arbitrary* $x \in C\ell_{p,q}$ *we have*

$$\text{Inv } R = \hat{x} = [x]_0 + [x]_1 - [x]_2 - [x]_3 + [x]_4 + \cdots,$$

i.e., for $x \in C\ell_{p,q}^k$ *we have:*

$$\hat{x} = x, \quad for \quad k \equiv 0, 1 \,(\text{mod}\, 4),$$
$$\hat{x} = -x, \quad for \quad k \equiv 2, 3 \,(\text{mod}\, 4).$$

Proof. The proof is left as an exercise to the reader (cf. Exercise 3.5.4). $\qquad\square$

For an arbitrary k-vector x we have

$$\text{Inv } R(x) = \hat{x} = (-1)^{\frac{k(k-1)}{2}} x.$$

It then follows for instance $\text{Inv } R(e_{i_1} e_{i_2} \cdots e_{i_k}) = e_{i_k} e_{i_{k-1}} \cdots e_{i_1}$, and therefore $\widehat{xy} = \hat{y}\,\hat{x}$, so that this is an antiautomorphism. Finally let us indicate the following relationships:

$$\hat{x} = \overline{\tilde{x}} = \tilde{\overline{x}} \quad \text{and} \quad \tilde{x} = \hat{\overline{x}} = \overline{\hat{x}}.$$

We want now to establish a connection with the already known vector products. Let x, y be vectors in $C\ell_{p,q}^1 \subset C\ell_{p,q}$. These have the representation

$$x = \sum_{i=1}^{n} x_i e_i \quad \text{and} \quad y = \sum_{i=1}^{n} y_i e_i.$$

We define the *inner product* or *scalar product* of x and y by means of

$$x \cdot y := -\sum_{i=1}^{p} x_i y_i + \sum_{i=p+1}^{n} x_i y_i,$$

which corresponds to the one of vector calculus. In the case $p = 0$ this can be written also in the form

$$x \cdot y := -\frac{xy + yx}{2},$$

using the multiplication in $C\ell_{p,q}$. From this it follows that

$$e_i \cdot e_j = -\frac{e_i e_j + e_j e_i}{2} = 0 \quad (i \neq j).$$

So the basis elements are orthogonal and we have

$$e_i \cdot e_i = -e_i^2 = 1, \quad i = 1, \ldots, n.$$

For $x = y$ precisely the quadratic form mentioned in the previous subsection arises,

$$x^2 = -x \cdot x = Q(x).$$

From the definition follows the commutativity of the inner product. Thus it represents the symmetric part of the Clifford product of two vectors. The Clifford product itself splits into a sum of this symmetric part and an antisymmetric part:

$$xy = \frac{(xy + yx)}{2} + \frac{(xy - yx)}{2}.$$

The antisymmetric part will be called the *outer product* or the *Graßmann product* (cf. Appendix 1, Example A.1.7 c). The scalar product defines the bilinear form $Q(x, y) := -x \cdot y$ on our vector space for which clearly $Q(x, x) = Q(x)$ holds. Besides we can obtain $Q(x, y)$ from $Q(x)$ by means of the equation

$$Q(x, y) = \frac{Q(x + y) - Q(x) - Q(y)}{2}.$$

It may be mentioned here that for paravectors x and y we have for their scalar product in $\mathbb{R}^{n+1}$ the expression

$$x \cdot y = \frac{1}{2}(x\overline{y} + y\overline{x}).$$

We should mention also that there are more possibilities to introduce inner products. F.R. HARVEY classifies in [60] eight types of inner product spaces. C. PERWASS introduces in his thesis [114] the commutator product $A \overline{\times} B = \frac{1}{2}(AB - BA)$ and the anti-commutator product $A \underline{\times} B = \frac{1}{2}(AB + BA)$ of two elements of a geometric algebra. By means of these products it becomes possible to consider several products including the scalar product and write them using its algebraic structure.

3.2.3 Modulus

For an element $x \in C\ell_{p,q}$ we denote as usual the *modulus* or *absolute value* by

$$|x| := \left(\sum_{A \in \mathcal{P}_n} x_A^2 \right)^{1/2}.$$

Then we can regard the Clifford algebra as a Euclidean space of dimension 2^n with Euclidean metric. We list at first the well-known rules for conjugation and for the modulus, inasmuch as here they are needed

Proposition 3.11. *The following relations hold, where* $\mathrm{Sc}(x) = [x]_0 = x_0$ *represents the scalar part of the Clifford number* x *and* $\lambda \in \mathbb{R}$:

(i) *in case x is a paravector:* $\mathrm{Sc}\, x = \frac{x + \overline{x}}{2}$, (ii) $\overline{x + y} = \overline{x} + \overline{y}$,

(iii) $\overline{\overline{x}} = x$,

(iv) $\overline{xy} = \overline{y}\,\overline{x}$,

(v) $|\overline{x}| = |-x| = |x|$,

(vi) $|\lambda x| = |\lambda||x|$,

(vii) $\big||x| - |y|\big| \le |x - y| \le |x| + |y|$.

3. Clifford numbers

The proofs are very simple and analogous to those in $\mathbb{C}$ and $\mathbb{H}$.

So far so good, but in $\mathbb{C}$ and $\mathbb{H}$ we had further rules, which unfortunately give difficulties here. Among the latter the existence of the inverse of the multiplication, which in $Cl_{p,q}$ is not always given:

Proposition 3.12. (i) *For $p > 0$ or for $p = 0$ and $q = n \geq 3$, $Cl_{p,q}$ contains zero divisors.*

(ii) *For $p = 0$ and for all n the paravectors, which are different from zero and consist only of scalars and vectors, possess the known inverse*

$$x^{-1} = \frac{\overline{x}}{x\overline{x}}, \quad x \neq 0.$$

(iii) *The positive and negative basis elements $\{\pm e_A : A \in \mathcal{P}_n\}$ constitute a group.*

Proof. (i) For $p > 0$ we just gave above the zero divisors:

$$(1 + e_1)(1 - e_1) = 0.$$

For $p = 0$ and $n = q \geq 3$ there is the zero divisor

$$(1 + e_{123})(1 - e_{123}) = 0,$$

since we have $e_{123}^2 = 1$.

(ii) For the paravectors we have

$$x = x_0 + \sum_{i=1}^{n} x_i e_i, \quad \overline{x} = x_0 - \sum_{i=1}^{n} x_i e_i$$

and therefore

$$\begin{aligned}
\overline{x}\,x = x\overline{x} &= x_0^2 - \sum_{i,k=1}^{n} x_i x_k e_i e_k \\
&= x_0^2 + \sum_{i=1}^{n} x_i^2 + \sum_{i<k} x_i x_k (e_i e_k + e_k e_i) = |x|^2,
\end{aligned}$$

so that $\overline{x}/|x|^2$ is the multiplicative inverse of x.

(iii) Independently of the sign, the product of two basis elements is again such an element. e_0 is naturally the unit element and the inverse of $\pm e_A$ is $\overline{e}_A$ or $-\overline{e}_A$, respectively. $\qquad\square$

The usual equation $x\overline{x} = |x|^2$ just proven for paravectors is unfortunately not right for all Clifford numbers. In $Cl_{0,n}$ we get

Proposition 3.13. (i) *For arbitrary Clifford numbers we have:*

$$\mathrm{Sc}(\overline{x}\,y) = \mathrm{Sc}(x\,\overline{y}) = x \cdot y$$

with the scalar product taken as the one for vectors x and y in $\mathbb{R}^{2^n}$.

(ii) *In particular we have* $\mathrm{Sc}(x\overline{x}) = |x|^2$ *and*

$$\overline{x}\,x = \sum_{e_A^2=1} (x \cdot (xe_A))e_A, \quad x\overline{x} = \sum_{e_A^2=1} (x \cdot (e_A x))e_A.$$

From (ii) we see that $\overline{x}\,x$ begins indeed with $x \cdot x = |x|^2$ but as a rule in the sum with $e_A \neq e_0$ further terms follow.

Proof. (i) We have

$$
\begin{aligned}
\overline{x}\,y &= \sum_{A,B} x_A y_B \overline{e}_A e_B = \sum_C \left(\sum_{\overline{e}_A e_B = \pm e_C} \pm x_A y_B \right) e_C \\
&= \sum_{\overline{e}_A e_B = \pm e_0} \pm x_A y_B + \sum_{C \neq \{0\}} \left(\sum_{\overline{e}_A e_B \neq \pm e_C} \pm x_A y_B \right) e_C,
\end{aligned}
$$

where $\overline{e}_A e_B = \pm e_0$ can hold only for $A = B$ and because $p = 0$ with $+e_0$ in view of the group property. We also have

$$\mathrm{Sc}(\overline{x}\,y) = \sum_A x_A y_A = x \cdot y.$$

The statement for $x\overline{y}$ follows in a similar way.

(ii) In view of $e_B e_A = \pm e_0$ for arbitrary y we have

$$\mathrm{Sc}\,(ye_A) = \mathrm{Sc}\left(\sum_B y_B e_B e_A \right) = y_A e_A^2$$

and

$$y = \sum_A \mathrm{Sc}(ye_A)\overline{e}_A$$

only in case $B = A$. In particular for $y = \overline{x}\,x$ it follows that

$$
\begin{aligned}
\overline{x}\,x &= \sum_A \mathrm{Sc}(\overline{x}\,xe_A)\overline{e}_A \\
&= \sum_A (x \cdot (xe_A))\overline{e}_A = \sum_{e_A^2=1} (x \cdot (xe_A))e_A.
\end{aligned}
$$

The last equality follows from the relation

$$x \cdot (xe_A) = (xe_A) \cdot x = 0$$

for all x and the A with $\overline{e}_A = -e_A$. This is obtained as follows:

$$
\begin{aligned}
x \cdot (xe_A) &= \mathrm{Sc}(\overline{x}xe_A) = \mathrm{Sc}(\overline{x}\overline{x}e_A) = \mathrm{Sc}(\overline{x}x\overline{e}_A) \\
&= x \cdot (x\overline{e}_A) = -x \cdot (xe_A),
\end{aligned}
$$

so that it is indeed $x \cdot (xe_A) = 0$.

$\square$

For the modulus of a product we prove now the following theorem, which does not give the best constant, but rather a usable estimate.

Theorem 3.14. (i) *For every n there is a constant K_n, so that for all x, y we have*

$$|xy| \leq K_n |x||y|.$$

In any case we have

$$K_n \leq 2^{n/2}.$$

(ii) *If y satisfies the relation $y\bar{y} = |y|^2$, then it possesses an inverse, and we have*

$$|xy| = |yx| = |x||y|.$$

In $\mathbb{C}$ and $\mathbb{H}$ we have $K_1 = K_2 = 1$. The estimate given in the theorem is therefore not sharp.

Proof. (i) From $xy = \sum_A xy_A e_A$, using the triangular inequality, it follows that

$$|xy| \leq \sum_A |xy_A e_A| = \sum_A |y_A||xe_A|.$$

In

$$xe_A = \sum_B x_B e_B e_A$$

both B and $e_B e_A$ run through all the subsets $C \subset \mathcal{P}_n$, so that

$$|xe_A|^2 = \sum_B |x_B|^2 = |x|^2$$

and with Schwarz's inequality we find

$$|xy|^2 \leq \left(\sum_A |y_A||x| \right)^2 \leq |x|^2 \left(\sum_A 1 \right) \left(\sum_A |y_A|^2 \right) = 2^n |x|^2 |y|^2.$$

(ii) $|xy|^2 = \mathrm{Sc}(xy \, \overline{xy}) = \mathrm{Sc}(xy \, \bar{y} \, \bar{x}) = \mathrm{Sc}(x\bar{x}|y|^2) = |y|^2 \, \mathrm{Sc}(x\bar{x}) = |y|^2 |x|^2$. For the case $|yx|^2$ the proof is similar. □

3.3 Geometric applications

3.3.1 Spin groups

In these considerations for simplicity's sake we assume the real Clifford algebra $C\ell_{0,n} = C\ell(n)$. We consider 1-vectors in $C\ell(n)$, thus vectors in $\mathbb{R}^n$. In $\mathbb{C}$ (Section 1.3) in $\mathbb{R}^3$ (Section 2.3) and in $\mathbb{H}$ we have already considered rotations. Here too we understand as *rotation* in $\mathbb{R}^n$ a linear transformation of the form

$$v = (v_1, \ldots, v_n) \rightarrow T(v) = \left(\sum_{i,j=1}^{n} A_{ij} v_j \right),$$

which preserves the scalar product of vectors and the orientation. The group of such transformations is called the *special orthogonal group* and is denoted by $SO(n)$. The matrices $A \in \mathbb{R}^{n \times n}$ are *orthogonal matrices* with $A^{-1} = A^{\top}$. In view of the preservation of orientation we must have $\det A = +1$.

We want first to address reflections in $\mathbb{R}^n$ from which we can construct rotations. A *reflection* about a plane $x \cdot u = 0$ determined by taking the unit vector u as normal vector is clearly a mapping by which the component of x orthogonal to u remains unchanged, while the one parallel to u gets a minus sign. In formula we split x accordingly,

$$x = x_u + (u \cdot x)u,$$

where x_u is orthogonal to u so that the reflected point x' is given by

$$x' := R_u(x) := x_u - (u \cdot x)u = x - 2(u \cdot x)u.$$

Since $u^2 = u \cdot u = -1$ it follows further that

$$R_u(x) = x + (ux + xu)u = uxu.$$

This map is isometric, since

$$
\begin{aligned}
R_u(x) \cdot R_u(y) &= (uxu) \cdot (uyu) = -\frac{1}{2}[uxuuyu + uyuuxu] \\
&= \frac{1}{2}u[xy + yx]u = x \cdot y.
\end{aligned}
$$

As matrix multiplication the corresponding matrix map is

$$A = E - 2uu^{\top}$$

with the identity matrix E and

$$A^{\top} = A = A^{-1},$$

the latter equality following since the inverse mapping is equal to the original. However $\det A = -1$ must hold, since the orientation is not preserved (a fact that can be calculated). If we finally carry out two reflections one after the other, we obtain a mapping with orthogonal matrices, but the determinant is $+1$, thus it is a rotation. In summary:

Proposition 3.15. *A reflection about an $(n-1)$-dimensional (hyper-) plane $u \cdot x = 0$ in $\mathbb{R}^n$ with a unit vector u is described by*

$$R_u(x) = x - 2(u \cdot x)u = uxu.$$

The composition of two reflections gives a rotation

$$T(x) = u_2 u_1 x u_1 u_2.$$

As already Hamilton determined, we can in fact obtain every rotation as composition of an even number of reflections. We now want to investigate this problem. To this end we first define

Definition 3.16. The *Spin group* Spin(n) is the totality of all products of an even number of unit vectors from $\mathbb{R}^n \cong C\ell^1(n)$.

Here the empty product is allowed, it is equal to the unit element e_0 of the Clifford algebra. We can show that in fact it is a group, since at most pairs of such vectors vanish, so that the even number of factors remains unchanged. For a product

$$s = u_1 u_2 \ldots u_{2k}$$

the inverse is given by

$$\bar{s} = \overline{u_{2k}} \ldots \overline{u_2}\, \overline{u_1} = u_{2k} \ldots u_2 u_1.$$

If we define now a map $h : \text{Spin}(n) \to SO(n)$ by $h(s)(x) := \bar{s}\, x\, s$, we can then ask what such map looks like. We will prove in the next subsection that it is surjective. It is however not injective: let $h(s)$ be equal to the identity map, then

$$h(s)(x) = \bar{s}\, x\, s = x$$

for all $x \in \mathbb{R}^n$, so that s commutes with all the basis elements e_j and thus belongs to the center of the algebra $C\ell(n)$. The spin group is the subset of the even subalgebra $C\ell(n)^+$, whose elements are not exchangeable with all e_j, for instance $(e_i e_j)e_j = -e_i \neq e_j(e_i e_j) = e_i$ for $i \neq j$. Thus s must be real, and this holds only for the identity apart from the sign. Therefore we have shown:

Theorem 3.17. *The map* $h : \text{Spin}(n) \to SO(n)$ *is a twofold covering, i.e., the complete preimage of a given element from $SO(n)$ consists exactly of two elements.*

3.3.2 Construction of rotations of $\mathbb{R}^n$

We want to treat the problem of the construction of rotations of $\mathbb{R}^n$ by means of Clifford algebraic concepts. Our presentation follows a work of H. KRÜGER (Kaiserslautern), which was kindly made available to us for our book project.

A general linear isometry or orthogonal map T will also be indicated from the fact that it maps an orthonormal basis $\{e_1, \ldots, e_n\}$ of $\mathbb{R}^n$ into another orthonormal basis $\{g_1, \ldots, g_n\}$ thanks to the prescription

$$g_k = T(e_k).$$

Clearly we can express the g_k by means of the e_k. The coefficients are then exactly the elements of the corresponding rotation matrix, if the orientation coincides, otherwise the matrix has the determinant -1. But we want to work here within the Clifford algebra. From the definition it follows directly that

$$g_j \cdot g_k = T(e_j) \cdot T(e_k) = e_j \cdot e_k = \delta_{jk}.$$

The linearity of T gives immediately the property for arbitrary $x, y \in \mathbb{R}^n$,

$$T(x) \cdot T(y) = x \cdot y.$$

The goal of our observations is constructing reflections R_u which transfer the orthonormal basis $\{e_1, \ldots, e_n\}$ successively into the orthonormal basis $\{g_1, \ldots, g_n\}$. At first we observe that for two unit vectors $x, y \in \mathbb{R}^n$ with $x \neq y$ there is always a unit vector u for which $R_u(x) = y$ holds. In fact we need only to choose $u = (x - y)/|x - y|$ since in view of $x^2 = y^2 = -1$ we have

$$
\begin{aligned}
R_u(x) &= \frac{1}{|x-y|^2}(x-y)x(x-y) = \frac{x^3 - x^2 y - yx^2 + yxy}{2(1 - x \cdot y)} \\
&= \frac{y(1 - x \cdot y)}{1 - x \cdot y} = y.
\end{aligned}
$$

This important mapping will be essential in the following.

Let now $\{e_1, \ldots, e_n\}$ and $\{g_1, \ldots, g_n\}$ be orthonormal bases, which we would like to send into each other. If $e_1 = g_1$, in the first step there is nothing to do, otherwise we define $u_1 := (e_1 - g_1)/|e_1 - g_1|$. Then R_{u_1} takes the basis vector e_1 into the new basis vector g_1. But R_{u_1} applies also to the other basis vectors. We set

$$e_k^{(1)} := R_{u_1}(e_k) \qquad (2 \leq k \leq n).$$

From the properties of R_{u_1} we have

$$e_j^{(1)} \cdot e_k^{(1)} = e_j \cdot e_k = \delta_{jk} \qquad (2 \leq j, k \leq n).$$

For $k = 2$ with $x = e_2^{(1)}$ and $y = g_2$ the consideration is repeated unless it happens that we have $e_2^{(1)} = g_2$. For $e_2^{(1)} \neq g_2$ let

$$u_2 := \frac{e_2^{(1)} - g_2}{|e_2^{(1)} - g_2|}.$$

R_{u_2} maps as desired $e_2^{(1)}$ into g_2. Luckily g_1 remains invariant under this reflection: at first it follows that

$$R_{u_2}(g_1) = u_2 g_1 u_2 = u_2(g_1 u_2 + u_2 g_1) - u_2^2 g_1 = -2u_2(g_1 \cdot u_2) + g_1.$$

We then have still to study the following scalar product:

$$g_1 \cdot u_2 = \frac{1}{|e_2^{(1)} - g_2|} g_1 \cdot (e_2^{(1)} - g_2).$$

We find $g_1 \cdot g_2 = 0$ from the assumption and

$$g_1 \cdot e_2^{(1)} = R_{u_1}(e_1) \cdot R_{u_1}(e_2) = e_1 \cdot e_2 = 0.$$

3. Clifford numbers

Thus g_1 remains in fact invariant under R_{u_2}.

We now use an induction argument on k. Let us construct

$$g_1 = R_{u_1}(e_1), \quad u_1 = \frac{e_1 - g_1}{|e_1 - g_1|},$$

$$g_2 = R_{u_2}(e_2^{(1)}), \quad u_2 = \frac{e_2^{(1)} - g_2}{|e_2^{(1)} - g_2|},$$

$$\ldots$$

$$g_{k-1} = R_{u_{k-1}}(e_{k-1}^{(k-2)}), \quad u_{k-1} = \frac{e_{k-1}^{(k-2)} - g_{k-1}}{|e_{k-1}^{(k-2)} - g_{k-1}|},$$

where eventually R_{u_j} is the identity, if $e_j^{(j-1)} = g_j$. Then the R_{u_j} leave the g_i with $i < j$ invariant. Finally for the induction step from $k-1$ to k we define again

$$e_j^{(k-1)} := R_{u_{k-1}}(e_j^{(k-2)}), \quad j = k, \ldots, n.$$

As before we set

$$u_k := \frac{e_k^{(k-1)} - g_k}{|e_k^{(k-1)} - g_k|}$$

and it then follows that

$$R_{u_k}(e_k^{(k-1)}) = g_k.$$

We should consider whether the previous g_j remain invariant with this procedure. In any case for $j < k$ we obtain that

$$R_{u_k}(g_j) = u_k g_j u_k = -u_k^2 g_j + u_k(u_k g_j + g_j u_k) = g_j - 2u_k(u_k \cdot g_j).$$

It remains to show the vanishing of the scalar product

$$u_k \cdot g_j = \frac{1}{|e_k^{(k-1)} - g_k|}(e_k^{(k-1)} - g_k) \cdot g_j.$$

From the assumption we have $g_k \cdot g_j = 0$. For the rest we can again write

$$e_k^{(k-1)} \cdot g_j = R_{u_{k-1}}(e_k^{(k-2)}) \cdot R_{u_{k-1}}(g_j) = e_k^{(k-2)} \cdot g_j$$

which is zero by inductive hypothesis.

Besides both e_n and g_n are determined by the other basis elements and the orientation. In case both orthonormal systems are oriented in the same way we can skip the last step.

Thus we have obtained the following theorem:

Theorem 3.18. *Every isometry of $\mathbb{R}^n$, in particular every rotation is the product of at most $n - 1$ reflections about the planes $u_k \cdot x = 0$, $k = 1, \ldots, n - 1$, defined by the vectors u_k. Thus the basis $\{e_1, e_2, \ldots, e_n\}$ is transformed into the basis $\{g_1, g_2, \ldots g_n\}$ with $g_k = T(e_k)$, where we have*

$$T = R_{u_{n-1}} \circ \cdots \circ R_{u_2} \circ R_{u_1}.$$

In particular every rotation is a composition of an even number of reflections.

Remark 3.19. In case $u \cdot v = 0$ the product $R_u \circ R_v$ is commutative, i.e.,

$$(R_u \circ R_v)(x) = (R_v \circ R_u)(x),$$

since we have $0 = u \cdot v = -\frac{1}{2}(uv + vu)$, and therefore $uv = -vu$ so that

$$(R_u \circ R_v)(x) = uvxvu = vuxuv = (R_v \circ R_u)(x).$$

In particular all compositions $R_{e_j} \circ R_{e_k}$ are commutative.

3.3.3   Rotations of $\mathbb{R}^{n+1}$

The rotations of $\mathbb{R}^{n+1}$ distinguish themselves from those of $\mathbb{R}^n$ in the following way:

Theorem 3.20. *The Clifford group Γ_{n+1} is the set of all finite products of nonzero paravectors. Rotations of $\mathbb{R}^{n+1}$ have always the form*

$$x' = u\, x\, \hat{u} = u\, x\, \tilde{u}^{-1},$$

where the $u \in \Gamma_{n+1}$ represent finite products of unit paravectors. Moreover $\tilde{u}$ denotes the inversion and $\hat{u}$ the reversion.

Proof. As described above the reflection about a plane with the unit paravector $\overline{u}$ as normal is given by

$$x' = x - 2(\overline{u} \cdot x)\overline{u} = x - (\overline{u}\,\overline{x} + xu)\overline{u} = -\overline{u}\,\overline{x}\,\overline{u}.$$

A second reflection with $u = 1$, $x'' = -\overline{x'}$, which is exchangeable with the first one leads to $x'' = uxu$. This is thus a rotation. The composition of more rotations leads to expressions of the form

$$x' = u_k \ldots u_1 x u_1 \ldots u_k.$$

With $u := u_k \ldots u_1$ we have $u_1 \ldots u_k = \hat{u}$, since $\hat{u}_j = u_j$. In view of

$$\tilde{u}\,\hat{u} = \tilde{u}\,\overline{\tilde{u}} = |u|^2 = |u_1|^2 \ldots |u_k|^2 = 1$$

the rotation can be written also in the form

$$x' = u\, x\, \tilde{u}^{-1} = u\, x\, \hat{u}.$$

Conversely from Theorem 3.18 every rotation T is the composition of an even number of reflections R_u, so that

$$T = R_{u_{2k}} \circ \cdots \circ R_{u_1}.$$

If we now premultiply by $2k$ reflections $R_i(x) = -\overline{x}$ on one side, then the latter together give the identity since they are even. On the other side they can be interchanged with the R_{u_i} and then give as above the desired form.   □

3.4 Representations

As in $\mathbb{C}$ and $\mathbb{H}$ we should investigate whether an isomorphic matrix algebra can be associated to every Clifford algebra. To this end we need to prove some fundamental results. At first we want to investigate the lower dimensional cases. Here it is useful to use the Pauli matrices already used in Section 2.3 in Remark 2.28

$$\sigma_0 := \begin{pmatrix} 1 & 0 \\ 0 & 1 \end{pmatrix}, \ \sigma_1 := \begin{pmatrix} 0 & 1 \\ 1 & 0 \end{pmatrix}, \ \sigma_2 := \begin{pmatrix} 0 & -i \\ i & 0 \end{pmatrix}, \ \sigma_3 := \begin{pmatrix} 1 & 0 \\ 0 & -1 \end{pmatrix}.$$

We see easily that $\sigma_k^2 = \sigma_0$ for $k = 1, 2, 3$. Further the Pauli matrices satisfy the anticommutator relations:

$$\sigma_j \sigma_k + \sigma_k \sigma_j = 0 \quad (j \neq k; \ j, k = 1, 2, 3).$$

With $\mathcal{A}^{d \times d}$ we denote the ring of all $(d \times d)$ matrices over the algebra $\mathcal{A}$. Up to now we can establish in detail the following representations:

Example 3.21. a) $C\ell_{0,0}$: an element $x \in C\ell_{0,0}$ has the form $x = x_0 e_0$ with $x_0 \in \mathbb{R}$, thus $C\ell_{0,0} \cong \mathbb{R} \cong \mathbb{R}^{1 \times 1}$.

b) $C\ell_{1,0}$: for the algebra of the dual numbers we have the basis $\{(1, 0), (0, 1)\}$. The $\mathbb{R}^{2 \times 2}$ matrices σ_0 and σ_1 constitute moreover an isomorphic basis. We can assign to an element of the algebra $x = x_0 + x_1 e_1$ the matrix $x_0 \sigma_0 + x_1 \sigma_1$ and have as representation the matrix

$$\begin{pmatrix} x_0 & x_1 \\ x_1 & x_0 \end{pmatrix}.$$

We can also write $C\ell_{1,0} \cong \mathbb{R} \oplus \mathbb{R}$.

c) $C\ell_{0,1}$: This algebra is exactly equal to $\mathbb{C}$ and its basis can be constructed from the matrices $\sigma_0, -i\sigma_2$. For a complex number $x = x_0 + ix_1$ we obtain the representation

$$\begin{pmatrix} x_0 & -x_1 \\ x_1 & x_0 \end{pmatrix},$$

which we already encountered in Proposition 1.13.

d) $C\ell_{2,0}$: the basis $e_0, e_1, e_2, e_1 e_2$ with $e_1^2 = e_2^2 = 1$ can be given by the matrices $\sigma_0, \sigma_1, \sigma_3, \sigma_1 \sigma_3$, so that for an element $x = x_0 e_0 + x_1 e_1 + x_2 e_2 + x_3 e_1 e_2$ we obtain the representation

$$\begin{pmatrix} x_0 + x_2 & x_1 - x_3 \\ x_1 + x_3 & x_0 - x_2 \end{pmatrix}.$$

This delivers the whole matrix algebra $\mathbb{R}^{2 \times 2}$.

e) $Cl_{1,1}$: the basis elements are again e_0, e_1, e_2, e_1e_2 with $e_1^2 = -e_2^2 = 1$, which can be given by the matrices $\sigma_0, \sigma_1, i\sigma_2, \sigma_1 i\sigma_2$. An arbitrary element $x = x_0e_0 + x_1e_1 + x_2e_2 + x_3e_1e_2$ has then the representation

$$\begin{pmatrix} x_0 - x_3 & x_1 + x_2 \\ x_1 - x_2 & x_0 + x_3 \end{pmatrix}.$$

This is again the whole matrix algebra $\mathbb{R}^{2\times 2}$.

f) $Cl_{0,2}$: here we have the quaternions for whose usual basis the matrices $\sigma_0, -i\sigma_1$, $-i\sigma_2, \sigma_1\sigma_2(= -i\sigma_3)$ can be used. An arbitrary element of $\mathbb{H}$ can be represented in the form

$$x = \begin{pmatrix} x_0 - ix_3 & -x_2 - ix_1 \\ x_2 - ix_1 & x_0 + ix_3 \end{pmatrix},$$

which we have calculated in Section 2.3 in Remark 2.28, and we find that it is not $\mathbb{C}^{2\times 2}$, rather it is $\mathbb{C} \oplus \mathbb{C}$.

g) $Cl_{3,0}$: the basis of this algebra can be given by the Pauli matrices. In view of $\sigma_1\sigma_2 = i\sigma_3$, $\sigma_2\sigma_3 = i\sigma_1$ and $\sigma_3\sigma_1 = i\sigma_2$ as well as $\sigma_1\sigma_2\sigma_3 = i\sigma_0$ an arbitrary element of the algebra

$$x = x_0e_0 + x_1e_1 + x_2e_2 + x_3e_3 + x_4e_1e_2 + x_5e_2e_3 + x_6e_3e_1 + x_7e_1e_2e_3$$

gets the complex representation

$$\begin{pmatrix} (x_0 + ix_7) + (x_3 + ix_4) & (x_1 + ix_5) + (x_6 - ix_2) \\ (x_1 + ix_5) + (x_6 + ix_2) & (x_0 + ix_7) - (x_3 + ix_4) \end{pmatrix},$$

i.e., $Cl_{3,0} \cong \mathbb{C}^{2\times 2}$. We can also construct the isomorphy with the complex quaternions, for which we need only to rewrite x in another way:

$$x = (x_0 + ix_7)\sigma_0 + (x_4 + ix_3)\sigma_1\sigma_2 + (x_5 + ix_1)\sigma_2\sigma_3 + (x_6 + ix_2)\sigma_1\sigma_3.$$

From this example it is clear that representations are in no way unique. The fact that the last representation is isomorphic to the complex quaternions is recognized by observing that the basis elements now correspond to those of the real quaternions.

We want now to show that all other Clifford algebras are tensor products of the ones given here. The next representation stems from results of F. Sommen in a joint work with the third author.

Theorem 3.22 (Dimension reduction). *The algebras $Cl_{p+1,q+1}$ and $Cl_{p,q} \otimes Cl_{1,1}$ are isomorphic to each other.*

3. Clifford numbers

Proof. The basis of the algebra $C\ell_{p+1,q+1}$ can be given in the form

$$\breve{e}_1, \ldots, \breve{e}_{p+1}, e_1, \ldots, e_{q+1},$$

with $\breve{e}_i{}^2 = 1$, $e_i^2 = -1$. Let us now set $e := e_{q+1}$, $\breve{e} := \breve{e}_{p+1}$. We easily see that the bivector $e\breve{e}$ commutes with all elements $\breve{e}_1, \ldots, \breve{e}_p, e_1, \ldots, e_q$ and possesses the square 1. From this the elements

$$\breve{E}_1 = \breve{e}_1 e\breve{e}, \ldots, \breve{E}_p = \breve{e}_p e\breve{e},$$
$$E_1 = e_1 e\breve{e}, \ldots, E_q = e_q e\breve{e}$$

generate also the algebra $C\ell_{p,q}$, and thus $C\ell_{p+1,q+1}$ is isomorphic to the one constructed from the elements $\breve{E}_1, \ldots, \breve{E}_p, E_1, \ldots, E_q, e, \breve{e}$. Further we can write the elements of $C\ell_{p+1,q+1}$ in the form $x_0 + x_1 e + x_2 \breve{e} + x_3 e\breve{e}$ with $x_i \in C\ell_{p,q}$. It follows then that $C\ell_{p+1,q+1} \cong C\ell_{p,q} \otimes C\ell_{1,1}$. $\square$

Since from the previous considerations $C\ell_{1,1}$ is isomorphic to the whole matrix algebra $\mathbb{R}^{2\times 2}$ we have a reduction of p and q by passing to the corresponding matrices.

Corollary 3.23. *The following relationships hold:*

$$C\ell_{p+1,q+1} \cong C\ell_{p,q}^{2\times 2}, \quad C\ell_{p+2,q+2} \cong C\ell_{p,q}^{4\times 4},$$

and therefore

$$C\ell_{p,q} \cong C\ell_{p-q,0}^{2^q \times 2^q} \quad for \quad p \geq q,$$
$$C\ell_{p,q} \cong C\ell_{0,q-p}^{2^p \times 2^p} \quad for \quad q \geq p.$$

We have still to investigate the algebras $C\ell_{p,0}$ and $C\ell_{0,q} = C\ell(q)$. We consider first two useful examples:

Example 3.24. We have (see Exercise 3.5.6):

a) $C\ell_{2,2} \cong C\ell_{1,1} \otimes C\ell_{1,1} \cong \mathbb{R}^{4\times 4}$.

b) $C\ell_{1,2} \cong C\ell_{0,1} \otimes C\ell_{1,1} \cong \mathbb{C} \otimes C\ell_{1,1} \cong \mathbb{C}^{2\times 2}$.

Theorem 3.25 (Exchange property). *We have $C\ell_{p+1,q} \cong C\ell_{q+1,p}$.*

Proof. Let the canonical basis of $C\ell_{p+1,q}$ be

$$\breve{e}_1, \ldots, \breve{e}_{p+1}, e_1, \ldots, e_q.$$

We set $\breve{e} := \breve{e}_{p+1}$. An isomorphic algebra can be constructed from the elements

$$\breve{E}_1 := e_1\breve{e}, \ldots, \breve{E}_q := e_q\breve{e}, \breve{E}_{q+1} := \breve{e}$$
$$E_1 = \breve{e}_1\breve{e}, \ldots, E_p = \breve{e}_p\breve{e},$$

which however are clearly the generators of an algebra $C\ell_{q+1,p}$. $\square$

For $p > 1$ in view of Theorems 3.22 and 3.25 as well as from Corollary 3.23 it follows that

$$Cl_{p,0} \cong Cl_{1,p-1} \cong Cl_{0,p-2} \otimes Cl_{1,1} \cong Cl_{0,p-2}^{2 \times 2}.$$

In summary we have

Proposition 3.26. *For the first eight algebras $Cl_{p,0}$ it follows that*

$$Cl_{0,0} \cong \mathbb{R}^{1 \times 1},$$
$$Cl_{1,0} \cong \mathbb{R} \oplus \mathbb{R},$$
$$Cl_{2,0} \cong \mathbb{R}^{2 \times 2},$$
$$Cl_{3,0} \cong \mathbb{C}^{2 \times 2} \cong \mathbb{R}^{4 \times 4},$$
$$Cl_{4,0} \cong Cl_{0,2} \otimes Cl_{1,1} \cong \mathbb{H}^{2 \times 2},$$
$$Cl_{5,0} \cong Cl_{0,3} \otimes Cl_{1,1} \cong (\mathbb{H} \otimes \mathbb{C})^{2 \times 2},$$
$$Cl_{6,0} \cong Cl_{0,4}^{2 \times 2},$$
$$Cl_{7,0} \cong Cl_{0,5}^{2 \times 2}.$$

Only the Clifford algebras $Cl_{0,p} = Cl(p)$ remain. In view of the last two rows which are not satisfactory, we still need a modification of the principle of dimension reduction. More precisely: we need a property which relates the algebra $Cl_{0,p}$ again with $Cl_{p,0}$.

Theorem 3.27. *The following isomorphism holds:*

$$Cl_{p,q+2} \cong Cl_{q,p} \otimes Cl_{0,2} \cong Cl_{q,p} \otimes \mathbb{H}.$$

Proof. In a similar way as in the proof of the last theorems, by a given basis we determine a new basis

$$\breve{e}_1, \ldots, \breve{e}_p, e_1, \ldots, e_q, e_{q+1}, e_{q+2}$$

by means of

$$\breve{E}_1 = e_1 e_{q+1} e_{q+2}, \ldots, \breve{E}_q = e_q e_{q+1} e_{q+2},$$
$$E_1 = \breve{e}_1 e_{q+1} e_{q+2}, \ldots, E_p = \breve{e}_p e_{q+1} e_{q+2}, E_{p+1} = e_{q+1}, E_{p+2} = e_{q+2}.$$

These elements produce also $Cl_{p,q+2}$, while the elements $\breve{E}_1, \ldots, E_p$ generate exactly the algebra $Cl_{q,p}$. We can thus take again e_{q+1}, e_{q+2} and $e_{q+1} e_{q+2}$ as bases with coefficients from $Cl_{q,p}$. But that is the stated isomorphism. $\square$

With these considerations we obtain the following representation, where for instance according to Theorem 3.27,

$$\mathbb{H} \otimes \mathbb{H} \cong Cl_{0,2} \otimes \mathbb{H} \cong Cl_{2,2} \cong \mathbb{R}^{4 \times 4},$$

is used according to Example 3.24 a).

Corollary 3.28. *Thus we have obtained:*

$$Cl_{0,1} \cong \mathbb{C},$$
$$Cl_{0,2} \cong \mathbb{H},$$
$$Cl_{0,3} \cong Cl_{1,0} \otimes \mathbb{H} \cong \mathbb{H} \oplus \mathbb{H},$$
$$Cl_{0,4} \cong Cl_{2,0} \otimes \mathbb{H} \cong \mathbb{H}^{2 \times 2},$$
$$Cl_{0,5} \cong Cl_{3,0} \otimes \mathbb{H} \cong \mathbb{R}^{4 \times 4} \otimes \mathbb{H} \cong \mathbb{H}^{4 \times 4},$$
$$Cl_{0,6} \cong Cl_{4,0} \otimes \mathbb{H} \cong \mathbb{H}^{2 \times 2} \otimes \mathbb{H} \cong \mathbb{R}^{8 \times 8},$$
$$Cl_{0,7} \cong Cl_{5,0} \otimes \mathbb{H} \cong (\mathbb{H} \otimes \mathbb{C})^{2 \times 2} \otimes \mathbb{H}.$$

As completion of the previous construction it follows that

$$Cl_{6,0} \cong Cl_{0,4}^{2 \times 2} \cong \mathbb{H}^{4 \times 4},$$
$$Cl_{7,0} \cong Cl_{0,5}^{2 \times 2} \cong \mathbb{H}^{8 \times 8}.$$

The remaining Clifford algebras can be calculated with the help of Bott's periodicity law:

Theorem 3.29 (Bott's periodicity law). *We have*

$$Cl_{p+8,q} \cong Cl_{p,q+8} \cong Cl_{p,q} \otimes \mathbb{R}^{16 \times 16} \cong Cl_{p,q}^{16 \times 16}.$$

Proof. From the previous theorems we obtain

$$Cl_{p+4,q} \cong Cl_{q+1,p+3} \cong Cl_{p+1,q+1} \otimes \mathbb{H}$$
$$\cong Cl_{p,q} \otimes Cl_{1,1} \otimes \mathbb{H} \cong Cl_{p,q} \otimes \mathbb{H}^{2 \times 2},$$

from which it follows that

$$Cl_{p+8,q} \cong Cl_{p+4,q} \otimes \mathbb{H}^{2 \times 2} \cong Cl_{p,q} \otimes \mathbb{H}^{2 \times 2} \otimes \mathbb{H}^{2 \times 2}$$
$$\cong Cl_{p,q} \otimes \mathbb{H}^{4 \times 4} \cong Cl_{p,q} \otimes \mathbb{R}^{16 \times 16}.$$

$\square$

3.5 Exercises

1. Prove that the set of all products of nonvanishing paravectors $C\ell(n)$ constitutes a group, the group Γ_{n+1}. For $a, b \in C\ell(n)$ and a from the Clifford group Γ_{n+1} the following relationship holds:

$$|ab| = |a||b|.$$

2. Let x be a paravector. Show the identity:

$$\sum_{i=0}^{n} e_i x e_i = -(n-1)\overline{x}.$$

3. Prove Proposition 3.4: the center of the algebra $C\ell(n)$ for even n consists only of the real numbers $\mathbb{R}$ and for odd n it can be obtained from 1 and $e_1 e_2 \ldots e_n$.

4. Prove Proposition 3.10: For an arbitrary $x \in C\ell_{p,q}$ we have

$$\operatorname{Inv} R = \hat{x} = [x]_0 + [x]_1 - [x]_2 - [x]_3 + [x]_4 + \cdots,$$

i.e., for $x \in C\ell_{p,q}^k$ we have:

$$\hat{x} = x, \quad \text{for} \quad k \equiv 0, 1 \ (\mathrm{mod}\, 4),$$
$$\hat{x} = -x, \quad \text{for} \quad k \equiv 2, 3 \ (\mathrm{mod}\, 4).$$

5. Prove the relationships (i) $C\ell_{2,2} \cong C\ell_{1,1} \otimes C\ell_{1,1} \cong \mathbb{R}^{4 \times 4}$,
   (ii) $C\ell_{1,2} \cong C\ell_{0,1} \otimes C\ell_{1,1} \cong \mathbb{C} \otimes C\ell_{1,1} \cong \mathbb{C}^{2 \times 2}$.

Chapter II

Functions

4 Topological aspects

4.1 Topology and continuity

We know already distances in $\mathbb{C}, \mathbb{H}$, and $C\ell(n)$. They all have the following properties and define therefore a *metric* in the respective sets: We have for a *distance* $d(z_1, z_2)$ for all z_1, z_2, z_3:

$$d(z_1, z_2) \geq 0, \quad d(z_1, z_2) > 0 \Leftrightarrow z_1 \neq z_2 \quad (\textit{positivity}),$$
$$d(z_1, z_2) = d(z_2, z_1) \qquad \qquad \qquad \quad (\textit{symmetry}),$$
$$d(z_1, z_2) \leq d(z_1, z_3) + d(z_3, z_2) \qquad \quad (\textit{triangle inequality}).$$

This has been shown in the preceding sections.

The topological fundamentals, as, e.g., convergence, will now be defined as usual in metric spaces (always of finite dimension). We start with a real vector space X with a given metric d without any further specification and remark that subsets of X are also metric spaces with the induced metric. Therefore subsets of $\mathbb{C}, \mathbb{H}$, or $C\ell(n)$ are always included.

Definition 4.1 (Sequence and limit). Let X be a vector space with metric d. A function from $\mathbb{N}$ into X, written $(z_n)_{n \in \mathbb{N}}$ or shortly (z_n), is called a *sequence in* X. A sequence (z_n) has the *limit* $a \in X$ or it *converges to* a, if for all $\varepsilon > 0$ an $N = N(\varepsilon)$ exists such that $d(z_n, a) < \varepsilon$ for all $n > N(\varepsilon)$. This is written shortly $z_n \to a$.

We know this definition from the reals, examples are not necessary. But we use the notion of Cauchy sequence as in the real numbers, and the Cauchy criterion is applicable also here. A sequence may start with an index other than 1.

Proposition 4.2. *A sequence (x_n) in $\mathbb{C}, \mathbb{H}$, or $C\ell(n)$ converges if and only if the real sequences given by the components converge.*

The proof is recommended as an exercise (see Exercise 4.4.1). Now we define further fundamental topological notions as in metric spaces:

Definition 4.3 (Topology). Let X be a vector space with metric d and $M \subset X$.

(i) A ball $B_\varepsilon(z_0) := \{z \in X : d(z, z_0) < \varepsilon\}$ is called an ε-*neighborhood* of z_0, a set $U \supset B_\varepsilon(z_0)$ is called a *neighborhood* of the point z_0.

(ii) A point a is called an *accumulation point* of the set M, if there exists a sequence of points in M, different from a, which converges to a. The set of all accumulation points of M is denoted by M'.

(iii) A set is called *closed*, if it contains all its accumulation points, i.e., $M' \subset M$. The set $\overline{M} := M \cup M'$ is called the *closure* of M.

(iv) The set $CM := X \backslash M$ is called the *complement of M relative to X*, and $\partial M := \overline{M} \cap \overline{CM}$ is the *boundary* of M.

(v) For a subset N of M all these notions may be understood *relative* to M; we then have a topology relative to M. So N is a *relative neighborhood* of z, if there exists a neighborhood $U = U(z)$ with $N = M \cap U$. The set N is called *relatively closed*, if $N = M \cap A$ for a closed set A.

Further definitions are necessary:

Definition 4.4 (Domain). Let X be a vector space and $M \subset X$.

(i) A set M is called *open*, if M contains an ε-neighborhood of z for every point $z \in M$. A set $N \subset M$ is called *relatively open with respect to* M, if N contains for every $z \in N$ a relative neighborhood of z .

(ii) A set M is called *connected*, if it is not possible to find two open sets M_1 and M_2, such that

$$M \subset M_1 \cup M_2, \ M \cap M_i \neq \emptyset \ (i = 1, 2), \ M \cap M_1 \cap M_2 = \emptyset.$$

If M is open, the definition simplifies to: There do not exist two open non-empty sets G_1 and G_2 such that $M = G_1 \cup G_2$, $G_1 \cap G_2 = \emptyset$.

(iii) A set M is called *polygonally connected*, if every two points in M may be connected by a polygon in M.

(iv) An open and connected set is called a *domain*.

In general we shall deal with domains and their boundaries. We need the following simple topological propositions; for the proofs we refer to real analysis:

Proposition 4.5. (i) *For an open set M the complement CM is closed; vice versa, CM is open if M is closed.*

(ii) *The intersection of finitely or infinitely many closed sets is closed; the union of finitely or infinitely many open sets is open.*

(iii) *The union of finitely many closed sets is closed; the intersection of finitely many open sets is open.*

We shall prove the following proposition as an exercise; it is useful for working with domains:

Proposition 4.6. *An open set in a vector space is connected if and only if it is polygonally connected.*

Proof. If in the open connected set M two points z_1 and z_2 exist that cannot be connected by a polygon in M, we divide M into two subsets M_1 and $M_2 := M \backslash M_1$. Here M_1 contains all points from M that can be connected with z_1 by a polygon. From our assumption it follows that M_1 and M_2 are nonempty and disjoint. They are moreover open: For $z_0 \in M_1$ an ε-neighborhood of z_0 exists in M, all of whose points can be connected with z_0 by a straight line, therefore they can be connected with z_1 by a polygon. So, the whole ε-neighborhood is contained in M_1 and M_1 is open. We may prove similarly that M_2 is open. In such a way we have divided M into two open, disjoint and nonempty subsets; that is a contradiction to our definition of a connected set.

Now the other way around: If M is polygonally connected and not connected, we have by definition a dissection of M into two open and disjoint sets M_1 and M_2. We choose $z_i \in M_i$, $i = 1, 2$. These two points can by assumption be connected by a polygon $\mathcal{P}$ in M. This polygon may be given by the straight lines $\overline{P_0 P_1}, \dots, \overline{P_{n-1} P_n}$, where the points P_j $(j = 0, \dots, n)$ are given by the vectors a_j.

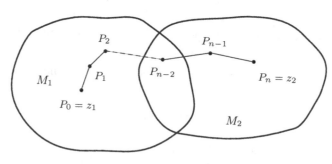

Figure 4.1

Then we have:

$$\overline{P_{j-1} P_j} : \quad z(t) = (a_j - a_{j-1})(t - j + 1) + a_{j-1}, \ j - 1 \le t \le j \ \ (j = 1, \dots, n),$$

in particular $z(0) = a_0 = z_1$ and $z(n) = a_n = z_2$. The variable t runs in the interval $[0, n]$. We define now $Q_i := \mathcal{P} \cap M_i$, $i = 1, 2$; Q_1 contains at least z_1. Let

$$t^* := \sup\{t \in [0, n] \ : \ z(t) \in Q_1\}.$$

If we have $z(t^*)$ in Q_1, then it also belongs to M_1. It would follow that a neighborhood of $z(t^*)$ is contained in M_1 and therefore all points $z(t)$ with $t^* - \delta < t < t^* + \delta$ for a sufficiently small δ are contained in M_1. That would contradict the definition of t^* as supremum. So, we have $z(t^*)$ in Q_2 and in M_2. As M_2 is open, we have a neighborhood of $z(t^*)$, which is contained in M_2. But, this contradicts also the definition of t^*, which says that there are points from $Q_1 \subset M_1$ as near as we wish to $z(t^*)$. So, we have a final contradiction and M has to be connected. $\qquad\square$

Our introductory topological definitions end with the notion of a compact set:

Definition 4.7 (Compactness).   (i) An arbitrary system of open sets is called a *covering* of a set M, if every point of M is contained in at least one of these open sets.

(ii) A set K is called *compact*, if an arbitrary covering of K can always be reduced to a finite covering.

Part (i) of the theorem below is named after H. E. HEINE (1821–1881) and E. BOREL (1871–1956) ; it is important for dealing with compact sets. The proof may be found, e.g., in [8], vol. I, Chapter III.3.

Theorem 4.8. (i) *A set K is compact if and only if it is closed and bounded.*

(ii) *A set K possesses a compact closure if and only if every sequence from K contains a convergent subsequence.*

Now, we define some notions connected with continuity of functions. Therefore let X and Y be vector spaces with metrics d_X, resp. d_Y. We deal with functions $f : M \to N$ for sets $M \subset X, N \subset Y$. The variable in the domain of definition (which may not be a domain in the sense of Definition 4.4) is called z, the variable in the range $w = f(z)$; here w is called the *image* of z and z is a *pre-image* of w. Similarly $f^{-1}(w) := \{z \in M : f(z) = w\}$ denotes the *set of pre-images* of the point w in the domain of definition and $f(M) := \{w : \exists z \in M, f(z) = w\}$ the *image set* of a set M.

Definition 4.9 (Continuity). (i) A function f is said to have the *limit* w_0 at a point $z_0 \in \overline{M}$, if for all $\varepsilon > 0$ a $\delta = \delta(\varepsilon, z_0)$ exists such that $d_Y(f(z), w_0) < \varepsilon$ for all z with $d_X(z, z_0) < \delta(\varepsilon)$ and $z \in M$, $z \neq z_0$.

(ii) A function f is said to be *continuous* at $z_0 \in M$, if it has the limit $w_0 = f(z_0)$ at z_0. A function is said to be *continuous on a set* if it is continuous at all points of M. The function is said to be *uniformly continuous on a set* M if the δ from (i) may be chosen independently of $z_0 \in M$.

(iii) A function that is bijective and continuous together with its inverse is called a *homeomorphism*.

(iv) A sequence of functions (f_n) defined on M is called *convergent to a limit function* f, if for all $\varepsilon > 0$ and for all $z \in M$ an $N(\varepsilon, z)$ exists such that $d_Y(f_n(z), f(z)) < \varepsilon$ for $n > N(\varepsilon, z)$. It is called *uniformly convergent*, if N may be chosen independently of z.

We have the usual theorems on continuity also for functions between vector spaces:

Theorem 4.10. (i) *A function f is continuous at $z_0 \in M$ if for every neighborhood V of the image point $w_0 = f(z_0)$ a relative neighborhood U of z_0 exists such that $f(U \cap M) \subset V$.*

(ii) *A function f is continuous on M if and only if the pre-images of open sets are open.*

(iii) *A function f is continuous on M if and only if the pre-images of closed sets are closed.*

(iv) *For a continuous function f the images of compact sets in M are compact.*

(v) *For a continuous function f the images of connected sets in M are connected.*

(vi) *The limit function of a uniformly convergent sequence of continuous functions is continuous.*

Proof. (i) Firstly we prove continuity: V is a neighborhood of w_0 and contains an ε-neighborhood V_ε of w_0. We assumed a neighborhood U_δ of z_0 with $f(U_\delta \cap M) \subset V_\varepsilon$ exists and that was just our definition of continuity. If continuity is given, then for a neighborhood V_ε in the range of f, a neighborhood U_δ exists such that $f(U_\delta \cap M) \subset V_\varepsilon$. That is the assertion of the other direction of our proof.

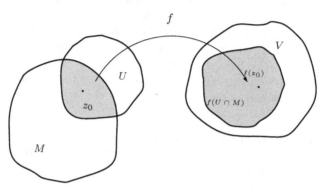

Figure 4.2

(ii) Let $f : M \to N$ be continuous on M, let G be a relative open set in N, and $z \in f^{-1}(G)$. Then $w = f(z)$ is contained in G and G is a neighborhood of w. Then it follows from (i) that a neighborhood U of z exists with $f(U \cap M) \subset G$, $U \cap M \subset f^{-1}(G)$, i.e., $f^{-1}(G)$ is relatively open. If for the other direction of our proof the pre-images of open sets are relatively open, then an open ε-neighborhood of w exists which has a relatively open pre-image. Therefore this pre-image contains a relative δ-neighborhood of all of its points, i.e., continuity.

(iii) The proof is analogous to (ii).

To prove (iv) we take an open covering of the set $f(K)$ by sets G_α. From (ii) it follows that all sets $f^{-1}(G_\alpha)$ are relatively open, and they form a covering of the compact set K. The definition gives us a finite covering, i.e.,

$$K \subset f^{-1}(G_{\alpha_1}) \cup \cdots \cup f^{-1}(G_{\alpha_m}).$$

From $f(f^{-1}(E)) = E$ for an arbitrary subset $E \subset f(X)$ it follows that $f(K) \subset G_{\alpha_1} \cup \cdots \cup G_{\alpha_m}$, i.e., the assertion.

To prove (v) we assume $f(M)$ not to be connected. Then there exist open sets G_1 and G_2 in Y which have a non-empty intersection with $f(M)$; these intersections are disjoint and their union contains $f(M)$. The pre-images $f^{-1}(G_i)$ $(i = 1, 2)$ are also relatively open in M and disjoint. Then we should have $M \subset f^{-1}(G_1) \cup f^{-1}(G_2)$, i.e., a contradiction with the connectivity of M.

We have to prove (vi): For all $n \geq N(\varepsilon/3)$ independently of $x \in M$ we have $d_Y(f(x), f_n(x)) < \varepsilon/3$; now let $n \geq N(\varepsilon/3)$ be fixed. For these n and ε, a $\delta(x) > 0$ exists such that for every $d_X(x, y) < \delta(x)$ we have always $d_Y(f_n(x), f_n(y)) < \varepsilon/3$. It follows that

$$d_Y(f(x), f(y)) \leq d_Y(f(x), f_n(x)) + d_Y(f_n(x), f_n(y)) + d_Y(f_n(y), f(y)) < \varepsilon. \qquad \square$$

4.2 Series

We shall define series in a vector space with Euclidean metric:

Definition 4.11 (Series). A pair of sequences $((a_n), (s_n))$ with $s_n := \sum_{j=1}^{n} a_j$ is called a *series*; shortly written also $\sum a_n$. The s_n are called *partial sums* of the series. The series $\sum a_n$ is *convergent to s*, or we say it has the *sum s* if $s_n \to s$. This is written

$$s = \sum_{n=1}^{\infty} a_n.$$

The series $\sum a_n$ is called *absolutely convergent*, if $\sum |a_n|$ is convergent.

A series may start with other indices, often with 0. We remark explicitly that $\sum a_n$ stands for the series and $\sum_{n=1}^{\infty} a_n$ for the sum of the series, the latter if the sum exists.

We will now ask whether a series in our algebras $\mathbb{C}, \mathbb{H}$ and $C\ell(n)$ has special properties. In view of the triangle inequality, absolute convergence implies always convergence. Similarly the Cauchy criterion is unchanged, this follows simply from the Cauchy criterion for sequences. The comparison test brings no problems as it is based also on the triangle inequality. The ratio and the root test are not so clear as the multiplicativity of the absolute value may be important, a fact which is not given in $C\ell(n)$. We deal firstly with the root test:

Proposition 4.12 (Root test). *Let $\sum a_k$ be a series from $\mathbb{C}, \mathbb{H}$ or $C\ell(n)$ and let $\limsup_{k \to \infty} \sqrt[k]{|a_k|} = R$ be given. Then we have:*

(i) *For $R < 1$ the series converges absolutely.*

(ii) *For $R > 1$ the series diverges.*

Proof. To prove (i) we take a real number r with $R < r < 1$. Then a natural number N exists such that

$$\sqrt[k]{|a_k|} < r$$

for $k \geq N$. So these elements of our series $\sum |a_k|$ are bounded from above by the elements of the real geometric series $\sum r^k$, and our series converges absolutely. If we have in (ii) $R > 1$, then we have infinitely many indices k with $\sqrt[k]{|a_k|} > 1$ resp. $|a_k| > 1$ implying divergence of the series. $\qquad \square$

As we see, the root test is not influenced by our algebras. Now we shall consider the ratio test:

Proposition 4.13 (Ratio test). *Let $\sum a_k$ be a series from $\mathbb{C}, \mathbb{H}$ or $C\ell(n)$ and let*

$$\limsup_{k \to \infty} \frac{|a_{k+1}|}{|a_k|} = Q.$$

For the R from the root test we have $R \leq Q$, thus the series converges absolutely for $Q < 1$. If instead

$$\lim_{k \to \infty} \frac{|a_{k+1}|}{|a_k|} = Q$$

with $Q > 1$, then the series diverges.

Proof. For $Q < 1$ let $Q < q < 1$; then it follows for $k \geq N$ with a suitable N that

$$|a_k| \leq q|a_{k-1}| \leq q^2|a_{k-2}| \leq \cdots \leq q^{k-N}|a_N|$$

and

$$|a_k| \leq q^k \frac{|a_N|}{q^N} \quad \text{as well as} \quad \sqrt[k]{|a_k|} \leq q \sqrt[k]{\frac{|a_N|}{q^N}}.$$

From the second inequality it follows that $R \leq Q$ and from the first we see that up to a factor our series is bounded above by $\sum q^k$, so we have convergence. For the second part of the proof let $Q > q > 1$, then we have for all sufficiently large k the inequality $|a_{k+1}| \geq q|a_k| \geq |a_k|$. Therefore the absolute values of the generic terms of our series do not converge to zero and the series diverges. $\qquad \square$

In the above proposition in $C\ell(n)$ it is important to have the quotients of the absolute values and not the absolute value of the quotients. Due to the inequality $|ab| \leq K|a||b|$ from Theorem 3.14 the following remark is needed:

Remark 4.14. If in $C\ell(n)$ instead of $|a_{k+1}|/|a_k|$ we take the quotient $|a_{k+1}/a_k|$, the a_k must not be zero divisors as otherwise the quotient may not be defined. Moreover we have to take into account the just cited multiplicativity inequality for the absolute value in $C\ell(n)$ and $Q < q < 1$:

$$|a_k| = \left| \frac{a_k}{a_N} a_N \right| \leq K|a_N| \left| \frac{a_k}{a_N} \right| = K|a_N| \left| \frac{a_k}{a_{k-1}} \frac{a_{k-1}}{a_{k-2}} \cdots \frac{a_{N+1}}{a_N} \right|$$

$$\leq K^{k-N}|a_N| \left| \frac{a_k}{a_{k-1}} \right| \cdots \left| \frac{a_{N+1}}{a_N} \right| \leq (Kq)^k (Kq)^{-N}|a_N|.$$

Up to a factor we get that our series is bounded above by $\sum (Kq)^k$, hence we have to assume $Q < 1/K$ for convergence.

The problems from the above remark in $C\ell(n)$ can be seen in a relatively simple example:

Example 4.15. For the *geometric series* $\sum x^k$ we have always

$$s_k = 1 + x + x^2 + \cdots + x^{k-1} = \frac{1 - x^k}{1 - x}.$$

The well-known sum

$$\sum_{k=0}^{\infty} x^k = \frac{1}{1 - x}$$

4. *Topological aspects*

follows as usual for $x^k \to 0$ if $k \to \infty$. In $\mathbb{C}$ and $\mathbb{H}$ this holds with $|x| < 1$ due to $|x^k| = |x|^k$. In $C\ell(n)$ we can derive convergence only for $|x| < 1/K$ since $|x^k| \leq K^{k-1}|x|^k$. But convergence follows in $C\ell(n)$ also for $|x| < 1$ if we deal with paravectors only.

Now we shall consider series of functions $\sum u_k(x)$ in our algebras:

Proposition 4.16 (Weierstraß). *Suppose the functions $u_k(x) : G \to C\ell(n)$ are given in a domain $G \subset C\ell(n)$ where they are bounded by*

$$|u_k(x)| \leq b_k.$$

If the real series $\sum b_k$ converges, then the series of functions converges in G absolutely and uniformly to a limit function

$$s(x) = \sum_{k=0}^{\infty} u_k(x).$$

Proof. Due to the triangle inequality

$$\left| \sum_{k=N}^{M} u_k(x) \right| \leq \sum_{k=N}^{M} |u_k(x)| \leq \sum_{k=N}^{M} b_k$$

and, taking into account the convergence of $\sum b_k$, the Cauchy criterion for the series of functions is fulfilled at every point, independently of the points of G. So the series of functions converges absolutely and uniformly in G. $\qquad \square$

According to Theorem 4.10 (vi) the limit function of a uniformly convergent series of continuous functions is itself continuous. An example is given by power series; in the following form they are important mainly in $\mathbb{C}$.

Definition 4.17 (Power series). A *power series* is a series $\sum a_k x^k$ where the a_k for $k \geq 0$ and x are elements of one of our algebras. If one uses powers of $x - x_0$ one speaks of a *power series with center x_0*.

The important theorem is::

Theorem 4.18. *Every power series $\sum a_k x^k$ possesses a circle of convergence $\{|x| < \rho/K\}$ for $\mathbb{C}$, resp. a ball of convergence for $\mathbb{H}$ or $C\ell(n)$, such that the series converges absolutely in the interior to a continuous function*

$$f(x) = \sum_{k=0}^{\infty} a_k x^k.$$

Here K is the factor in $|xy| \leq K|x||y|$, i.e., $K = 1$ in $\mathbb{C}$, $\mathbb{H}$ or if x is a paravector in $C\ell(n)$. In $\mathbb{C}$ and $\mathbb{H}$ the series diverges if $|x| > \rho$; if $|x| = \rho$ the convergence is *not determined and may differ from point to point. The number $\rho \neq 0$ is called* radius of convergence *and is given by:*

$$\frac{1}{\rho} = \limsup_{k \to \infty} |a_k|^{1/k}.$$

In case this limit is zero we define $\rho := \infty$. If this limit equals infinity we define $\rho := 0$. The power series converges uniformly in every smaller circle or ball $\{|x| \leq \frac{\rho}{K} - \varepsilon\}$.

Proof. The sum function f is continuous by the uniform convergence. To prove the latter we assume firstly $0 < \rho < \infty$, then from the above definition of ρ it follows that

$$\frac{1}{\rho - \varepsilon} > |a_k|^{1/k}$$

for all $\varepsilon > 0$ with $k > N(\varepsilon)$. For these k we have

$$|a_k x^k| \leq K^k |a_k| |x|^k \leq \left(\frac{K|x|}{\rho - \varepsilon}\right)^k ;$$

the right-hand side shows a convergent series for $|x| \leq (\rho - 2\varepsilon)/K$. So by comparison we have uniform convergence of our power series for $|x| \leq (\rho - 2\varepsilon)/K$. As $\varepsilon > 0$ is arbitrary we have convergence for $|x| < \rho/K$.

To prove divergence for $|x| > \rho$ in $\mathbb{C}$ or $\mathbb{H}$ we remark that, for infinitely many $k \in \mathbb{N}$,

$$\frac{1}{\rho + \varepsilon} \leq |a_k|^{1/k},$$

so it follows that

$$|a_k| |x|^k \geq \left(\frac{|x|}{\rho + \varepsilon}\right)^k .$$

As we find for every $|x| > \rho$ an ε such that $|x|/(\rho + \varepsilon) > 1$, the power series diverges for all x with $|x| > \rho$. In $C\ell(n)$ the situation is more difficult and we skip it here. But, if we deal only with paravectors the assertion is also correct.

In case of $\rho = 0$ in $\mathbb{C}$ or $\mathbb{H}$ we can conclude just as before and the series diverges for all $x \neq 0$. Finally we have for $\rho = \infty$ in all three algebras

$$\lim_{k \to \infty} |a_k|^{1/k} = 0;$$

it follows for all $\varepsilon > 0$ and sufficiently large k that the inequality $|a_k x^k| \leq K^k |a_k| |x|^k \leq (\varepsilon K |x|)^k$. On the right-hand side we have for $\varepsilon K |x| < 1$ a convergent comparison series that holds for all $|x|$ with sufficiently small ε. $\square$

Example 4.19. a) Later we shall use also power series with negative powers of x:

$$\sum a_{-k} x^{-k}.$$

These series converge in $\mathbb{C}$, $\mathbb{H}$, and for paravectors in $C\ell(n)$ in the exterior of a circle, resp. ball, $\{|x| < \rho\}$ with

$$\frac{1}{\rho} = \limsup_{k \to \infty} |a_{-k}|^{1/k}.$$

b) a) gives us in $\mathbb{C}$, $\mathbb{H}$, and for paravectors a second power series for $1/(1-x)$:

$$\frac{1}{1-x} = -\frac{1}{x}\frac{1}{1-\frac{1}{x}} = -\sum_{k=1}^{\infty} x^{-k};$$

this series converges for $|x| > 1$ and complements the geometric series in the exterior of the unit circle, resp. ball.

c) If the limit

$$\lim_{k \to \infty} \frac{|a_k|}{|a_{k+1}|}$$

exists it equals the radius of convergence of the power series $\sum a_k x^k$. The reader may prove this and may calculate the radius of convergence for the power series $\sum x^k/k!$ (see Exercise 4.4.5).

d) In $\mathbb{H}$ and $C\ell(n)$ one could investigate series of the form

$$\sum a_{k0} x a_{k1} x a_{k2} x \ldots a_{k(k-1)} x a_{kk}$$

due to the noncommutativity. With the estimate

$$|a_{k0} x a_{k1} x \ldots a_{kk}| \leq K^{2k} |a_{k0}||a_{k1}| \ldots |a_{kk}||x|^k$$

this may be reduced to the above cases. We do not carry this out here in detail, as we shall define later on different and in some sense more convenient power series in $\mathbb{H}$ and $C\ell(n)$.

4.3 Riemann spheres

4.3.1 Complex case

BERNHARD RIEMANN introduced a completion of the complex numbers, which closes or compactifies the complex plane $\mathbb{C}$ by a point $z = \infty$. This is called *one point compactification* of the complex plane or *closed complex plane* and will be denoted by $\hat{\mathbb{C}} := \mathbb{C} \cup \{\infty\}$. Relative to the notion *compact* we refer to Subsection 4.1 above. The point $z = \infty$ is at first to be seen as an ideal element which is not a point of the complex plane $\mathbb{C}$. We deal here only with the plane case, as we can illustrate this in $\mathbb{R}^3$; the generalizations to $\mathbb{H}$ and $C\ell(n)$ will follow.

The ideal point $z = \infty$ can be illustrated by the *Riemann sphere* in the following manner: We put a ball of radius $1/2$ on the complex plane, such that it touches the plane at the origin. This point is called the *south pole* of the Riemann sphere, the diametral point is called the *north pole N*.

BERNHARD RIEMANN (1826–1866) studied at the universities of Berlin and Göttingen and became Privatdozent in Göttingen. After the death of JOHANN PETER GUSTAV LEJEUNE DIRICHLET he inherited the chair of GAUSS. Riemann investigated already in his doctoral thesis the foundations of function theory which are still in use today. We shall meet his name very often. RIEMANN made important contributions to many fields in mathematics. He died of tuberculosis when he was 39 years old during a stay in Italy.

Bernhard Riemann

In space we introduce a coordinate system (ξ, η, ζ), the ξ-axis and the x-axis coinciding as well as the η-axis and the y-axis. Then the north pole has the coordinates $(0, 0, 1)$. Now every point $z \in \mathbb{C}$ will be connected with N by a straight line. The point where this line cuts the sphere is the image point of z on the sphere. As all these lines start at one point N, this mapping is called *stereographic projection*. Every point in $\mathbb{C}$ has an image on the sphere, every point on the sphere has a pre-image except N. This motivates us to define $z = \infty$ as the image of N. So we have found a concrete illustration of $\hat{\mathbb{C}}$.

This mapping transforms the southern half-sphere onto the interior of the unit circle, while the northern half-sphere is mapped onto the exterior of this circle. The Riemann sphere is given by the equation

$$\xi^2 + \eta^2 + \left(\zeta - \frac{1}{2}\right)^2 = \frac{1}{4}$$

or by

$$\xi^2 + \eta^2 + \zeta^2 - \zeta = 0.$$

Figure 4.3 shows a view into the Riemann sphere, here r is the absolute value of z and ρ that of $\xi + i\eta$.

As the occuring triangles are similar and the arguments of z and $\xi + i\eta$ are equal we have

$$\frac{r}{\rho} = \frac{1}{1 - \zeta}, \quad x = \frac{\xi}{1 - \zeta}, \quad y = \frac{\eta}{1 - \zeta}.$$

In view of the above equation of the sphere it follows

$$z = \frac{\xi + i\eta}{1 - \zeta}, \quad z\overline{z} = \frac{\rho^2}{(1 - \zeta)^2} = \frac{\zeta}{1 - \zeta}$$

and

$$\xi = \frac{x}{1 + z\overline{z}}, \quad \eta = \frac{y}{1 + z\overline{z}}, \quad \zeta = \frac{z\overline{z}}{1 + z\overline{z}}.$$

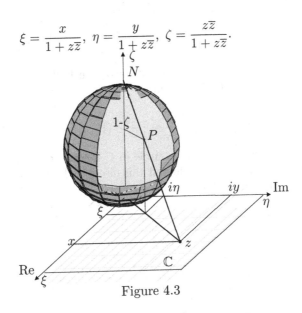

Figure 4.3

We are now able to introduce the Euclidean distance in $\mathbb{R}^3$ on the Riemann sphere as a new distance of two complex numbers called *chordal distance*. We get by some tedious calculations

$$d_{ch}(z_1, z_2) := \sqrt{(\xi_1 - \xi_2)^2 + (\eta_1 - \eta_2)^2 + (\zeta_1 - \zeta_2)^2} = \frac{|z_1 - z_2|}{\sqrt{1 + |z_1|^2}\sqrt{1 + |z_2|^2}}.$$

This distance may be used especially for $z = \infty$. Indeed, if $|z_2|$ goes to ∞ we get for $z_1 = z$,

$$d_{ch}(z, \infty) = \frac{1}{\sqrt{1 + |z|^2}}.$$

We remark that $\mathbb{C}$ is not compact either in the Euclidean or the chordal metric, but $\hat{\mathbb{C}}$ is compact in the chordal metric. For example we have $\overline{\mathbb{C}} = \mathbb{C}$ in the Euclidean metric, but $\overline{\mathbb{C}} = \hat{\mathbb{C}}$ in the chordal metric.

Now we shall look at the images of lines and circles under the stereographic map. A straight line in $\mathbb{C}$ is mapped onto the sphere by the lines which connect the points of the given line with N; these lines span a plane. This plane cuts the sphere in a circle such that the straight lines in $\mathbb{C}$ are mapped onto circles on the sphere going through N. If we now look at a circle in $\mathbb{C}$ given by the equation

$$z\overline{z} - z\overline{z_0} - \overline{z}z_0 + z_0\overline{z_0} = R^2$$

and map it by the above equations to the Riemann sphere, we get with $\tau := \xi + i\eta$,

$$\tau\bar{\tau} - (\tau\bar{\tau_0} + \bar{\tau}\tau_0)\frac{1-\zeta}{1-\zeta_0} + \zeta_0\frac{(1-\zeta)^2}{1-\zeta_0} = R^2(1-\zeta)^2.$$

If we substitute here $\tau\bar{\tau}$ using the equation of the ball $\tau\bar{\tau} = -\zeta^2 + \zeta$, we may divide the result by $1-\zeta$ (N is not on the circle) and we get a linear equation in ξ, η, ζ,

$$\zeta - (\tau\bar{\tau_0} + \bar{\tau}\tau_0)\frac{1}{1-\zeta_0} + \frac{\zeta_0}{1-\zeta_0}(1-\zeta) = R^2(1-\zeta).$$

That means the images of our plane circle in $\mathbb{C}$ lie on a plane in space, the latter cuts the Riemann sphere in a circle and this circle is the image of our circle in $\mathbb{C}$. So we have proved (Figure 4.4):

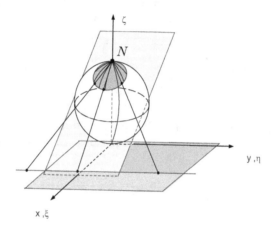

Figure 4.4

Proposition 4.20. *Circular and straight lines in $\mathbb{C}$ correspond to circles on the Riemann sphere. Circles in $\mathbb{C}$ correspond to circles not through N, straight lines correspond to circles through N.*

One sees that *parallels* on the sphere correspond to circles with center at the origin in the plane, the *meridians* to the straight lines through the origin. It does also not constitute a problem to find the plane circles which correspond to *great circles* on the sphere, i.e., the shortest lines. For these great circles the defining plane has to go through the center of the Riemann sphere $(0, 0, \frac{1}{2})$. If we put this into the equation for the plane it follows that

$$1 + \frac{\zeta_0}{1-\zeta_0} = R^2$$

or

$$1 + |z_0|^2 = R^2 :$$

a circle in the complex plane with center z_0 and this Radius R corresponds to a great circle on the Riemann sphere.

4.3.2 Higher dimensions

We start with the generalization of the Riemann sphere to the case of real quaternions. Let $\{e_0 = 1, \ldots, e_4\}$ be an orthonormal coordinate system in $\mathbb{R}^5$ and we identify $\mathbb{R}^5$ with $\mathbb{H} \times \mathbb{R}$. The points in $\mathbb{R}^5$ are given by $\xi = \sum_{j=0}^{4} \xi_j e_j$; we identify $\hat{\xi} = \sum_{j=0}^{3} \xi_j e_j$ with $x \in \mathbb{H}$. The generalized Riemann sphere is now defined by the sphere

$$\xi_0^2 + \xi_1^2 + \xi_2^2 + \xi_3^2 + \left(\xi_4 - \frac{1}{2} \right)^2 = \frac{1}{4}$$

or

$$\xi_0^2 + \xi_1^2 + \xi_2^2 + \xi_3^2 + \xi_4(\xi_4 - 1) = 0.$$

The point $N = (0,0,0,0,1)$ is called the *north pole*, the origin is the south pole. Every point $x \in \mathbb{H}$ can be connected with N by a straight line which meets the sphere at the image point ξ. Adding an ideal point $x = \infty$ we get the *one point compactification* $\hat{\mathbb{H}} = \mathbb{H} \cup \{\infty\}$ of $\mathbb{H}$, completely analogous to $\hat{\mathbb{C}}$. This point $x = \infty$ will be appointed to the north pole which has no pre-image in $\mathbb{H}$. So $\hat{\mathbb{H}}$ becomes a compact set which we can equip with a chordal metric.

The relations between ξ on the sphere and $x \in \mathbb{H}$ are similar to the complex case. From the theorem of intersecting lines we get, with $\rho = |\hat{\xi}|$,

$$\frac{x_0}{\xi_0} = \frac{x_1}{\xi_1} = \frac{x_2}{\xi_2} = \frac{x_3}{\xi_3} = \frac{|x|}{\rho} = \frac{1}{1 - \xi_4} \; ;$$

it follows that

$$x = \frac{\hat{\xi}}{1 - \xi_4} \; .$$

The reverse direction is a bit more difficult. From $\overline{x} = \overline{\hat{\xi}}/(1 - \xi_4)$ it follows by the sphere equation that

$$x\overline{x} = |x|^2 = \frac{|\hat{\xi}|^2}{(1 - \xi_4)^2} = \frac{(1 - \xi_4)\xi_4}{(1 - \xi_4)^2} = \frac{\xi_4}{1 - \xi_4}$$

or

$$\xi_4 = \frac{|x|^2}{1 + |x|^2} \quad \text{and} \quad \hat{\xi} = \frac{x}{1 + |x|^2} \; .$$

As for $\mathbb{C}$ we define now the chordal distance

$$d_{\text{ch}}(x, x') = |\xi - \xi'|$$

to be the Euclidean distance on the sphere. Calculating as in $\mathbb{C}$ (we have to substitute only iy by $\mathbf{x}$) we get

$$d_{\text{ch}}(x, x') = \frac{|x - x'|}{\sqrt{1 + |x|^2}\sqrt{1 + |x'|^2}} \ .$$

This metric allows us to include the point at ∞:

$$d_{\text{ch}}(x, \infty) = \frac{1}{\sqrt{1 + |x|^2}} \ .$$

Naturally the topological properties in $\mathbb{H}$ and $\mathbb{C}$ are the same.

It is not necessary to do this again in $C\ell(n)$. One has to take an analogous ball of diameter 1 in $\mathbb{R}^{n+2}$. The point $(0, \ldots, 0, 1)$ defines the north pole and corresponds to the ideal point $x = \infty$ which is added to $\mathbb{R}^{n+1}$. The same formulas as in $\mathbb{H}$ hold in this case too, and the corresponding chordal distance allows us to include the ideal point $x = \infty$.

4.4 Exercises

1. Let (x_n) be a sequence in $\mathbb{C}$, $\mathbb{H}$ or $C\ell(n)$. Prove:
   a) (x_n) converges if and only if the real sequences of the components of x_n converge.
   b) If x_n are complex numbers the sequence (x_n) converges if and only if the sequences of the absolute values and arguments converge. Are there exceptions?

2. Describe geometrically those sets which have the following definition in the plane: a) $|z| < 1$, b) $\text{Im } z \leq 0$.

3. Do both series $\sum -x^{-k}$ and $\sum x^k$ have the same values on the unit circle of $\mathbb{C}$, resp. the unit ball of $\mathbb{H}$?

4. Find examples of convergent power series with radius of convergence $\rho = 0$ and $\rho = \infty$.

5. a) Prove that if the limit

$$\lim_{k \to \infty} \frac{|a_k|}{|a_{k+1}|}$$

   exists it equals the radius of convergence for the power series $\sum a_k x^k$.
   b) Calculate the radius of convergence for the power series $\sum x^k/k!$.

4. *Topological aspects*

6. Calculate the chordal distance $d_{ch}(x, x')$ on the Riemann sphere in $\mathbb{H}$ and $C\ell(n)$.

7. Study the images of spheres and planes in $\mathbb{H}$ on the 4-dimensional Riemann sphere under the stereographic map.

5 Holomorphic functions

5.1 Differentiation in $\mathbb{C}$

We now look at functions

$$f(z) = u(z) + iv(z) = u(x,y) + iv(x,y),$$
$$u = \mathrm{Re}f, \ v = \mathrm{Im}f,$$

in a domain $G \subset \mathbb{C}$ with values also in $\mathbb{C}$. We know the notion of partial differentiation of the real functions u and v from analysis, which functions depend on the two real variables x and y. We assume u and v to be once continuously differentiable to simplify the proceeding. There exists extensive research related to weaker assumptions for holomorphic functions which is known under the names of H. LOOMAN and D. MENCHOV. Generally the set of k-times continuously differentiable real functions is denoted by $C^k(G)$. The set $C^0(G)$ contains the continuous functions in G. For complex functions, $f \in C^k(G)$ means that the components of f are in $C^k(G)$. With the assumption $f \in C^1(G)$ we are allowed to work with differentials (for differentials see Appendix 1):

$$df = du + idv = (u_x + iv_x)dx + (u_y + iv_y)dy.$$

We denote by ∂_x, ∂_y the partial derivatives to x and y and we define further:

$$\partial_z := \frac{1}{2}(\partial_x - i\partial_y), \ \partial_{\bar{z}} := \frac{1}{2}(\partial_x + i\partial_y).$$

Moreover we have for $z = x + iy$,

$$dx = \frac{1}{2}(dz + d\bar{z}), \ \ dy = \frac{1}{2i}(dz - d\bar{z}).$$

Besides this we introduce the notation

$$\partial := 2\partial_z, \ \ \bar{\partial} := 2\partial_{\bar{z}}$$

to hint already of higher dimensions. So, we are able to express df as follows:

$$\begin{aligned} df &= \frac{1}{2}\left(u_x + v_y + i(v_x - u_y)\right)dz + \frac{1}{2}\left(u_x - v_y + i(v_x + u_y)\right)d\bar{z} \\ &= (\partial_z u + i\partial_z v)dz + (\partial_{\bar{z}} u + i\partial_{\bar{z}} v)d\bar{z} \end{aligned}$$

and

$$df = (\partial_z f)dz + (\partial_{\bar{z}} f)d\bar{z}.$$

This rearrangement shows more explicitly the dependence of f on z and $\bar{z}$, a fact which we soon shall need. Sometimes also the short notation

$$f_z := \partial_z f, \ \ f_{\bar{z}} := \partial_{\bar{z}} f$$

5. Holomorphic functions

is used. We shall introduce now the notion of *holomorphic function*; for this purpose we use approximation by linear functions following the real case. This may be used also with success in higher dimensions and in function spaces. A *linear function* or *linear form* in $\mathbb{C}$ has the form

$$L(z) = ax + by$$

with $a, b \in \mathbb{C}$. This may be written also as

$$L(z) = Az + B\bar{z} \quad (A, B \in \mathbb{C}).$$

Firstly this is an $\mathbb{R}$-linear function; we have with real α, α',

$$L(\alpha z + \alpha' z') = \alpha L(z) + \alpha' L(z').$$

Stressing the complex variable z requires a $\mathbb{C}$-linear function L which has to fulfil for complex λ, λ' the relation

$$L(\lambda z + \lambda' z') = \lambda L(z) + \lambda' L(z').$$

Obviously this is possible only for an $L(z)$ given by

$$L(z) = Az, \quad A \in \mathbb{C},$$

otherwise we have $B\lambda z = B\bar{\lambda}z$ for all z and λ thus implying $B = 0$. Now we are able to define:

Definition 5.1 (Holomorphic function). A function $f \in C^1(G)$ in a domain $G \subset \mathbb{C}$ is called *holomorphic*, if for each point $z \in G$ a complex number $f'(z)$ exists, such that for $h \to 0$,

$$f(z + h) = f(z) + f'(z)h + o(h).$$

The number $f'(z)$ is called the *(complex) differential quotient* or the *(complex) derivative* of f in z.

The difference $f(z + h) - f(z)$ is approximated in h by the linear function $f'(z)h$. The literature often uses the notion *regular* instead of holomorphic, or also *monogenic*. For consistency we shall use only the word "holomorphic". For the *Bachmann–Landau symbol* o(h) see Definition A.1.11 in Appendix 1.

We have reached an important point: *function theory* deals mainly with holomorphic functions. We have chosen for the definition the linear approximation since this is possible also in higher dimensions. In the sequel we shall get to know further equivalent properties of holomorphic functions which are sometimes easier to show than the definition. We remark explicitly that a holomorphic function is defined in a domain. The approximation itself is defined pointwise.

We know from the reals that a function can be approximated by a linear function if and only if it possesses a limit of the difference quotient. Unfortunately this does not hold in higher dimensions, but in the plane we have:

Definition 5.2 (Complex differentiability). A function $f \in C^1(G)$ in a domain $G \subset \mathbb{C}$ is called *complex differentiable*, if the limit

$$\lim_{h \to 0} \frac{f(z+h) - f(z)}{h}$$

with $h \in \mathbb{C}$ exists for all $z \in G$. This limit is called the *complex derivative* if it exists and is denoted by $f'(z)$.

A complex differentiable function is obviously continuous. We prove:

Proposition 5.3. *A complex function f in a domain G is complex differentiable if and only if it is holomorphic in G.*

Proof. It is easy to convert the linear approximation

$$f(z+h) = f(z) + f'(z)h + o(h)$$

into the difference quotient and vice versa:

$$\frac{f(z+h) - f(z)}{h} = f'(z) + o(1).$$

For $h \to 0$, $f'(z)$ is also the limit of the difference quotient. □

Complex differentiability is a stronger assumption for complex functions than is real differentiability. The reason is that h may tend to zero from all directions in the plane while on the real axis only the two directions on the real axis are taken into account. Therefore we assumed for the definition of a holomorphic function the approximation by a *complex* linear function. We shall show now a second equivalent property for a function to be holomorphic, one that will fulfil a system of partial differential equations. And this is possible also in higher dimensions:

Theorem 5.4 (Cauchy–Riemann differential equations). *A function $f \in C^1(G)$ in a domain $G \subset \mathbb{C}$ is holomorphic in G if and only if one has*

$$\bar{\partial} f = 2\partial_{\bar{z}} f = u_x - v_y + i(u_y + v_x) = 0.$$

The equations

$$\bar{\partial} f = 2\partial_{\bar{z}} f = 0$$

or

$$u_x - v_y = 0, \quad u_y + v_x = 0$$

are called Cauchy–Riemann differential equations (CRD) *and $\partial_{\bar{z}}$ is called the* Cauchy–Riemann operator.

Before we start with the proof we give some remarks about CAUCHY; the reader may find remarks about RIEMANN in Subsection 4.3.

5. Holomorphic functions

Augustin-Louis Cauchy (1789–1857) was after 1822 one of the most eminent mathematicians of his time. He was professor at the Ecole Polytechnique in Paris and member of the French Academy of Science. His textbooks have been standard literature in different languages and for several decades. His enormous productivity is mirrored in his extensive collected works. He laid important foundations in several fields of mathematics.

A.-L. Cauchy (1789–1867)

Proof. From the existence of real derivatives of the functions u and v with $h =: h_1 + ih_2$, it follows that

$$u(x + h_1, y + h_2) - u(x, y) = u_x(x, y)h_1 + u_y(x, y)h_2 + o(h),$$
$$v(x + h_1, y + h_2) - v(x, y) = v_x(x, y)h_1 + v_y(x, y)h_2 + o(h).$$

We define $u(x, y) =: u(z)$ and $v(x, y) =: v(z)$. Both equations for u and v, because of $h_1 = (h + \overline{h})/2$ and $h_2 = (h - \overline{h})/2i$, are contained in the following:

$$
\begin{aligned}
f(z + h) - f(z) &= (u_x(z) + iv_x(z))h_1 + (u_y(z) + iv_y(z))h_2 + o(h) \\
&= \frac{1}{2}(u_x(z) + v_y(z) + i(v_x(z) - u_y(z)))\,h \\
&\quad + \frac{1}{2}(u_x(z) - v_y(z) + i(v_x(z) + u_y(z)))\,\overline{h} + o(h) \\
&= \partial_z f(z)h + \partial_{\overline{z}} f(z)\overline{h} + o(h) \qquad (*).
\end{aligned}
$$

If f fulfils the CRD we have just the complex approximation of f,

$$f(z + h) - f(z) = (\partial_z f)(z)h + o(h),$$

which means $f'(z) = (\partial_z f)(z)$ and this is one direction of our proof.

For the other direction of proof if f is holomorphic, we have

$$f(z + h) - f(z) = f'(z)h + o(h),$$

and together with (*) it follows that

$$f'(z) = \partial_z f(z) + \partial_{\overline{z}} f(z)\frac{\overline{h}}{h} + o(1).$$

The last term on the right-hand side vanishes for $h = |h|(\cos\varphi + i\sin\varphi) \to 0$; the central term has the factor $(\cos\varphi - i\sin\varphi)^2$ which may take all values on the unit circle. As the other terms do not depend on h our equation holds only for $(\partial_{\overline{z}} f)(z) = 0$, i.e., the CRD are fulfilled. $\qquad\square$

Proposition 5.5 (Differentiation rules). *The differentiation rules are the same as in the real case:*

$$
\begin{aligned}
(f+g)'(z) &= f'(z) + g'(z), \\
(fg)'(z) &= f'(z)g(z) + f(z)g'(z), \\
\left(\frac{f}{g}\right)'(z) &= \frac{f'(z)g(z) - f(z)g'(z)}{g^2(z)}, \quad g(z) \neq 0, \\
(f(g))'(z) &= f'(g(z))g'(z) \quad (\textit{chain rule}).
\end{aligned}
$$

Also the rule for the inverse function is true:.

$$
(f^{-1})'(z) = \frac{1}{f'(f^{-1}(z))}.
$$

But in this case some assumptions have to be made, in particular the function f' must not vanish in a neighborhood of the point of differentiation.

Proof. We shall prove only two of the rules, firstly the product rule:

$$
\begin{aligned}
f(z+h)g(z+h) &= (f(z) + f'(z)h + o(h))(g(z) + g'(z)h + o(h)) \\
&= f(z)g(z) + (f'(z)g(z) + f(z)g'(z))h + o(h),
\end{aligned}
$$

as the product of a bounded function with a o(1) is again a o(1); secondly the chain rule:

$$
\begin{aligned}
f(g(z+h)) &= f(g(z)) + f'(g(z))[g(z+h) - g(z)] + o(h) \\
&= f(g(z)) + f'(g(z))g'(z)h + o(h),
\end{aligned}
$$

as the sum of the different remainders is a o(h). $\qquad\square$

Example 5.6. a) As an example we differentiate $f(z) = z^n$, $n \in \mathbb{N}$:

$$
f(z+h) - f(z) = (z+h)^n - z^n = nz^{n-1}h + o(h)
$$

for $h \to 0$, as in the binomial formula all terms but the first one contain a higher power of h. *Rational functions*, i.e., quotients of polynomials, require the use of the quotient rule. So we have, e.g.,

$$
\frac{dz^{-n}}{dz} = \frac{0 - nz^{n-1}}{z^{2n}} = -nz^{-n-1}, \quad z \neq 0;
$$

the differentiation rule for positive exponents holds analogously also for negative ones.

This direct reference to known derivatives makes it sometimes easier to use differentiation than to use the CRD.

b) As an example for a non-holomorphic function we look at $f(z) = \bar{z}$:

$$
f(z+h) - f(z) = \bar{z} + \bar{h} - \bar{z} = \frac{\bar{h}}{h}h,
$$

and the quotient $\overline{h}/h$ has no limit for $h \to 0$.

c) A non-constant real function $f(x, y)$ is not complex differentiable, as $f = u$ reduces the CRD to

$$f_x = 0, \quad f_y = 0,$$

thus f has to be constant.

d) As we may differentiate with a real h we get for a holomorphic function

$$f'(z) = f_x(z),$$

and similarly with an imaginary h,

$$f'(z) = -if_y(z).$$

Remark 5.7. At the beginning of this section we had written a differentiable complex function in the form

$$f(z + h) = f(z) + (\partial_z f)(z)h + (\partial_{\overline{z}} f)(z)\overline{h} + \mathrm{o}(h).$$

If h varies on a small circle around 0 we have, with $h = |h|(\cos \varphi + i \sin \varphi)$,

$$f(z + h) - f(z) = h[(\partial_z f)(z) + (\partial_{\overline{z}} f)(z)(\cos 2\varphi - i \sin 2\varphi)] + \mathrm{o}(h).$$

If f is holomorphic then $f(z + h) - f(z)$ behaves like h up to a factor, if we abandon the inessential remainder term. One says that $w = f(z)$ maps infinitesimal circles into infinitesimal circles. The dilation of $f(z+h) - f(z)$ by $|f'(z)|$ is in all directions the same, therefore this behavior has been called *monogenic*.

But if we have $(\partial_z f)(z) = 0$ and $(\partial_{\overline{z}} f)(z) \neq 0$, one speaks of an *antiholomorphic function*. The infinitesimal behavior is similar to the holomorphic case, however the image circles are crossed in the opposite direction relative to the pre-image circles. These are the functions depending on $\overline{z}$, e.g., $f(z) = \overline{z}$. This function gives the reflection about the real axis, as for all z the imaginary part is multiplied by -1. This reflection generates the inversion of the crossing direction through the infinitesimal circles.

If finally both derivatives do not vanish, i.e., $\partial_z f(z) \neq 0$ and $\partial_{\overline{z}} f(z) \neq 0$, then $f(z+h) - f(z)$ runs through an infinitesimal ellipse if h runs through an infinitesimal circle. In this case the dilation of $f(z+h) - f(z)$ varies from one direction of h to another one. Therefore such functions have been called *polygenic*. Also these functions contain interesting classes, but this is not the place to deal with them. One may study the behavior of f at every single point, but for holomorphic functions we assume always the same behavior in a domain G.

5.2 Differentiation in $\mathbb{H}$

As has been shown in Section 2, the algebra of quaternions is a division algebra just as $\mathbb{C}$ and $\mathbb{R}$. Therefore one may introduce differentiation also for functions $f = f(x)$, $x \in \mathbb{H}$, with values in $\mathbb{H}$ via a difference quotient

$$[f(x + h) - f(x)]h^{-1} \quad \text{or} \quad h^{-1}[f(x + h) - f(x)]$$

with $h \in \mathbb{H}$ and to define all functions which have such a limit for $h \to 0$ as $\mathbb{H}$-differentiable. We have already mentioned that h varies for real functions f at a point x only from two directions, while h in $\mathbb{C}$ may vary from all plane directions to zero, and this causes severe restrictions for the existence of the limit, namely the complex functions must satisfy the Cauchy–Riemann differential equations. The higher dimension in $\mathbb{H}$ brings still more freedom for h, so we have to expect a more restrictive system of differential equations for the existence of a differential quotient.

For this reason we have already used in $\mathbb{C}$ the linear approximation for the definition of a holomorphic function. We shall carry out this idea also in $\mathbb{H}$, but some further problems have to be solved here. From the historical point of view the understanding of this situation is very recent. Twenty years ago the existence of a useful notion of differentiation in the quaternions was thought to be impossible.

We shall show first that the classical notion of differentiability via the difference quotient is possible only for trivial cases.

5.2.1 Mejlikhzhon's result

Already in 1947 N.M. KRYLOV [87] and his pupil A.S. MEJLIKHZHON [107] could prove the following theorem which makes impossible the definition of an $\mathbb{H}$-holomorphic function via the existence of the limit of the difference quotient:

Theorem 5.8 (Krylov, Mejlikhzhon). *Let $f \in C^1(G)$ be a function given in a domain $G \subset \mathbb{H}$ with values in $\mathbb{H}$. If for all points in G the limit*

$$\lim_{h \to 0} h^{-1}[f(x+h) - f(x)] =: {}'f(x)$$

exists, then in G the function f has the form

$$f(x) = a + x\,b \qquad (a, b \in \mathbb{H}).$$

We have an analogous result if the difference quotient with h^{-1} on its right-hand side possesses a limit. We see also that the approximation of a function f by $f(x_0) + xb$, resp. by $f(x_0) + ax$, is too strong an assumption for the function f. We have to think of this fact while defining an $\mathbb{H}$-holomorphic function.

Proof. (Following [152].) We choose for h the special values h_0, $h_1 e_1$, $h_2 e_2$, $h_3 e_3$ with real $h_i, i = 0, 1, 2, 3$. Using $\partial_i = \partial/\partial x_i$ for $i = 0,1,2,3$ we get for the different $h_i \to 0$ in x the identities

$$'f = \partial_0 f = -e_1 \partial_1 f = -e_2 \partial_2 f = -e_3 \partial_3 f. \tag{$*$}$$

Now we identify e_1 with the complex unit i, and we split also the function f by $F_1 := f_0 + i f_1$, $F_2 := f_2 - i f_3$ in the following way:

$$f(x) = F_1(x) + e_2 F_2(x).$$

It then follows from $(*)$ that

$$\partial_0(F_1 + e_2 F_2) = -i\partial_1(F_1 + e_2 F_2) = -e_2\partial_2(F_1 + e_2 F_2) = -e_3\partial_3(F_1 + e_2 F_2).$$

5. Holomorphic functions

As 1 and e_2 are complex linearly independent, the complex parts and those with the factor e_2 are separately equal:

$$\partial_0 F_1 = -i\partial_1 F_1 = \partial_2 F_2 = i\partial_3 F_2,$$
$$\partial_0 F_2 = i\partial_1 F_2 = -\partial_2 F_1 = i\partial_3 F_1.$$

By convenient summation it follows that

$$(\partial_0 + i\partial_1)F_1 = (\partial_2 - i\partial_3)F_2 = 0,$$
$$(\partial_0 - i\partial_1)F_2 = (\partial_2 + i\partial_3)F_1 = 0.$$

That means that F_1 is a holomorphic function of the complex variables $z_1 := x_0 + ix_1$ and $z_2 := x_2 + ix_3$, while this is correct for F_2 relative to the conjugate complex variables $\bar{z}_1$ and $\bar{z}_2$. Therefore all functions are differentiable infinitely often relative to the real variables x_0 and x_1, resp. x_2 and x_3. We see further that

$$(\partial_0 - i\partial_1)F_1 = -2i\partial_1 F_1 = 2\partial_2 F_2 = (\partial_2 + i\partial_3)F_2,$$
$$(\partial_2 - i\partial_3)F_1 = -2i\partial_3 F_1 = -2i\partial_1 F_2 = -(\partial_0 + i\partial_1)F_2.$$

In addition we can conclude that $(\partial_0 - i\partial_1)F_1$ is also holomorphic relative to z_2, so that the mixed derivative $\partial_{\bar{z}_2}\partial_{z_1}F_1$ also exists and is continuous. From the theorem of Schwarz it follows that we can change the order of differentiation and so

$$(\partial_0 - i\partial_1)^2 F_1 = (\partial_0 - i\partial_1)(\partial_2 + i\partial_3)F_2 = (\partial_2 + i\partial_3)(\partial_0 - i\partial_1)F_2 = 0,$$
$$(\partial_2 - i\partial_3)^2 F_1 = -(\partial_2 - i\partial_3)(\partial_0 + i\partial_1)F_2 = -(\partial_0 + i\partial_1)(\partial_2 - i\partial_3)F_2 = 0.$$

The last two equations show that F_1 depends only linearly on the two complex variables as the mixed derivative vanishes also:

$$(\partial_0 - i\partial_1)(\partial_2 - i\partial_3)F_1 = -(\partial_0 - i\partial_1)(\partial_0 + i\partial_1)F_2 = -(\partial_0 + i\partial_1)(\partial_0 - i\partial_1)F_2 = 0.$$

Quite analogously the same follows for F_2, so f depends only linearly on the variables:

$$f(x) = a + \sum_{k=0}^{3} b_k x_k$$

with quaternions a and b_k. From the equation $(*)$ it follows finally that

$$b_0 = -ib_1 = -e_2 b_2 = -e_3 b_3$$

or

$$f(x) = a + \left(\sum_{k=0}^{3} e_k x_k\right) b_0 = a + xb_0. \qquad \Box$$

5.2.2 $\mathbb{H}$-holomorphic functions

One can proceed in analogy to the complex case also for a continuously differentiable function $f(x)$ in a domain $G \subset \mathbb{H}$ with values in $\mathbb{H}$, but the calculation of the quaternionic form of the total differential depending on x and $\bar{x}$ is more difficult. The differentials dx_k commute with quaternions, but one should observe that quaternionic expressions such as df or dx do not commute with quaternions. Regarding differentials we refer to Appendix 1. Let us define $\partial_k := \partial/\partial x_k$, we then have

$$df = \partial_0 f \, dx_0 + \sum_{k=1}^{3} \partial_k f \, dx_k = dx_0 \partial_0 f + \sum_{k=1}^{3} dx_k \partial_k f.$$

From

$$dx = dx_0 + \sum_{k=1}^{3} e_k dx_k \ \text{ and } \ d\bar{x} = dx_0 - \sum_{k=1}^{3} e_k dx_k$$

and an appropriate multiplication with the e_j from the left and the right the simple formulas

$$dx_0 = \frac{1}{2}(dx + d\bar{x}) \ \text{ as well as } \ dx_j = \frac{1}{2}(e_j d\bar{x} - dx \, e_j), \ \ j = 1, 2, 3,$$

follow, and by substituting this into df we get

$$df = \frac{1}{2}\partial_0 f \, (dx + d\bar{x}) + \frac{1}{2} \sum_{k=1}^{3} \partial_k f \, (e_k \, d\bar{x} - dx \, e_k)$$

$$= \frac{1}{2}\left(\sum_{k=0}^{3} \partial_k f \, e_k\right) d\bar{x} + \frac{1}{2}\left(\partial_0 f \, dx - \sum_{k=1}^{3} \partial_k f \, dx \, e_k\right).$$

A comparison with the formula for df in $\mathbb{C}$ shows that here also the differential of the conjugate variable $d\bar{x}$ has been isolated with a differential operator of the form

$$\bar{\partial} := \frac{\partial}{\partial x_0} + \frac{\partial}{\partial x_1}e_1 + \frac{\partial}{\partial x_2}e_2 + \frac{\partial}{\partial x_3}e_3$$

as a foregoing factor. One sees easily that $\frac{1}{2}\bar{\partial}$ is a generalization of the complex differential operator

$$\partial_{\bar{z}} := \frac{1}{2}(\partial_x + i\partial_y).$$

One can further see that the differential operator $\frac{1}{2}\partial$ with

$$\partial := \frac{\partial}{\partial x_0} - \frac{\partial}{\partial x_1}e_1 - \frac{\partial}{\partial x_2}e_2 - \frac{\partial}{\partial x_3}e_3$$

corresponding to

$$\partial_z := \frac{1}{2}(\partial_x - i\partial_y)$$

5. *Holomorphic functions*

occurs only entangled with the differentials. That corresponds to the above remark that ax or xb are not the correct approximations for a function f.

Before proceeding we remark that the operators $\overline{\partial}$ and ∂ are often used in the literature with interchanged meaning. We have decided to use the above notation in agreement with the operator $\partial_{\overline{z}}$. The operator $\overline{\partial}$ is also in $\mathbb{H}$ and $C\ell(n)$, just as in $\mathbb{C}$, the central operator for the whole theory.

The variable x is not appropriate for the approximation and it also does not satisfy the equation $\overline{\partial}x = 0$, which happens in $\mathbb{C}$. This may advise us to look for linear expressions which satisfy the equation $f\overline{\partial} = 0$ or $\overline{\partial}f = 0$. So, the following expressions satisfying the condition we wanted — also called *Fueter variables* — are helpful:

$$z_k := -\frac{1}{2}(e_k x + x e_k) = x_k - x_0 e_k, \quad (k = 1, 2, 3).$$

It is also important that all calculations have to take care of the fact that the differential operators may act upon f from the left, resp. from the right. Normally we shall work with operators acting from the left, a fact which moreover is the common writing style for operators. But for all statements we have a corresponding one from the right.

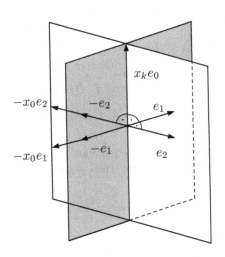

Figure 5.1

The new variables z_k satisfy both the conditions $z_k\overline{\partial} = 0$, resp. $\overline{\partial}z_k = 0$. If we substitute them into the formula for df, because of $dx_k = dz_k + dx_0 e_k$ we have easily

$$df = (f\overline{\partial})dx_0 + \sum_{k=1}^{3}(\partial_k f)dz_k = (\overline{\partial}f)dx_0 + \sum_{k=1}^{3} dz_k(\partial_k f).$$

If we now assume $f\bar{\partial} = 0$, resp. $\bar{\partial}f = 0$, this corresponds to the linear approximation by

$$\ell(x) = \sum_{k=1}^{3} a_k z_k, \quad \text{resp.} \quad \ell(x) = \sum_{k=1}^{3} z_k a_k.$$

As Malonek [100] showed this leads successfully to

Definition 5.9 (ℍ-holomorphic function). Let the function $f \in C^1(G)$ be given in the domain $G \subset \mathbb{H}$ with values in $\mathbb{H}$. Then f is called *right-ℍ-holomorphic* in G, if for every point $x \in G$ and for appropriate quaternions a_k depending on x,

$$f(x + h) = f(x) + \sum_{k=1}^{3} a_k(h_k - h_0 e_k) + \mathrm{o}(h)$$

holds for $h \to 0$, resp. *left-ℍ-holomorphic*, if

$$f(x + h) = f(x) + \sum_{k=1}^{3} (h_k - h_0 e_k) a_k + \mathrm{o}(h).$$

Here the h_k are the coordinates of h.

If no confusion is possible we shall speak simply of holomorphic functions or a bit more precisely of right- resp. left-holomorphic functions. Without difficulties we can prove the equivalence of our definition with one given via differential equations:

Theorem 5.10 (CRD in ℍ). *A function $f \in C^1(G)$ in a domain $G \subset \mathbb{H}$ and with values in $\mathbb{H}$ is there right- resp. left-ℍ-holomorphic if and only if*

$$f\bar{\partial} = 0, \quad \text{resp.} \quad \bar{\partial}f = 0.$$

We call these differential equations also Cauchy–Riemann differential equations (CRD) *(in ℍ).*

These Cauchy–Riemann differential equations are sometimes called *generalized Cauchy–Riemann differential equations*, resp. *Cauchy–Fueter differential equations*.

Proof. We restrict ourselves to left-holomorphic functions. Following A.1.14 and the above considerations we get

$$f(x + h) - f(x) = df[h] + \mathrm{o}(h) = (\bar{\partial}f)h_0 + \sum_{k=1}^{3} (h_k - h_0 e_k)(\partial_k f) + \mathrm{o}(h).$$

For $\bar{\partial}f = 0$ we have explicitly the desired linear approximation. If vice versa the linear approximation is given we see firstly $a_k = \partial_k f$, as for $h_0 = 0$ the h_k are independent variables with uniquely determined coefficients. If we compare this with the second formula in Definition 5.9 we get

$$0 = (\bar{\partial}f)h_0 + \mathrm{o}(h),$$

and that is possible only for $\bar{\partial}f = 0$. □

5. *Holomorphic functions*

5.2.3   Holomorphic functions and differential forms

It is possible to express the $\mathbb{H}$-holomorphy very shortly by differential forms, a fact that will be useful later for the integral theorems.

Theorem 5.11. *A function $f \in C^1(G)$ in a domain $G \subset \mathbb{H}$ with values in $\mathbb{H}$ is left-$\mathbb{H}$-holomorphic if and only if*

$$d(dx^* f) = 0,$$

analogously it is right-$\mathbb{H}$-holomorphic if and only if

$$d(f dx^*) = 0.$$

For the Hodge operator $*$ and dx^* see Definition A.1.15 in Appendix 1.

Proof. We prove only the statement for left-$\mathbb{H}$-holomorphic functions, the proof for right-$\mathbb{H}$-holomorphic functions is completely analogous. Together with $d\sigma := dx_0 \wedge dx_1 \wedge dx_2 \wedge dx_3$ we have

$$
\begin{aligned}
d(dx^* f) &= (-1)^3 dx^* \wedge df = -\sum_{j,k=0}^{3} e_j dx_j^* \wedge (\partial_k f) dx_k \\
&= -\sum_{j=0}^{3} e_j (\partial_j f) dx_j^* \wedge dx_j \\
&= \sum_{j=0}^{3} e_j (\partial_j f) d\sigma = (\overline{\partial} f) d\sigma,
\end{aligned}
$$

as the wedge product vanishes if two dx_j, dx_k are equal. Therefore $d(dx^* f) = 0$ if and only if $\overline{\partial} f = 0$, i.e., the statement. $\square$

We remark that this theorem is also correct in $\mathbb{C}$ since there we have $dx^* = dy, dy^* = -dx$ and $dz^* = dy - i dx = -i dz$, and it follows that

$$d(dz^* f) = -(dy - i dx) \wedge (\partial_x f \, dx + \partial_y f \, dy) = 2(\partial_{\overline{z}} f) dx \wedge dy,$$

i.e., the same formula as in $\mathbb{H}$ from which we concluded that $d(dz^* f)$ and $\partial_{\overline{z}} f$ vanish simultaneously.

Moreover we are able to characterize $\mathbb{H}$-holomorphic functions by differential forms having a smaller degree, see A. SUDBERY [152].

Theorem 5.12. *A function $f \in C^1(G)$ in a domain $G \subset \mathbb{H}$ is left-$\mathbb{H}$-holomorphic in G if and only if*

$$\frac{1}{2} d(dx \wedge dx f) = dx^*(Df)$$

where D denotes the so-called DIRAC operator

$$D = \sum_{k=1}^{3} e_k \partial_k.$$

Analogously a function is right-$\mathbb{H}$-holomorphic if and only if

$$\frac{1}{2}d(fdx \wedge dx) = (fD)dx^*.$$

The Dirac operator acts only upon the three variables x_1, x_2, x_3; it has important applications in physics and is named after the English physicist PAUL A.M. DIRAC (1902–1984). For our basic operator $\overline{\partial}$ we have obviously

$$\overline{\partial} = \partial_0 + D.$$

Proof. We shall firstly convert $dx \wedge dx$ in a form convenient for higher dimensions:

$$\begin{aligned}
\frac{1}{2}\,dx \wedge dx &= \frac{1}{2}\sum_{i,j=0}^{3} e_i e_j dx_i \wedge dx_j = \sum_{i<j;i,j=1}^{3} e_i e_j dx_i \wedge dx_j \\
&= e_1 e_2 dx_1 \wedge dx_2 + e_1 e_3 dx_1 \wedge dx_3 + e_2 e_3 dx_2 \wedge dx_3 \\
&= \sum_{i=1}^{3}(-1)^{i-1}e_i d\hat{x}_{0,i} =: d\tau.
\end{aligned}$$

Here the two subscripts at $d\hat{x}$ mean that the two corresponding differentials have to be cancelled. For this $d\tau$ it follows $(x = x_0 + \mathbf{x})$ that

$$dx_0 \wedge d\tau = \sum_{i=1}^{3}(-1)^{i-1}e_i d\hat{x}_i = -\sum_{i=1}^{3} e_i dx_i^* = -dx^* + dx_0^* = -d\mathbf{x}^*.$$

Now keeping in mind that $\partial_0 = \overline{\partial} - D$ it can easily be seen that

$$\begin{aligned}
\frac{1}{2}d(dx \wedge dx\,f) &= d(d\tau\,f) = (-1)^2 d\tau \wedge df \\
&= d\tau \wedge dx_0\,\partial_0 f + \sum_{i=1}^{3}(-1)^{i-1}e_i d\hat{x}_{0,i} \wedge dx_i\,\partial_i f \\
&= (-dx^* + dx_0^*)(\overline{\partial}f - Df) + \sum_{i=1}^{3} d\hat{x}_0\,e_i \partial_i f \\
&= (-dx^* + dx_0^*)(\overline{\partial}f - Df) + dx_0^*\,Df \\
&= -d\mathbf{x}^*\overline{\partial}f + dx^*\,Df.
\end{aligned}$$

So, $\overline{\partial}f = 0$ is equivalent to the theorem's assertion. $\qquad\square$

We shall present now three other attempts to approach differentiation in the quaternions to show that, over the years, quite different research has been done in this direction.

Remark 5.13. In applications, maps from $\mathbb{R}^3$ into $\mathbb{R}^4$ are often used. If we identify $\mathbb{R}^3$ with $\operatorname{Vec}\mathbb{H}$ and $\mathbb{R}^4$ with $\mathbb{H}$ (regarding them as vector spaces) we can call a function $f : \operatorname{Vec}\mathbb{H} \to \mathbb{H}$ holomorphic if $Df = 0$. Then it follows that

$$d(d\mathbf{x}^* f) = -(Df)dx_1 \wedge dx_2 \wedge dx_3.$$

We may hint of another possibility, sometimes necessary, to identify $\mathbb{R}^3$ with the space $\mathbb{R}$-span$\{e_0, e_1, e_2\}$, which means with the paravectors in $C\ell(2)$.

Remark 5.14. Using the notation $(x)^l := -e_l x e_l$, $l = 1, 2, 3$, V. Souček found in 1983 [143] another form for the total differential df. The operator $(x)^l$ is the principal involution introduced in Definition 2.17 (ii).

One can easily calculate the formulas

$$(x)^l (y)^l = (xy)^l, \quad \sum_{l=1}^{3} (x)^l = 2x_0 + \overline{x} = 4x_0 - x.$$

Then following Souček the decomposition

$$d\overline{x}\,\overline{\partial} + \sum_{l=1}^{3} (d\overline{x}\,\overline{\partial})^l = 4(d\overline{x}\,\overline{\partial})_0 = 4d$$

is given for the total differential operator d. If one calls the first term in this decomposition $\mathcal{D}$, then $\mathcal{D}f = 0$ is equivalent to $\overline{\partial}f = 0$ or to $\mathbb{H}$-holomorphy.

Souček calls the second part in this decomposition $\vec{\mathcal{D}}$ and shows that $\vec{\mathcal{D}}f = 0$ is equivalent to the existence of the limit of the difference quotient, using however $\overline{h}$ instead of h. So, $\vec{\mathcal{D}}f = 0$ is equivalent to the linearity of f.

Following the definition, $\vec{\mathcal{D}}f = 0$ is equivalent to $4df = d\overline{x}\,\overline{\partial}f$. As one may put $dx_j = \delta_{jk}$ this is equivalent to the four equations

$$4\partial_0 f = \overline{\partial}f, \quad 4\partial_1 f = -e_1\overline{\partial}f, \quad 4\partial_2 f = -e_2\overline{\partial}f, \quad 4\partial_3 f = -e_3\overline{\partial}f$$

or

$$\partial_0 f = e_1\partial_1 f = e_2\partial_2 f = e_3\partial_3 f = \frac{1}{4}\overline{\partial}f.$$

Following Theorem 5.8 this is equivalent to the existence of the limit

$$\lim_{h \to 0} \overline{h}^{-1}[f(x+h) - f(x)]$$

and therefore to $f(x) = a + \overline{x}b$. It is interesting that one here has a decomposition of df in part related to holomorphy and in part related to classical differentiation.

Remark 5.15. M.S. Marinov [106] introduced a quite differently structured notion of a so-called S-differentiability. Firstly a new multiplication is defined in $\mathbb{H}$:

$$x * y = (z_1 + z_2 e_2) * (w_1 + w_2 e_2) := (z_1 w_1 - z_2 w_2) + (z_1 w_2 + z_2 w_1)e_2,$$

which differs only slightly from the usual Hamilton multiplication

$$xy = (z_1 + z_2 e_2)(w_1 + w_2 e_2) := (z_1 w_1 - z_2 \overline{w}_2) + (z_1 w_2 + z_2 \overline{w}_1)e_2.$$

But this multiplication is commutative, so a new algebra is defined which contains zero divisors due to the Theorem of Frobenius 2.10. Marinov puts $h = h_1 + h_2 e_2$ and $h^* = \overline{h}_1 - h_2 e_2$ and defines then:

Let $G \subset \mathbb{H}$ be a domain and $f : G \to \mathbb{H}$. The function f is called S-differentiable in $x \in G$ if the limit

$$\lim_{h \to 0} \frac{h^* * (f(x+h) - f(h))}{h_1^2 + h_2^2}, \qquad h_1^2 + h_2^2 \neq 0,$$

exists.

For this version of differentiability it is remarkable that $f * g$ is again S-differentiable, moreover the classical Leibniz rule holds due to commutativity. Marinov is able to prove that this S-differentiability is equivalent to CRD of the form

$$\partial_* * f = 0, \qquad \partial_* := \partial_{z_1} + \partial_{z_2} e_2.$$

We shall not go into details here.

Remark 5.16. In [130] M.V. Shapiro and N.I. Vasilevski introduced the notion of a ψ-hyperholomorphic function. Corresponding differential operators of Cauchy–Riemann type

$$^\psi Du := \sum_{k=4-m}^{3} \psi^k \partial_k u,$$

$$D^\psi u := \sum_{k=4-m}^{3} \partial_k \psi^k u$$

have been studied where $\psi \in \mathbb{H}^m$, $m = 1, 2, 3, 4$. Necessary and sufficient conditions are formulated such that the m-dimensional Laplace operator can be split up into the composition of two operators of type $^\psi D$, resp. D^ψ. Further facts are to be found in the article cited above.

5.3 Differentiation in $C\ell(n)$

For the n-dimensional case the same difficulties come up as in the quaternions, so we may proceed analogously. We deal with functions $f \in C^1(G)$ in a domain $G \subset \mathbb{R}^{n+1}$, the variable x will be identified with the paravector $x = x_0 + x_1 e_1 + \cdots + x_n e_n$. The values of f are in $C\ell(n)$. Again we have to observe the non-commutativity in $C\ell(n)$; we always have a left and a right version of the definitions and theorems. From

$$df = \sum_{k=0}^{n} \partial_k f \, dx_k = \sum_{k=0}^{n} dx_k \partial_k f$$

with

$$dx_0 = \frac{1}{2}(dx + d\bar{x}) \quad \text{as well as} \quad dx_k = \frac{1}{2}(e_k d\bar{x} - dx \, e_k), \ k = 1, \ldots, n,$$

similarly to the quaternions we get

$$df = \frac{1}{2} \left(\sum_{k=0}^{n} \partial_k f \, e_k \right) d\bar{x} + \frac{1}{2} \left(\partial_0 f \, dx - \sum_{k=1}^{n} \partial_k f \, dx \, e_k \right).$$

5. *Holomorphic functions*

This formula is quite analogous to those in $\mathbb{C}$ and $\mathbb{H}$. Again the operator

$$\overline{\partial} := \sum_{k=0}^{n} \partial_k e_k$$

is important and basic for the whole theory. As in $\mathbb{H}$ it is not possible to isolate the conjugate operator

$$\partial := \partial_0 - \sum_{k=1}^{n} \partial_k e_k.$$

We introduce again new variables, the *Fueter variables*, which are solutions of the equations $\overline{\partial} f = 0 = f\overline{\partial}$:

$$z_k := -\frac{1}{2}(e_k x + x e_k) = x_k - x_0 e_k, \quad k = 1, \ldots, n.$$

Substituting these variables into the formula for df, since $dx_k = dz_k + dx_0 e_k$, $k = 1, \ldots, n$, it follows that

$$df = (f\overline{\partial})dx_0 + \sum_{k=1}^{n} (\partial_k f)dz_k = (\overline{\partial}f)dx_0 + \sum_{k=1}^{n} dz_k(\partial_k f).$$

If we now assume $f\overline{\partial} = 0$, resp. $\overline{\partial}f = 0$, this corresponds to linear approximation as in $\mathbb{H}$ via

$$\ell(x) = \sum_{k=1}^{n} a_k z_k, \quad \text{resp.} \quad \ell(x) = \sum_{k=1}^{n} z_k a_k.$$

We follow again Malonek and define:

Definition 5.17 (Clifford holomorphic functions). Let the function $f \in C^1(G)$ have values in $C\ell(n)$ in a domain $G \subset \mathbb{R}^{n+1}$. The function f is called *right-Clifford holomorphic* in G if at every point $x \in G$ for $h \to 0$ and for convenient Clifford numbers a_k depending on x,

$$f(x + h) = f(x) + \sum_{k=1}^{n} a_k(h_k - h_0 e_k) + o(h)$$

holds. It is *left-Clifford holomorphic* if

$$f(x + h) = f(x) + \sum_{k=1}^{n} (h_k - h_0 e_k)a_k + o(h).$$

Here the h_k are the components of h.

If no confusion is possible we shall also speak here simply of holomorphic or right- resp. left-holomorphic functions. With the same proof as for the quaternions, see Theorem 5.10, the equivalence with the corresponding differential equations follows:

Theorem 5.18 (CRD in $C\ell(n)$). *A function $f \in C^1(G)$ in a domain $G \subset \mathbb{R}^{n+1}$ with values in $C\ell(n)$ is right- resp. left-holomorphic if and only if*

$$f\overline{\partial} = 0 \text{ resp. } \overline{\partial}f = 0.$$

We shall call these differential equations again Cauchy–Riemann differential equations (CRD).

The equivalence with a condition for holomorphic functions using differential forms is given as in $\mathbb{H}$:

Theorem 5.19. *A function $f \in C^1(G)$ in a domain $G \subset \mathbb{R}^{n+1}$ with values in $C\ell(n)$ is left-holomorphic if and only if*

$$d(dx^* f) = 0$$

in G. Similarly it is right-holomorphic if and only if

$$d(f dx^*) = 0.$$

Proof. The proof is quite analogous to $\mathbb{H}$ with $d\sigma = dx_0 \wedge \cdots \wedge dx_n$, only the sums run from 0 to n. □

We have just as in $\mathbb{H}$ the equivalence of holomorphy with an equation using differential forms with a degree lower than n, as it was shown in 1999 in [53].

Theorem 5.20. *A function $f \in C^1(G)$ in a domain $G \subset \mathbb{R}^{n+1}$ is left-holomorphic in G if and only if*

$$d(d\tau\, f) = dx^*(Df)$$

holds with the Dirac *operator*

$$D = \sum_{k=1}^{n} e_k \partial_k$$

and the differential form

$$d\tau = \sum_{k=1}^{n} (-1)^{k-1} e_k d\hat{x}_{0,k}.$$

Analogously a function f is right-holomorphic if and only if

$$d(f\, d\tau) = (fD)dx^*.$$

Proof. The proof is again quite analogous to the one in $\mathbb{H}$, only the summation runs from 1 to n. □

5.4 Exercises

1. Find a product rule for $\overline{\partial}(fg)$!

2. Prove in $\mathbb{C}$ the differentiation rule for the inverse function

$$(f^{-1})'(z) = \frac{1}{f'(f^{-1}(z))}.$$

3. Compute in $C\ell(n)$ the expressions

$$\overline{\partial} x^m \quad \text{and} \quad \partial x^m, \quad m \in \mathbb{N}.$$

4. Show the identities (cf. Remark 5.14)

$$(x)^l (y)^l = (xy)^l, \quad \sum_{l=1}^{3} (x)^l = 2x_0 + \overline{x} = 4x_0 - x.$$

6   Powers and Möbius transforms

6.1   Powers

6.1.1   Powers in $\mathbb{C}$

We have already seen that powers

$$w = z^n, \quad n \in \mathbb{N} \cup \{0\},$$

are holomorphic in $\mathbb{C}$. In polar coordinates, using the formula of de Moivre, we have

$$w = r^n (\cos n\varphi + \sin n\varphi)$$

and

$$|w| = r^n, \quad \arg w = n \arg z.$$

The image w winds around the origin precisely n-times if z winds around the origin once. It is interesting to look at the inverse functions:

We divide the complex plane into n sectors:

$$S_k \; : \; \frac{2k\pi}{n} \le \arg z < \frac{2(k+1)\pi}{n} \quad (k = 0, 1, \dots, n-1).$$

The mapping $w = z^n$ transforms each of these sectors onto a complete copy W_k of the w-plane. We assume these planes W_k to be cut along the positive real axis. Then we have upper and lower boundaries; the upper boundary of a cut is always the image of the bounding half line of the sector S_k with the smaller argument. If we cross from S_k to S_{k+1}, the w of the lower boundary cut of W_k crosses to the upper boundary cut of W_{k+1}. Therefore we "glue" the lower boundary cut of W_k to the upper one of W_{k+1}. There emerges a "spiral staircase" which closes when crossing from S_{n-1} to S_0; the lower boundary cut of W_{n-1} has to be glued to the upper boundary cut of W_0. We get a geometric structure over the w-plane and call it a *Riemann surface* $\mathcal{F}(w^{1/n})$ of $w^{1/n}$. Upon this surface the inverse mapping $z = w^{1/n}$ of $w = z^n$ is uniquely defined: On the copy W_k of the w-plane, $z = w^{1/n}$ has to be defined by

$$z = |w|^{1/n} \left(\cos \left(\frac{\arg w + 2k\pi}{n} \right) + i \left(\frac{\arg w + 2k\pi}{n} \right) \right),$$

here $0 \le \arg w < 2\pi$ is assumed. The copies W_k of the w-plane are called *sheets* of the Riemann surface. The points $z = 0$ and $w = 0$ are the unique images of each other, the same with the points $z = \infty$ and $w = \infty$, the latter if we consider the situation on the Riemann spheres in z and w. Both these points are called an *algebraic singularity* or a *winding point* of the Riemann surface.

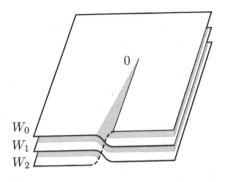

W_0
W_1
W_2

Figure 6.1

As differentiation is a local property we may differentiate $z = w^{1/n}$ without difficulties, and using the differentiation rule for the inverse function we get

$$\frac{dz}{dw} = \frac{1}{n} w^{\frac{1}{n}-1},$$

naturally with the exception of the point $w = 0$. The case $n = 2$ and the Riemann surface for $w^{1/2}$ make it possible to understand the connection between the positive and the negative root in the reals: the transition from one root to the other corresponds to one winding around of the origin on the Riemann surface.

6.1.2 Powers in higher dimensions

The polynomials we deal with are either defined in $\mathbb{H}$ with values in $\mathbb{H}$ or in $\mathbb{R}^{n+1}$ with values in $C\ell(n)$. Unfortunately it is difficult to transfer the notion of holomorphic polynomials to polynomials in $\mathbb{H}$ or $C\ell(n)$ as $\overline{\partial}x = x\overline{\partial} = 1 - n \neq 0$. Moreover we have no convenient product rule for the differentiation in view of the non-commutativity. For the differentiation we could overcome these problems by introducing new variables, which we call *Fueter variables*:

$$z_j = x_j - x_0 e_j \qquad (j = 1, \ldots, n).$$

For them we have

$$\overline{\partial}\, z_j = z_j \overline{\partial} = 0 \qquad (j = 1, \ldots, n);$$

but regrettably already

$$\overline{\partial}(z_j z_i) \neq 0 \qquad (i, j = 1, \ldots, n,\ i \neq j),$$

as one can calculate. RUDOLF FUETER found in the thirties of the last century a method to get over these difficulties: He symmetrized products of his variables in the sense of Appendix A.1.3 and defined homogeneous holomorphic polynomials of arbitrary degree in $\mathbb{H}$ and $C\ell(n)$.

KARL RUDOLF FUETER was born in 1880 in Basel (Switzerland) and studied mathematics in Basel and Göttingen. He got his doctorate in 1903 in Göttingen as a pupil of D. HILBERT. After his habilitation in Marburg he worked as professor in Clausthal, Basel, Karlsruhe and since 1916 onwards at the university in Zürich. In the thirties of the last century together with his pupils he developed function theory in the quaternions. He made important progress in the theory and publicized it. After 1940 his group started also to construct a function theory in Clifford algebras. Rudolf Fueter died in 1950 in Brunnen (Switzerland).

Fueter introduced the polynomials named after him in 1936 [43]. They were used by his pupils and later on, in particular, R. DE-LANGHE proved that they are left- and right-holomorphic in $C\ell(n)$. Much later in 1987 H. MALONEK [101] showed that the Fueter polynomials have values only in the paravectors. We now define the Fueter polynomials. We remark that we have to choose $n = 3$ if the domain of definition is in $\mathbb{H}$ where we have to use the three variables z_1, z_2, z_3, in which case the polynomials are defined in $\mathbb{R}^4$. If we on the contrary understand $\mathbb{H}$ in the sense of $C\ell(2)$ we have only to use the two variables z_1, z_2, i.e., $n = 2$ and the polynomials are defined then in $\mathbb{R}^3$.

Rudolf Fueter

Definition 6.1 (Fueter polynomials). Let x be in $\mathbb{H}$ or $\mathbb{R}^{n+1}$.

(i) We call $\mathbf{k} := (k_1, \ldots, k_n)$ with integer k_i a *multiindex*; for multiindices with non-negative components let us take

$$k := |\mathbf{k}| := \sum_{i=1}^{n} k_i, \quad \mathbf{k}! := \prod_{i=1}^{n} k_i!.$$

We call $k = |\mathbf{k}|$ the *degree* of the multiindex $\mathbf{k}$.

(ii) For a multiindex $\mathbf{k}$ with at least one negative component we define

$$\mathcal{P}_{\mathbf{k}}(x) := 0.$$

For the degree $k = 0$ we write shortly $\mathbf{k} = (0, \ldots, 0) = \mathbf{0}$ and define

$$\mathcal{P}_{\mathbf{0}}(x) := 1.$$

(iii) For a $\mathbf{k}$ with $k > 0$ we define the *Fueter polynomial* $\mathcal{P}_{\mathbf{k}}(x)$ as follows: For each $\mathbf{k}$ let the sequence of indices $j_1, \ldots, j_k$ be given such that the first k_1

indices equal 1, the next k_2 indices equal 2 and, finally the last k_n indices equal n. We put

$$\mathbf{z}^{\mathbf{k}} := z_{j_1} z_{j_2} \ldots z_{j_k} = z_1^{k_1} \ldots z_n^{k_n};$$

this product contains z_1 exactly k_1-times and so on. Then

$$\mathcal{P}_{\mathbf{k}}(x) := \frac{1}{k!} \sum_{\sigma \in \mathrm{perm}(k)} \sigma(\mathbf{z}^{\mathbf{k}}) := \frac{1}{k!} \sum_{\sigma \in \mathrm{perm}(k)} z_{j_{\sigma(1)}} \ldots z_{j_{\sigma(k)}}.$$

Here $\mathrm{perm}(k)$ is the permutation group with k elements (see Definition A.1.1).

This symmetrization compensates in some respects the non-commutativity in $\mathbb{H}$ resp. $C\ell(n)$; for $n = 1$ in $\mathbb{C}$ we get nothing new but for the variable z/i. We show the most important properties of Fueter polynomials:

Theorem 6.2. (i) *The Fueter polynomials satisfy the following recursion formula where $\varepsilon_i := (0, \ldots, 0, 1, 0, \ldots, 0)$ with one 1 in position i:*

$$k\mathcal{P}_{\mathbf{k}}(x) = \sum_{i=1}^{n} k_i \mathcal{P}_{\mathbf{k}-\boldsymbol{\varepsilon}_i}(x) z_i = \sum_{i=1}^{n} k_i z_i \mathcal{P}_{\mathbf{k}-\boldsymbol{\varepsilon}_i}(x).$$

This gives also

$$\sum_{i=1}^{n} k_i \mathcal{P}_{\mathbf{k}-\boldsymbol{\varepsilon}_i}(x) e_i = \sum_{i=1}^{n} k_i e_i \mathcal{P}_{\mathbf{k}-\boldsymbol{\varepsilon}_i}(x).$$

(ii) *We get for the derivatives with $j = 1, \ldots, n$,*

$$\partial_j \mathcal{P}_{\mathbf{k}}(x) = k_j \mathcal{P}_{\mathbf{k}-\boldsymbol{\varepsilon}_j}(x).$$

(iii) *Finally we have*

$$\partial_0 \mathcal{P}_{\mathbf{k}}(x) = -\sum_{j=1}^{n} k_j e_j \mathcal{P}_{\mathbf{k}-\boldsymbol{\varepsilon}_j}(x) = -\sum_{j=1}^{n} e_j \partial_j \mathcal{P}_{\mathbf{k}}(x)$$

and

$$\mathcal{P}_{\mathbf{k}}(x) \partial_0 = -\sum_{j=1}^{n} k_j \mathcal{P}_{\mathbf{k}-\boldsymbol{\varepsilon}_j}(x) e_j = -\sum_{j=1}^{n} \partial_j \mathcal{P}_{\mathbf{k}}(x) e_j,$$

i. e. the $\mathcal{P}_{\mathbf{k}}$ are right- and left-holomorphic as the last two equations imply

$$\overline{\partial} \mathcal{P}_{\mathbf{k}}(x) = \mathcal{P}_{\mathbf{k}}(x) \overline{\partial} = 0.$$

Proof. (i) We prove only the left formula, the proof of the right one is completely analogous. Every summand of $\mathcal{P}_{\mathbf{k}}$ contains one of the z_i as its last factor, so, if we order with respect to this last factor we get an expression of the form

$$k\mathcal{P}_{\mathbf{k}}(x) = \sum_{i=1}^{n} Q_{i,\mathbf{k}}(x) z_i.$$

In such an expression $Q_{i,\mathbf{k}}(x)$ the sum runs over all permutations of the other elements z_j. The z_i in the last position is fixed, and we have ahead of that fixed element all permutations of the other $k-1$ elements for the index $\mathbf{k}-\boldsymbol{\varepsilon}_i$. Moreover the factor before the sum is just $1/(k-1)!$ as the k has been multiplied on the left side. That means the polynomial $\mathcal{P}_{\mathbf{k}-\boldsymbol{\varepsilon}_i}(x)$ is the factor before the fixed z_i. But we have to take into account that each of the k_i equal factors z_i can take the last position without any change in the previous factors. So we have in $Q_{i,\mathbf{k}}$ exactly k_i-times the same expression $\mathcal{P}_{\mathbf{k}-\boldsymbol{\varepsilon}_i}(x)$:

$$Q_{i,\mathbf{k}}(x) := k_i \mathcal{P}_{\mathbf{k}-\boldsymbol{\varepsilon}_i}(x).$$

That is the proof. We get the second equation if we substitute $z_i = x_i - x_0 e_i$. The real factors x_i commute with $\mathcal{P}_{\mathbf{k}}$ so that these summands can be cancelled. In the remaining equation

$$\sum_{i=1}^{n} k_i \mathcal{P}_{\mathbf{k}-\boldsymbol{\varepsilon}_i}(x) x_0 e_i = \sum_{i=1}^{n} k_i x_0 e_i \mathcal{P}_{\mathbf{k}-\boldsymbol{\varepsilon}_i}(x),$$

one can divide by x_0 and the assertion is proved.

(ii) Mathematical induction with respect to k: If $k = 0$, then the $\mathbf{k}-\boldsymbol{\varepsilon}_j$ contains a negative component, therefore $\mathcal{P}_{\mathbf{k}-\boldsymbol{\varepsilon}_j} = 0$ and $0 = \partial_k \mathcal{P}_\mathbf{0}$.

To conclude from $k - 1$ to k we apply the recursion formulas from (i): Following the assumption for $k-1$ and (i) we have

$$
\begin{aligned}
k\partial_j \mathcal{P}_{\mathbf{k}}(x) &= \sum_{i=1}^{n} k_i \partial_j (\mathcal{P}_{\mathbf{k}-\boldsymbol{\varepsilon}_i}(x) z_i) \\
&= \sum_{i=1}^{n} k_i k_j \mathcal{P}_{\mathbf{k}-\boldsymbol{\varepsilon}_i-\boldsymbol{\varepsilon}_j}(x) z_i + \sum_{i=1}^{n} k_i \mathcal{P}_{\mathbf{k}-\boldsymbol{\varepsilon}_i}(x) \partial_j z_i \\
&= (k-1) k_j \mathcal{P}_{\mathbf{k}-\boldsymbol{\varepsilon}_j}(x) + k_j \mathcal{P}_{\mathbf{k}-\boldsymbol{\varepsilon}_j}(x) = k k_j \mathcal{P}_{\mathbf{k}-\boldsymbol{\varepsilon}_j}(x).
\end{aligned}
$$

A division by k gives the assertion.

(iii) We apply also mathematical induction with respect to k and prove simultaneously the formulas for the differentiation from the left and from the right. If $k = 0$ we have $\mathcal{P}_{\mathbf{0}-\boldsymbol{\varepsilon}_j} = 0$ so we get $0 = \partial_0 \mathcal{P}_\mathbf{0}(x)$. To go from $k - 1$ to k we apply again the recursion formula from (i) and the assumption for $k-1$ for the derivative from the right:

$$
\begin{aligned}
k\partial_0 \mathcal{P}_{\mathbf{k}}(x) &= \sum_{i=1}^{n} k_i (\partial_0 \mathcal{P}_{\mathbf{k}-\boldsymbol{\varepsilon}_i}(x)) z_i - \sum_{i=1}^{n} k_i \mathcal{P}_{\mathbf{k}-\boldsymbol{\varepsilon}_i}(x) e_i \\
&= -\sum_{i,j=1}^{n} k_i k_j e_j \mathcal{P}_{\mathbf{k}-\boldsymbol{\varepsilon}_i-\boldsymbol{\varepsilon}_j}(x) z_i - \sum_{i=1}^{n} k_i \mathcal{P}_{\mathbf{k}-\boldsymbol{\varepsilon}_i}(x) e_i \\
&= -(k-1) \sum_{j=1}^{n} k_j e_j \mathcal{P}_{\mathbf{k}-\boldsymbol{\varepsilon}_j}(x) - \sum_{i=1}^{n} k_i e_i \mathcal{P}_{\mathbf{k}-\boldsymbol{\varepsilon}_i}(x) \\
&= -k \sum_{i=1}^{n} k_i \mathcal{P}_{\mathbf{k}-\boldsymbol{\varepsilon}_i}(x) e_i.
\end{aligned}
$$

According to (ii) we may substitute $k_i \mathcal{P}_{\mathbf{k}-\boldsymbol{\varepsilon}_i} = \partial_i \mathcal{P}_{\mathbf{k}}$. The same has to be done for the right version starting with the derivative from the left, concluding the proof. □

A simple corollary is:

Corollary 6.3. *Fueter polynomials are right- and left-$C\ell(n)$-linearly independent.*

Proof. The polynomials $\mathcal{P}_\mathbf{k}$ are right-$C\ell(n)$-linearly independent if from

$$\sum_\mathbf{k} \mathcal{P}_\mathbf{k}(x)\lambda_\mathbf{k} = 0$$

with $\lambda_\mathbf{k} \in C\ell(n)$ it follows that all $\lambda_\mathbf{k} = 0$. This can be seen at once as every $\mathcal{P}_\mathbf{k}$ contains at least one of the x_i with a power differing from the other $\mathcal{P}_\mathbf{k}$, so that linear independence may be proven by real differentiation. □

Remark 6.4. All the polynomials $\mathcal{P}_\mathbf{k}\left(\frac{x}{|x|}\right)$ with $|\mathbf{k}| = k$ constitute on the unit sphere S^n the so-called *spherical polynomials of degree k*.

Corollary 6.5. *We have the following estimate for the Fueter polynomials:*

$$|\mathcal{P}_\mathbf{k}(x)| \leq |z_1|^{k_1} \ldots |z_n|^{k_n} = |\mathbf{z}|^\mathbf{k} \leq |x|^{|\mathbf{k}|}.$$

The proof may be an exercise for the reader (see Exercise 6.3.1).

Example 6.6. a) The Fueter polynomials of degree 1 are just the z_i:

$$\mathcal{P}_{\boldsymbol{\varepsilon}_i}(x) = z_i.$$

b) For degree 2 obviously the simple squares of the z_i are the polynomials of the form

$$\mathcal{P}_{2\boldsymbol{\varepsilon}_i}(x) = z_i^2.$$

For mixed indices we have $(i < j)$

$$\mathcal{P}_{\boldsymbol{\varepsilon}_i + \boldsymbol{\varepsilon}_j}(x) = \frac{1}{2}(z_i z_j + z_j z_i).$$

c) For degree 3 the expressions become longer, e.g., for $i < j$,

$$\mathcal{P}_{2\boldsymbol{\varepsilon}_i + \boldsymbol{\varepsilon}_j}(x) = \frac{1}{3}(z_i^2 z_j + z_i z_j z_i + z_j z_i^2).$$

We remark finally that the $\mathcal{P}_\mathbf{k}$ have values in $\mathbb{R}^{n+1}$, but we shall not prove this here (see Proposition 10.6).

Remark 6.7. R. Delanghe introduced in 1970 [31] the notion of a totally analytic variable. He calls a variable $z \in \mathbb{H}$ *totally analytic* if together with z also each power z^j is $\mathbb{H}$-holomorphic. For this purpose he defines

$$z = \sum_{i=0}^{3} x_i a_i$$

with convenient $a_i \in \mathbb{H}$ $(i = 0, 1, 2, 3)$. These a_i are not necessarily $\mathbb{R}$-linearly independent. This variable is totally analytic if and only if $a_i a_j = a_j a_i$ (see [54]). Simple examples are the Fueter variables z_i introduced above.

Between the system $\{a_0, a_1, a_2, a_3\}$ and the Euclidean base $\{e_0, e_1, e_2, e_3\}$ we have the usual connection

$$\begin{pmatrix} a_0 \\ a_1 \\ a_2 \\ a_3 \end{pmatrix} = A \begin{pmatrix} e_0 \\ e_1 \\ e_2 \\ e_3 \end{pmatrix}$$

with $A \in \mathbb{R}^{4 \times 4}$. We cancel now the first column in A and call the resulting matrix A', let rank $A' < 2$ and let z be a totally analytic variable. For $u = (u_1, \ldots, u_k) \in \mathbb{H}^k$ we define the following function with values in $\mathbb{H}$:

$$(L_k u)(x) = \sum_{l=1}^{k} \left\{ \prod_{j \neq l} \{[(z(x) - z(a_j)] [z(a_k) - z(a_j)]^{-1} u_l\} \right\}.$$

Here we assume $a_j \in \mathbb{H}$ $(j = 1, \ldots, k)$ and $z(a_k) \neq z(a_j)$ for $k \neq j$. This function L_k then takes just the values u_j at the points a_j:

$$\begin{aligned} &\text{(i)} \quad (L_k u)(a_j) = u_j \quad (j = 1, \ldots, k), \\ &\text{(ii)} \quad (L_k u)^p \in \ker \overline{\partial} \quad (p = 1, 2, \ldots) \ (holomorphy). \end{aligned}$$

A polynomial with such (interpolation) properties is called a *Lagrange polynomial*.

6.2 Möbius transformations

6.2.1 Möbius transformations in $\mathbb{C}$

In this section we shall investigate the properties of a simple but interesting class of mappings: The function $w = z + b$ translates all points of the Gaussian plane by b, circles and straight lines are mapped onto circles and straight lines. This map is called *translation*.

The function $f(z) = az$ has the derivative a; one sees from

$$w = f(z) = az =: |a|(\cos \alpha + i \sin \alpha)z$$

that we have a *rotation* combined with a *dilation* of the plane. Every point is rotated by the angle α around the origin and dilated by the factor $|a|$.

The function $f(z) = 1/z$ has the derivative $f'(z) = -1/z^2$. Points in the open unit disc are mapped into the exterior of the disc and vice versa, therefore this mapping is called *reflection about the unit circle*. In particular the origin is mapped to the point $z = \infty$. At the end of Section 1.3 we had an equation for circles and straight lines,

$$A z\overline{z} + 2\mathrm{Re}(\overline{B}z) + C = 0$$

with A, C real; $AC < |B|^2$, if we apply our mapping $z = 1/w$ we get

$$A + 2\mathrm{Re}(\overline{B}\overline{w}) + Cw\overline{w} = A + 2\mathrm{Re}(Bw) + Cw\overline{w} = 0.$$

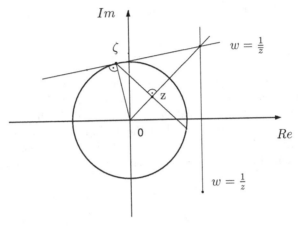

Figure 6.2

We see that circles and straight lines are mapped onto the same geometric structures. More precisely we can say that straight lines through the origin are mapped onto themselves ($A = C = 0$); straight lines not through the origin are mapped onto circles through the origin and vice versa ($A = 0$, $C \neq 0$ resp. $A \neq 0$, $C = 0$). At last circles not through the origin are mapped onto the same circles ($AC \neq 0$).

Definition 6.8 (Möbius transformation). A mapping $w = f(z)$, given by

$$f(z) = \frac{az + b}{cz + d}$$

with $ad - bc \neq 0$ is called a *Möbius transformation*. The matrix

$$\begin{pmatrix} a & b \\ c & d \end{pmatrix}$$

is associated with the Möbius transformation. We remark that an analogous matrix in higher dimensions is called a *Vahlen matrix*. The mapping has to be conveniently defined in $\hat{\mathbb{C}}$ to have a continuous function in the chordal metric.

The mapping is named after August Ferdinand Möbius (1790–1868) and the matrix after Karl Theodor Vahlen (1869–1945).

The Möbius transformations are also called *(fractional) linear maps* despite the presence of the denominator. The derivative is

$$f'(z) = \frac{ad - bc}{(cz + d)^2};$$

it shows the Möbius transformation to be constant if the determinant $ac - bd$ of the associated matrix is zero. For this reason we have excluded this case.

From $w = f(z)$ it is easy to calculate the inverse function

$$z = f^{-1}(w) = \frac{dw - b}{-cw + a}.$$

Its associated matrix has the same determinant as f and is also a Möbius transformation. Consequently a Möbius transformation is a one-to-one mapping even from $\hat{\mathbb{C}}$ onto $\hat{\mathbb{C}}$; in case $c \neq 0$ the point $z = -d/c$ will be mapped to $w = \infty$ and $z = \infty$ to $w = a/c$. If $c = 0$ then the Möbius transformation $f(z) = (az + b)/d$ is the composition of a translation, a rotation, and a dilation. If $c \neq 0$ we have a composition of a translation, a rotation, a dilation, and the mapping $w = 1/z$, as we may write f in the form

$$f(z) = \frac{a}{c} - \frac{ad - bc}{c^2} \frac{1}{z + d/c}.$$

All these simple maps transform circles and straight lines into the same geometric structures, so we have proved:

Proposition 6.9. *A Möbius transformation maps circles and straight lines in the z-plane onto circles and straight lines in the w-plane.*

This is the reason why circles and straight lines are both called "circles" in a more general sense: *Möbius transformations are circle-preserving.*

It is easy to see that the Möbius transformations form a group relative to composition. Indeed, the mapping $f : z \to z$ is the neutral element of this group, the inverse mapping has been calculated above, and the composition of two Möbius transformations is again a Möbius transformation, as may be easily calculated: From

$$\zeta = \frac{a'w + b'}{c'w + d'}; \quad w = \frac{az + b}{cz + d}$$

it follows that

$$\zeta = \frac{(a'a + b'c)z + (a'b + b'd)}{(c'a + d'c)z + (c'b + d'd)}.$$

The matrix of the composition is just the product of the matrices of the factors. So we have proved:

Proposition 6.10. *The Möbius transformations form a group relative to composition, called the* Möbius group.

The Möbius transformations possess many interesting geometric properties, we shall now prove some of them:

Definition 6.11. The *cross-ratio* of four different points in $\mathbb{C}$ is defined as follows:

$$[z_1, z_2, z_3, z_4] := \frac{z_3 - z_1}{z_3 - z_2} : \frac{z_4 - z_1}{z_4 - z_2} = \frac{z_3 - z_1}{z_3 - z_2} \frac{z_4 - z_2}{z_4 - z_1}.$$

If one of the four points is $z = \infty$ we have to take the corresponding limit.

We show some properties of the cross-ratio and of the Möbius transformations:

Proposition 6.12 (Cross-ratio). (i) *Möbius transformations preserve the cross-ratio of four points.*

(ii) *Let z_1, z_2, z_3 and w_1, w_2, w_3 be given points, different in pairs, then the Möbius transformation implicitly defined by the equation*

$$[w_1, w_2, w_3, w] = [z_1, z_2, z_3, z]$$

transforms the sequence of points z_1, z_2, z_3 into the sequence w_1, w_2, w_3.

(iii) *The cross-ratio of four points z_1, z_2, z_3, z_4 is real if and only if the points are lying on a straight line or on a circle (i.e., on a circle in the above generalized sense).*

Proof. (i) We have seen above that a Möbius transformation may be split into a translation, a rotation, a dilation, and eventually a reflection in the unit circle $w = 1/z$. For a translation $w = z + a$ the a vanishes in the differences, so the cross-ratio is invariant. For a rotation and a dilation $w = az$ the a vanishes by a division, the invariance is given. It remains the reflection about the unit circle $w = 1/z$, for which it follows that

$$[w_1, w_2, w_3, w_4] = \frac{z_1 - z_3}{z_2 - z_3} \frac{z_2 z_3}{z_1 z_3} \frac{z_2 - z_4}{z_1 - z_4} \frac{z_1 z_4}{z_2 z_4} = [z_1, z_2, z_3, z_4],$$

i.e., we also have an invariant cross-ratio.

(ii) w, resp. z, can be calculated uniquely from the cross-ratio, in particular one sees from

$$[z_1, z_2, z_3, z] = \frac{z_3 - z_1}{z_3 - z_2} \frac{z - z_2}{z - z_1}$$

that

$$[z_1, z_2, z_3, z_1] = \infty, \quad [z_1, z_2, z_3, z_2] = 0, \quad [z_1, z_2, z_3, z_3] = 1.$$

This is correct also for the w-part so that the transformation given in the proposition indeed transforms z_1 into w_1, z_2 into w_2, and z_3 into w_3.

(iii) The proof is recommended as an exercise (see exercise 6.3.4). $\qquad \square$

Example 6.13. a) We look for a transformation which transforms the unit circle $\mathbb{D} = B_1(0)$ onto the upper half plane. Using the proposition above we choose the points $-1, -i, 1$ on the unit circle and transform them to the points $-1, 0, 1$ on the real axis. In the given sequence we run the boundaries in a direction such that the unit circle and the upper half plane are both on the left side; it then follows that

$$\frac{1+1}{1} \frac{w}{w+1} = \frac{1+1}{1+i} \frac{z+i}{z+1}$$

or

$$\frac{w}{w+1} = \frac{1}{1+i} \frac{z+i}{z+1}$$

and at last

$$w = \frac{z+i}{iz+1}.$$

This is the transformation named after the English mathematician ARTHUR CAYLEY (1821–1895).

b) The Möbius group has many interesting subgroups, so the Möbius transformations

$$w = e^{i\theta} \frac{z - z_0}{1 - \overline{z_0} z}, \quad |z_0| < 1, \theta \in \mathbb{R},$$

form a group. These are the transformations which map the interior of the unit circle $\mathbb{D} = \{z : |z| < 1\}$ onto itself (see Exercise 6.3.5).

6.2.2 Möbius transformations in higher dimensions

Unfortunately we meet in $\mathbb{H}$ and $C\ell(n)$ some difficulties while defining Möbius transformations, in part from the non-commutativity. The first published results in this direction are due to Vahlen [156], and later to Maaß [99]; and in the eighties of the last century Ahlfors studied these transformations in a series of papers [2]–[4], see also the dissertation of Zöll [161].

At the end of Section 1.3 we gave a consistent equation for circles and straight lines in the plane, which we shall repeat in higher dimensions. An n-dimensional (hyper)plane in $\mathbb{R}^{n+1}$ is given by the equation

$$x \cdot B = c$$

or

$$x \overline{B} + B \overline{x} = 2c =: -C$$

with real c and a paravector B. A sphere, an n-dimensional surface of a ball in $\mathbb{R}^{n+1}$, is described by the equation

$$(x + B)(\overline{x} + \overline{B}) = r^2,$$

$-B$ is the center and r the radius of the sphere. This gives

$$x\,\overline{x} + x\,\overline{B} + B\overline{x} = r^2 - |B|^2 =: -C.$$

In summary we have:

Proposition 6.14. *Planes and spheres of dimension n in $\mathbb{R}^{n+1}$ are described by the equation*

$$Ax\,\overline{x} + x\,\overline{B} + B\overline{x} + C = 0;$$

here A and C are real and B is an arbitrary paravector and we have to assume that

$$|B|^2 - AC > 0.$$

For $A = 0$ the equation defines a plane, for $A \neq 0$ a sphere.

As in the plane we can speak simply only of spheres if we deal with planes or spheres. A translation

$$x' = x + a$$

with $a \in \mathbb{R}^{n+1}$ transforms the above equation into one of the same type, which means planes and spheres or simply spheres are preserved. The same is true for rotations and dilations of the form

$$x' = axb.$$

We saw in Cayley's Theorem 2.27 that in $\mathbb{H}$ this form describes rotations and an additional factor gives the dilation. In $\mathbb{R}^{n+1}$ we have to satisfy the assumptions of Theorem 3.20, we then get $x' \in \mathbb{H}$, resp. $\in \mathbb{R}^{n+1}$. The simple calculations are recommended to the reader. Finally we deal in an analogous way with the plane by means of mappings of the form

$$x' = \frac{1}{x} = \frac{\overline{x}}{|x|^2}.$$

Also here x' is in $\mathbb{H}$ resp. in $\mathbb{R}^{n+1}$, and we see easily that planes are transformed into spheres and vice versa. In a similar way as for the plane this transformation is called *reflection about the unit sphere*. We state

Proposition 6.15. *Translations, rotations, dilations, and the reflection about the unit sphere $x' = 1/x$ preserve general spheres.*

We shall define now the Möbius transformations. To be clearer we deal with both the cases $\mathbb{H}$ and $\mathbb{R}^{n+1}$ separately since there are some differences:

Definition 6.16. Let $a, b, c, d \in \mathbb{H}$. The mapping

$$f(x) = (ax + b)(cx + d)^{-1} \quad \left(f(x) = (xc + d)^{-1}(xa + b)\right)$$

with

$$H := cac^{-1}d - cb \neq 0 \quad \text{for} \quad c \neq 0 \quad \text{and} \quad ad \neq 0 \quad \text{for} \quad c = 0,$$
$$\left(H := dc^{-1}ac - bc \neq 0 \quad \text{for} \quad c \neq 0 \quad \text{and} \quad da \neq 0 \quad \text{for} \quad c = 0\right)$$

is called a *Möbius transformation in the left representation* (*Möbius transformation in the right representation*). The associated matrix

$$\begin{pmatrix} a & b \\ c & d \end{pmatrix}$$

is called a *Vahlen matrix*. Möbius transformations may be continued continuously to $\hat{\mathbb{H}}$ in the chordal metric.

H corresponds to the determinant in $\mathbb{C}$; if c commutes with a this is also formally correct. We have a theorem analogous to $\mathbb{C}$:

Theorem 6.17.   (i) *The Möbius transformations preserve spheres.*

(ii) *The left representation may be transformed into a right one and vice versa.*

(iii) *The Möbius transformations form a group, the* Möbius group.

As in the plane for each mapping we have a whole class of representations.

Proof. (i) For $c \neq 0$ we calculate as follows:

$$\begin{aligned}
f(x) = (ax + b)(cx + d)^{-1} &= (a(x + c^{-1}d) + b - ac^{-1}d)(x + c^{-1}d)^{-1}c^{-1} \\
&= ac^{-1} - c^{-1}H(x + c^{-1}d)^{-1}c^{-1},
\end{aligned}$$

and similarly for the right representation. Then we see that the mapping is composed of translations, rotations, dilations, and a reflection in the unit sphere $x' = 1/x$. So, the last proposition shows that planes and spheres are preserved. For $c = 0$ we see at once that

$$f(x) = (ax + b)d^{-1}, \quad \text{resp.} \quad f(x) = d^{-1}(xa + b),$$

and the assertion follows. Clearly a and d may not be zero, which means $H \neq 0$. Incidentally, one sees that $H = 0$ would correspond to the constant map, but we have excluded that case.

(ii) We start with the left representation: for $c = 0$ it is necessary to have $a \neq 0$ and we get the desired right representation with

$$f(x) = axd^{-1} + bd^{-1} = (a^{-1})^{-1}(xd^{-1} + a^{-1}bd^{-1}).$$

For $c \neq 0$ we use the formula from (i):

$$\begin{aligned}
f(x) &= ac^{-1} - c^{-1}H(x + c^{-1}d)^{-1}c^{-1} = ac^{-1} - [(x + c^{-1}d)H^{-1}c]^{-1}c^{-1} \\
&= (xH^{-1}c + c^{-1}dH^{-1}c)^{-1}(xH^{-1}cac^{-1} + c^{-1}dH^{-1}cac^{-1} - c^{-1}).
\end{aligned}$$

If we denote the new coefficients with an index r, for $c \neq 0$ they read:

$$a_r = H^{-1}cac^{-1}, \quad b_r = c^{-1}dH^{-1}cac^{-1} - c^{-1},$$
$$c_r = H^{-1}c, \quad d_r = c^{-1}dH^{-1}c.$$

For $c = 0$ they read

$$a_r = d^{-1}, \ b_r = a^{-1}bd^{-1}, \ c_r = 0, \ d_r = a^{-1}.$$

The new determinant for the right representation has to be different from zero, as the mapping is not constant. The calculations from the right to the left representation are quite analogous.

(iii) The inverse mapping to $x' = (ax + b)(cx + d)^{-1}$ is easily calculated,

$$x = (x'c - a)^{-1}(-x'd + b),$$

we get a right representation, the determinant is the same. The neutral element of the group is naturally the identity. The composition with a second Möbius transformation $x'' = (a'x' + b')(c'x' + d')^{-1}$ gives analogously to $\mathbb{C}$,

$$x'' = ((a'a + b'c)x + (a'b + b'd))((c'a + d'c)x + (c'b + d'd))^{-1}.$$

The latter is again a Möbius transformation with an associated Vahlen matrix, which is the product of both the single Vahlen matrices. The determinant cannot be zero as otherwise one of the single maps would be constant. Therefore we have all the necessary properties of a group. $\qquad\square$

In $C\ell(n)$ we have an analogous situation with the additional difficulty that the Möbius transformation really has to map into $\mathbb{R}^{n+1}$. As in the quaternions we have to exclude that the mapping is constant.

Definition 6.18. Also in $\mathbb{R}^{n+1}$ we call the mappings

$$f(x) = (ax + b)(cx + d)^{-1}, \ \text{resp.} \ f(x) = (xc + d)^{-1}(xa + b)$$

with $x \in \mathbb{R}^{n+1}$ and $a, b, c, d \in \Gamma'_{n+1} := \Gamma_{n+1} \cup \{0\}$ *Möbius transformations*; the first form is called the *left representation*, the second one the *right representation*.

$$\begin{pmatrix} a & b \\ c & d \end{pmatrix}$$

is called again the *Vahlen matrix*, Γ_{n+1} is the Clifford group from Theorem 3.20. We assume for the left representation that

$$H^* := a\hat{d} - b\hat{c} \in \mathbb{R}_0 := \mathbb{R} \setminus \{0\},$$

where $\hat{d}$ is the reversion from Definition 3.9. For $c \neq 0$ we assume moreover

$$ac^{-1}, c^{-1}d \in \mathbb{R}^{n+1}$$

and for $c = 0$ only $bd^{-1} \in \mathbb{R}^{n+1}$. For the right representation this reads

$$H^* := \hat{d}a - \hat{c}b \in \mathbb{R}_0,$$

for $c \neq 0$ let

$$c^{-1}a, dc^{-1} \in \mathbb{R}^{n+1}$$

and at last for $c = 0$ let $d^{-1}b \in \mathbb{R}^{n+1}$. The mapping has to be continued to the one-point-compactification $\hat{\mathbb{R}}^{n+1}$ continuously with the chordal metric.

Here H^* is nearer to the determinant in $\mathbb{C}$, at least formally, than the determinant H in the quaternions. We have an analogous theorem to the previous one:

Theorem 6.19. (i) *Möbius transformations map $\hat{\mathbb{R}}^{n+1}$ onto itself and preserve spheres in the general sense.*

(ii) *The left representation may be transformed into a right representation and vice versa.*

(iii) *The Möbius transformations form a group, the Möbius group.*

Also in the case of the Clifford algebra one has many different possibilities to represent a Möbius transformation.

Proof. (i) As in $\mathbb{H}$ we have for $c \neq 0$,

$$f(x) = ac^{-1} + (b - ac^{-1}d)(x + c^{-1}d)^{-1}c^{-1}.$$

As we assumed $ac^{-1}, c^{-1}d \in \mathbb{R}^{n+1}$ both these summands generate only translations in $\mathbb{R}^{n+1}$. Moreover $(x + c^{-1}d)^{-1}$ is composed by a translation and the reflection about the unit sphere $x' = 1/x$, which map also into $\mathbb{R}^{n+1}$. Finally we have to study the mapping

$$x' = (b - ac^{-1}d)xc^{-1}.$$

Since $c^{-1}d \in \mathbb{R}^{n+1}$ we have $c^{-1}d = (c^{-1}d)\hat{} = \hat{d}\hat{c}^{-1}$, and it then follows that

$$c^{-1}H = b - ac^{-1}d = b - a\hat{d}\hat{c}^{-1} = (b\hat{c} - a\hat{d})\hat{c}^{-1} = \lambda\hat{c}^{-1},$$

as the expression in brackets is assumed to be real and not 0. So we get

$$x' = \lambda\hat{c}^{-1}xc^{-1};$$

following Theorem 3.20 this is a rotation for $\lambda = 1$, otherwise an additional dilation. So, the transformation maps into the $\mathbb{R}^{n+1}$, and Proposition 6.15 gives the assertion.

For $c = 0$ we have $f(x) = axd^{-1} + bd^{-1}$. Because of $bd^{-1} \in \mathbb{R}^{n+1}$ this summand gives a translation. From $H^* = a\hat{d} \in \mathbb{R}$ it follows further with a real $\lambda \neq 0$ that

$$a = \lambda\hat{d}^{-1},$$

i.e., again the desired rotation composed with a dilation in $\mathbb{R}^{n+1}$.

The proofs of (ii) and (iii) are the same as in $\mathbb{H}$, so it is not necessary to repeat them. $\square$

As our last proposition we shall differentiate the Möbius transformations. In contrast to $\mathbb{C}$ for $n > 2$ we shall see that they are not holomorphic functions in $\mathbb{H}$ and $C\ell(n)$.

Proposition 6.20. *For $f(x) = (ax + b)(cx + d)^{-1}$ we have with $h \in \mathbb{H}$, resp. $h \in \mathbb{R}^{n+1}$,*

(i) $f'(x)[h] = (xc_r + d_r)^{-1}h(cx + d)^{-1}$,

(ii) $(\bar{\partial}f)(x) = -(n-1)\overline{(xc_r + d_r)}^{-1}(cx + d)^{-1}$,

(iii) $(f\bar{\partial})(x) = -(n-1)(xc_r + d_r)^{-1}\overline{(cx+d)}^{-1}$.

We see as stated that a Möbius transformation is neither left- nor right-holomorphic. The factor $n-1$ shows why the derivative is zero in the plane, where we have a holomorphic function. But that does not matter, the Möbius transformations are important for us when we have to transform variables of integration. The derivative f' is to be understood in the sense of a mapping between vector spaces. The additional argument h is necessary. The derivative may be calculated without too many difficulties.

Proof. (i) We assume firstly $c \neq 0$ and calculate

$$
\begin{aligned}
f(x) - f(y) &= (ax+b)(cx+d)^{-1} - (yc_r + d_r)^{-1}(ya_r + b_r) \\
&= (yc_r + d_r)^{-1}[(yc_r + d_r)(ax+b) - (ya_r + b_r)(cx+d)](cx+d)^{-1} \\
&= (yc_r + d_r)^{-1}[y(c_r a - a_r c)x + y(c_r b - a_r d) \\
&\quad + (d_r a - b_r c)x + (d_r b - b_r d)](cx+d)^{-1}.
\end{aligned}
$$

Surprisingly this simplifies with the above calculated expressions for the coefficients of the right representation (proof of Theorem 6.17 (ii)):

$$
c_r a - a_r c = 0, \quad d_r b - b_r d = 0,
$$
$$
d_r a - b_r c = 1, \quad c_r b - a_r d = -1.
$$

This gives us

$$
f(x) - f(y) = (yc_r + d_r)^{-1}(x-y)(cx+d)^{-1},
$$

also for $c = c_r = 0$ we may calculate this without difficulties. We now substitute x by $x+h$ and y by x; it then follows that

$$
f(x+h) - f(x) = (xc_r + d_r)^{-1} h\,(c(x+h) + d)^{-1}.
$$

If we apply the same formula to $g(x) = (cx+d)^{-1}$ we get

$$
g(x+h) = (cx+d)^{-1} + |h|O(1).
$$

This gives for our $f(x)$,

$$
f(x+h) = f(x) + (xc_r + d_r)^{-1} h\,(cx+d)^{-1} + |h|^2 O(1),
$$

i.e., the assertion for $f'(x)$.

(ii) Because of $\partial_i f(x) = f'(x)[e_i] = (xc_r + d_r)^{-1} e_i (cx+d)^{-1}$ we have

$$
(\bar{\partial}f)(x) = \sum_{i=0}^{n} e_i(xc_r + d_r)^{-1} e_i (cx+d)^{-1} = -(n-1)\overline{(xc_r + d_r)}^{-1}(cx+d)^{-1}.
$$

Here we have used the formula from Exercise 3.5.5

(iii) The proof is quite analogous to (ii). $\qquad\square$

6.3   Exercises

1. Prove the estimate

$$|\mathcal{P}_\mathbf{k}(x)| \le |z_1|^{k_1} \ldots |z_n|^{k_n} = |\mathbf{z}|^\mathbf{k} \le |x|^{|\mathbf{k}|}.$$

2. Prove the *binomial formula* for Fueter polynomials: We have for all x and y in $\mathbb{R}^{n+1}$ ($\mathbf{k}, \mathbf{i}, \mathbf{j}$ denoting multiindices)

$$\mathcal{P}_\mathbf{k}(x + y) = \sum_{\mathbf{i}+\mathbf{j}=\mathbf{k}} \frac{\mathbf{k}!}{\mathbf{i}!\,\mathbf{j}!} \mathcal{P}_\mathbf{i}(x) \mathcal{P}_\mathbf{j}(y).$$

Here one should use mathematical induction relative to $|\mathbf{k}|$, the conclusion from $|\mathbf{k}| - 1$ to $|\mathbf{k}|$ should be proved by differentiation with respect to x_i.

3. Show that the Möbius transformations of

$$w = e^{i\theta} \frac{z - z_0}{1 - \overline{z_0}\, z}, \quad |z_0| < 1,\ \theta \in \mathbb{R},$$

form a group and map the interior of the unit circle onto itself.

4. Prove that the cross-ratio of four points z_1, z_2, z_3, z_4 is real if and only if the points are lying on a straight line or a circle. (Proposition 6.12 (iii)).

5. Calculate a Möbius transformation which maps the interior of the unit circle onto the right half-plane.

6. Prove that

$$x' = \frac{r^2}{\overline{x}}$$

is a reflection about the sphere $\{|x| = r\}$ in the following sense: the points x and x' should lie on the same half-line starting from the origin, and x' should be the polar point of x. Nevertheless $x' = 1/x$ is also called reflection about the unit ball; here a reflection about the plane $x_0 = 0$ is added.

7. Calculate the Möbius transformation in $\mathbb{R}^{n+1}$ which maps the interior of the sphere S^n onto the half-space $\{x_0 > 0\}$.

8. Prove that translations, rotations, dilations, and the reflection about the unit sphere $x' = 1/x$ preserve the set of general circles, i.e., n-dimensional planes and spheres in $\mathbb{R}^{n+1}$.

Chapter III

Integration and integral theorems

7   Integral theorems and integral formulae

7.1   Cauchy's integral theorem and its inversion

The Cauchy integral theorem belongs to the central results of complex analysis and tells us in its classical formulation that, for a holomorphic function f in a domain G, the integral along a sufficiently smooth closed curve which is located in G has always the value zero.

Cauchy (see Subsection 5.1) proved his famous theorem in 1825 (cf. [23]). He formulated it for rectangle domains and used methods of variational calculus. In 1883 EDOUARD GOURSAT (1858–1936) considered domains with more general boundaries and weakened the assumption of the continuity of the derivative f'. It was, in the end, ALFRED PRINGSHEIM (1850–1941) who brought the Cauchy integral theorem to the form used today with a method based on a triangular decomposition of the domain.

The Cauchy integral theorem is for us an easy consequence of Gauß' theorem:

Theorem 7.1 (Cauchy's integral theorem in $C\ell(n)$). *Let $f, g \in C^1(\overline{G})$, where G is a bounded domain of finite connectivity with a sufficiently smooth boundary ∂G, so that always the normal points outwards. Let f be right-holomorphic and g be left-holomorphic in G. We then have:*

$$\int\limits_{\partial G} f(x)dx^* g(x) = 0.$$

The classical proof uses the method of Goursat. B. Wirthgen [104] has transferred it to higher dimensions, so that also this proof would be possible, but with considerable effort. We therefore restrict ourselves to the proposed method.

For $n = 1$ we have $C\ell(n) = \mathbb{C}$. Because of the commutativity we can work in the plane with one holomorphic function f. The theorem in $\mathbb{C}$ reads as follows:

Theorem 7.2 (Cauchy's integral theorem in $\mathbb{C}$). *Let G be a bounded domain with boundary ∂G consisting of a finite number of piecewise smooth curves. The boundary curves may be oriented so that G is always on the left side. The function f is assumed to be holomorphic in G and continuous in $\overline{G}$. We then have*

$$\int\limits_{\partial G} f(z)dz = 0.$$

Now we want to prove the theorem of the Italian mathematician GIACINTO MORERA (1856–1909), a kind of inversion of the integral theorem which shows the equivalence of Cauchy's integral theorem and the definition of holomorphy.

Theorem 7.3 (Morera' s theorem in $C\ell(n)$). *Let $f \in C^1(G)$ be a function in a domain $G \subset \mathbb{R}^{n+1}$. If we have for all balls $B_r(x) \subset G$,*

$$\int_{\partial B_r(x)} dy^* f(y) = 0,$$

then $\overline{\partial} f = 0$ in G, thus f is a left-holomorphic function.

Proof. Let $x \in G$ be an arbitrary point. Then Gauß' theorem in $C\ell(n)$ A.2.22 leads to

$$0 = \int_{\partial B_r(x)} dy^* f(y) = \int_{B_r(x)} \overline{\partial} f(y) \, d\sigma_y.$$

Because of the continuity on the one hand and the theorem of Lebesgue from integral calculus on the other, with the volume V_r of a ball of radius r we obtain

$$\lim_{r \to 0} \frac{1}{V_r} \int_{B_r(x)} \overline{\partial} f d\sigma = \overline{\partial} f(x) = 0,$$

so that f is a left-holomorphic function in G. □

In $\mathbb{C}$ the theorem was proved by Morera in 1886. There a primitive function exists and the assertion can be formulated differently and under weaker assumptions:

Theorem 7.4 (Morera' s theorem in $\mathbb{C}$). *Let $G \subset \mathbb{C}$ be a domain and let $f : G \to \mathbb{C}$ be continuous in G. If*

$$\int_{\Pi} f(z) dz = 0$$

for any closed polygon Π in G, then f is holomorphic and possesses in G a primitive function F with $F'(z) = f(z)$.

Proof. One of the difficulties of the assumptions of this theorem lies in the "holes" eventually existing in the domain. One has to be able to integrate around them in every direction. This is not always given as one can see for the integral of $1/z$ over the unit circle (cf. Exercise 7.4.3).

Let now z_0 be an arbitrary, but fixed point in G. We define

$$F(z) := \int_{z_0}^{z} f(\zeta) d\zeta$$

and integrate over an arbitrary polygon Π connecting z_0 and z in G. Because of the conditions in Morera's theorem the function F is uniquely defined: the integral over another polygon Π_1 has no different value as the integral over $\Pi + (-\Pi_1)$ has the value zero.

We examine the differentiability of F:

$$\frac{F(z+h) - F(z)}{h} = \frac{1}{h} \int_z^{z+h} f(\zeta) d\zeta.$$

If $|h|$ is small enough, $z + h$ is contained in a small disk around z which belongs to G. There we can integrate over the straight line from z to $z + h$. The latter requires again the path independence of the integral. As f is continuous we have $f(\zeta) = f(z) + \mathrm{o}(1)$ for any ζ on the line from z to $z + h$, thus

$$\frac{1}{h} \int_z^{z+h} f(\zeta)d\zeta = \frac{1}{h} f(z)h + R(h)$$

with the remainder term

$$R(h) := \frac{1}{h} \int_z^{z+h} \mathrm{o}(1)d\zeta.$$

Because of the continuity we have $|\mathrm{o}(1)| < \varepsilon$ for $|h| < \delta$ and $|R(h)| \leq \varepsilon$ follows. Then $R(h) = \mathrm{o}(1)$ and we have proved

$$F'(z) = f(z).$$

Consequently, F is a holomorphic function and as we shall see later, also f is.   □

A domain is called *star-shaped* with regard to a point z_0 if any point $z \in G$ can be connected with z_0 by a straight line contained in G. For such a star-shaped domain it is sufficient to assume in Morera' s theorem that the curve integral vanishes for all triangles with one corner in z_0 .

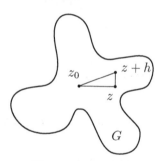

Figure 7.1

A property of a function in a domain G is said to hold *locally* if for every point $x \in G$ a small ball (or disc) with center x exists which is contained in G and in which the property in question holds. To give an example, typical properties of this kind are the holomorphy and the uniform convergence of a sequence of functions. The latter is given mostly only locally as the convergence generally becomes worse near the boundary of G. We want to formulate a supplement of Cauchy's theorem which shows the equivalence with holomorphy:

Theorem 7.5. *A continuous function f in a domain $G \subset \mathbb{C}$ is holomorphic if and only if locally the integral over the boundary of all triangles is zero.*

Proof. If we restrict ourselves to a small disc in G the assertion follows immediately from Morera's theorem in $\mathbb{C}$, proven above.   □

7.2 Formulae of Borel–Pompeiu and Cauchy

7.2.1 Formula of Borel–Pompeiu

This section is devoted to the treatment of an integral formula which is named after the French mathematician ÉMILE BOREL (1871–1956) and the Romanian DIMITRIE POMPEIU (1873–1954). An application to holomorphic functions proves Cauchy's integral formula which is the road into classical complex analysis. Of course, Cauchy's formula was obtained earlier than Borel–Pompeiu's. We do not treat the statement in $\mathbb{C}$ separately, the proof is included in the corresponding theorem in $C\ell(n)$.

For the integral formulae we need the so-called *Cauchy kernel*. In some sense it represents the weakest isolated singularity a holomorphic function may have at a point:

Definition 7.6. We call the function

$$E_n(x) = \frac{1}{\sigma_n} \frac{\overline{x}}{|x|^{n+1}} \quad (x \neq 0),$$

defined in $\mathbb{R}^{n+1} \setminus \{0\}$, the *Cauchy kernel*. Here σ_n is the surface area of the unit ball S^n in $\mathbb{R}^{n+1}$. According to Example A.2.17 with the Gamma function from Section 13.1 we have

$$\sigma_n = 2 \frac{\Gamma^{n+1}\left(\frac{1}{2}\right)}{\Gamma\left(\frac{n+1}{2}\right)} = \frac{2\pi^{(n+1)/2}}{\Gamma\left(\frac{n+1}{2}\right)}.$$

In $\mathbb{C}$ the Cauchy kernel simplifies to

$$E_1(z) = \frac{1}{2\pi} \frac{1}{z}.$$

Proposition 7.7. *The Cauchy kernel is right- and left-holomorphic.*

Proof. We have

$$\overline{\partial} \frac{\overline{x}}{|x|^{n+1}} = (\overline{\partial}\overline{x}) \frac{1}{|x|^{n+1}} + \left(\overline{\partial}\left(|x|^2\right)^{-(n+1)/2}\right) \overline{x},$$

and with

$$\overline{\partial}\overline{x} = \sum_{i=0}^{n} e_i \overline{e}_i = n + 1, \quad \overline{\partial}|x|^2 = 2 \sum_{i=0}^{n} x_i e_i = 2x$$

the desired relation

$$\overline{\partial} \frac{\overline{x}}{|x|^{n+1}} = 0$$

follows (proceed analogously for the right-holomorphy). $\qquad\square$

Theorem 7.8 (Formula of Borel–Pompeiu in $C\ell(n)$). *Let $G \subset \mathbb{R}^{n+1}$ be a bounded domain with sufficiently smooth boundary and an outwards pointing normal. Then we have for any $f \in C^1(\overline{G})$,*

$$\int_{\partial G} E_n(y - x) dy^* f(y) - \int_G E_n(y - x)(\overline{\partial} f)(y) d\sigma_y = \begin{cases} f(x), & x \in G, \\ 0, & x \in \mathbb{R}^{n+1} \setminus \overline{G}. \end{cases}$$

Proof. Cutting out from G the ball $B_\varepsilon(x)$ we obtain the domain $G_\varepsilon := G \backslash \overline{B_\varepsilon(x)}$ with the boundary $\partial G_\varepsilon = \partial G \cup (-S_\varepsilon)$. S_ε is the sphere with radius ε and center x. The orientation of the sphere's normal to the exterior of the domain G_ε is taken into account by the minus sign. Now we apply Gauß' Theorem A.2.22 to the domain G_ε and the functions $E_n(y-x)$ and $f(y)$, y being the variable of integration. As $E_n(y-x)$ is a holomorphic function we obtain:

$$\int\limits_{\partial G} E_n(y-x)dy^* f(y) - \int\limits_{S_\varepsilon} E_n(y-x)dy^* f(y) = \int\limits_{G_\varepsilon} E_n(y-x)(\overline{\partial}f)(y)d\sigma_y.$$

For the second integral in agreement with the special cases A.2.17 b) and c) with $\tilde{y} = (y-x)/|y-x|$ it follows that

$$\int\limits_{S_\varepsilon} E_n(y-x)dy^* f(y) = \frac{1}{\sigma_n} \int\limits_{S_\varepsilon} \frac{\overline{y-x}}{|y-x|^{n+1}} \frac{y-x}{|y-x|} f(y)\varepsilon^n |do_{\tilde{y}}|$$

$$= \frac{1}{\sigma_n} \int\limits_{S_1} f(x+\varepsilon\tilde{y})|do_{\tilde{y}}|.$$

Because of the continuity of f for $\varepsilon \to 0$ we have

$$\lim_{\varepsilon \to 0} \frac{1}{\sigma_n} \int\limits_{S_1} f(x+\varepsilon\tilde{y})|do_{\tilde{y}}| = \frac{1}{\sigma_n} \int\limits_{S_1} |do_{\tilde{y}}|f(x) = f(x).$$

The volume integral over G_ε causes no difficulties for $\varepsilon \to 0$ although $E_n(y-x)$ becomes singular. In new coordinates $y - x =: rt$ with $|t| = 1$, from the special case A.2.17 b) we deduce

$$E_n(y-x) = \frac{1}{\sigma_n}\frac{\overline{t}}{r^n}, \quad d\sigma_y = r^n dr|do_t|,$$

so that the singularity disappears and the volume integral converges for $\varepsilon \to 0$. $\qquad\square$

For $n = 1$ the algebra $C\ell(n)$ is just $\mathbb{C}$. So we obtain the Borel–Pompeiu formula of complex analysis:

Theorem 7.9 (Formula of Borel–Pompeiu in $\mathbb{C}$). *Let G be a domain of finite connectivity with piecewise smooth boundary which is oriented so that the domain lies on the left-hand side. Let the function $f \in C^1(\overline{G})$ have complex values. Then we have*

$$\frac{1}{2\pi i} \int\limits_{\partial G} \frac{f(\zeta)}{\zeta - z}d\zeta - \frac{1}{\pi} \int\limits_{G} \frac{\partial_{\overline{\zeta}}f(\zeta)}{\zeta - z}d\sigma_\zeta = \begin{cases} f(z), & z \in G, \\ 0, & z \notin \overline{G}. \end{cases}$$

For the notion *piecewise smooth* we refer to Example A.2.10 a and for *finite connectivity* to Definition A.2.19.

Important conclusions can be drawn from the formula of Borel–Pompeiu, but we shall introduce here only some notation which has become standard today, and with which we shall deal later on.

Definition 7.10. Let $f \in C^1(\overline{G})$ and ∂G be a sufficiently smooth manifold. The operator $F_{\partial G}$, defined by

$$(F_{\partial G} f)(x) := \int_{\partial G} E_n(y - x) dy^* f(y),$$

is called the *Cauchy–Bitsadze operator*. The operator T_G, defined by

$$(T_G f)(x) = - \int_G E_n(y - x) f(y) d\sigma_y ,$$

is called the *Teodorescu transform*.

Remark 7.11. With this new notation the Borel–Pompeiu formula can be written in the form

$$(F_{\partial G} f)(x) + (T_G(\overline{\partial} f))(x) = \begin{cases} f(x), & x \in G, \\ 0, & x \in \mathbb{R}^n \setminus \overline{G}. \end{cases}$$

7.2.2 Formula of Cauchy

Now it is easy to deduce Cauchy's integral formula which turns out to be the way into important parts of hypercomplex analysis. We shall draw the first consequences of this formula in the next subsection. To prove Cauchy's formula we have only to use a left-holomorphic function in the formula of Borel–Pompeiu, so that the volume integral vanishes:

Theorem 7.12 (Cauchy's integral formula). *Let* $G \in \mathbb{R}^{n+1}$ *be a bounded domain with sufficiently smooth boundary and outwards oriented normal. For a left-holomorphic function* $f \in C^1(\overline{G})$ *we have*

$$\int_{\partial G} E_n(y - x) dy^* f(y) = \begin{cases} f(x) , & x \in G, \\ 0 , & x \in \mathbb{R}^{n+1} \setminus \overline{G}. \end{cases}$$

For a right-holomorphic function, f and $E_n(y - x)$ have to be interchanged. In particular, we have for $n = 1$:

Theorem 7.13 (Cauchy's integral formula in $\mathbb{C}$). *Let* G *be a bounded domain of finite connectivity with piecewise smooth boundary, which is oriented so that the domain* G *lies on the left-hand side. Let the function* f *be holomorphic in* G *and continuous in* $\overline{G}$. *We then have*

$$\frac{1}{2\pi i} \int_{\partial G} \frac{f(\zeta)}{\zeta - z} d\zeta = \begin{cases} f(z), & z \in G, \\ 0, & z \notin \overline{G}. \end{cases}$$

Cauchy's integral formula shows that the function f is completely defined by its boundary values on ∂G. This strong inner connection of holomorphic functions is the reason for many interesting properties. For the history of the formulae of Cauchy and Borel–Pompeiu we refer to Subsection 7.2.4.

Now we want to prove a further variant of Cauchy's integral formula over a so-called *exterior domain*. For this purpose we consider a *Jordan surface* Γ, which is a piecewise smooth bounded manifold whose complement relative to $\mathbb{R}^{n+1}$ consists only of two domains. Thus, $\mathbb{R}^{n+1}$ is split by Γ into two domains, one of them having the point ∞ as boundary point: this is called the *exterior domain* G^- of Γ. Correspondingly $G^+ = \mathbb{R}^{n+1} \setminus (\Gamma \cup G^-)$ is called the *interior domain* of Γ. We then have

Theorem 7.14 (Cauchy's integral formula for the exterior domain). *Let Γ be a Jordan surface with the exterior domain G^- and the interior domain G^+. The orientation of Γ is to be chosen so that the normal points toward G^-. The function f is assumed to be left-holomorphic in G^- and continuously differentiable in $G^- \cup \Gamma$ and to have a limit value $f(\infty)$ at $x = \infty$ using the chordal metric. We then have*

$$\int_{\Gamma} E_n(y-x)dy^*f(y) = \begin{cases} -f(x) + f(\infty), & x \in G^-, \\ f(\infty), & x \in G^+. \end{cases}$$

Proof. We choose a sphere $\Gamma_\rho = \{x : |x| = \rho\}$ with sufficiently large radius ρ which contains Γ and G^+, and consider the domain $G_\rho := G^- \cap \{|x| < \rho\}$. Its boundary is $\partial G_\rho = \Gamma_\rho \cup (-\Gamma)$ taking into account the orientation of Γ. Then Cauchy's integral formula for $x \in G_\rho$ yields

$$\begin{aligned} f(x) &= -\int_{\Gamma} E_n(y-x)dy^*f(y) + \int_{|y|=\rho} E_n(y-x)dy^*(f(y)-f(\infty)) \\ &\quad + f(\infty) \int_{|y|=\rho} E_n(y-x)dy^* \\ &= -\int_{\Gamma} E_n(y-x)dy^*f(y) + f(\infty) + R, \end{aligned}$$

$$R := \int_{|y|=\rho} E_n(y-x)dy^*(f(y)-f(\infty)).$$

From our assumption the inequality $|f(y)-f(\infty)| < \varepsilon$ follows for sufficiently large ρ. If we assume moreover $\rho > 2|x|$ we get

$$|E_n(y-x)| = \frac{1}{\sigma_n} \frac{1}{|y-x|^n} \leq \frac{1}{\sigma_n} \frac{2^n}{\rho^n}$$

because of $|y - x| \geq |y| - |x| > \rho/2$. Finally we obtain with Theorem 3.14 (ii)

$$|R| \leq \int_{|y|=\rho} |E_n(y-x)dy^*|\varepsilon \leq \frac{\varepsilon 2^n}{\sigma_n} \int_{|t|=1} |do_t| = \varepsilon 2^n.$$

For $\rho \to \infty$ the value ε can be chosen arbitrarily small and the assertion is proved.
Our proof is also valid in the inner domain G^+ as then the right-hand side contains the value 0 instead of $f(x)$ in the formulas above. $\qquad\square$

7.2.3 Formulae of Plemelj–Sokhotski

We have proved in the previous section that a holomorphic function is determined completely by its boundary values and that it can be described with the help of Cauchy's integral. Obviously the Cauchy integral can be written down formally for any integrable function on Γ, and it defines a holomorphic function outside Γ. Then naturally the question arises on the boundary values of this integral. The following theorem and all questions around it were first considered by the Slovenian mathematician JOSEF PLEMELJ (1873–1967) and the Russian YULIAN VASILIEVICH SOKHOTSKI (1842–1927).

First of all we prove an auxiliary formula:

Proposition 7.15. *Let G be a domain with a sufficiently smooth boundary Γ, i.e., $\Gamma \in C^2$ at least, and let $x \in \mathbb{R}^{n+1}$ be an arbitrary fixed point. Then we have*

$$\int\limits_{\Gamma} E_n(y - x)dy^* = \begin{cases} 1, & x \in G, \\ \frac{1}{2}, & x \in \Gamma, \\ 0, & x \in \mathbb{R}^{n+1} \setminus \overline{G}. \end{cases}$$

The integral is singular, i.e., it exists in the sense of *Cauchy's principal value*. Cauchy's principal value is defined in the following way: By a ball $B_\varepsilon(x)$ one cuts out from Γ a neighborhood $\Gamma_\varepsilon(x)$. Then the integral over $\Gamma' := \Gamma \setminus \Gamma_\varepsilon(x)$ should converge for $\varepsilon \to 0$.

Proof. For $x \in G$ the statement follows from Cauchy's integral formula Theorem 7.12 using the function $f = 1$; for $x \in \mathbb{R}^{n+1} \setminus \overline{G}$ the result follows from the integral formula in the exterior domain Theorem 7.14. For $x \in \Gamma$ we cut out the neighborhood $\Gamma_\varepsilon(x)$ by a ball $B_\varepsilon(x)$ with surface $S_\varepsilon(x)$ obtaining Γ'. The part of the sphere within G is $S_\varepsilon(x) \cap G$. The kernel function $E_n(y - x)$ is holomorphic with regard to y in the domain bounded by $\Gamma' \cup (S_\varepsilon(x) \cap G)$ and thus the integral over this boundary is zero. Taking into account the orientation we obtain

$$\int\limits_{\Gamma'} E_n(y - x)dy^* = \int\limits_{S_\varepsilon(x) \cap G} E_n(y - x)dy^*.$$

On $S_\varepsilon(x) \cap G$ we have as usual $dy^* = \nu|do|$ (see Example A.2.17 c) with the unit vector of the outer normal $\nu = (y - x)/|y - x|$, thus the integrand reads

$$\frac{\overline{y - x}}{|y - x|^{n+1}} \frac{y - x}{|y - x|}|do| = |do_1|$$

with the surface element $|do_1|$ of the unit sphere S^n.

Our integral over Γ' turns out to be the area of $S_\varepsilon(x) \cap G$ divided by the area of the whole sphere. Because of the differentiability, the surface $S_\varepsilon(x) \cap G$ converges for $\varepsilon \to 0$ to the hemisphere. Thus we get in the limit the value $1/2$. $\qquad\square$

$$B_\varepsilon(x)$$

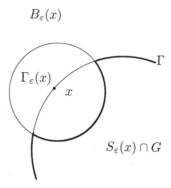

Figure 7.2

Remark 7.16. The proof remains valid if there are the so-called irregular subsets on Γ (edges and corners). Let x be an irregular point of Γ, which is located on an edge or is a corner (e.g., on a cuboid). If one measures the apex angle $\gamma(x)$ of Γ in x by the limit for $\varepsilon \to 0$ of the area of the sphere's part $S_\varepsilon(x) \cap G$ divided by the area of the whole sphere, one has for $x \in \Gamma$ simply

$$\int_\Gamma E_n(y - x)dy^* = \gamma(x).$$

This is a natural generalisation of the value $\frac{1}{2}$ at a regular point. Nevertheless, we remark that Γ should fulfil at least the *cone property*, which says that a circular cone exists with the apex in x which belongs to G excluding the apex. This means that cusps and cuts are not admitted.

The integral

$$(S_\Gamma u)(x) := 2 \int_\Gamma E_n(y - x)dy^* u(y), \quad x \in \Gamma,$$

is an important singular integral obtained from the Cauchy–Bitsadze integral

$$(F_\Gamma u)(x) = \int_\Gamma E_n(y - x)dy^* u(y), \quad x \in \mathbb{R}^{n+1} \setminus \Gamma.$$

The proposition just proved is related to $(F_\Gamma 1)(x)$ for $x \notin \Gamma$ and to $\frac{1}{2}(S_\Gamma 1)(x)$ for $x \in \Gamma$. For Hölder continuous functions u the following decomposition follows immediately:

$$(S_\Gamma u)(x) = 2 \int_\Gamma E_n(y - x)dy^* [u(y) - u(x)] + u(x), \quad x \in \Gamma,$$

as this integral exists as an improper integral because of the Hölder continuity. For the concept of Hölder continuity we refer the reader to Appendix 3, Definition A.3.1 and Proposition A.3.2.

Now we turn to the announced formula of Plemelj–Sokhotski:

Theorem 7.17 (Plemelj–Sokhotski formula). *Let u be Hölder continuous on a sufficiently smooth surface $\Gamma = \partial G$. Then at any regular point $x \in \Gamma$ we have*

$$\text{n.t.-}\lim_{t \to x} (F_\Gamma u)(t) = \frac{1}{2} \left[\pm u(x) + (S_\Gamma u)(x) \right],$$

where $t \in G^\pm$ with $G^+ = G$ and $G^- = \mathbb{R}^{n+1} \setminus \overline{G}$. The notation $\text{n.t.-}\lim\limits_{t \to x}$ means that the limit should be taken non-tangential, i.e., t tends to x within a circular cone whose symmetry axis points toward the normal direction of Γ at the point x. The apex angle of the cone should be smaller than π.

Proof. Let x be a fixed point on Γ. Proposition 7.15 yields immediately for $t \in G$,

$$(F_\Gamma u)(t) = \int_\Gamma E_n(y - t) dy^* [u(y) - u(x)] + u(x)$$

and for $t \in \mathbb{R}^{n+1} \setminus \overline{G}$,

$$(F_\Gamma u)(t) = \int_\Gamma E_n(y - t) dy^* [u(y) - u(x)].$$

We set

$$a(t) := \int_\Gamma E_n(y - t) dy^* [u(y) - u(x)]$$

and intend to show that

$$a(t) \to a(x) = \int_\Gamma E_n(y - x) dy^* [u(y) - u(x)] = \frac{1}{2}(S_\Gamma u)(x) - \frac{1}{2} u(x)$$

for the non-tangential convergence $t \to x$. Hence, the assertion

$$(F_\Gamma u)(t) \to \frac{1}{2}[\pm u(x) + (S_\Gamma u)(x)]$$

would be proved. In the integrand of $a(t) - a(x)$ the absolute value can be drawn to the factors as E_n and dy^* are paravector-valued, thus

$$|a(t) - a(x)| \le \int_\Gamma |E_n(y - t) - E_n(y - x)||u(y) - u(x)||dy^*|$$

follows. Because of the Hölder continuity of u we have $|u(y) - u(x)| \le L|y - x|^\mu$, and from the triangle inequality we obtain the estimate (see Exercise 7.4.1)

$$|E_n(y - t) - E_n(y - x)| \le \frac{1}{\sigma_n} |x - t| \frac{2|y - x|^n + |y - x|^{n-1}|y - t| + \cdots + |y - t|^n}{|y - x|^n |y - t|^{n+1}}.$$

Now we choose an $\varepsilon > 0$ and a circular cone C_ε with symmetry axis in the normal direction, having its apex in x, and having a slightly greater apex angle (but $< \pi$) than that cone in which t runs such that the following condition holds: In the interior of $B_\varepsilon(x)$ the surface Γ lies in the exterior of C_ε. Now let $|t - x| < \varepsilon/2$, inside $B_\varepsilon(x)$ the non-tangential limit comes to play a decisive role.

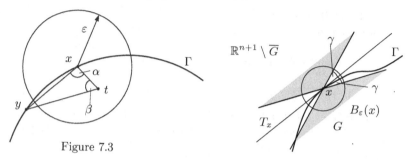

Figure 7.3

Figure 7.4

We apply the sine theorem (cf. Figures 7.3 and 7.4) in the triangle with corners x, t, y and angles α, β, γ:

$$\frac{|y - t|}{|y - x|} = \frac{\sin \alpha}{\sin \beta} \geq \sin \alpha \geq c > 0.$$

We have also

$$\frac{|x - t|}{|y - t|} = \frac{\sin \gamma}{\sin \alpha} \leq \frac{1}{c}.$$

Therefore, we are able to use in the numerator of the above estimate for the difference $E_n(y - t) - E_n(y - x)$ the inequality $|y - x| \leq \frac{1}{c}|y - t|$ and in the denominator the inequality $|y - t| \geq c|y - x|$. With a suitable constant K_1 as an upper bound for the integrand in $a(t) - a(x)$ we have the expression

$$K_1 \frac{|x - t|^{1-\delta}}{|y - x|^{n+1-\mu-\delta}}.$$

Outside $B_\varepsilon(x)$ we get an analogous estimate from $|y - x| \geq \varepsilon \geq 2|t - x|$, i.e., $|y - t| \geq |y - x| - |t - x| \geq \frac{1}{2}|y - x|$. We can thus estimate as before with a suitable constant K_2. Actually, with $K = \max\{K_1, K_2\}$ we get

$$|a(t) - a(y)| \leq \frac{K|t - x|^{1-\delta}}{\sigma_n} \int\limits_\Gamma \frac{|dy^*|}{|y - x|^{n+1-\mu-\delta}}.$$

For $\mu + \delta > 1$ this integral exists as an improper integral, and $a(t)$ converges to $a(x)$. $\square$

Remark 7.18. Analogously to Remark 7.16 a suitable statement holds for an x on edges or at corners if the non-tangential convergence is defined accordingly. We may work with the apex angle $\gamma(x)$ as in Remark 7.16. Then the operator S_Γ has to be replaced by the operator

$$(S_\Gamma^\gamma)(u) = \frac{\sigma_n - 2\gamma(x)}{\sigma_n} u(x) + (S_\Gamma u)(x).$$

Corollary 7.19. *Let G be a domain with a C^2-boundary. The relation*

$$(S_\Gamma u)(x) = u(x) \quad \text{for all} \quad x \in \Gamma$$

is necessary and sufficient so that u represents the boundary values of a holomorphic function $(F_\Gamma u)$ defined in G. On the other hand, the condition

$$(S_\Gamma u)(x) = -u(x) \quad \text{for all} \quad x \in \Gamma$$

is necessary and sufficient so that $(F_\Gamma u)$ is the holomorphic continuation of u into the domain G^- which vanishes at $x = \infty$.

Proof. At first let U be the holomorphic continuation into the domain G of the $C\ell(n)$-valued function $u(x)$ given on Γ. Then Cauchy's integral formula yields $U(x) = (F_\Gamma u)(x)$. Therefore, the non-tangential boundary values of U are u. Using the formulae of Plemelj–Sokhotski we get $u(x) = \frac{1}{2}[u(x) + (S_\Gamma u)(x)]$, which leads to $(S_\Gamma u)(x) = u(x)$.

If vice versa we have $(S_\Gamma u)(x) = u(x)$ for all $x \in \Gamma$, then $(F_\Gamma u)(x)$ has the boundary values u, thus it is the holomorphic continuation of u into G. The proof for the exterior domain proceeds analogously. $\qquad\square$

Corollary 7.20. *Let u be Hölder continuous on Γ, we then have the algebraic identity $S_\Gamma^2 u = Iu$ where I is the identity operator.*

Proof. Setting

$$\text{n.t.-} \lim_{t \to x, t \in G^+} (F_\Gamma u)(t) = F^+(x)$$

and

$$\text{n.t.-} \lim_{t \to x, t \in G^-} (F_\Gamma u)(t) = F^-(x)$$

we get from the formula of Plemelj–Sokhotski and Corollary 7.19

$$(S_\Gamma u)(x) = F^+(x) + F^-(x).$$

Here F^+ represents the boundary values of the function $F_\Gamma u$, holomorphic in G. Applying the last corollary we find $S_\Gamma F^+ = F^+$. Similarly we get $S_\Gamma F^- = -F^-$, thus

$$(S_\Gamma^2 u)(x) = (S_\Gamma F^+)(x) + (S_\Gamma F^-)(x) = F^+(x) - F^-(x) = u(x). \qquad\square$$

Definition 7.21. The operators $P_\Gamma := \frac{1}{2}(I + S_\Gamma)$ and $Q_\Gamma := \frac{1}{2}(I - S_\Gamma)$ are called *Plemelj projections.*

Corollary 7.22. *The operator P_Γ is the projection onto the space of all functions defined on Γ which are holomorphically continuable into G^+. The operator Q_Γ is the projection onto the space of all functions which are holomorphically continuable into the exterior domain G^- vanishing at ∞. The following algebraic properties are fulfilled:*

$$P_\Gamma{}^2 = P_\Gamma, \quad Q_\Gamma{}^2 = Q_\Gamma, \quad P_\Gamma Q_\Gamma = Q_\Gamma P_\Gamma = 0.$$

Proof. This is an immediate conclusion from the definition as well as from the Plemelj–Sokhotski formula with its consequences. $\qquad\square$

7.2.4 History of Cauchy and Borel–Pompeiu formulae

In 1831 Cauchy proved his integral formula during his stay in Torino in Italy. The result was published within a treatise on celestial mechanics. In [24] the result is formulated in an unusual way for today's way of writing

$$f(x) = \frac{1}{2\pi} \int\limits_{-\pi}^{\pi} \frac{\overline{x} f(\overline{x})}{\overline{x} - x} dp, \quad \overline{(x)} = |\overline{(x)}| e^{p\sqrt{(-1)}},$$

where the integration runs over the unit circle and x is inside the unit circle. Today dp would be replaced by $d\varphi$ and $\overline{x}$ would be ζ.

In 1905 D. Pompeiu recognized in his thesis that the set of non-holomorphic points of a continuous function f can already be characterized by the values of

$$\int\limits_{\Gamma} f(z) dz.$$

Here Γ is a piecewise smooth closed curve in the domain of f.

In 1912 he introduced the *areolar derivative* of a function f as a measure of its non-holomorphy at a given point $z_0 \in G$:

$$\frac{Df}{D\sigma}(z_0) := \lim_{G \to \{z_0\}} \frac{\int_{\partial G} f(z) dz}{2i \int_G d\sigma}.$$

The notation $G \to \{z_0\}$ means that G shrinks to the point z_0 if $diam\, G \to 0$ (cf. Exercise 7.4.9).

D. Pompeiu (1873–1954) was born in Broscauti in the Roumanian province of Moldova. He graduated in 1893 from the "Scoala Nationala de Institutori" and got a position as a primary schoolteacher in Ploiesti. In 1898 he moved to Paris where he enrolled in the Sorbonne. There he dealt, above all, with complex analysis and mechanics. In 1905 he finished his thesis *Sur la continuité des fonctions de deux variable complexes* [117]. His results were taken up very sceptically and L. Zoretti, a pupil of Émile Borel, tried to disprove them. Later D. Pompeiu worked as a professor in Bucharest and Cluj.

Dimitrie Pompeiu

The first two summands in the following representation (cf. Section 5.1, Remark 5.7)

$$f(z) = f(z_0) + \partial_z f(z_0)(z - z_0) + \partial_{\overline{z}} f(z_0)(\overline{z} - \overline{z_0}) + o(|z - z_0|)$$

vanish by Cauchy's theorem while integrating along a piecewise smooth curve which encloses the point z_0. We are left with

$$\int_\Gamma f(z)dz = (\partial_{\bar z}f)(z_0)\int_\Gamma \bar z dz + \int_\Gamma o(|z-z_0|)dz.$$

According to Exercise A.2.3.8 the middle integral is just the area $2i\sigma(G)$ (up to the factor $2i$) of G bounded by Γ, such that

$$\lim_{G\to\{z_0\}} \frac{\int_\Gamma f(z)dz}{2i\sigma(G)} = (\partial_{\bar z}f)(z_0)$$

and thus

$$(\partial_{\bar z}f)(z_0) = \frac{Df}{D\sigma}(z_0).$$

Hence one obtains a coordinate free representation for the complex derivative $\partial_{\bar z}$. In addition, a definition of $\partial_{\bar z}f$ in a weak sense [118] is possible.

In 1909 Pompeiu proved that a function which is holomorphic in the whole complex plane and vanishes at ∞ has the representation

$$f(z) = \frac{1}{2\pi i}\int_{\mathbb{C}} \frac{Df}{D\sigma}(\zeta)\frac{1}{\zeta-z}d\sigma_\zeta.$$

At first this result remained unnoticed, it became known by an analogous, but much more special result by É. Borel [13], who lectured on this topic in 1912 in Cambridge (U.K.) during the International Congress of Mathematicians. In the end, in 1912 Pompeiu was able to show the formula

$$f(z) = \frac{1}{2\pi i}\int_{\partial G} \frac{f(\zeta)}{\zeta-z}d\zeta - \frac{1}{\pi}\int_G \frac{Df}{D\sigma}(\zeta)\frac{1}{\zeta-z}d\sigma_\zeta \quad (z\in G)$$

for continuous functions f with continuous areolar derivative in a neighborhood of $\overline{G}$. He thought this formula to be an analogy of the main theorem of calculus

$$f(x) = f(x_0) + \int_{x_0}^{x} f'(t)dt.$$

In his thesis [154] N. Teodorescu transferred Pompeiu's results to the case of quaternions. He introduced the so-called *volume derivative* for quaternion-valued functions f in a domain $G\subset\mathbb{R}^3$, which is given by

$$\frac{Df}{D\sigma}(x) := -\lim_{G\to\{x\}} \frac{\int_{\partial G} f\,do}{\int_G d\sigma} \quad (x\in G).$$

It can be shown that the volume derivative coincides with the application of the Hamilton operator ∇ (see Definition 7.24). The first generalization into space of Borel–Pompeiu's formula was proved in 1930 by G.C. Moisil [110]:

$$f(x) = \frac{1}{4\pi} \int\limits_{\partial G} \frac{\overline{y-x}}{|y-x|^3} do_y f(y) + \frac{1}{4\pi} \int\limits_{G} \frac{Df}{D\sigma}(y) \frac{\overline{y-x}}{|y-x|^3} d\sigma_y.$$

This formula holds for functions f continuous in $\overline{G}$ with continuous and bounded volume derivative in G. The reader will find further statements in [109].

A. GROTHENDIECK used an analogous formula within the context of complex analysis of several variables. Such a formula is proved in the monograph of H. CARTAN (1961) [21] as well as in the book of A. W. BITSADZE (1973) [11]. Based on R. DELANGHE's results (see [31]) a Borel–Pompeiu formula in real Clifford algebras could be obtained.

In 1975 E. M. SAAK [127] considered a system of differential equations of first order constructed as follows:

One takes a family of orthogonal $n \times n$ matrices $\{e_1, \ldots, e_n\}$, whose entries are only the numbers $0, 1, -1$, with the following property:

$$e_i^\top e_j + e_j^\top e_i = 0 \quad (i \neq j), \quad i, j = 1, \ldots, n.$$

One then represents n-dimensional vector-valued functions by the e_i as a basis and defines a differential operator of the system by $\sum_{i=1}^{n} e_i \partial_i$. In 1978 an n-dimensional matrix analogy to the formula of Borel–Pompeiu was developed for such a system (cf. [144]).

Another interesting example of a Borel–Pompeiu formula in $\mathbb{R}^3$ goes back to A. DZHURAEV. He considers in his article [37] the following matrix differential operator with three real variables x_1, x_2, x_3 in a domain $G \subset \mathbb{R}^3$,

$$\overline{\partial_x} = \begin{pmatrix} \partial_{x_1} & \partial_z \\ -\partial_{\bar{z}} & \partial_{x_1} \end{pmatrix},$$

and defines $u = (u_1, u_2)$ by the complex numbers $u := u_1 + iu_2$ as well as $z := x_2 + ix_3$ and $\zeta := y_2 + iy_3$. He studies then the differential equation $\overline{\partial_x} u = f$ by setting

$$E(y-x) = -\frac{1}{|y-x|^3} \begin{pmatrix} y_1 - x_1 & -(\overline{\zeta} - \overline{z}) \\ \zeta - z & y_1 - x_1 \end{pmatrix}, \quad n = \begin{pmatrix} n_1 & n_2 - in_3 \\ -(n_2 + in_3) & n_1 \end{pmatrix}$$

with the outer normal unit vector n on ∂G. Then he proves a formula of Borel–Pompeiu type:

$$u(x) = \frac{1}{\sigma_2} \int\limits_{\partial G} E(y-x) n(y) f(y) |do_y| - \frac{1}{\sigma_2} \int\limits_{G} E(y-x)(\overline{\partial_y} u)(y) d\sigma_y.$$

Formulae of Borel–Pompeiu type open new possibilities for the treatment of partial differential equations by complex methods ([9]). Today analogies to the Borel–Pompeiu formula also exist on Riemannian manifolds [20], in particular on the sphere (cf. [157]) as well as in complex Clifford algebras [126] and for several classes of differential operators [80]. Discrete Borel–Pompeiu formulae have also been proved.

7.3 Consequences of Cauchy's integral formula

7.3.1 Higher order derivatives of holomorphic functions

An important consequence shows us a decisive property of holomorphic functions, namely that these are real continuously differentiable not only once, but infinitely many times. Thus all derivatives of a holomorphic function are again holomorphic: Because of Schwarz' theorem the derivatives with respect to the variables x_i commute if they are continuous, consequently, the equation $\overline{\partial} f = 0$ remains valid for all derivatives of f.

At first, we formulate the theorem in $\mathbb{C}$, because it is simpler and more precise:

Corollary 7.23 (Cauchy's integral formula for derivatives in $\mathbb{C}$). *Let f be holomorphic in the disc $\{z : |z - z_0| < R\}$, then f is infinitely often complex differentiable in it and for any ρ and z with $|z - z_0| < \rho < R$ we have*

$$f^{(n)}(z) = \frac{n!}{2\pi i} \int\limits_{|\zeta - z_0| = \rho} \frac{f(\zeta)}{(\zeta - z)^{n+1}} d\zeta.$$

Moreover, if $|f(z)| \leq M$ on the circle $|z - z_0| = \rho$, we have the estimate

$$\left| f^{(n}(z_0) \right| \leq \frac{n!}{\rho^n} M.$$

In view of Cauchy's integral theorem the radius does not matter. The function f has only to be holomorphic in the given disc. Hence, one may choose ρ as large as possible to come close to the next singular point of f.

Proof. The function f is holomorphic for $|z - z_0| \leq \rho + \varepsilon < R$. So we can apply Cauchy's integral formula. We prove the first part of the theorem by mathematical induction. For $n = 0$ we have Cauchy's integral formula. For the induction step from n to $n + 1$ we have to exchange differentiation and integration (differentiation of parameter integrals). Thus we obtain

$$\frac{d}{dz} f^{(n)}(z) = \frac{n!}{2\pi i} \int\limits_{|\zeta - z_0| = \rho} \frac{(n+1)f(\zeta)}{(\zeta - z)^{n+2}} d\zeta,$$

i.e., just the assertion for $n + 1$.

The second part of the theorem follows directly from the formula for the derivatives:

$$\left| f^{(n)}(z_0) \right| \leq \frac{n!}{2\pi} \int_{|\zeta - z_0| = \rho} \frac{M}{\rho^{n+1}} |d\zeta| = \frac{n!M}{\rho^n},$$

because the integral generates the factor $2\pi\rho$ once again.   □

In particular the real differentiability as well as the holomorphy of all partial derivatives follow.

To prove an analogous result in $\mathbb{R}^{n+1}$ we need some preliminaries:

Definition 7.24 (Nabla and Delta). We call the operator

$$\nabla := (\partial_0, \partial_1, \ldots, \partial_n)$$

Nabla, it corresponds to our vector operator $\overline{\partial}$. With the multiindex $\mathbf{k} = (k_0, \ldots, k_n)$ the symbol

$$\nabla^{\mathbf{k}} := \partial_0^{k_0} \partial_1^{k_1} \ldots \partial_n^{k_n}$$

is defined. Furthermore, we introduce the operator *Delta*

$$\Delta := \nabla \cdot \nabla = \partial\overline{\partial} = \sum_{i=0}^{n} \partial_i^2.$$

A solution of the equation $\Delta f = 0$ is called a *harmonic function.*

We shall deal with harmonic functions later on, but we have immediately:

Proposition 7.25. *The coordinate functions of a holomorphic function are harmonic.*

Proof. For a holomorphic function, $\overline{\partial} f = 0$ holds by definition. As the derivatives of any order exist (see Corollary 7.28) $\Delta f = \partial\overline{\partial} = 0$ is given. The operator Δ is a real operator, it acts separately on every coordinate function f_A of f, so we have $\Delta f_A = 0$: the f_A are harmonic functions.   □

The symbol of the operator ∇ was introduced by Hamilton already during the fourties of the 19-th century in his quaternionic analysis. Firstly he wrote it lying on its side with the peak to the right. JAMES CLERK MAXWELL recommended in 1870 the name *Atled*, suggesting a Delta standing on its head. Shortly after ROBERTSON SMITH suggested the name *Nabla*, which was used by PETER G. TAIT in the second edition of his book [153]. The reason for this naming was probably the resemblance to the figure of an assyric string instrument, since it was at that time that the first findings of the mesopotamic excavations reached the British museum (cf. C.G. KNOTT [70]). The name might be borrowed from the Greek where it means "string instrument".

With the operator ∇ we define now

Definition 7.26.

$$\mathcal{Q}_{\mathbf{k}}(x) := \frac{(-1)^{|\mathbf{k}|}}{\mathbf{k}!} \nabla^{\mathbf{k}} \sigma_n E_n(x) = \frac{(-1)^{|\mathbf{k}|}}{\mathbf{k}!} \nabla^{\mathbf{k}} \frac{\overline{x}}{|x|^{n+1}},$$

in particular

$$\mathcal{Q}_0(x) = \frac{\overline{x}}{|x|^{n+1}} = \sigma_n E_n(x).$$

We have to consider these functions, which we shall use often, in more detail: Generally, we shall work with the multiindex $\mathbf{k}$ where $k_0 = 0$, so that we shall differentiate only with respect to $x_1, \ldots, x_n$.

The functions $\mathcal{Q}_{\mathbf{k}}$ replace the negative powers of z in $\mathbb{C}$: For $n = 1$ we have $\mathcal{Q}_0 = 1/z$ and in this case we have to use $\mathbf{k} = (0, k_1)$ as well as $z = x_0 + ix_1$. In view of $\partial_1 f = if'$ we get, with $k_1 =: k$,

$$\mathcal{Q}_{\mathbf{k}}(z) = \frac{(-1)^k}{k!} \partial_1^k \mathcal{Q}_0(z) = \frac{i^k}{z^{k+1}}.$$

Those are the negative powers of z. We need some more properties of the functions $\mathcal{Q}_{\mathbf{k}}$:

Proposition 7.27. *We have*

$$\mathcal{Q}_{\mathbf{k}}(x) = \frac{q_{\mathbf{k}}(x)}{|x|^{n+2|\mathbf{k}|+1}}$$

with a homogeneous polynomial $q_{\mathbf{k}}$ of degree $|\mathbf{k}| + 1$ which takes only values in $\mathbb{R}^{n+1}$, i.e., in the paravectors. Furthermore, constants $C_{n,\mathbf{k}}$ exist such that

$$|\mathcal{Q}_{\mathbf{k}}(x)| \leq \frac{C_{n,\mathbf{k}}}{|x|^{n+|\mathbf{k}|}}.$$

Proof. We use mathematical induction with respect to $|\mathbf{k}|$. For $\mathbf{k} = \mathbf{0}$ the assertion is given by $\sigma_n E_n(x) = \mathcal{Q}_0(x)$. To conclude from $|\mathbf{k}|$ to $|\mathbf{k}| + 1$ we have to differentiate the expression for $\mathcal{Q}_{\mathbf{k}}$ once:

$$\partial_i \mathcal{Q}_{\mathbf{k}}(x) = \frac{|x|^2 \partial_i q_{\mathbf{k}}(x) - (n + 2|\mathbf{k}| + 1) x_i q_{\mathbf{k}}}{|x|^{n+2|\mathbf{k}|+3}}.$$

The differentiation of a homogeneous polynomial with respect to one of its variables yields either zero or it reduces the degree by 1, so that the first term in the numerator has the right degree $|\mathbf{k}| + 2$. For the second term this is also clear, and the homogeneity is not altered by the differentiation or the multiplication with x_i. With such a step the numerator remains a paravector too.

Moreover a homogeneous polynomial $q_{\mathbf{k}}$ can be estimated by a suitable constant, just $C_{n,\mathbf{k}}$, with the factor $|x|^{|\mathbf{k}|+1}$, which is the assertion. $\qquad\square$

Now we are able to formulate and prove a theorem for higher derivatives of a holomorphic function in $C\ell(n)$:

Corollary 7.28 (Cauchy's integral formula for derivatives in $C\ell(n)$). *Let f be holomorphic in the ball $B_R(x_0)$. Then f is infinitely often real continuously differentiable and we have, for all ρ with $|x - x_0| < \rho < R$,*

$$\nabla^{\mathbf{k}} f(x) = \frac{k!}{\sigma_n} \int_{|y-x_0|=\rho} \mathcal{Q}_{\mathbf{k}}(y - x) dy^* f(y).$$

If $|f(x)| \leq M$ for $|x - x_0| = \rho$ we have moreover

$$\left|\nabla^{\mathbf{k}} f(x_0)\right| \leq \frac{M C_{n,\mathbf{k}} \mathbf{k}!}{\rho^{|\mathbf{k}|}}$$

with the same constants of the last proposition.

Proof. The Cauchy kernel $Q_0(y - x)$ is arbitrarily often continuously differentiable with respect to the x_i, so one can differentiate under the integral sign and get the integral for the derivatives of f as given. Because the functions $Q_{\mathbf{k}}(y - x)$ and dy^* have only values in $\mathbb{R}^{n+1}$ we may take the absolute value under the integral and further on into the product. The integration yields the value $\sigma_n \rho^n$, which implies the estimate as desired.   □

Thus an important property of holomorphic functions has been proved:

Corollary 7.29. *A holomorphic function in $C\ell(n)$ is real continuously differentiable arbitrarily many times and all its derivatives are also holomorphic.*

Another quick consequence is the equivalence of Cauchy's integral formula with the holomorphy:

Corollary 7.30. *A continuous function f in a domain $G \subset \mathbb{R}^{n+1}$ is holomorphic if and only if Cauchy's integral formula holds locally.*

Proof. If f is holomorphic in G then Cauchy's integral formula holds locally. If vice versa this formula holds locally then we can differentiate and f is locally holomorphic, but this means it is in the whole domain G.   □

7.3.2   Mean value property and maximum principle

In this subsection we shall deal with an important set of consequences of Cauchy's integral formula. The first one is the so-called *mean value property* which was proved in case $n = 1$ by *Poisson* in 1823:

Corollary 7.31 (Mean value property). *A holomorphic function f possesses the mean value property, i.e., for all x_0 in its domain of holomorphy G and for all balls (disks) $\{x : |x - x_0| \leq \rho\} \subset G$ we have*

$$f(x) = \frac{1}{\sigma_n} \int\limits_{|y|=1} f(x_0 + \rho y) |do_y|.$$

That means that the value of f in the center of the ball is equal to the integral of f over the boundary of the ball, hence the name mean value property.

This theorem expresses once more the close relationship of the values of a holomorphic function.

Proof. From Cauchy's integral formula we get

$$f(x_0) = \int\limits_{|t - x_0| = \rho} E_n(t - x_0) dt^* f(t).$$

Substituting $E_n(t - x_0)$ from its definition and considering the special cases A.2.17 b) and c) with $t = x_0 + \rho y$ we find

$$E_n(t - x_0)dt^* = \frac{1}{\sigma_n}|do_y|.$$

This is the assertion.   □

An especially important consequence of Cauchy's integral formula is the following:

Theorem 7.32 (Maximum principle). *Let f be holomorphic and bounded in a do-main $G \subset \mathbb{R}^{n+1}$, i.e., $\sup_{x \in G} |f(x)| = M < \infty$. If $|f|$ attains the value M at a point of G then f is constant in G with $|f(x)| = M$.*

This shows also the close inner connection of the values of f. If f is not constant $|f|$ has to tend to the supremum M at a sequence of points converging to the boundary of the domain. For a bounded domain G and f continuous in $\overline{G}$ the function $|f|$ has to take the value M on the boundary.

A corresponding *minimum principle* in $\mathbb{C}$ reads as follows: :

Let f be holomorphic in $G \subset \mathbb{C}$ and $f(z) \neq 0$ for all $z \in G$. Then f is constant or we have for each $z \in G$,

$$|f(z)| > \inf_{\zeta \in \partial G} |f(\zeta)|$$

(see Exercise 7.4.6).

Proof. We want to show that the relation $|f(x)| = M$ is true in the whole domain G. If this is not the case there would exist a point x_1 with $|f(x_1)| < M$, and because of the assumption a point x_0 exists with $|f(x_0)| = M$. The points x_0 and x_1 can be connected by a polygon. Running this polygon from x_1 to x_0, let x_2 be the first point with $|f(x_2)| = M$ ($x_2 = x_0$ is not excluded). Such a point exists because of the continuity of $|f|$.

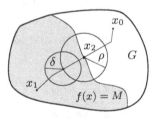

Figure 7.5

Now we consider a sufficiently small ball in G with center x_2 and radius ρ, use in it the mean value property from the last theorem, and obtain

$$M = |f(x_2)| \leq \frac{1}{\sigma_n} \int_{|y|=1} |f(x_2 + \rho y)||do_y|.$$

On the sphere $|x - x_2| = \rho$ at least one point with $|f(x)| < M$ exists, that is where the polygon leaves the ball in the direction toward x_1. Because of the continuity of f on a small cap of the sphere of area δ we have also the inequality $|f(x)| \leq M - \varepsilon$ with suitably chosen ε and δ. For the mean value integral this means

$$M = |f(x_2)| \leq \frac{1}{\sigma_n} \left(M(\sigma_n - \delta) + (M - \varepsilon)\delta \right) < M,$$

and this is a contradiction. Thus $|f|$ is constant in $|x - x_2| \leq \rho$. The procedure has to be repeated starting from a boundary point of the small ball. We obtain that $|f|$ is constant in G. Now it remains only to show that also f is constant. With $|f|^2 = \sum_A |f_A|^2 = \text{const}$ the derivatives of $|f|$ with respect to x_i are zero, so

$$\sum_A f_A (\partial_i f_A) = 0.$$

A second differentiation with respect to x_i and summation over i yield

$$\sum_A \left(\sum_i (\partial_i f_A)^2 + f_A \Delta f_A \right) = 0.$$

From Proposition 7.25 we know $\Delta f_A = 0$, hence we find

$$\sum_A \sum_i (\partial_i f_A)^2 = 0.$$

But this means that all partial derivatives of all f_A vanish, so that f must be constant. $\quad\square$

7.3.3 Liouville's theorem

We want to add another consequence of Cauchy's integral formula, a theorem first proved by the French mathematician JOSEPH LIOUVILLE (1809–1882) who formulated it for complex functions. This theorem is an easy and far-reaching tool in the theory of holomorphic functions. As an application we shall give a simple proof of the fundamental theorem of algebra 1.12. The second part of the theorem is a generalization.

Proposition 7.33.   (i) *A function holomorphic and bounded in $\mathbb{R}^{n+1}$ is constant.*

(ii) *If f is holomorphic in $\mathbb{R}^{n+1}$ and if*

$$|f(x)| \leq M \, |x|^m,$$

then f is a polynomial of degree at most m.

Proof. (i) Let $|f(x)| \leq M$ for all $x \in \mathbb{R}^{n+1}$. We then use the formula of Corollary 7.28 for the first derivatives of f with $\boldsymbol{\varepsilon}_i = (\delta_{0i}, \ldots, \delta_{ni})$:

$$|\partial_i f(x_0)| \leq \frac{M C_{n,\varepsilon_i}}{\rho}.$$

As ρ is here an arbitrary real number, for $\rho \to \infty$ we can conclude that $|\partial_i f(x_0)| = 0$. Because of the arbitrary choice of x_0 all derivatives $\partial_i f$ are zero at all points in $\mathbb{R}^{n+1}$ and f is constant.

The proof of (ii) is recommended as an exercise (see Exercise 7.4.7). $\quad\square$

Now we are able to prove the fundamental theorem of algebra Theorem 1.12 without any difficulty:

Corollary 7.34 (Fundamental theorem of algebra). *A polynomial of degree $n > 0$ in $\mathbb{C}$ has exactly n zeros in $\mathbb{C}$ if every zero is counted according to its multiplicity.*

Proof. If

$$P(z) = a_n z^n + a_{n-1} z^{n-1} + \cdots + a_0$$

had no zero in $\mathbb{C}$, then $1/P(z)$ would be a holomorphic function in $\mathbb{C}$. However, we have

$$|P(z)| \geq |z|^n \left(|a_n| - \frac{|a_{n-1}|}{|z|} - \cdots - \frac{|a_0|}{|z|^n} \right) \geq \frac{1}{2} |a_n| |z|^n$$

for $|z|$ large enough. Hence $1/|P(z)| \to 0$ would follow for $|z| \to \infty$ and $P(z)$ would be bounded in $\mathbb{C}$. From Liouville's theorem it would follow that f is constant. This is not true as $P(0) = a_0$ and $P(z) \to \infty$ for $z \to \infty$. So P has to have at least one zero, let us call it z_1. By division of P by $(z - z_1)$ we get $P(z) = (z - z_1) P_1(z)$ with a polynomial P_1 of degree $n - 1$. An n-times iteration of the procedure gives the assertion. $\square$

Remark 7.35. Unfortunately, this proof does not work in $\mathbb{R}^{n+1}$ because the reciprocal value of a holomorphic function is not holomorphic in general. The question about the zeros of a polynomial is much more complicated in higher dimensions, as zeros do not have necessarily to be isolated. Later on we shall prove this statement in $\mathbb{C}$.

7.3.4 Integral formulae of Schwarz and Poisson

As usual let $B_r(0) =: B_r$ be the disk with center at the origin and radius r in $\mathbb{C}$. H.A. SCHWARZ (see Section 2.4) succeeded in 1869 to solve the following boundary value problem with the help of an integral formula which bears his name:

We seek a complex function u holomorphic and bounded in B_1 whose real part has as boundary values a continuous function g given on the boundary ∂B_1.

The integral formula developed for this purpose by Schwarz reads as follows:

Theorem 7.36 (Integral formula of Schwarz). *Let $f = u + iv$ be a function holomorphic in $B_r \subset \mathbb{C}$ and continuous in $\overline{B_r}$. Then for $z \in B_r$ we have*

$$f(z) = \frac{1}{2\pi i} \int_{\partial B_r} \frac{\zeta + z}{\zeta - z} u(\zeta) \frac{d\zeta}{\zeta} + i \, v(0).$$

Analogously the formula

$$f(z) = \frac{1}{2\pi} \int_{\partial B_r} \frac{\zeta + z}{\zeta - z} v(\zeta) \frac{d\zeta}{\zeta} + u(0)$$

holds as well.

Of course one can transfer these formulae easily to circles with other centers by substituting z and ζ by $z - a$ resp. $\zeta - a$.

Proof. If we reflect a point z about the circle ∂B_r then, following Exercise 6.3.6, we get the reflected point z^* by $z^* z = r^2$. Moreover, we have $\zeta\bar\zeta = r^2$ using the parametrization $\zeta = r(\cos\varphi + i\sin\varphi)$ of the circle. Furthermore,

$$d\zeta = i\zeta d\varphi, \quad d\bar\zeta = -i\bar\zeta d\varphi = -\frac{\bar\zeta}{\zeta}d\zeta.$$

As z^* is located in the exterior of the disk B_r we obtain using Cauchy's integral formula

$$\int_{\partial B_r} \frac{f(\zeta)}{\zeta - z^*}d\zeta = 0.$$

If we conjugate this equation and if we use

$$\frac{1}{\bar\zeta - \frac{r^2}{z}} = \frac{\zeta z}{r^2(z - \zeta)} = \frac{z}{\bar\zeta(z - \zeta)} = \frac{1}{\bar\zeta} + \frac{\zeta}{\bar\zeta(z - \zeta)}$$

we get the expressions

$$0 = \int_{\partial B_r} \frac{\overline{f(\zeta)}}{\bar\zeta - \frac{r^2}{z}}d\bar\zeta = \overline{\int_{\partial B_r} \frac{f(\zeta)}{\zeta}}d\zeta + \int_{\partial B_r} \frac{\overline{f(\zeta)}}{z - \zeta}\frac{\zeta d\bar\zeta}{\bar\zeta}$$

$$= 2\pi i\overline{f(0)} - \int_{\partial B_r} \frac{\overline{f(\zeta)}}{\zeta - z}d\zeta.$$

Therefore the Cauchy integral of the conjugated function $\bar f$ does not depend on the variable point z. We obtain

$$\frac{1}{2\pi i}\int_{\partial B_r} \frac{\overline{f(\zeta)}}{\zeta - z}d\zeta = \overline{f(0)}$$

and further

$$f(z) + \overline{f(0)} = \frac{1}{\pi i}\int_{\partial B_r} \frac{u(\zeta)d\zeta}{\zeta - z}.$$

For the value of the real part of f at the origin this gives

$$u(0) = \frac{1}{2\pi i}\int_{\partial B_r} \frac{u(\zeta)}{\zeta}d\zeta,$$

and thus we get

$$f(z) = \frac{1}{\pi i}\int_{\partial B_r} \frac{u(\zeta)}{\zeta - z}d\zeta - \frac{1}{2\pi i}\int_{\partial B_r} \frac{u(\zeta)}{\zeta}d\zeta + i\,\mathrm{Im}\,f(0)$$

$$= \frac{1}{2\pi i}\int_{\partial B_r} \frac{\zeta + z}{\zeta - z}u(\zeta)\frac{d\zeta}{\zeta} + i\,\mathrm{Im}\,f(0).$$

One gets easily from this formula the representation of f by its imaginary part by considering $-if(z)$. $\square$

By decomposition into real and imaginary part one gets the well-known *Poisson integral formula*:

Theorem 7.37. *Let u be harmonic in B_ρ and continuous in $\overline{B_\rho}$. With $z = r(\cos \varphi + i \sin \varphi)$ and $\zeta = \rho(\cos \theta + i \sin \theta)$ we then have*

$$u(z) = \frac{1}{2\pi} \int_0^{2\pi} \frac{\rho^2 - r^2}{\rho^2 + r^2 - 2r\rho \cos (\varphi - \theta)} u(\zeta) d\theta.$$

Proof. The integral formula of Schwarz for a given u yields a holomorphic function f with $\operatorname{Re} f = u$ as the boundary values determine u uniquely. Then one takes simply the real part of Schwarz' integral formula to prove the assertion. $\qquad\square$

Poisson's integral formula is a proven device for harmonic continuation of continuous functions into the interior of a ball, i.e., the solution of a *boundary value problem*.

7.4 Exercises

1. Prove the following estimate for paravectors a and b:

$$\left| a|b|^{n+1} - b|a|^{n+1} \right| \leq |b||a - b|[2|b|^n + |b|^{n-1}|a| + \cdots + |a|^n].$$

2. Prove Cauchy's integral theorem for triangles in the plane. How can one extend such a statement to an arbitrary domain with sufficiently smooth boundary? (Advice: Dissect the domain into finer and finer triangles.)

3. Formulate and prove in $\mathbb{R}^{n+1}$ an analogy to Theorem 7.5 about a local condition for holomorphy.

4. Calculate the integral

$$\int_{|z|=1} \frac{dz}{z}$$

in $\mathbb{C}$. Why is it not zero?

5. Calculate in $C\ell(n)$ the integral

$$\int_{|x|=1} \frac{\overline{x}}{|x|^{n+1}} dx^*.$$

6. Prove the minimum principle in $\mathbb{C}$: *Let f be holomorphic in $G \subset \mathbb{C}$ and $f(z) \neq 0$ for any $z \in G$. Then either f is constant or we have*

$$|f(z)| > \inf_{\zeta \in \partial G} |f(\zeta)|$$

for all $z \in G$.

7. Prove part (ii) of Liouville' s Theorem 7.33 : If f is holomorphic in $\mathbb{R}^{n+1}$ and if $|f(x)| \leq M|x|^m$, then f is a polynomial of degree at most m.

8. Let a sequence of holomorphic functions (f_k) be given which are defined in a domain $G \subset \mathbb{C}$ and which converges at a fixed point $z = a \in G$ to zero. Furthermore, let the sequence of its real parts converge uniformly to zero in G. Show that the sequence (f_k) tends uniformly to zero on every compact subset $K \subset G$.

9. Show that the limit value

$$\lim_{G \to \{z_0\}} \frac{\int_{\partial G} f(z)dz}{2i \int_G d\sigma}$$

exists, independently of the contraction $G \to \{z_0\}$.

8 Teodorescu transform

Paraphrasing the Borel–Pompeiu formula in the domain G we are led to

$$(T_G(\overline{\partial}f))(x) = f(x) - (F_{\partial G}f)(x), \ x \in G.$$

i.e., the application of the Teodorescu transform to the image of the Cauchy–Riemann operator reproduces the source function up to a boundary integral which is the well-known Cauchy integral. If f has boundary values which allow a holomorphic continuation into the exterior domain, then the Cauchy integral disappears and T_G works like an inverse operator for $\overline{\partial}$. Thus the question to study $\overline{\partial}T_G$ arises and this study is undertaken in this section. Such considerations require the knowledge of suitable function spaces. $T_G f$ should at least be partially differentiable in order to give the expression $\overline{\partial}T_G f$ some sense. We have put together an introduction to suitable function spaces in Appendix 3.

8.1 Properties of the Teodorescu transform

First we deal with the definition of the *Teodorescu transform*

$$(T_G u)(x) = (Tu)(x) = -\frac{1}{\sigma_n} \int\limits_G \mathcal{Q}_0(y - x)u(y)d\sigma_y$$

introduced in 7.10 and of the *Cauchy–Bitsadze operator*

$$(F_\Gamma u)(x) = (Fu)(x) = \frac{1}{\sigma_n} \int\limits_\Gamma \mathcal{Q}_0(y - x)dy^* u(y),$$

where G is a bounded domain in $\mathbb{R}^{n+1}$, $\sigma_n = 2\pi^{(n+1)/2}/\Gamma((n+1)/2)$ is the area of the n-dimensional unit sphere S^n in $\mathbb{R}^{n+1}$ following Example A.2.17 a, and $\Gamma = \partial G$ is the sufficiently smooth (in general twice continuously differentiable) boundary surface. We should recall that $\mathcal{Q}_0(x) := \overline{x}/|x|^{n+1} = \sigma_n E_n(x)$.

Proposition 8.1. *Let be $u \in L^p(G)$ for $p > n + 1$.*

(i) *The integral $(Tu)(x)$ exists everywhere in $\mathbb{R}^{n+1}$ and tends to zero for $|x| \to \infty$; in addition, $T_G u$ is holomorphic in $\mathbb{R}^{n+1} \setminus \overline{G}$. Further we have for a bounded domain G,*

$$\|T_G u\|_p \le C_1(G, p, n)\|u\|_p.$$

(ii) *For $x, z \in \mathbb{R}^{n+1}$ and $x \ne z$ we get the inequality*

$$|(T_G u)(x) - (T_G u)(z)| \le C_2(G, p, n)\|u\|_p |x - z|^{\frac{p-n-1}{p}}.$$

Although the constants can be estimated explicitly the corresponding expressions are not very informative.

Proof. (i) For $|x| \to \infty$ we always have $x \neq y$. Recall that $L^p(G) \subset L^1(G)$ for a bounded domain G and $p > n + 1$. Then the estimate

$$|(T_G u)(x)| \leq \frac{1}{\sigma_n} \int\limits_G \frac{1}{|x-y|^n} |u(y)| d\sigma_y \leq \frac{1}{\sigma_n} \max_{z \in G} \frac{1}{|x-z|^n} \int\limits_G |u(y)| d\sigma_y.$$

follows. So we get $|(T_G u)(x)| \to 0$ for $|x| \to \infty$. Because $\mathcal{Q}_0$ is holomorphic up to the singularity, $T_G u$ is holomorphic in $\mathbb{R}^{n+1} \setminus \overline{G}$. The Hölder inequality leads to

$$|(T_G u)(x)| \leq \frac{1}{\sigma_n} \int\limits_G |\mathcal{Q}_0(y-x) u(y)| d\sigma_y \leq \frac{1}{\sigma_n} \|u\|_p \left(\int\limits_G |\mathcal{Q}_0(y-x)|^q d\sigma_y \right)^{1/q}.$$

The last integral is considered in Exercise 8.3.4. Repeated integration over G yields the statement of the theorem.

(ii) Now let $x, z \in \mathbb{R}^{n+1}$ and $x \neq z$. We restrict ourselves to the integral $T'u$ over $G' =: G \setminus B_\varepsilon(x)$. Furthermore, let $|z - x| < \frac{\varepsilon}{2}$, then $|y - z| \geq |y - x|/2$ and we obtain

$$\left| \frac{\overline{y-x}}{|y-x|^{n+1}} - \frac{\overline{y-z}}{|y-z|^{n+1}} \right| = \left| \frac{\overline{z-x}}{|y-x|^{n+1}} + (\overline{y-z}) \frac{|y-z|^{n+1} - |y-x|^{n+1}}{|y-x|^{n+1}|y-z|^{n+1}} \right|$$

$$\leq \frac{|z-x|}{|y-x|^{n+1}} + |y-z| \frac{|z-x|(|y-z|^n + |y-z|^{n-1}|y-x| + \cdots)}{|y-x|^{n+1}|y-z|^{n+1}}$$

$$\leq |z-x| \frac{2^{n+1}}{|y-x|^{n+1}}.$$

We apply Hölder's inequality and get

$$|(T'u)(z) - (T'u)(x)| \leq \frac{2^{n+1}|z-x|}{\sigma_n} \left(\int\limits_{G'} |u(y)|^p d\sigma_y \right)^{1/p} \left(\int\limits_{G'} \frac{d\sigma_y}{|y-x|^{(n+1)q}} \right)^{1/q}.$$

The last integral can be estimated in spherical coordinates by

$$\sigma_n \int\limits_\varepsilon^R r^{n-(n+1)q} dr.$$

Let G be contained in a ball of radius R with center x. With a suitable constant C we obtain

$$|(T'u)(z) - (T'u)(x)| \leq C\|u\|_p |z-x| \varepsilon^{(n+1)(1-q)\frac{1}{q}} \leq C(G, p, n) \|u\|_p |z-x|^{1-\frac{n+1}{p}}.$$

The limit $\varepsilon \to 0$ gives the assertion. $\square$

For $u \in L^1(G)$ it can be shown that the integral $(T_G u)(x)$ exists everywhere in $\mathbb{R}^{n+1}$. This proof will be left to the reader (Exercise 8.3.1). Now the proof of a very important property of the Teodorescu transform follows, namely that it is the right-inverse of the $\overline{\partial}$–operator :

Theorem 8.2. *Let the function u be continuously differentiable in G. Then $T_G u$ is also differentiable for all $x \in G$ with*

$$\partial_i (T_G u)(x) = -\frac{1}{\sigma_n} \int_G \partial_{i,x} \mathcal{Q}_0(y - x) u(y) d\sigma_y + \overline{e_i} \frac{u(x)}{n+1}.$$

In particular, we have the identity

$$\overline{\partial} (T_G u)(x) = u(x).$$

The integral over G is strongly singular and exists only as Cauchy's principal value. That means one has to cut out from G the ball $B_\varepsilon(x)$ and then to consider the limit of the integral over $G' := G \setminus B_\varepsilon(x)$ for $\varepsilon \to 0$.

Proof. First we deal with real-valued functions u, the general case of algebra-valued functions can be done by addition. Let $x \in G$, in a similar way as in the preceding proof we consider

$$(T_G u)(x) = (T_{G \setminus B_\varepsilon(x)} u)(x) + (T_{B_\varepsilon(x)} u)(x).$$

Let ε be small enough so that the ball $B_\varepsilon(x)$ together with its boundary is contained in G. The integral $(T_{G \setminus B_\varepsilon(x)} u)(x)$ can be transformed by Gauß' theorem which we use in the form (cf. Exercise A.2.3.9)

$$\int_\partial f(y) d\overline{y}^* = \int_{G'} (\partial_y f)(y) d\sigma_y.$$

It is evident that

$$\partial_{i,y} \frac{1}{|y - x|^{n-1}} = -(n-1)\frac{y_i - x_i}{|y - x|^{n+1}} = -\partial_{i,x} \frac{1}{|y - x|^{n-1}},$$

$$\partial_y \frac{1}{|y - x|^{n-1}} = -(n-1)\mathcal{Q}_0(y - x),$$

which leads to

$$
\begin{aligned}
(T_{G \setminus B_\varepsilon(x)} u)(x) &= \frac{1}{(n-1)\sigma_n} \int_{G \setminus B_\varepsilon(x)} \left(\partial_y \frac{1}{|y - x|^{n-1}} \right) u(y) d\sigma_y \\
&= -\frac{1}{(n-1)\sigma_n} \int_{G \setminus B_\varepsilon(x)} \frac{1}{|y - x|^{n-1}} \partial_y u(y) d\sigma_y \\
&\quad + \frac{1}{(n-1)\sigma_n} \left(\int_{\partial G} - \int_{\partial B_\varepsilon(x)} \right) \frac{u(y)}{|y - x|^{n-1}} d\overline{y}^*.
\end{aligned}
$$

For $\varepsilon \to 0$ the integral over $\partial B_\varepsilon(x)$ tends to zero, because $d\overline{y}^*$ includes the factor $|y - x|^n$. Hence we obtain for $\varepsilon \to 0$,

$$(T_G u)(x) = -\frac{1}{(n-1)\sigma_n} \int_G \frac{1}{|y - x|^{n-1}} (\partial_y u)(y) d\sigma_y + \frac{1}{(n-1)\sigma_n} \int_{\partial G} \frac{u(y)}{|y - x|^{n-1}} d\overline{y}^*.$$

According to Exercise 8.3.3 the first integral can be differentiated under the integral sign. The second integral is a parametric integral without singularities in the integrand and may be differentiated very easily:

$$\partial_{i,x}(T_G u)(x) = -\frac{1}{\sigma_n}\int_G \frac{y_i - x_i}{|y - x|^{n+1}}(\partial_y u)(y)d\sigma_y + \frac{1}{\sigma_n}\int_{\partial G} \frac{y_i - x_i}{|y - x|^{n+1}}u(y)d\overline{y}^*.$$

All integrals exist as improper integrals. Now we want to transform the integral over ∂G again by using Gauß' theorem:

$$\left(\int_{\partial G} - \int_{\partial B_\varepsilon(x)}\right)\frac{y_i - x_i}{|y - x|^{n+1}}u(y)d\overline{y}^*$$

$$= \int_{G\backslash B_\varepsilon(x)}\left[\left(\partial_y \frac{y_i - x_i}{|y - x|^{n+1}}\right)u(y) + \frac{y_i - x_i}{|y - x|^{n+1}}(\partial_y u)(y)\right]d\sigma_y.$$

This leads to

$$\partial_{i,x}(T_G u)(x) = -\frac{1}{\sigma_n}\int_{B_\varepsilon(x)}\frac{y_i - x_i}{|y - x|^{n+1}}(\partial_y u)(y)d\sigma_y$$

$$-\frac{1}{\sigma_n}\int_{G\backslash B_\varepsilon(x)}\partial_{i,x}\mathcal{Q}_0(y - x)u(y)d\sigma_y + \frac{1}{\sigma_n}\int_{\partial B_\varepsilon(x)}\frac{y_i - x_i}{|y - x|^{n+1}}u(y)d\overline{y}^*,$$

where we used the relation

$$\partial_y \frac{y_i - x_i}{|y - x|^{n+1}} = \partial_y \partial_{i,x}\frac{1}{(n-1)|y - x|^{n-1}} = -\partial_{i,x}\mathcal{Q}_0(y - x).$$

In the coordinates $y - x = rt$ the first integral has the form

$$-\frac{1}{\sigma_n}\int_{B_\varepsilon(x)}\frac{y_i - x_i}{|y - x|^{n+1}}(\partial_y u)(y)d\sigma_y = -\frac{1}{\sigma_n}\int_{r=0}^{\varepsilon}\int_{|t|=1}t_i(\partial u)(x + rt)dr|do_1|$$

and converges to zero for $\varepsilon \to 0$. The second integral converges for $\varepsilon \to 0$ to the desired Cauchy's principal value as stated (cf. Exercise 8.3.2). For the third integral we may decompose as follows:

$$\int_{\partial B_\varepsilon(x)}\frac{y_i - x_i}{|y - x|^{n+1}}u(y)d\overline{y}^* = \int_{\partial B_\varepsilon(x)}\frac{y_i - x_i}{|y - x|^{n+1}}[u(y) - u(x)]d\overline{y}^*$$

$$+u(x)\int_{\partial B_\varepsilon(x)}\frac{y_i - x_i}{|y - x|^{n+1}}d\overline{y}^*;$$

because of $d\overline{y}^* = \overline{(y - x)}|y - x|^{n-1}|do_1|$ on $\partial B_\varepsilon(x)$ and $u(y) \to u(x)$ the first integral on the right-hand side tends to zero for $\varepsilon \to 0$. For the second integral on the right with $y - x = \varepsilon t$ we obtain

$$u(x)\int_{\partial B_\varepsilon(x)}\frac{y_i - x_i}{|y - x|^{n+1}}d\overline{y}^* = u(x)\int_{|t|=1}t_i d\overline{t}^*,$$

and using Gauß' theorem (see Exercise A.2.3.9)

$$= u(x) \int\limits_{|t|<1} (\partial t_i)d\sigma_t = u(x)\overline{e}_i\sigma_n \int\limits_0^1 r^n dr = \frac{\sigma_n}{n+1}u(x)\overline{e}_i.$$

This proves our first assertion. Let us now study $\overline{\partial}(Tu)$. Because of the holomorphy of $\mathcal{Q}_0$ the integral over $\overline{\partial}\mathcal{Q}_0$ is zero and we find

$$\overline{\partial}(Tu)(x) = \sum_{i=0}^n e_i\partial_i(Tu)(x) = u(x). \qquad \square$$

The theorem just proved remains correct if only the continuity of u is assumed. Starting from the result above we have to approximate the continuous function u by a sequence of continuously differentiable functions and to prove the uniform boundedness of $T_G u$.

It should be noted that the differentiability of Γ does not matter. One can cut out a fixed ball B from G and use the above proof. The integral over $G \setminus B$ can be differentiated as a non–singular parametric integral and added to the result in the ball.

We are able to show even the sharper result:

Theorem 8.3. *The operator*

$$\partial_k T_G : L^p(G) \to L^p(G)$$

is continuous and in the bounded domain G it fulfils the estimate

$$\|\partial_k T_G u\|_p \le C_3(G,p,n)\|u\|_p \quad (p > n+1)$$

with a suitable constant $C_3(G,p,n)$.

Proof. Theorem 8.2 yields for $u \in C^1(G)$,

$$\partial_k T_G u(x) = -\frac{1}{\sigma_n} \int\limits_G \partial_{k,x}\mathcal{Q}_0(x-y)u(y)d\sigma_y + \overline{e}_k\frac{u(x)}{n+1}.$$

Now we apply the theorem of Calderon–Zygmund ([108] XI, §3). It follows with $\frac{1}{p} + \frac{1}{q} = 1$ that

$$\left\|\frac{1}{\sigma_n} \int\limits_G \partial_{k,x}\mathcal{Q}_0(y-x)u(y)d\sigma_y\right\|_p \le C\|f\|_{L^q(S^{n+1})}\|u\|_p,$$

if only (with $\omega := (y-x)/|y-x|$)

$$f(\omega) := \frac{1}{\sigma_n}|y-x|^{n+1}\partial_{k,x}\mathcal{Q}_0(y-x) = \frac{1}{\sigma_n}\left(-\overline{e}_k - (n+1)\omega_k\overline{\omega}\right)$$

has on S^n a bounded q-norm with respect to ω. However, this is the case since

$$\int\limits_{S^n} |f(\omega)|^q |do_y| \leq \int\limits_{S^n} \left(\frac{n+2}{\sigma_n}\right)^q |do_y| = \sigma_n^{1-q}(n+2)^q,$$

and so

$$\|f(\omega)\|_{L^q(S^n)} \leq (n+2)\sigma_n^{-1/p}.$$

It is rather difficult to obtain an explicit estimate for the constant C. Because $C^1(G)$ is dense in $L^p(G)$, the inequality can be extended to $L^p(G)$, i.e., it follows that

$$\left\|\frac{1}{\sigma_n} \int\limits_G \partial_{k,x} Q_0(x-y)u(y)dy\right\|_p \leq C(n+2)\sigma_n^{-1/p}\|u\|_p.$$

Moreover we have

$$\left\|\overline{e}_k \frac{u}{n+1}\right\|_p \leq \frac{1}{n+1}\|u\|_p.$$

Adding up both the inequalities we conclude

$$\|\partial_k T_G u\|_p \leq C_3(G,p,n)\|u\|_p$$

with $C_3(G,p,n) = C(n+2)\sigma_n^{-1/p} + \frac{1}{n+1}$.   □

Actually, it follows

Theorem 8.4. *If G is a bounded domain, then*

$$T_G : L^p(G) \to W^{1,p}(G)$$

is continuous.

Proof. Because of Proposition 8.1 and Theorem 8.3 we have

$$\|T_G u\|_{1,p} \leq (C_1(G,p,n) + (n+1)C_3(G,p,n)) \|u\|_p$$

and the assertion is proved.   □

8.2 Hodge decomposition of the quaternionic Hilbert space

8.2.1 Hodge decomposition

Here we limit ourselves to $\mathbb{R}^3 = \operatorname{Vec}\mathbb{H}$; the functions should have values in $\mathbb{H}$. Then the Cauchy–Riemann operator $\overline{\partial}$ is transformed into the Dirac operator D according to Remark 5.13. Let further $X := \ker D \cap L^2(G)$ be the set of all holomorphic functions in $L^2(G)$. At first it will be shown that X is a closed subspace in $L^2(G)$.

Proposition 8.5. *Let $G \subset \mathbb{R}^3$ be a bounded domain with sufficiently smooth boundary Γ. Then $X = \ker D \cap L^2(G)$ is a right-linear subspace in $L^2(G)$.*

Proof. Let $(\phi_i)_{i=1}^{\infty}$ be a Cauchy sequence in X. Because of the completeness of $L^2(G)$ a function $\phi \in L^2(G)$ exists with $\phi_i \to \phi$ in $L^2(G)$. The mean value theorem Corollary 7.31 for r small enough yields

$$|\phi_i - \phi_j|(x) \le \frac{1}{\sigma_3} \int\limits_{|y|=1} |\phi_i - \phi_j|(x+ry)|do_y|\,,$$

from which it follows that

$$|\phi_i - \phi_j|(x) \le C_r \,\|\phi_i - \phi_j\|_2$$

and

$$\sup_{x \in G(r)} |\phi_i - \phi_j|(x) \le C_r \|\phi_i - \phi_j\|_2$$

with $G(r) = \{x \in G : \text{dist}\,(x,\Gamma) > r\}$. Hence, we know that the sequence (ϕ_i) converges uniformly on compact subsets. With Corollary 7.28 we obtain also that all-first order partial derivatives on compact subsets converge, i.e., also $(D\phi_i)$ converges (to zero). In this way the assertion is verified. $\qquad\square$

Remark 8.6. An analogous result is valid in $L^p(G)$ for $(1 < p < \infty)$ (cf. [55]).

We now study a basic theorem for all further considerations.

Theorem 8.7 (Hodge decomposition). *The $\mathbb{H}$-valued Hilbert space $L^2(G)$ allows the orthogonal decomposition*

$$L^2(G) = (L^2(G) \cap \ker D) \oplus DW_0^{1,2}(G)$$

with respect to the inner product in $L^2(G)$ which is defined in Appendix A.3.3.

Proof. We write briefly $X := L^2(G) \cap \ker D$ and $Y := L^2(G) \ominus X$ for the orthogonal complement to X in $L^2(G)$. For $u \in Y$ we have $v = Tu \in W^{1,2}(G)$ by using Theorem 8.4. It is clear that $u = Dv$ and, in addition, for any holomorphic $\phi \in L^2(G)$ we find

$$\int\limits_G \overline{u}\phi d\sigma = \int\limits_G \overline{Dv}(y)\phi(y)d\sigma = 0.$$

We select points $x^{(l)}$ in $\mathbb{R}^3 \setminus \overline{G}$ and associate to them the singular functions

$$\phi_l(x) = \frac{1}{\sigma_2} \frac{x - x^{(l)}}{|x - x^{(l)}|^3}.$$

The latter are $\mathbb{H}$-holomorphic in G. Hence,

$$\int\limits_G (\overline{Dv})(y)\phi_l(y)d\sigma = 0.$$

Since $\overline{\phi_l} = -\phi_l$ and using Gauß' theorem the representation

$$0 = \int\limits_G \overline{Dv(y)\phi_l(y)}d\sigma = \int\limits_G \overline{\phi_l}(y)(Dv)(y)d\sigma = \int\limits_G (\phi_l D)(y)v(y)d\sigma$$

$$\int\limits_\Gamma \phi_l(y)dy^* v(y) = -(F_\Gamma v)(x^{(l)})$$

follows. If we choose $\{x^{(l)}\}$ to be a dense subset in $\mathbb{R}^3 \setminus \overline{G}$, the function $(F_\Gamma v)(x)$ has to be 0 on $\mathbb{R}^3 \setminus \overline{G}$ for continuity reasons. So the trace $\mathrm{tr}_\Gamma v$ is holomorphically extendable into the domain G. The continuation is denoted by h. Then we have $\mathrm{tr}_\Gamma h = \mathrm{tr}_\Gamma v$ and the *trace operator* tr_Γ describes just the restriction onto the boundary Γ. Now we set $w := v - h$. Obviously, w has the boundary value zero and we get $w \in W_0^{1,2}(G)$. It is now clear that

$$u = Dv = Dw,$$

which was to be proven.   □

Remark 8.8. An orthogonal decomposition generates two orthoprojections on the corresponding subspaces X and Y, i.e., we have

$$\mathbf{P} : L^2(G) \to L^2(G) \cap \ker D,$$
$$\mathbf{Q} : L^2(G) \to DW_0^{1,2}(G).$$

Moreover, the operator $\mathbf{P}$ can be seen as a generalisation of the classical *Bergman projection*, which maps the functions from $L^2(G)$ onto the holomorphic ones in $L^2(G)$.

We have tied together the Borel–Pompeiu formula with the behavior of a function in the domain G by using Cauchy's integral with its boundary values. Interestingly the boundary values of $T_G f$ are connected with the above mentioned orthoprojections. The following proposition gives us an entire characterization of the image of $\mathbf{Q}$.

Proposition 8.9. *A function u belongs to im $\mathbf{Q}$ if and only if $\mathrm{tr}_\Gamma Tu = 0$, where im $\mathbf{Q}$ is the image of the operator $\mathbf{Q}$ and $\mathrm{tr}_\Gamma f$ is again the trace or restriction of f onto Γ.*

Proof. At first let us take $u \in$ im $\mathbf{Q}$. Then a function $w \in W_0^{1,2}(G)$ exists such that u has the representation

$$u = Dw.$$

Then the formula of Borel–Pompeiu Theorem 7.8 leads to

$$Tu = TDw = w - F_\Gamma w = w,$$

and therefore $\mathrm{tr}_\Gamma Tu = \mathrm{tr}_\Gamma w = 0$. Vice versa we assume that $\mathrm{tr}_\Gamma Tu = 0$ and so

$$\mathrm{tr}_\Gamma T\mathbf{Q}u + \mathrm{tr}_\Gamma T\mathbf{P}u = 0 \ .$$

It is clear that $\mathrm{tr}_\Gamma T\mathbf{Q}u = 0$, so we find $\mathrm{tr}_\Gamma T\mathbf{P}u = 0$. Using Theorem 8.4, $T\mathbf{P}u$ belongs to $W^{1,2}(G)$, therefore $T\mathbf{P}u \in W_0^{1,2}(G)$. Then we get $D(T\mathbf{P}u) \in \mathrm{im}\,\mathbf{Q}$. However, at the same time we have $D(T\mathbf{P})u = \mathbf{P}u \in \mathrm{im}\,\mathbf{P}$. From $\mathbf{P}u \in \mathrm{im}\,\mathbf{P} \cap \mathrm{im}\,\mathbf{Q}$ it follows $\mathbf{P}u = 0$, and the assertion is proven. $\qquad\square$

8.2.2 Representation theorem

We already know that each component of a holomorphic function is harmonic. In the following we look for a qualitative description of harmonic functions by expressions as holomorphic functions. Thereby we also clarify the relation between these important function classes.

Theorem 8.10 (Representation theorem). *Let G be a domain in $\mathbb{R}^3$ with sufficiently smooth boundary Γ, $g \in W^{k+\frac{3}{2},2}(\Gamma)$ and $k \geq 0$, $k \in \mathbb{N}$. Every solution $u \in W^{k+2,2}(G)$ of the Dirichlet problem*

$$\Delta u = 0 \quad \text{in} \quad G, \tag{8.1}$$

$$u = g \quad \text{on} \quad \Gamma,$$

is determined by functions $\phi_1 \in W^{k+2,2}(G)$ and $\phi_2 \in W^{k+1,1}(G)$ of the form

$$u = \phi_1 + T\,\phi_2,$$

where ϕ_1 solves the boundary value problem

$$D\phi_1 = 0 \quad \text{in} \quad G, \tag{8.2}$$

$$\mathrm{tr}_\Gamma \phi_1 = \mathbf{P}_\Gamma g \quad \text{on} \quad \Gamma,$$

and ϕ_2 is a solution of the boundary value problem

$$D\phi_2 = 0 \quad \text{in} \quad G, \tag{8.3}$$

$$\mathrm{tr}_\Gamma T\phi_2 = \mathbf{Q}_\Gamma g \quad \text{on} \quad \Gamma.$$

The functions ϕ_i $(i = 1, 2)$ are uniquely defined.

Proof. Let $u \in W^{k+2,2}(G)$ be a solution of the boundary value problem (8.1). Then the Borel–Pompeiu formula leads to

$$u = F_\Gamma \mathrm{tr}_\Gamma u + TDu = F_\Gamma g + TDu.$$

We know that $F_\Gamma g \in \ker D$ and $Du \in \ker D$. Setting now $\phi_1 := F_\Gamma g$ and $\phi_2 := Du$ from $u \in W^{k+2,2}(G)$ it follows immediately that $\phi_2 \in W^{k+1,2}(G)$. Using the mapping properties of the Teodorescu transform we get $F_\Gamma g = u - TDu \in W^{k+2,2}(G)$, so that the regularity statement of the theorem is proven.

With $\phi_1 = F_\Gamma g$ we have $\mathrm{tr}_\Gamma \phi_1 = \mathrm{tr}_\Gamma F_\Gamma g = \mathbf{P}_\Gamma g$, and from $\mathrm{tr}_\Gamma T\phi_2 = \mathrm{tr}_\Gamma u - F_\Gamma g = g - \mathbf{P}_\Gamma g = \mathbf{Q}_\Gamma g$, we obtain that ϕ_1 and ϕ_2 really solve the boundary value problems (8.2) and (8.3).

It remains to prove the uniqueness which is done as usual indirectly. The assumption of two representations $u = \phi_1 + T\phi_2 = \psi_1 + T\psi_2$ leads to $0 = (\phi_1 - \psi_1) + T(\phi_2 - \psi_2)$. The action of D from the left gives directly $\phi_2 = \psi_2$; using this $\phi_1 = \psi_1$ follows. $\qquad\square$

8.3   Exercises

1. Prove that the Teodorescu transform

$$(T_G u)(x) = -\frac{1}{\sigma_n} \int\limits_G \mathcal{Q}_0(y - x)u(y)d\sigma_y$$

   exists everywhere in $\mathbb{R}^{n+1}$ under the condition $u \in L^1(G)$.

2. Prove that the integrals

$$\int\limits_{G\setminus B_\varepsilon(x)} \partial_{i,x}\mathcal{Q}_0(y - x)u(y)d\sigma_y \quad \text{and} \quad \int\limits_{|y-x|=\varepsilon} \mathcal{Q}_0(y - x)dy_i^* u(y),$$

   appearing in the differentiation of the Teodorescu transform in Theorem 8.2, converge for $\varepsilon \to 0$ locally uniformly in x.

3. Let the continuous function u be given in $G \subset \mathbb{R}^{n+1}$. Prove that

$$\partial_{i,x} \int\limits_G \frac{1}{|y - x|^{n-1}} u(y)d\sigma_y = \int\limits_G \partial_{i,x}\left(\frac{1}{|y - x|^{n-1}}\right)u(y)d\sigma_y.$$

   Is it possible to weaken the condition for u?

4. For $1 \le q < (n + 1)/n$ prove the inequality

$$\int\limits_G |\mathcal{Q}_0(x - y)|^q d\sigma_y \le \sigma_n \frac{(\operatorname{diam} G)^{n+1-qn}}{n + 1 - qn}.$$

Chapter IV

Series expansions and local behavior

9 Power series

9.1 Weierstraß' convergence theorems, power series

9.1.1 Convergence theorems according to Weierstraß

In this section we will use Cauchy's integral theorem and the integral formula to derive results concerning the convergence behavior of function sequences.

These theorems were discovered by KARL WEIERSTRASS, who played an important role in the development of complex analysis.

Karl Weierstraß

KARL THEODOR WILHELM WEIERSTRASS (1815–1897) left the University of Bonn, where he was preparing for a career in Prussian administration, to dedicate himself fully to mathematics. One year later he already received his certification as a teacher in Münster. After that he held teaching positions in Deutsch-Krone (West Prussia) and from 1848 until 1855 in Braunsberg (East Prussia). In 1854 the University of Königsberg awarded him an honorary doctor's degree for his paper "Zur Theorie der abelschen Funktionen" (On the theory of Abelian functions). In June 1856 he was offered a chair at the Industry Institute of Berlin, later the Berlin Institute of Technology, since October 1856 he held a professorship at the Berlin University and in 1867 he was elected fellow of the Berlin Academy of Sciences. In 1861 with the help of E.E. KUMMER he started the first research seminar for mathematics at a German University. His lectures attracted numerous students from all around the world and had a big influence on the development of mathematics, inside and outside of Germany. He worked in the field of complex analysis and developed — in competition with B. RIEMANN — a closed theory of complex analytic functions.

At first we recall the following proposition from real analysis which will serve as a basic tool in the proofs:

Proposition 9.1. *Let (f_m) be a sequence of continuous functions in a domain $G \subset \mathbb{R}^{n+1}$ that converges uniformly to a function f. Then the limit function f is continuous in G, and for a piecewise smooth manifold $\Gamma \subset G$ of dimension n and a function g continuous in G the following equation holds:*

$$\lim_{m \to \infty} \int_{\Gamma} f_m(x) dx^* g(x) = \int_{\Gamma} f(x) dx^* g(x).$$

The statement will remain true if $f(x, s)$ depends on a real or paravector-valued parameter s and if for $s \to s_0$ uniformly in x we have

$$f(x, s) \to f(x).$$

Since the functions can be split into its component functions, the proof is a direct consequence of the real case. Now we will prove the following theorem covering differentiation under the integral sign:

Theorem 9.2. *Let the sequence of differentiable functions (f_m) converge pointwise to a function f in a domain $G \subset \mathbb{R}^{n+1}$; let the partial derivatives $\partial_i f_m$ of f_m be continuous in G and converge in the local uniform sense to the functions g_i. Then f is partially differentiable in G and g_i is the corresponding derivative $\partial_i f$. A similar statement holds for a sequence of holomorphic functions f_m.*

Again the sequence of functions can be replaced by a parameter-depending function set, e.g., $f(x, s)$ with the parameter s.

Proof. We will use the preceding proposition; differentiation with respect to x_i at a point x yields

$$f_m(\ldots, x_i + h, \ldots) - f_m(\ldots, x_i, \ldots) = \int_0^h \partial_i f_m(\ldots, x_i + t, \ldots) dt.$$

Recall that it is sufficient to consider small neighborhoods of x — differentiability is a local property. Taking the limit $m \to \infty$ the sequences on the left-hand side converge and on the right-hand side we can interchange the limit and the integration signs because of the uniform convergence property. We obtain

$$f(\ldots, x_i + h, \ldots) - f(\ldots, x_i, \ldots) = \int_0^h g_i(\ldots, x_i + t, \ldots) dt.$$

Thus f is differentiable with respect to x_i and g_i is the corresponding derivative.

Let f_m be holomorphic in $\mathbb{C}$. In a small disc with center z_0 consider the integral

$$f_m(z) - f_m(z_0) = \int_{z_0}^z f'_m(\zeta) d\zeta.$$

This integral is uniquely defined since the f'_m are also holomorphic. In addition the Cauchy integral theorem yields that integrals over triangles with one corner at z_0 vanish. This still holds after taking the limit $m \to \infty$, which we can exchange with the integral sign:

$$f(z) - f(z_0) = \int_{z_0}^z g(\zeta) d\zeta.$$

We can differentiate as in the proof of Morera's theorem and finally obtain $f'(z) = g(z)$. $\square$

Now we want to prove a different type of theorem where only the functions have to converge uniformly:

Theorem 9.3. *Suppose that the functions f_m, $m \in \mathbb{N}$, are holomorphic in a domain $G \subset \mathbb{R}^{n+1}$ and the sequence (f_m) is convergent in G to a function f in the locally uniform sense. Then f is holomorphic in G.*

Proof. Let $x_0 \in G$ and the ball (disc) $\{|x - x_0| \leq \rho\}$ be a subset of G. We can then use Cauchy's integral formula,

$$f_m(x) = \int_{|y-x_0|=\rho} E_n(y - x) dy^* f_m(y).$$

Because of the uniform convergence we can exchange the limit $m \to \infty$ and the integration. So we obtain:

$$f(x) = \int_{|y-x_0|=\rho} E_n(y - x) dy^* f(y).$$

Thus Cauchy's integral formula holds locally and because of Corollary 7.30, f is holomorphic. $\qquad\square$

We just showed that the set of holomorphic functions is closed with respect to locally uniform convergence. The following corollary stresses this fact:

Corollary 9.4. *Suppose that the functions f_m are holomorphic in a domain $G \subset \mathbb{R}^{n+1}$ and the sequence (f_m) converges to f in G in the locally uniform sense. Then for every multiindex $\mathbf{k}$ the sequence of derivatives $\nabla^{\mathbf{k}} f_m$ converges to $\nabla^{\mathbf{k}} f$ in the locally uniform sense.*

It is recommended to prove this as Exercise 9.4.1.

9.1.2 Power series in $\mathbb{C}$

Power series constitute a third approach to complex analysis, which was mainly developed by Weierstraß. At the end of the 19th and the beginning of the 20th century the Riemannian and the Weierstrassian approaches to complex analysis were competing, with some mathematicians claiming one or the other to be the "proper" way. Fortunately this conflict has long since faded away and both viewpoints are sound and indispensable, the question about the better approach is superfluous. At first we will consider the complex case, in which the main properties are clearer to see.

Power series were introduced in Definition 4.17 as series of the type $\sum a_n z^n$. It was also shown that power series possess a disc of convergence $|z| < \rho$. Inside this disc they converge absolutely to a continuous limit function. The radius of convergence can be computed by

$$\frac{1}{\rho} = \limsup_{n \to \infty} |a_n|^{1/n}$$

with the identifications $\frac{1}{\infty} := 0$, $\frac{1}{0} := \infty$.

One could think of differentiating the power series *term by term* inside the disc of convergence:

$$f(z) = \sum_{n=0}^{\infty} a_n z^n \;\Rightarrow\; f'(z) = \sum_{n=1}^{\infty} n a_n z^{n-1} = \sum_{j=0}^{\infty} = (j+1) a_{j+1} z^j. \tag{9.1}$$

We shall now prove that this is indeed possible:

Theorem 9.5. *In the interior of the disc of convergence a power series can be differentiated termwise (in the complex sense). The termwise differentiated series is the derivative of the given series and has the same radius of convergence. Thus, power series are holomorphic inside their discs of convergence.*

Proof. At first we want to show that the termwise differentiated series has the same radius of convergence as the original series: The central series of formula (9.1) is multiplied by z and this has no influence on convergence issues. Then the coefficients are $b_n := n a_n$ and

$$|b_n|^{1/n} = n^{1/n} \, |a_n|^{1/n}.$$

Because of $n^{1/n} \to 1$ if $n \to \infty$ we obtain

$$\limsup_{n \to \infty} |b_n|^{1/n} = \limsup_{n \to \infty} |a_n|^{1/n} = \frac{1}{\rho}$$

with the original radius of convergence ρ. Now we have to show that $f'(z)$ is indeed the derivative of $f(z)$ inside the disc of convergence. The partial sum $s_n(z)$ of f and the term by term differentiated partial sums $s_n'(z)$ converge inside the disc of convergence in a locally uniform sense. Using Theorem 9.3, f is holomorphic inside the disc of convergence and the termwise differentiated series is the derivative of f:

$$f'(z) = \sum_{n=1}^{\infty} n a_n z^{n-1} = \sum_{k=0}^{\infty} (k+1) a_{k+1} z^k. \qquad \square$$

Obviously the differentiation process can be repeated and inverted — we will include this in two corollaries. The first one will also describe the relation of f to the coefficients of the power series.

Corollary 9.6. *A power series*

$$f(z) = \sum_{n=0}^{\infty} a_n (z - z_0)^n$$

is differentiable infinitely often inside its disc of convergence and for all $n \geq 0$,

$$a_n = \frac{1}{n!} f^{(n)}(z_0).$$

The a_n are called the Taylor coefficients of f at z_0.

Proof. Differentiating n times gives

$$f^{(n)}(z) = n! \sum_{j=n}^{\infty} \binom{j}{n} a_j (z - z_0)^{j-n}.$$

Substituting $z = z_0$ completes the proof. $\qquad\qquad\qquad\qquad\qquad\qquad\qquad\square$

Corollary 9.7 (Primitive of a power series). *Inside its disc of convergence a power series*

$$f(z) = \sum_{n=0}^{\infty} a_n (z - z_0)^n$$

has a holomorphic primitive, i.e., a function F satisfying $F' = f$, which is unique up to a constant:

$$F(z) = c + \sum_{n=0}^{\infty} \frac{1}{n+1} a_n (z - z_0)^{n+1} = c + \sum_{k=1}^{\infty} \frac{1}{k} a_{k-1} (z - z_0)^k.$$

The proof is obvious. Before we multiply or divide power series, we want to prove the uniqueness of the power series expansion:

Corollary 9.8 (Uniqueness of power series). *A holomorphic function f can admit only one power series expansion at a point.*

Proof. Assume that a function f admits two different power series expansions, e.g., at the origin,

$$f(z) = \sum_{n=0}^{\infty} a_n z^n = \sum_{n=0}^{\infty} b_n z^n.$$

Let n_0 denote the smallest index with $a_n \neq b_n$. It follows that

$$0 = z^{n_0} \sum_{n=n_0}^{\infty} (a_n - b_n) z^{n-n_0}.$$

For $z \neq 0$ we can divide by z^{n_0}. Considering $z \to 0$ leads to the contradiction $a_{n_0} = b_{n_0}$. $\qquad\qquad\qquad\square$

Using the multiplication formula for absolutely converging series we can derive a multiplication formula for power series:

Proposition 9.9 (Power series multiplication). *Suppose that*

$$f(z) = \sum_{n=0}^{\infty} a_n z^n, \quad g(z) = \sum_{n=0}^{\infty} b_n z^n$$

are two power series with radii of convergence ρ_f and ρ_g, respectively, with $0 < \rho_f \leq \rho_g$. Then for $|z| < \rho_f$ we have:

$$f(z)g(z) = \sum_{n=0}^{\infty} \left(\sum_{k=0}^{n} a_k b_{n-k} \right) z^n.$$

Proof. For $|z| < \rho_f$ we can use the multiplication formula for absolutely converging series

$$f(z)g(z) = \sum_{n=0}^{\infty} a_n z^n \sum_{k=0}^{\infty} b_k z^k = \sum_{n=0}^{\infty} \sum_{k=0}^{n} a_k z^k b_{n-k} z^{n-k}. \qquad \square$$

Dividing power series proves to be a little more difficult:

Proposition 9.10 (Power series division). *Suppose that*

$$f(z) = \sum_{n=0}^{\infty} a_n z^n, \quad g(z) = \sum_{n=0}^{\infty} b_n z^n$$

are two power series with radii of convergence ρ_f and ρ_g, respectively, with $0 < \rho_f \leq \rho_g$ and $g(z) \neq 0$ for $|z| < \rho$. Then for $|z| < \min\{\rho_f, \rho\}$ the equality

$$\frac{f(z)}{g(z)} = h(z) = \sum_{n=0}^{\infty} c_n z^n$$

is satisfied with the recursion formula

$$c_n = \frac{1}{b_0}\left(a_n - \sum_{k=0}^{n-1} c_k b_{n-k}\right)$$

for the coefficients c_n.

Proof. The preceding proposition gives

$$\sum_{n=0}^{\infty} a_n z^n = f(z) = h(z)g(z) = \sum_{n=0}^{\infty}\left(\sum_{k=0}^{n} c_k b_{n-k}\right) z^n.$$

The uniqueness of the coefficients of the power series allows coefficients comparison:

$$a_n = \sum_{k=0}^{n} c_k b_{n-k},$$

which leads us — taking into account $b_0 = g(0) \neq 0$ — to

$$c_n = \frac{1}{b_0}\left(a_n - \sum_{k=0}^{n-1} c_k b_{n-k}\right), \quad n \geq 0.$$

Now it is an easy task to determine the c_n; $c_0 = a_0/b_0$ is the first value. $\qquad \square$

9.1.3 Power series in $C\ell(n)$

In $C\ell(n)$ or $\mathbb{R}^{n+1}$ the definition of the term power series and its convergence is more complicated than in the complex case. Incidentally, functions admitting a power series expansion are called *analytic*. But this term is used under many different aspects, so we want to avoid it if possible.

Definition 9.11 (Power series in $C\ell(n)$). Generalizing Definition 4.17, a *power series in $C\ell(n)$ with variables in $\mathbb{R}^{n+1}$* is a series of the form

$$\sum P_k(x)$$

with homogeneous polynomials

$$P_k(x) = \sum_{|\mathbf{k}|=k} a_{\mathbf{k}} x^{\mathbf{k}}.$$

Here $\mathbf{k} = (k_0, \ldots, k_n)$ denotes a multiindex and $x^{\mathbf{k}} := x_0^{k_0} \cdots x_n^{k_n}$. This series is said to *converge absolutely* if the series

$$\sum \widetilde{P}_k(x)$$

with

$$\widetilde{P}_k(x) := \sum_{|\mathbf{k}|=k} |a_{\mathbf{k}}| |x^{\mathbf{k}}|$$

converges.

We should explicitly point out that different combinations of the terms of such a *multi-infinite series* are possible and lead to different definitions of convergence. It is difficult to determine the domain of convergence and absolute convergence in detail. Since $|x_i| \leq |x|$ is true, an estimation of the following kind holds:

$$\widetilde{P}_k(x) \leq |x|^k \left(\sum_{|\mathbf{k}|=k} |a_{\mathbf{k}}| \right) =: A_k |x|^k.$$

For the series $\sum_k A_k |x|^k$ we can use the results of the complex case. In this way one obtains a *ball of convergence* $\{|x| < \rho\}$ with

$$\frac{1}{\rho} = \limsup_{k \to \infty} |A_k|^{1/k}.$$

We now show the following result:

Theorem 9.12. *Inside its ball of convergence a power series can be differentiated term by term and the differentiated series is the derivative of the original series.*

Proof. The degree of a homogeneous polynomial in x is decreased by one (or becomes zero) after differentiation with respect to x_i. Hereby the factors $k_i \leq |\mathbf{k}| = k$ appear, so that one obtains

$$\left| \widetilde{\partial_i P_k}(x) \right| \leq k |x|^{k-1} A_k.$$

Similarly to the complex case the convergence for $|x| < \rho$ follows.

Thus the partial sums of the original series $s_m(x)$ and the $\partial_i s_m(x)$ of the termwise differentiated series converge uniformly to the functions

$$f(x) = \sum_{k=0}^{\infty} P_k(x) \text{ resp. } g(x) = \sum_{k=0}^{\infty} \partial_i P_k(x).$$

Using Theorem 9.2 we have $g(x) = \partial_i f(x)$. $\qquad\qquad\qquad\qquad\qquad\qquad$ □

We shall soon encounter examples of these multi-infinite series.

Theorem 9.13. *The power series expansion of a function f (which admits such an expansion) is uniquely determined.*

Proof. Suppose that P_k and Q_k denote the first homogeneous polynomials that differ in the two series for a function f. Since $f(0)$ determines the values P_0 and Q_0, we have $k > 0$. A contradiction follows after differentiating with respect to a proper x_i. $\qquad$ □

9.2 Taylor and Laurent series in $\mathbb{C}$

9.2.1 Taylor series

We shall again start with the complex case since it is relatively simple and leads to important results. We shall again draw conclusions from Cauchy's integral formula and show that a function is holomorphic if and only if it admits a power series expansion. Particularly Weierstraß preferred to use power series for his considerations.

Theorem 9.14 (Taylor expansion). *Let f be holomorphic in the disc $B_R(z_0) = \{|z - z_0| < R\}$. Then f admits a converging power series expansion*

$$f(z) = \sum_{n=0}^{\infty} a_n (z - z_0)^n$$

with

$$a_n = \frac{1}{n!} f^{(n)}(z_0) = \frac{1}{2\pi i} \int_{|\zeta - z_0| = \rho} \frac{f(\zeta)}{(\zeta - z_0)^{n+1}} d\zeta,$$

where ρ is arbitrary with $0 < \rho < R$.

Thus a function is holomorphic in a domain G if and only if it allows a converging power series expansion at any point $z_0 \in G$. Consequently power series include all holomorphic functions. Nevertheless for different matters, different representations can be useful. We already deduced that the geometric series for $1/(1-z)$ at $z_0 = 0$ is only a viable representation in the unit circle, although the rational representation describes the complete behavior in $\hat{\mathbb{C}}$ without problems.

Proof. We use Cauchy's integral formula in $B_{\rho+\varepsilon}$ with $\rho < \rho + \varepsilon < R$,

$$f(z) = \frac{1}{2\pi i} \int_{|\zeta-z_0|=\rho+\varepsilon} \frac{f(\zeta)}{\zeta - z} d\zeta,$$

and expand the Cauchy kernel $1/(\zeta - z)$ into a converging geometric series with respect to $(z - z_0)$:

$$\frac{1}{\zeta - z} = \frac{1}{\zeta - z_0} \frac{1}{1 - \frac{z-z_0}{\zeta-z_0}} = \sum_{n=0}^{\infty} \frac{(z - z_0)^n}{(\zeta - z_0)^{n+1}}.$$

In the closed disc $\overline{B}_\rho(z_0)$ this series converges uniformly. Thus it can be integrated term by term. We obtain

$$f(z) = \sum_{n=0}^{\infty} a_n (z - z_0)^n, \quad a_n = \frac{1}{2\pi i} \int_{|\zeta-z_0|=\rho+\varepsilon} \frac{f(\zeta)}{(\zeta - z_0)^{n+1}} d\zeta.$$

Cauchy's integral theorem yields that the integral is independent of the radius, so that we can substitute $|\zeta - z_0| = \rho$. The radius ρ can be chosen arbitrarily close to R, so that the representation is correct in the whole disc $B_R(z_0)$. The calculation of the coefficients by using derivatives of f has already been discussed in Corollary 9.6. $\qquad\square$

We will continue with an important theorem for holomorphic functions. Such a function is already determined uniquely by its values on a sequence of points.

Theorem 9.15 (Uniqueness theorem). *Let f be holomorphic in a domain G and let us suppose that (z_n) is a sequence of points from G with $z_n \to z_0 \in G$. If $f(z_n) = 0$ for all z_n, then $f = 0$.*

We can also reformulate the statement as follows:

The zeros of a holomorphic function, different from the zero function, are isolated.

Or alternatively:

Suppose that the two functions f and g are holomorphic in G and identical on a sequence of points with an accumulation point in G. Then the functions are identical.

Proof. Using Corollary 9.8 a holomorphic function admits a uniquely determined power series expansion. Since f is continuous in G, $f(z_0) = 0$ holds and we obtain for the power series in z_0,

$$f(z) = \sum_{n=0}^{\infty} a_n (z - z_0)^n$$

that $a_0 = 0$. Now we define

$$f_1(z) := \frac{f(z) - f(z_0)}{z - z_0} = \sum_{n=0}^{\infty} a_{n+1} (z - z_0)^n.$$

Since $f(z_n) = f(z_0) = 0$, this function also fulfils the requirements, we thus deduce $a_1 = 0$. Induction with respect to n yields the assertion inside the convergence disc of the power series. If there was another point z^* in G with $f(z^*) \neq 0$, we would use the already well known proving principle: We connect z_0 and z^* by a polygonal curve. Let : z^{**} be the "last" point between z_0 and z^* satisfying $f(z^{**}) = 0$; it has points z' with $f(z') \neq 0$ in any small neighborhood of z^{**}. We repeat the same at z^{**}. Compactness arguments lead to a contradiction. □

The following definition is a direct consequence of this theorem:

Definition 9.16 (Holomorphic continuation). Let G_1 and G_2 be two domains with non-empty intersection, let f and g be holomorphic in G_1 and in G_2, respectively. If $f(z) = g(z)$ in $G_1 \cap G_2$ (or on a sequence of points with an accumulation point in $G_1 \cap G_2$), g is called a *holomorphic continuation* of f to G_2 (and f a holomorphic continuation of g to G_1).

Because of the uniqueness theorem there can only be one holomorphic continuation of that type. Thus the term is defined properly. Recall the geometric series in the unit circle: its holomorphic continuation to $\mathbb{C} \setminus \{1\}$ is $1/(1 - z)$.

One can imagine the holomorphic continuation in the following way: Consider a function f defined on a disc $B_{R_0}(z_0)$ and find a point z_1, where the disc of convergence of the Taylor series in z_1, e.g., $B_{R_1}(z_1)$, extends $B_{R_0}(z_0)$. By repeating this process we obtain a sequence of discs; the original function f is continued by using this process. Each disc and the corresponding holomorphic function is called a *function element*. If the centers of such a *chain of discs* are connected by a polygonal curve (which does not leave the discs), f is said to possess a holomorphic continuation *along a polygonal curve*.

Unfortunately it is not clear that one obtains the original function by continuing f along a closed polygonal curve. Therefore the following result will be useful (the term simply connected domain is explained in Definition A.2.19):

Theorem 9.17 (Holomorphic continuation to simply connected domains). *Consider a holomorphic function element f in $B_{R_0}(z_0) \subset G$, where G denotes a simply connected domain and $z_0 \in G$. If f admits a holomorphic continuation to G, the obtained function will be uniquely defined and holomorphic in G.*

Considering the example $\sqrt{z}$, it is easy to see that in a not simply connected domain the theorem does not necessarily hold. By continuing $\sqrt{z}$ along the unit circle starting from $z_0 = 1$ around the origin, we obtain a function that differs from the original one by a minus sign.

Proof. We will only sketch the proof and leave the details to the reader.

Let us assume the function obtained by holomorphic continuation not to be uniquely defined. Then there is a closed polygonal curve, so that we obtain a function not equal to the original one after holomorphic continuation along the path. We can assume the polygonal curve to be a Jordan curve: If the path contained a self-intersection we could split it into a closed polygonal curve and a second curve with one self-intersection less, and so on. Holomorphic continuation along at least one of them must not be unique. Otherwise the continuation along the original path would be unique, too.

Now we cover the plane with a sufficiently close grid parallel to the axes. For continuity reasons the original polygon can now be substituted by a polygonal curve parallel to the axes. The small squares of the grid which are adjacent to the boundary of the polygonal curve should of course be inside the discs on which our function elements used for holomorphic continuation are defined. The interior of the closed polygonal curve parallel to the axes cannot contain exterior points of G: Either there are no exterior points of G or there is exactly one, so we can assume it to be at $z = \infty$, or there are more exterior points, so we can assume one to be at $z = \infty$. In the last case no point can belong to the interior of the polygonal curve, because in this case the boundary points covered by the open sets of the interior or the exterior of the polygonal curve would be part of the exterior. But these open sets would be disjoint, so that the boundary would not be connected and G would not be simply connected.

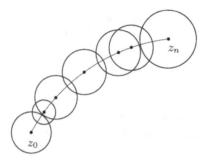

Figure 9.1

Now we can get rid of the interior squares of the polygonal curve one by one until there is only one remaining. Since each of these squares is a subset of one of the function elements' discs, the continuation along its boundary cannot lead to a different function. Thus this property has to be true also for the original polygonal curve – the holomorphic continuation of a function element results in a uniquely defined function f in G.   □

We shall draw one final conclusion from the Taylor series expansion:

Corollary 9.18. *If a holomorphic function has the power series expansion*

$$f(z) = \sum_{n=0}^{\infty} a_n (z - z_0)^n$$

with radius of convergence $\rho > 0$, it cannot be holomorphic at all boundary points of the disc, i.e., of the circle $|z - z_0| = \rho$.

Proof. If the function f was holomorphic at every point ζ of the circle $|z - z_0| = \rho$, it would also be — by definition — holomorphic in a small (open) disc around ζ. The circle is a compact set covered by these discs, so that we can select a finite number of these discs still covering the circle. This finite number of open discs overlap each other and cover a small annulus $\rho - \varepsilon \le |z - z_0| \le \rho + \varepsilon$. Thus one could apply Cauchy's integral formula to a bigger disc $B_{\rho+\varepsilon}(z_0)$ and one would obtain a converging power series in this disc. This is a contradiction with the assumption of the theorem.   □

9.2.2 Laurent series

We can extend the Taylor series principle to functions holomorphic only in an annulus, the behavior inside the interior disc is arbitrary. We have the following theorem.

Theorem 9.19 (Laurent series). *Suppose that f is holomorphic in the annulus $G = \{z : R_1 < |z - z_0| < R_2\}$, in which $R_1 = 0$ and/or $R_2 = \infty$ is possible. Then f can be represented by a Laurent series*

$$f(z) = \sum_{n=-\infty}^{\infty} a_n (z - z_0)^n$$

in the annulus. The coefficients are given by

$$a_n = \frac{1}{2\pi i} \int_{|\zeta - z_0| = \rho} \frac{f(\zeta)}{(\zeta - z_0)^{n+1}} d\zeta,$$

where ρ is arbitrary with $R_1 < \rho < R_2$. The convergence of the two-sided infinite series is defined by the convergence of the two series with positive and negative indices, respectively.

The series of negative powers

$$\sum_{n=-\infty}^{-1} a_n (z - z_0)^n$$

is called the principal part *of the function f in z_0, the series of positive powers is called the* Taylor part.

If the function f is holomorphic in the whole disc $|z - z_0| < R_2$, all coefficients with negative indices vanish and the coefficient formula coincides with the one of the Taylor series. Hence Laurent series can be interpreted as extensions of Taylor series.

Proof. In the smaller annulus $R_1 < \rho_1 < |z - z_0| < \rho_2 < R_2$ the function f is also holomorphic on the boundary. So we can use Cauchy's integral formula

$$f(z) = \frac{1}{2\pi i} \int_{|\zeta - z_0| = \rho_2} \frac{f(\zeta)}{\zeta - z} d\zeta - \frac{1}{2\pi i} \int_{|\zeta - z_0| = \rho_1} \frac{f(\zeta)}{\zeta - z} d\zeta.$$

In order to have the annular region on the left we have to integrate over the inner circle clockwise, so that a minus sign shows up in front of the second integral. Analogously with the proof of the Taylor series we expand $1/(\zeta - z)$ into a geometric series: This leads to

$$\frac{1}{\zeta - z} = \frac{1}{\zeta - z_0} \frac{1}{1 - \frac{z - z_0}{\zeta - z_0}} = \sum_{n=0}^{\infty} \frac{(z - z_0)^n}{(\zeta - z_0)^{n+1}},$$

for the outer circle and to

$$\frac{1}{\zeta - z} = -\frac{1}{z - z_0} \frac{1}{1 - \frac{\zeta - z_0}{z - z_0}} = -\sum_{n=0}^{\infty} \frac{(\zeta - z_0)^n}{(z - z_0)^{n+1}}$$

for the inner one. Both series converge in a little smaller, resp. greater, disc uniformly, so that we can substitute integration and summation:

$$
f(z) = \sum_{n=0}^{\infty} (z - z_0)^n \frac{1}{2\pi i} \int_{|\zeta - z_0|\rho_2} \frac{f(\zeta)}{(\zeta - z_0)^{n+1}} d\zeta
$$
$$
+ \sum_{n=0}^{\infty} (z - z_0)^{-(n+1)} \frac{1}{2\pi i} \int_{|\zeta - z_0| = \rho_1} f(\zeta)(\zeta - z_0)^n d\zeta.
$$

The first sum represents the Taylor part of the Laurent series, the second one can be put in its final form by substituting the summation index $n =: -m - 1$:

$$
f(z) = \sum_{n=-\infty}^{\infty} a_n (z - z_0)^n, \quad a_n = \frac{1}{2\pi i} \int_{|\zeta - z_0| = \rho} \frac{f(\zeta)}{(\zeta - z_0)^{n+1}} d\zeta.
$$

Because of Cauchy's integral theorem the circle of integration with radius between R_1 and R_2 is arbitrary. Since the circles of integration can be chosen sufficiently close to the boundaries of the annulus, the formula remains true in the whole annular domain $R_1 < |z - z_0| < R_2$.   □

Example 9.20. a) In example 4.19 b we introduced a series which now is recognized as being the Laurent series in $1 < |z| < \infty$:

$$
\frac{1}{1 - z} = -\sum_{n=1}^{\infty} z^{-n}.
$$

b) The function

$$
f(z) = \frac{1}{z}
$$

is also a (very simple) example of a Laurent series in the annulus $0 < |z| < \infty$.

c) Let the function f be holomorphic in the annulus $1 - \varepsilon < |z| < 1 + \varepsilon$ where it possesses the Laurent series expansion

$$
f(z) = \sum_{n=-\infty}^{\infty} a_n z^n.
$$

On the boundary of the unit disc we have $z = e^{i\varphi}$ and the Laurent series transforms into

$$
f(e^{i\varphi}) = \sum_{n=-\infty}^{\infty} a_n e^{in\varphi}
$$

with coefficients

$$a_n = \frac{1}{2\pi i} \int_{|\zeta|=1} \frac{f(\zeta)}{\zeta^{n+1}} d\zeta = \frac{1}{2\pi} \int_{-\pi}^{\pi} f(e^{i\varphi}) e^{-in\varphi} d\varphi.$$

This turns out to be the *Fourier series* of the periodic complex function $f(e^{i\varphi})$, which can be decomposed into sine and cosine series in the usual way.

As for the Taylor series the function f cannot be holomorphic on the whole boundary of the annulus of convergence. Otherwise our annulus could be expanded.

9.3 Taylor and Laurent series in $C\ell(n)$

9.3.1 Taylor series

In this section we want to generalize the results of the complex plane to higher dimensional spaces. The Fueter polynomials introduced and discussed in Section 6.2 will serve as generalizations of the usual powers z^n in the complex plane.

Let us recall *Euler's formula* for *homogeneous functions*:

Suppose that the function f is homogeneous of degree k in $\mathbb{R}^{n+1}$. Then we have

$$x \cdot \nabla f(x) = k f(x).$$

This is proved by differentiating $f(tx) = t^k f(x)$ with respect to t and substituting $t = 1$. Now we are able to prove a first statement concerning the representation by Fueter polynomials:

Proposition 9.21. *Every homogeneous holomorphic polynomial of degree k can be written as a $C\ell(n)$-linear combination of Fueter polynomials:*

$$P(x) = \sum_{|\mathbf{k}|=k} \frac{1}{\mathbf{k}!} \mathcal{P}_{\mathbf{k}}(x)(\nabla^k P)(0), \quad \mathbf{k} = (0, k_1, ..., k_n).$$

Proof. Let P be a left-holomorphic homogeneous polynomial of degree k. We obtain the formulae

$$\partial_0 P(x) + \sum_{i=1}^{n} e_i \partial_i P(x) = 0$$

and

$$x_0 \partial_0 P(x) + \sum_{i=1}^{n} x_i \partial_i P(x) = k P(x),$$

since P satisfies the Cauchy–Riemann equations and Euler's formula. These yield

$$k P(x) = \sum_{i=1}^{n} (x_i - x_0 e_i) \partial_i P(x) = \sum_{i=1}^{n} z_i \partial_i P(x).$$

This can also be regarded as a justification for introducing the variables $z_i = x_i - x_0 e_i$ in Section 5.2 when we defined holomorphic functions. Because of Corollary 7.29 each derivative $\partial_i P$ is holomorphic. $\partial_i P$ is a homogeneous polynomial of degree $k - 1$. After k steps one gets

$$k! P(x) = \sum_{i_1, \ldots, i_k = 1}^{n} z_{i_1} \ldots z_{i_k} \partial_{i_1} \ldots \partial_{i_k} P(x).$$

Now we combine all derivatives belonging to a multiindex $\mathbf{k}$. The order of differentiation does not play any role in computing the derivatives, but in computing the z_i. Every distribution of the z_i occurs only once in the k components. These are also to be permuted with each other in the $\mathcal{P}_{\mathbf{k}}$. We get a factor $\mathbf{k}!$ in the denominator and obtain

$$
\begin{aligned}
P(x) &= \sum_{|\mathbf{k}|=k} \frac{1}{k! \, \mathbf{k}!} \sum_{\sigma \in \mathrm{perm}(k)} z_{i_{\sigma(1)}} \ldots z_{i_{\sigma(k)}} \nabla^{\mathbf{k}} P(x) \\
&= \sum_{|\mathbf{k}|=k} \frac{1}{k!} \mathcal{P}_{\mathbf{k}}(x) \nabla^{\mathbf{k}} P(x) \quad \left(\mathbf{k} = (0, k_1, \ldots, k_n) \right).
\end{aligned}
$$

This relation can be seen as the justification for introducing Fueter polynomials. Since we obtain constants by differentiating a polynomial of degree k exactly k times, the desired representation follows:

$$P(x) = \sum_{|\mathbf{k}|=k} \mathcal{P}_{\mathbf{k}}(x) a_{\mathbf{k}}, \quad a_{\mathbf{k}} = \frac{(\nabla^{\mathbf{k}} P)(0)}{\mathbf{k}!}. \qquad \square$$

Let us now consider the *Taylor series* in $C\ell(n)$. For left-holomorphic functions it is of the form

$$\sum_{k=0}^{\infty} \sum_{|\mathbf{k}|=k} \mathcal{P}_{\mathbf{k}}(x) a_{\mathbf{k}},$$

for right-holomorphic functions the $a_{\mathbf{k}}$ are on the left side. If this series converges absolutely, we can exchange summation and differentiation. Thus a series of the given form is holomorphic in case it converges absolutely. This condition can be checked by using the estimates of Corollary 6.5.

The further considerations require the introduction of another kind of special polynomials.

Definition 9.22. The function

$$C_k^{\mu}(s) := \sum_{m=\left[\frac{k}{2}\right]}^{k} \binom{-\mu}{m} \binom{m}{2m-k} (-2s)^{2m-k}$$

is called a *Gegenbauer polynomial* (due to LEOPOLD BERNHARD GEGENBAUER (1849–1903)).

Special cases are $C_0^{\mu}(s) = 1$, $C_1^{\mu}(s) = \binom{-\mu}{1} \binom{1}{1} (-2s)^1 = 2\mu s$.

Proposition 9.23. (i) *The polynomial* $C_k^\mu(s) = \sum_{j=0}^{\left[\frac{k}{2}\right]} \binom{-\mu}{k-j}\binom{k-j}{k-2j}(-2s)^{k-2j}$

contains only powers of s with degrees $k, k-2, \ldots$.

(ii) *The Taylor series expansion for* $(1 - 2st + t^2)^{-\mu}$ *with respect to t converges for* $|t| < 1$ *and* $-1 \le s \le 1$ *in the locally uniform sense.*

Proof. (i) With $k - m =: j$ we get $0 \le j \le \frac{k}{2}$, the result follows. $\left[\frac{k}{2}\right]$ denotes the largest integer smaller than or equal to $\frac{k}{2}$.

(ii) For small t the binomial series (see Exercise 9.4.4) for $(1 - 2st + t^2)^{-\mu}$ converges in the locally uniform sense. For $0 < t < 1, 0 \le s \le 1$ we obtain the strict inequality

$$1 \ge t(t - 2s) = (t - s)^2 - s^2 > -1,$$

so that the series converges for these s and $0 < t < 1$ in the locally uniform sense. According to Exercise 9.4.2 the radius of convergence is 1. In a similar way one can prove the convergence for negative s with negative t, so that for all s the series in t has the radius of convergence 1. $\qquad\square$

Theorem 9.24 (Taylor series). *Let the function f be left-holomorphic for* $|x| < R$ *in* $\mathbb{R}^{n+1}$, *where then it can be expanded into the converging Taylor series*

$$f(x) = \sum_{k=0}^{\infty} \sum_{|\mathbf{k}|=k} \mathcal{P}_{\mathbf{k}}(x)a_{\mathbf{k}}, \quad \mathbf{k} = (0, k_1, \ldots, k_n)$$

with

$$a_{\mathbf{k}} = \frac{\nabla^{\mathbf{k}} f(0)}{\mathbf{k}!} = \frac{1}{\sigma_n} \int_{|y|=\rho} \mathcal{Q}_{\mathbf{k}}(y)dy^* f(y).$$

ρ *is arbitrary with* $0 < \rho < R$, *the* $\mathcal{Q}_{\mathbf{k}}$ *are given in Definition 7.26 as derivatives of* $\mathcal{Q}_0(x) = \sigma_n E_n(x)$. *For right-holomorphic functions we need to interchange the terms* $a_{\mathbf{k}}$ *with* $\mathcal{P}_{\mathbf{k}}$ *and the* $\mathcal{Q}_{\mathbf{k}}$ *with* f.

Proof. As usual Cauchy's integral formula is the starting point. For $|x| < \rho < R$ we have

$$f(x) = \int_{|y|=\rho} E_n(y - x)dy^* f(y).$$

The Cauchy kernel $E_n(y - x)$ or rather $\mathcal{Q}_0(y - x) = \sigma_n E_n(y - x)$ needs to be expanded into a power series: We choose ρ arbitrarily with $0 < \rho < R$, so that the expansion converges in $|x| < R$ due to Cauchy's integral theorem. We calculate as follows:

$$\begin{aligned} \partial_x \frac{1}{|y - x|^{n-1}} &= -\frac{n-1}{2} \frac{1}{|y-x|^{n+1}} \partial_x |y - x|^2 \\ &= (n-1)\frac{y-x}{|y-x|^{n+1}} = (n-1)\mathcal{Q}_0(y - x), \end{aligned}$$

i.e., we find the Cauchy kernel with a factor. Consequently it is sufficient to determine the series expansion of $|y - x|^{-(n-1)}$. Defining $x =: \omega_x |x|$, $y =: \omega_y |y|$ we get

$$|y - x|^2 = |y|^2 \left(1 - \frac{\overline{y}\,x}{|y|^2}\right)\left(1 - \frac{\overline{x}\,y}{|y|^2}\right) = |y|^2 \left(1 - 2(\omega_x \cdot \omega_y)\frac{|x|}{|y|} + \frac{|x|^2}{|y|^2}\right).$$

Finally setting

$$t := \frac{|x|}{|y|}, \quad s := \omega_x \cdot \omega_y$$

we get $0 \le t < 1$, $-1 \le s \le 1$ and with $\mu := (n-1)/2$ we obtain

$$\frac{1}{|y - x|^{n-1}} = \frac{1}{|y|^{n-1}} \frac{1}{(1 - 2st + t^2)^\mu}.$$

The binomial series expansion yields

$$\frac{1}{(1 - 2st + t^2)^\mu} = \sum_{m=0}^\infty \binom{-\mu}{m}(-2st + t^2)^m.$$

This series converges for $|-2st+t^2| < 1$ in the locally uniform sense. Thus it also converges for sufficiently small t in s and t in the locally uniform sense. Since $(-2st + t^2)^m$ leads only to a finite number of terms, we can reorder by powers of t:

$$\frac{1}{(1 - 2st + t^2)^\mu} = \sum_{m=0}^\infty \sum_{j=0}^m \binom{-\mu}{m}\binom{m}{j}(-2st)^j t^{2(m-j)}.$$

The factor preceding t^k can only be non-zero if $2m - j = k$ or $2m \ge k \ge m$, meaning for those m satisfying $\left[\frac{k}{2}\right] \le m \le k$ that

$$\frac{1}{(1 - 2st + t^2)^\mu} = \sum_{k=0}^\infty t^k \sum_{m \ge \frac{k}{2}}^k \binom{-\mu}{m}\binom{m}{2m - k}(-2s)^{2m-k}.$$

Since the series converges in the locally uniform sense, we can differentiate term by term. Taking into account $\partial_x C_0^\mu(s) = 0$ this leads to

$$\mathcal{Q}_0(y - x) = \frac{\overline{y - x}}{|y - x|^{n+1}} = \sum_{k=1}^\infty \frac{1}{n-1} \partial_x C_k^\mu(\omega_x \cdot \omega_y)\frac{|x|^k}{|y|^{n-1+k}}.$$

By defining

$$\frac{1}{n-1}\partial_x C_{k+1}^\mu(\omega_x \cdot \omega_y)\frac{|x|^{k+1}}{|y|^{n+k}} =: P_k(x, y),$$

one gets finally

$$\mathcal{Q}_0(y - x) = \frac{\overline{y - x}}{|y - x|^{n+1}} = \sum_{k=0}^\infty P_k(x, y).$$

Due to the preceding Proposition 9.23 $P_k(x, y)$ contains only powers of x having the form

$$(\omega_x \cdot \omega_y)^{k+1-2j}|x|^{k+1},$$

so it is a polynomial in x with terms (see Proposition 9.23)

$$\partial_x \left((x \cdot \omega_y)^{k+1-2j} |x|^{2j} \right).$$

These are homogenous polynomials in x of degree k. Since the series for $\mathcal{Q}_0(y-x)$ is bilaterally holomorphic, this property must be fulfilled by the terms. These are homogeneous polynomials of degree k, so that differentiating by $\overline{\partial}_x$ no terms cancel each other. Using Proposition 9.21 the $P_k(x,y)$ can be decomposed into linear combinations of Fueter polynomials $\mathcal{P}_\mathbf{k}$ with $|\mathbf{k}| = k$:

Definition 9.25. The functions $\tilde{Q}_\mathbf{k}(y)$ are defined by

$$\sum_{|\mathbf{k}|=k} \mathcal{P}_\mathbf{k}(x)\tilde{Q}_\mathbf{k}(y) = \sum_{|\mathbf{k}|=k} \tilde{Q}_\mathbf{k}(y)\mathcal{P}_\mathbf{k}(x) = P_k(x,y).$$

We have

$$\tilde{Q}_\mathbf{k}(y) = \frac{1}{(n-1)\mathbf{k}!} \nabla_x^\mathbf{k} \partial_x \left[C_{k+1}^\mu (\omega_x \cdot \omega_y) \frac{|x|^{k+1}}{|y|^{n+k}} \right]\Bigg|_{x=0}.$$

The $Q_\mathbf{k}$ are thus homogeneous of degree $-(n+k)$ in y. In the series expansion of $\mathcal{Q}_0$ no terms cancel the functions $P_k(x,y)$ by differentiation with respect to y, so that $\tilde{Q}_\mathbf{k}$ is bilaterally holomorphic. Now we want to show their equality to

$$\mathcal{Q}_\mathbf{k}(y) = \frac{(-1)^{|\mathbf{k}|}}{\mathbf{k}!} \nabla^\mathbf{k} \mathcal{Q}_0(y),$$

introduced in Definition 7.26: For arbitrary $\mathbf{j}$ we get

$$\mathcal{Q}_\mathbf{j}(y-x) = \frac{(-1)^{|\mathbf{j}|}}{\mathbf{j}!} \nabla_y^\mathbf{j} \mathcal{Q}_0(y-x) = \frac{1}{\mathbf{j}!} \nabla_x^\mathbf{j} \mathcal{Q}_0(y-x) = \sum_{k=0}^{\infty} \frac{1}{\mathbf{j}!} \nabla_x^\mathbf{j} P_k(x,y)$$

$$= \sum_{k=0}^{\infty} \sum_{|\mathbf{k}|=k} \frac{1}{\mathbf{j}!} \nabla_x^\mathbf{j} \mathcal{P}_\mathbf{k}(x)\tilde{Q}_\mathbf{k}(y).$$

Setting $x = 0$, $\nabla^\mathbf{j}\mathcal{P}_\mathbf{k}(0) = \mathbf{j}! \, \delta_{\mathbf{j}\,\mathbf{k}}$ (see Exercise 9.4.5) yields

$$\mathcal{Q}_\mathbf{j}(y) = \frac{1}{\mathbf{j}!} \nabla_x^\mathbf{j} \mathcal{P}_\mathbf{j}(0)\tilde{Q}_\mathbf{j}(y) = \tilde{Q}_\mathbf{j}(y).$$

Consequently $\tilde{Q}_\mathbf{k} = \mathcal{Q}_\mathbf{k}$ holds and for $|x| < \rho < R$ we obtain the series expansion for f,

$$f(x) = \frac{1}{\sigma_n} \int\limits_{|y|=\rho} \left(\sum_{k=0}^{\infty} \sum_{|\mathbf{k}|=k} \mathcal{P}_\mathbf{k}(x)\mathcal{Q}_\mathbf{k}(y) \right) dy^* f(y).$$

Since it converges in the locally uniform sense we can integrate term by term and get

$$f(x) = \sum_{k=0}^{\infty} \sum_{|\mathbf{k}|=k} \mathcal{P}_\mathbf{k}(x)a_\mathbf{k}$$

with

$$a_\mathbf{k} = \frac{\nabla^\mathbf{k} f(0)}{\mathbf{k}!} = \frac{1}{\sigma_n} \int\limits_{|y|=\rho} \mathcal{Q}_\mathbf{k}(y)dy^* f(y).$$

This is completely similar to the expansion in $\mathbb{C}$; the negative powers of z are replaced by $\mathcal{Q}_\mathbf{k}$. $\square$

In analogy to the complex case we want to prove a uniqueness theorem. But it will have a different form than in $\mathbb{C}$ since the zeros of a holomorphic function in $\mathbb{R}^{n+1}$ may not be isolated. Consider the simple example $f(x) = z_1$, where the $(n-1)$-dimensional plane $x_0 = x_1 = 0$ equals the set of zeros of f. By combining the z_i every lower-dimensional plane can be obtained as the set of zeros. We will start with a proposition used in the proof of the uniqueness theorem in $C\ell(n)$:

Proposition 9.26. *Suppose that f is holomorphic in a domain $G \subset \mathbb{R}^{n+1}$ and the equality $f(x) = 0$ is valid for all x from a ball $\{|x - x_0| < \rho\} \subset G$. Then $f = 0$ in G.*

Proof. x_0 can be connected to every $x^* \in G$ by a polygonal curve Π in G. As a compact set the polygonal curve has a distance $\delta > 0$ to ∂G. Thus at an arbitrary point of Π the Taylor series converges at least in a ball with radius δ. On Π we choose a point x_1 with $\delta/2 < |x_1 - x_0| < \delta$. In a small neighborhood of x_1 we have $f(x) = 0$, so that all Taylor coefficients in x_1 equal zero and $f(x) = 0$ for $|x - x_1| < \delta$. Repeating this process a finite number of times will lead us finally to x^*, so that $f(x^*) = 0$ holds. Thus we have $f = 0$ in G. $\square$

A possible uniqueness theorem reads as follows:

Theorem 9.27 (Identity theorem). *If a function f is holomorphic in a domain $G \subset \mathbb{R}^{n+1}$ and equals zero on an n-dimensional smooth manifold $M \subset G$, the equation $f = 0$ is satisfied in G.*

The dimension has to be at least n, as the example $f(x) = z_1$ shows. The manifold can be composed from an arbitrarily small piece, the focus lies on the dimension n. A more uniqueness-theorem-like formulation is the following:

Two functions holomorphic in G coinciding on an n-dimensional smooth manifold in G are identical.

Obviously this theorem remains true in $\mathbb{C}$, but since the zeroes are isolated the statement can be more general.

Proof. Let x^* denote an arbitrary point in M and $x(t_1, \ldots, t_n) =: x(t)$ be a parametrization of M in a neighborhood of x^* with $x(0) = x^*$. Since $f(x(t)) = 0$ holds for all t we have

$$\sum_{i=0}^{n} \frac{\partial x_i}{\partial t_j} \partial_i f(x^*) = 0, \quad j = 1, \ldots, n.$$

These are the tangential derivatives of f in the direction of the manifold. For our manifold we assume the matrix

$$\left(\frac{\partial x_i}{\partial t_j} \right)$$

to be of rank n. Thus in the mentioned system of equations we can express the $\partial_i f(x^*)$ by the $\partial_0 f(x^*)$ with the help of suitable real-valued a_i,

$$\partial_i f(x^*) = a_i \partial_0 f(x^*), \quad i = 1, \ldots, n.$$

Substituting into the Cauchy–Riemann equations yields

$$\left(\sum_{i=1}^{n} e_i a_i + 1\right) \partial_0 f(x^*) = 0.$$

The term in brackets is obviously not zero, consequently $\partial_0 f(x^*) = 0$ holds. This yields $\partial_i f(x^*) = 0$, $i = 0, \ldots, n$. Taking into account the correctness for all $x^* \in M$ all first-order derivatives of f vanish in M. The $\partial_i f$ are also holomorphic, so that we can expand our considerations to higher-order derivatives of f with the use of induction. Thus the coefficients of the Taylor series in x^* are zero, so that the function f equals zero in the ball of convergence of the Taylor series. The preceding proposition gives $f = 0$ in G. $\square$

9.3.2 Laurent series

After considering the Taylor series for functions holomorphic in a ball, we want to consider functions holomorphic in a *ballshell*, i.e., the multi-dimensional analogue of an annulus. A corresponding Laurent expansion will be derived proceeding similarly to the complex case.

Theorem 9.28 (Laurent series). *Let f be left-holomorphic in a ballshell domain $G = \{r < |x| < R\}$ with $0 \le r < R \le \infty$. Then f admits the following Laurent series expansion in G,*

$$f(x) = f_1(x) + f_2(x) = \sum_{k=0}^{\infty} \sum_{|\mathbf{k}|=k} \mathcal{P}_\mathbf{k}(x) a_\mathbf{k} + \sum_{k=0}^{\infty} \sum_{|\mathbf{k}|=k} \mathcal{Q}_\mathbf{k}(x) b_\mathbf{k}$$

with

$$a_\mathbf{k} = \frac{1}{\sigma_n} \int\limits_{|y|=\rho} \mathcal{Q}_\mathbf{k}(y) dy^* f(y),$$

$$b_\mathbf{k} = \frac{1}{\sigma_n} \int\limits_{|y|=\rho} \mathcal{P}_\mathbf{k}(y) dy^* f(y).$$

ρ is arbitrary with $r < \rho < R$; the series converge uniformly in any closed sub-ballshell. Furthermore f_1 is holomorphic in $B_R(0)$ and f_2 is holomorphic in $\mathbb{R}^{n+1} \setminus \overline{B_r(0)}$ with

$$\lim_{|x| \to \infty} f_2(x) = 0.$$

f_1 is called the Taylor part and f_2 is called the principal part of the Laurent series.

Proof. In the ballshell $r < r' < |x| < R' < R$ Cauchy's integral formula yields

$$f(x) = \frac{1}{\sigma_n} \int\limits_{|y|=R'} \mathcal{Q}_0(y - x) dy^* f(y) - \frac{1}{\sigma_n} \int\limits_{|y|=r'} \mathcal{Q}_0(y - x) dy^* f(y).$$

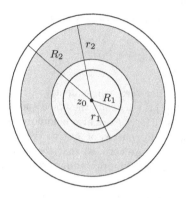

Figure 9.2

The sign in front of the second integral appears reversed due to the orientation of the boundary of the ballshell. The first integral, denoted by f_1, possesses the already known Taylor expansion

$$f_1(x) = \sum_{k=0}^{\infty} \sum_{|\mathbf{k}|=k} \mathcal{P}_{\mathbf{k}}(x) a_{\mathbf{k}}$$

with

$$a_{\mathbf{k}} = \frac{1}{\sigma_n} \int\limits_{|y|=\rho} \mathcal{Q}_{\mathbf{k}}(y) dy^* f(y).$$

ρ is an arbitrary number between r and R in view of Cauchy's integral theorem. Now we will focus on

$$f_2(x) = -\frac{1}{\sigma_n} \int\limits_{|y|=r'} \mathcal{Q}_0(y-x) dy^* f(y).$$

Similarly to the Taylor series

$$-\mathcal{Q}_0(y-x) = \partial_y \frac{1}{n-1} \frac{1}{|y-x|^{n-1}}$$

has to be expanded into a series. Since $|x| > |y|$, we have

$$\frac{1}{|y-x|^{n-1}} = \frac{1}{|x|^{n-1}} \frac{1}{(1-2st+t^2)^\mu}$$

with $t = |y|/|x|$ and $s = (\omega_x \cdot \omega_y)$, $\omega_x = x/|x|$, $\omega_y = y/|y|$, $\mu = \frac{n-1}{2}$. One gets

$$\mathcal{Q}_0(y-x) = -\sum_{k=1}^{\infty} \frac{1}{n-1} \partial_y C_{k+1}^\mu(\omega_x \cdot \omega_y) \frac{|y|^{k+1}}{|x|^{n-1+k}},$$

and with

$$R_k(x,y) := \frac{1}{n-1} \partial_y C_{k+1}^\mu(\omega_x \cdot \omega_y) \frac{|y|^{k+1}}{|x|^{n+k}}$$

we have

$$\mathcal{Q}_0(y - x) = - \sum_{k=0}^{\infty} R_k(x, y).$$

The R_k have the same properties as the P_k from the proof of the Taylor series after exchanging x and y. We obtain

$$R_k(x, y) = \sum_{|\mathbf{k}|=k} \mathcal{P}_\mathbf{k}(y)\mathcal{Q}_\mathbf{k}(x) = \sum_{|\mathbf{k}|=k} \mathcal{Q}_\mathbf{k}(x)\mathcal{P}_\mathbf{k}(y)$$

and hence we have

$$f_2(x) = \sum_{k=0}^{\infty} \sum_{|\mathbf{k}|=k} \mathcal{Q}_\mathbf{k}(x)b_\mathbf{k}$$

with

$$b_\mathbf{k} = \frac{1}{\sigma_n} \int\limits_{|y|=\rho} \mathcal{P}_\mathbf{k}(y)dy^* f(y).$$

Again ρ is arbitrary with $r < \rho < R$ (Cauchy's integral theorem). The statement $f_2(x) \to 0$ for $|x| \to \infty$ is left as Exercise 9.4.7. □

As a conclusion we want to point out some orthogonality relations of the $\mathcal{P}_\mathbf{k}$ and the $\mathcal{Q}_\mathbf{k}$ which correspond to similar properties of the $z^{\pm k}$ in $\mathbb{C}$:

Proposition 9.29. *For any positive ρ the Fueter polynomials $\mathcal{P}_\mathbf{k}$ and the $\mathcal{Q}_\mathbf{k}$ satisfy for arbitrary $\mathbf{j}$ and $\mathbf{k}$ the following orthogonality relations:*

$$\frac{1}{\sigma_n} \int\limits_{|x|=\rho} \mathcal{Q}_\mathbf{k}(x)dx^*\mathcal{P}_\mathbf{j}(x) = \delta_{\mathbf{jk}},$$

$$\frac{1}{\sigma_n} \int\limits_{|x|=\rho} \mathcal{P}_\mathbf{j}(x)dx^*\mathcal{Q}_\mathbf{k}(x) = \delta_{\mathbf{jk}},$$

$$\frac{1}{\sigma_n} \int\limits_{|x|=\rho} \mathcal{P}_\mathbf{j}(x)dx^*\mathcal{P}_\mathbf{k}(x) = 0,$$

$$\frac{1}{\sigma_n} \int\limits_{|x|=\rho} \mathcal{Q}_\mathbf{j}(x)dx^*\mathcal{Q}_\mathbf{k}(x) = 0.$$

Proof. The first line follows from the Taylor expansion of a polynomial $\mathcal{P}_\mathbf{j}$. All Taylor coefficients of $\mathcal{P}_\mathbf{j}$ have to vanish for $\mathbf{k} \neq \mathbf{j}$ and $a_\mathbf{j} = 1$. The second line is the corresponding statement for right-holomorphic functions since $\mathcal{Q}_\mathbf{k}$ and f are exchanged. The third line represents the $b_\mathbf{k}$ of the Laurent expansion of $\mathcal{P}_\mathbf{j}$, they have to vanish. Finally the fourth line shows the coefficients $a_\mathbf{k}$ of the Laurent expansion of $\mathcal{Q}_\mathbf{j}$, they have to equal zero, too. □

The Fueter polynomials are recognized to be left- and right-holomorphic generalizations of the positive powers z^k, but constructing higher-dimensional generalizations of the negative powers proves to be a more difficult task. We chose the derivatives of the Cauchy kernel, but other approaches are possible. The powers x^{-n} introduced by H. Malonek in 1990 [101] are not left-holomorphic. In [14] higher-dimensional left-holomorphic powers with negative exponents are computed similarly to our way. It was also shown that the values of $\mathcal{Q}_\mathbf{k}$ do not leave the paravector space. After tedious computation in [73] a recursion formula was obtained. It was also possible for R. S. Kraußhar to get an estimation for the $\mathcal{Q}_\mathbf{k}$ (see Proposition 7.27).

9.4   Exercises

1. Suppose that a sequence of functions (f_m) is holomorphic in a domain $G \subset \mathbb{R}^{n+1}$ and converges to a function f in G in the local uniform sense. Show that for every multi-index $\mathbf{k}$ the sequence of derivatives $(\nabla^\mathbf{k} f_m)$ converges to $\nabla^\mathbf{k} f$ in the local uniform sense (see Corollary 7.28).

2. Assume a power series (at the origin) converges at a fixed point z_0. Show that it does converge absolutely for all $|z| < |z_0|$. Explain the connection with the radius of convergence.

3. Obtain the two possible power series for the function $f(z) = \frac{1}{z-3i}$ at the origin.

4. Show that

$$(1+z)^a = \sum_{j=1}^{\infty} \binom{a}{j} z^j \quad (a \in \mathbb{C})$$

is a Taylor series. Use this to show

$$(1+z)^a = \exp(a \log(1+z))$$

with the principal argument of the logarithm that equals zero for $z = 1$ (see Section 11.4 and Definition 11.8). Evaluate the radius of convergence.

5. Show
   a)
$$\nabla^\mathbf{j} \mathcal{P}_\mathbf{k} = \frac{k!}{(\mathbf{k} - \mathbf{j})!} \mathcal{P}_{\mathbf{k}-\mathbf{j}} \ \text{ for } \ \mathbf{j} \le \mathbf{k},$$
   b)
$$\nabla^\mathbf{j} \mathcal{P}_\mathbf{k} = \mathbf{k}! \, \delta_{\mathbf{k}\mathbf{j}} \ \text{ for } \ |\mathbf{k}| = |\mathbf{j}|$$
   for Fueter polynomials with multiindices $\mathbf{j}$ and $\mathbf{k}$. $\mathbf{j} \le \mathbf{k}$ denotes $j_\mu \le k_\mu$ for all $\mu \in \{0, \ldots, n\}$.

6. Show the following uniqueness theorem: Suppose that a function is holomorphic in a domain $G \subset \mathbb{R}^{n+1}$ and different from the zero function. Then it does not have a *zero of infinite degree*. A function f is said to have a zero of infinite degree in x_0 if

$$\lim_{|x-x_0| \to 0} \frac{|f(x)|}{|x - x_0|^k} = 0$$

holds for all $k \in \mathbb{N}$. (Hint: Taylor coefficients.)

7. Show that the principal part $f_2(x)$ of a Laurent series converges to 0 for $|x| \to \infty$.

10　Orthogonal expansions in $\mathbb{H}$

The approximation, resp. the representation, of quadratically integrable functions by Fourier series requires the knowledge of complete or closed function systems in the Hilbert space as well as of the general representation of functionals.

10.1　Complete $\mathbb{H}$-holomorphic function systems

As the quaternions are non-commutative we have to define a new closedness and completeness. First we prove the following version of the Hahn–Banach theorem:

Theorem 10.1. *Let X be a right vector space with norm over $\mathbb{H}$ and let $X_0 \subset X$ be a closed subspace. Let f be a bounded, right-linear, $\mathbb{H}$-valued functional on X_0. Then there is an $\mathbb{H}$-valued right-linear extension F of f to X which is bounded on X.*

This theorem was proven first in 1938 by G. A. Suchumlinov [151]. A corresponding theorem for locally convex spaces with values in a real Clifford algebra was proven in 1982 [14].

Proof. The idea of the proof is that an arbitrary $\mathbb{H}$-right-linear functional always has the form

$$f(x) = f_0(x) - \sum_{i=1}^{3} f_0(xe_i)e_i.$$

To prove this we start with the representation

$$f(x) = f_0(x) + f_1(x)e_1 + f_2(x)e_2 + f_3(x)e_3.$$

Since $f(xe_i) = f(x)e_i$ $(i = 0, 1, 2, 3)$ we have the relations

$$f_1(x) = -f_0(xe_1), f_2(x) = -f_0(xe_2), f_3(x) = -f_0(xe_3).$$

Let now $X_{0,\mathbb{R}} \subset X_0$ be a real subspace and f_0 the restriction of f to $X_{0,\mathbb{R}}$. Following the classical Hahn–Banach theorem, f_0 can be extended to an $\mathbb{R}$-linear bounded functional F_0 on X. The needed $\mathbb{H}$-valued functional F is then given by

$$F(x) := F_0(x) - F_0(xe_1)e_1 - F_0(xe_2)e_2 - F_0(xe_3)e_3.$$

Finally we just need to prove the corresponding properties: We obviously have

$$|F(x)| \leq 4\|F_0\|\|x\|,$$

i.e., F is bounded. The additivity follows from the additivity of F_0. The proof of right homogeneity has to be done individually for each component. We now calculate the norm of F: we choose a point $x^* \in X$ so that $\|x^*\| = 1$ and $\|F\| \leq |F(x^*)| + \varepsilon$. Since $\|F\| \neq 0$ we can assume that $F(x^*) \neq 0$. We define $\Theta := F(x^*)/|F(x^*)|$ and $x^{**} = x^*\overline{\Theta}$ and get

$$F(x^{**}) = F\left(x^* \frac{\overline{F(x^*)}}{|F(x^*)|}\right) = F(x^*)\frac{\overline{F(x^*)}}{|F(x^*)|} = |F(x^*)|.$$

From Sc $F(x) = F_0(x)$ it follows $F_0(x^{**}) = F(x^{**})$.

So we have

$$F_0(x^{**}) \leq \|F_0\| \|x^{**}\| = \|F_0\|$$

and

$$\|F\| \leq \|F_0\| + \varepsilon.$$

Since $\varepsilon > 0$ is arbitrary,

$$\|F\| = \|F_0\|$$

holds. We still have to prove $\|F\| = \|f\|$ but this follows from $\|F_0\| = \|f_0\| = \|f\|$. Therefore F is a norm-preserving $\mathbb{H}$-right-linear extension of f. $\square$

Definition 10.2. Let X be a right vector space with norm over $\mathbb{H}$. A set $\{x^{(i)}\} \subset X$, $i \in \mathbb{N}$, of elements of X is called $\mathbb{H}$-*complete* if and only if for all $x \in X$ and for an arbitrary $\varepsilon > 0$ there is a finite right-linear combination $R_\varepsilon(x)$ of the $\{x^{(i)}\}$ so that

$$\|x - R_\varepsilon(x)\| < \varepsilon$$

holds. The set $\{x^{(i)}\}$ is called $\mathbb{H}$-*closed in* X, if for every bounded $\mathbb{H}$-right-linear functional F on X with values in $\mathbb{H}$ always from $F(x^{(i)}) = 0$ $(i \in \mathbb{N})$ it follows that $F = 0$.

Analogously with the real analysis a set is $\mathbb{H}$-closed if and only if it is $\mathbb{H}$-complete. The proof of this statement is left to the reader, with a strong recommendation to become familiar with the special right-linear structure of the spaces considered (see Exercise 10.4.2).

Theorem 10.3. *Let F be an $\mathbb{H}$-right-linear functional over $L^p(G)$. It then has the representation*

$$F(u) = \int\limits_G \overline{f(x)} u(x) d\sigma$$

with $f \in L^q(G)$ and $\frac{1}{p} + \frac{1}{q} = 1$.

For the proof for real Clifford algebras we refer to [14], in which the proof is related to the classical proof for complex-valued linear functionals over $L^p(G)$. For $p = 2$ we simply write $F(u) := (f, u)$.

A first and simple complete system of holomorphic functions can be given in $\mathbb{R}^3$ with the help of "shifted" fundamental solutions of the Dirac operator:

Theorem 10.4. *Let G, G_ε be bounded domains in $\mathbb{R}^3$ whose boundaries Γ and Γ_1 are at least C^2-surfaces. Moreover let $\overline{G} \subset G_\varepsilon$ and let $\{x^{(i)}\}$ be a dense subset of Γ_1. The system $\{\phi_i\}_{i=1}^{\infty}$ with $\phi_i(x) = (x - x^{(i)})/|x - x^{(i)}|^3$ is then $\mathbb{H}$-complete in $L^2(G) \cap \ker D$, where $D = \sum\limits_{i=1}^{3} e_i \partial_i$ is the Dirac operator.*

Proof. The proof is given by contradiction. We assume there is a function $u \in L^2(G) \cap$ ker D, different from the zero function, with $(\phi_i, u) = 0$ $(i = 1, 2, \ldots)$. It then follows that for the Teodorescu transform the relation

$$(T_G u)(x^{(i)}) = -\frac{1}{\sigma_3} \int\limits_G \frac{\overline{x - x^{(i)}}}{|x - x^{(i)}|^3} u(x) d\sigma_x = 0$$

holds on the set $\{x^{(i)}\}$, dense in Γ_1. Since $T_G u$ is continuous in $\mathbb{R}^3 \backslash G$ it follows $\text{tr}_{\Gamma_1} T_G u = 0$. Further, $T_G u$ is holomorphic in $\mathbb{R}^3 \backslash \overline{G}$, thus from Proposition 8.1 it follows that $\lim\limits_{|x| \to \infty} (T_G u)(x) = 0$. From theorem 7.14 we have $(T_G u)(x) = 0$ in the exterior of Γ_1. Being a holomorphic function in $\mathbb{R}^3 \setminus \overline{G}$, the Teodorescu transform $T_G u$ has to be zero in the exterior of G by the identity Theorem 9.27. Because of the continuity it holds also for the boundary values on the boundary Γ of G,

$$\text{tr}_\Gamma T_G u = 0.$$

From proposition 8.9 we have that $u \in \text{im } \mathbf{Q}$. We had assumed that $u \in \text{ker } D = \text{im } \mathbf{P}$. Since $\text{im } \mathbf{P} \cap \text{im } \mathbf{Q} = \{0\}$ it follows that $u = 0$, this is the contradiction. $\square$

The theorem just proved ensures that we can approximate with arbitrary accuracy in the L^2-norm the solutions of the Dirac equation by special simple solutions. Our function systems are arbitrarily differentiable, but they unfortunately are not orthogonal. It is easy to see that an orthogonalization would destroy the simple structure, not to speak of the difficulties coming from the numerical instability. In the next subsection we shall try to construct polynomial systems which do not possess these disadvantages.

10.1.1 Polynomial systems

It is this section's goal to provide orthogonal polynomial systems for approximation purposes. Following the results we proved in Section 9.3, the shifted fundamental solutions $\mathcal{Q}_0(x - a)$ of the generalized Cauchy–Riemann operator in $C\ell(n)$ can be regarded as analogues of the functions $1/(z - a)$ in $\mathbb{C}$. Our statements on approximation and completeness are then related to the rational approximation in complex analysis and lead us to the theorems of Walsh and Runge.

Up to now we have not clarified completely the question about analogues in $C\ell(n)$ of the positive powers of z in complex analysis. The results of the development into Taylor series of left-holomorphic functions in Theorem 9.24 give us the first ideas. We had proved that a function f, left-holomorphic in the unit ball $B_{n+1} := B_1(0)$ in $\mathbb{R}^{n+1}$, can be developed into a convergent Taylor series

$$f(x) = \sum_{k=0}^{\infty} \sum_{|\mathbf{k}|=k} \mathcal{P}_{\mathbf{k}}(x) a_{\mathbf{k}}$$

with

$$a_{\mathbf{k}} = \frac{\nabla^{\mathbf{k}} f(0)}{\mathbf{k}!} = \frac{1}{\sigma_n} \int\limits_{|y|=\rho} \mathcal{Q}_{\mathbf{k}}(y) dy^* f(y).$$

From this development we can conclude that f is locally approximable in the sense of locally uniform convergence. But a global completeness statement, e.g., in $L^2(B_{n+1})$, does not follow from this theorem.

Let us first have a look at the complex case. We assume a holomorphic function f to be in $L^2(B_2)$. This function then possesses a locally uniformly convergent Taylor series in B_2,

$$f(z) = \sum_{k=0}^{\infty} \frac{f^{(k)}(0)}{k!} z^k.$$

The functions z^k constitute an orthogonal system in the space $L^2(B_2)$, and after a normalization an orthonormal system in the form $\sqrt{\frac{k+1}{\pi}} z^k$. A function f can be developed into a Fourier series relative to this orthonormal system,

$$f(z) \mapsto \sum_{k=0}^{\infty} (f, \sqrt{\frac{k+1}{\pi}} z^k) \sqrt{\frac{k+1}{\pi}} z^k = \sum_{k=0}^{\infty} (f, z^k) \frac{k+1}{\pi} z^k.$$

With the usual assumption that there is a function $g \in L^2(B_2)$, which is orthogonal to all z^k, it follows from the above considerations that the system $\left\{ \sqrt{\frac{k+1}{\pi}} z^k \right\}_{k \in \mathbb{N}}$ is complete.

The reason for this easy result is the fact that the functions used for the Taylor series in $\mathbb{C}$ are "spontaneously" orthogonal, though orthogonality played no role for the Taylor series and we only used differentiation properties. Before we seek convenient higher dimensional generalizations of the powers z^k we will first state the properties of the complex powers as functions of the real variables x and y to get a frame of reference:

- The functions $(x + iy)^k$ are $\mathbb{R}$-homogeneous (holomorphic) polynomials.

- $\mathbb{R}$-homogeneous holomorphic polynomials of different order are orthogonal.

- We have $z^k \in \ker \partial_{\bar{z}}^{k+1} \setminus \ker \partial_{\bar{z}}^k$.

- Moreover $\partial_z z^k = k z^{k-1}$, $\int z^k = \frac{1}{k+1} z^{k+1}$ holds, i.e., differentiation and integration of basis functions deliver again basis functions. These properties ensure that the derivative of a power series is again a power series, and that the derivative of a Fourier series is at least formally a Fourier series.

Unfortunately these properties are not self-evident in the higher dimensional case. If we restrict ourselves for better understanding to functions $f : \mathbb{R}^3 \mapsto \mathbb{H}$, for the degree k, we then have just $k+1$ linearly independent homogeneous $\mathbb{H}$-holomorphic polynomials. This satisfies indeed the requirements of a Taylor expansion as there are also $k + 1$ partial derivatives of degree k,

$$\partial_1^{k_1} \partial_2^{k_2}, \qquad k_1 + k_2 = k.$$

But the requirements of a Fourier development are not satisfied as the $k+1$ Fueter polynomials $\mathcal{P}_{\mathbf{k}}$ from Definition 6.1 are not in any case orthogonal. But we can show that homogeneous holomorphic polynomials of different degree are indeed orthogonal, so we have to orthogonalize "only" in the subspaces $\mathcal{H}_k^+$ of the $\mathbb{H}$-holomorphic homogeneous polynomials of degree k. But as desired the derivatives $\partial_i \mathcal{P}_{\mathbf{k}}$ of the Fueter polynomials of degree $|\mathbf{k}| = k$ from Theorem 6.2 are again Fueter polynomials

$$\partial_i \mathcal{P}_{\mathbf{k}} = k_i\, \mathcal{P}_{\mathbf{k}-\boldsymbol{\varepsilon}_i}$$

with the vector $\boldsymbol{\varepsilon}_i$ having a 1 in place i and zeros otherwise.

Another attempt would be to use instead of the complex variable z, in $\mathbb{R}^3$ the quaternionic variable $x = x_0 + x_1 e_1 + x_2 e_2$, and in $\mathbb{R}^4$ the variable $x = x_0 + x_1 e_1 + x_2 e_2 + x_3 e_3$. Unfortunately both variables are not $\mathbb{H}$-holomorphic. If we assume symmetry between the "imaginary" variables x_1, x_2, x_3 the next candidates would then be the $\mathbb{H}$-holomorphic variables $x = x_0 + \frac{1}{2}(x_1 e_1 + x_2 e_2)$, resp. $x = x_0 + \frac{1}{3}(x_1 e_1 + x_2 e_2 + x_3 e_3)$. Such variables have been studied, e.g., in [100] (see Exercise 10.4.4).

In view of these problems we start with the Fueter variables $z_i = x_i - x_0 e_i$. These variables are equal to the complex variable z up to an isomorphism. A simple calculation shows that the positive powers z_i^k are $\mathbb{H}$-holomorphic as well. Away from the zeros of the z_i also the negative integer powers are $\mathbb{H}$-holomorphic (see Exercise 10.4.5). Such variables, which are called *totally analytic* by Delanghe in [31], have been studied in Remark 6.7, where we pointed out the fact that the $\mathbb{H}$-holomorphic variable

$$z = \sum_{k=0}^{3} a_k x_k, \quad a_k \in \mathbb{H},$$

is totally analytic if and only if $a_i a_j = a_j a_i$, $i, j = 0, 1, 2, 3$. If we now examine the functions $u(x) = \sum_{i=0}^{3} u_i(x) a_i$ we have the product rule for the Cauchy–Riemann operator $\overline{\partial}$ in its classical form $\overline{\partial}(uv) = (\overline{\partial}u)v + u(\overline{\partial}v)$. Restricting ourselves to functions with range $\mathrm{span}\{a_i\}$, we then can carry over almost the complete classical function theory. For a better understanding of this fact one should show (see Exercise 10.4. 6) that the condition $a_i a_j = a_j a_i$, $i, j = 0, 1, 2, 3$, means that the range of $z(x)$ is only a two-dimensional plane.

However this simple construction has significance in principle. We have defined in Remark 6.7 the following interpolation polynomial:

$$L_k u(x) := \sum_{j=1}^{k} \Big(\prod_{i \neq j} [(z(x) - z(b_i))(z(b_j) - z(b_i))^{-1}] \Big) u_j. \tag{10.1}$$

This polynomial generalizes the classical Lagrange interpolation polynomial. The significance is that we can now show that for k arbitrarily given different points

b_j and k given values u_j at these points, there is always a holomorphic function which takes the given values at the given points. For $\mathbb{H}$-holomorphic functions this statement is in no case obvious. It would still be desirable to show that always a totally analytic variable could be defined such that for the given points also the condition $z(b_i) \neq z(b_j)$ for $i \neq j$ is satisfied.

The above Lagrange interpolation polynomial is not appropriate to interpolate general $\mathbb{H}$-holomorphic functions, since in view of the restricted range of $z(x)$ no complete polynomial system is given.

10.1.2 Inner and outer spherical functions

Before we seek complete $\mathbb{H}$-holomorphic systems of possibly orthonormal polynomials, we will state some properties of Fueter polynomials which are already known. A question is whether the orthogonality relations from Proposition 9.29 can be useful in our context. We start with a definition:

Definition 10.5. (i) Let $\mathcal{H}_k^+$ be the space of holomorphic, paravector-valued, homogeneous polynomials of degree k. An arbitrary element P_k of this space is called an *inner spherical polynomial* of degree k.

(ii) Let $\mathcal{H}_k^-$ be the space of holomorphic, paravector-valued, homogeneous functions in $\mathbb{R}_0^{n+1} = \mathbb{R}^{n+1} \setminus \{0\}$ of degree $-(k+n)$. An arbitrary element Q_k of this space is called an *outer spherical function* of degree k.

(iii) Let $\mathcal{H}_k = \mathcal{H}_k^+ \cup \mathcal{H}_k^-$; the unions over all k are denoted by $\mathcal{H}^+, \mathcal{H}^-$, resp. $\mathcal{H}$.

In Proposition 9.21 we have already proved that these P_k are spanned by the Fueter polynomials $\mathcal{P}_\mathbf{k}$. In view of the Laurent expansion the Q_k are spanned by the $\mathcal{Q}_\mathbf{k}$. In Proposition 7.27 we have shown that the $\mathcal{Q}_\mathbf{k}$ are paravector-valued, this has to be proved also for the $\mathcal{P}_\mathbf{k}$:

Proposition 10.6. *The Fueter polynomials* $\mathcal{P}_\mathbf{k}$ *are paravector-valued, they span the space of the inner spherical polynomials. The* $\mathcal{Q}_\mathbf{k}$ *are also paravector-valued and span the space of the outer spherical functions.*

Proof. We need only to prove that the Fueter polynomials are paravector-valued. We use induction on the degree $k = |\mathbf{k}|$. For $k = 0$ or $k = 1$ we have only the functions 1, resp. z_i, which are paravector-valued. To prove the assertion for k we use Theorem 6.2 (ii),

$$\partial_j \mathcal{P}_\mathbf{k} = k_j \mathcal{P}_{\mathbf{k}-\boldsymbol{\varepsilon}_j},$$

with k_j from $\mathbf{k}$ and $\boldsymbol{\varepsilon}_j$ which has a 1 in place j and otherwise only zeros. If $\mathcal{P}_\mathbf{k}$ had a term not in $\mathbb{R}^{n+1}$, then this term upon suitable differentiation would not be zero, but it would not change the property not to be in $\mathbb{R}^{n+1}$, as the latter is not affected by convenient differentiation. This would be a contradiction with the property stating that the differentiated polynomial is paravector-valued. $\qquad\square$

It remains to consider whether the relations of Proposition 9.29 are orthogonality relations relative to $L^2(B_{n+1})$. To investigate this we define a mapping between $\mathcal{H}^+$ and $\mathcal{H}^-$: For $x \in \mathbb{R}^{n+1}$, $x \neq 0$, and f holomorphic in $\mathbb{R}_0^{n+1}$ we let

$$\mathbf{I}f(x) := \mathcal{Q}_0(x)f\left(x^{-1}\right).$$

The reader is left to prove in Exercise 10.4.7 that $\mathbf{I}f$ is a bijective map between $\mathcal{H}^+$ and $\mathcal{H}^-$. We now can show:

Proposition 10.7. *The spaces $\mathcal{H}^+$ and $\mathcal{H}^-$ are orthogonal in $L^2(S^n)$.*

Proof. Let P_j and Q_k be arbitrary elements from both spaces. Then there is a P_k with

$$Q_k(x) = \mathbf{I}P_k(x) = \mathcal{Q}_0(x)P_k\left(\frac{\overline{x}}{|x|^2}\right).$$

We are interested in the scalar product in $L^2(S^n)$, where $\mathcal{Q}_0(y) = \overline{y}$ and therefore

$$\int\limits_{S^n} \overline{Q_k(y)}P_j(y)|do_1| = \int\limits_{S^n} \overline{\overline{y}P_k(\overline{y})}P_j(y)|do_1| = \int\limits_{S^n} \overline{P_k(\overline{y})}dy^* P_j(y) = 0.$$

The last statement follows from the right-holomorphy of $\overline{P_k(\overline{y})}$ and the Cauchy integral theorem. Moreover for the Fueter polynomials we have $\overline{\mathcal{P}_{\mathbf{k}}(\overline{x})} = \mathcal{P}_{\mathbf{k}}(x)$, as only the order of the products of the z_i is inverted. But as we sum over all orders nothing changes. The individual z_i's will be conjugated twice, they do not change either. We then see that the last integral is just the third one given in Proposition 9.29.   □

We see that the statements about Fueter polynomials proved up to now are not very helpful for the orthogonality problem. We therefore have to go deeper into this question, and we restrict ourselves finally to holomorphic functions in $C\ell(2)$, which map $\mathbb{R}^3$ into $\mathbb{H}$. We then have only the two variables z_1 and z_2, and we are able to write down all formulas explicitly to make things clearer. We look for convenient $\mathbb{H}$-holomorphic polynomials which constitute a complete system of functions. Malonek in [100] and [101] has done this with symmetric polynomials which differ only slightly from the Fueter polynomials.

In this case we have z_1 and z_2 as the Fueter polynomials of degree 1; those of degree 2 are:

$$\mathcal{P}_{(2,0)}(x) = z_1^2, \ \mathcal{P}_{(1,1)}(x) = \frac{1}{2}(z_1z_2 + z_2z_1), \ \mathcal{P}_{(0,2)}(x) = z_2^2.$$

We already mentioned that the Fueter polynomials constitute a basis for the holomorphic polynomials in $\mathcal{H}_k^+$. To get a statement about the completeness of the system of the Fueter polynomials we need an assertion regarding the decomposition or the splitting of the space $L^2 \cap \ker \overline{\partial}$.

Theorem 10.8. *We have*

$$L^2(B_3) \cap \ker \overline{\partial} = \bigoplus_{k=0}^{\infty} \mathcal{H}_k^+.$$

Proof. The orthogonality of the subspaces of different degrees follows from the orthogonality of the homogeneous harmonic polynomials of different degree, which we have yet to show. If we now assume that the asserted splitting does not exhaust $L^2 \cap \ker \overline{\partial}$, then there must be a function f which is orthogonal to all subspaces. This function can be developed into a Taylor series in a neighborhood of the origin. In some ball this Taylor series converges uniformly, and in a smaller ball it also converges in the L^2-norm. Let this ball have the radius $\varepsilon < 1$; we then study the function $f^* \in L^2(B_{2/\varepsilon}(0)) \cap \ker \overline{\partial}$ with $f^*(x) = f(\frac{1}{2}\varepsilon x)$. On the one hand the Taylor series of f^* converges in $L^2(B_3)$. On the other one we have from the orthogonality to the subspaces that

$$\int\limits_{B_3} \overline{f\left(\frac{\varepsilon x}{2}\right)} \mathcal{P}_{\mathbf{k}}(x) d\sigma = c \int\limits_{B_3} \overline{f\left(\frac{\varepsilon x}{2}\right)} \mathcal{P}_{\mathbf{k}}\left(\frac{\varepsilon x}{2}\right) d\sigma = 0,$$

thus with a suitable c the function f^* would be orthogonal to all subspaces $\mathcal{H}_k^+$. From both statements it follows that $f^* = f = 0$. $\qquad\square$

We have thus also proved the needed completeness theorem:

Theorem 10.9. *The Fueter polynomials are complete in* $L^2(B_3) \cap \ker \overline{\partial}$.

The above results show that we are able to develop holomorphic functions into a Taylor series using Fueter polynomials, but we do not know precisely the radius of convergence of the power series. In the sense of best approximation any $\mathbb{H}$-holomorphic function, which is quadratically integrable in a ball, can be approximated with arbitrary precision in that ball. Unfortunately this is not sufficient for practical purposes, as the best approximation by arbitrary complete systems may not be stable numerically. For theoretical questions one often uses the Schmidt orthogonalization procedure, but this is also not numerically stable for such a system. So, we have to deal with the explicit construction of complete orthogonal systems. The system of Fueter polynomials, so important for the Taylor development, is not appropriate for such a construction. To show the non-orthogonality of the Fueter polynomials we cite a result of the article [18] (without the extensive proof), from which we also take some of the following constructions:

Theorem 10.10. *For* $|\nu| = |\mu| = n$ *we have*

$$\left(\mathcal{P}_{(\nu_1,\nu_2)}, \mathcal{P}_{(\mu_1,\mu_2)}\right)_{L^2(S^2)} = a,$$

where $a = a_0 + a_1 e_1 + a_2 e_2 + a_3 e_3 \in \mathbb{H}$ *as well as*

(i) $a_0 = a_1 = a_2 = 0$ *and* $a_3 \neq 0$ *for* $|\nu_1 - \mu_1|$ *and* $|\nu_2 - \mu_2|$ *odd,*

(ii) $a_1 = a_2 = a_3 = 0$ *and* $a_0 \neq 0$ *for* $|\nu_1 - \mu_1|$ *and* $|\nu_2 - \mu_2|$ *even,*

(iii) $a \in \mathbb{R}^+$ *for* $\nu_1 - \mu_1 = 0$.

Here we can see that the Fueter polynomials are in general not orthogonal. But if we work with the scalar part of the scalar product at least some of the Fueter polynomials are orthogonal:

Corollary 10.11. [19] *If $|\nu_1 - \mu_1|$ and $|\nu_2 - \mu_2|$ are odd we then have*

$$\left(\mathcal{P}_{(\nu_1,\nu_2)}, \mathcal{P}_{(\mu_1,\mu_2)}\right)_{0, L_2(S^2)} = 0.$$

Here $(.,.)_0$ is the scalar part of the scalar product.

As a shorthand we will denote both these scalar products by

$$(f,g) = \int_{B_3} \overline{f(x)} g(x) dx$$

and

$$(f,g)_0 = \text{Sc} \int_{B_3} \overline{f(x)} g(x) dx. \qquad (10.2)$$

We refer explicitly to Exercise 10.4.8, which shows that the scalar products relative to B_3 and S^2 are equal up to a factor. For approximation statements both scalar products are equivalent as they generate the same norm and so the same notion of convergence. But if we consider the structure of the space more in detail we then see that both the above scalar products are very different. $(f,g)_0$ generates a finer structure for the space; more functions are orthogonal, as we have seen already with the example of the Fueter polynomials. Viewed from the constructive side it seems to be easier to look first for orthonormal systems relative to the real scalar product. We shall do that and shall see later that the result delivers quite easily also orthonormal systems relative to the quaternion-valued scalar product.

If we have constructed these orthogonal polynomials, we are able to work in complete analogy to the complex system $\{\sqrt{\frac{n+1}{\pi}} z^n\}$. Due to the higher dimension the calculations are much more complicated, but they are necessary. For the practical application it is useful to think about a software solution of this formal problem. The CD, attached to the book, contains a Maple package QUATPACKAGE, which solves the generation problem of the polynomial basis for the 3-dimensional case.

10.1.3 Harmonic spherical functions

The idea of the following construction is to start with harmonic spherical functions, for which we have relatively simple explicit formulas. If we apply the operator ∂ to these functions we then get homogeneous holomorphic polynomials because of the factorization $\Delta = \overline{\partial} \partial$ of the Laplace operator. We introduce spherical coordinates as follows:

$$\begin{aligned}
x_0 &= r \cos\theta, \\
x_1 &= r \sin\theta \cos\varphi, \\
x_2 &= r \sin\theta \sin\varphi,
\end{aligned}$$

where $0 < r < \infty, 0 < \theta \leq \pi, 0 < \varphi \leq 2\pi$. Every $x = (x_0, x_1, x_2) \in \mathbb{R}^3 \setminus \{0\}$ possesses the representation already repeatedly used,

$$x = r\omega, \quad |x| = r, \quad |\omega| = 1.$$

If we transform also the Cauchy–Riemann operator we then have

$$\bar{\partial} = \frac{1}{r} L + \omega\, \partial_\omega$$

with

$$L = \sum_{i=0}^{2} e_i(r\partial_i - x_i\partial_\omega), \quad \partial_\omega = \sum_{i=0}^{2} \omega_i\partial_i.$$

The reader is asked to show that $\partial_\omega = \partial_r$ in Exercise 10.4.9; the *adjoint Cauchy–Riemann operator* ∂ can now be written in the form

$$\partial = \bar{\omega}\, \partial_r + \frac{1}{r}\overline{L},$$

where

$$\overline{L} = (-\sin\theta - e_1 \cos\theta\cos\varphi - e_2\cos\theta\sin\varphi)\partial_\theta + \frac{1}{\sin\theta}(e_1\sin\varphi - e_2\cos\varphi)\partial_\varphi. \quad (10.3)$$

Every homogeneous harmonic polynomial P_n of degree n can be represented in spherical coordinates in the form

$$P_n(x) = r^n P_n(\omega), \quad \omega \in S^2. \quad (10.4)$$

The restriction to the boundary of the unit ball $P_n(\omega)$ is called a *harmonic spherical function*. Analogously and according to Definition 10.5 we will denote $\mathbb{H}$-holomorphic homogeneous polynomials by $H_n(x)$ and the restrictions to the boundary $H_n(\omega)$ by *holomorphic spherical functions* or by *inner spherical polynomials*.

Our starting point is the well-known complete orthogonal system of *harmonic spherical functions* ([128]),

$$U_{n+1}^0(\theta, \varphi) = P_{n+1}(\cos\theta), \quad\quad\quad\quad\quad\quad\quad\quad\quad\quad\quad (10.5)$$
$$U_{n+1}^m(\theta, \varphi) = P_{n+1}^m(\cos\theta)\cos m\varphi, \quad\quad\quad\quad\quad\quad\quad\quad (10.6)$$
$$V_{n+1}^m(\theta, \varphi) = P_{n+1}^m(\cos\theta)\sin m\varphi, \quad n = 0, \ldots, \infty; \; m = 1, \ldots, n+1. \quad (10.7)$$

Here P_{n+1} denotes the *Legendre polynomial* of degree $n+1$, given by

$$P_{n+1}(t) = \sum_{k=0}^{[\frac{n+1}{2}]} a_{n+1,k}\, t^{n+1-2k}, \quad P_0(t) = 1, \quad\quad\quad t \in [-1, 1],$$

with

$$a_{n+1,k} = (-1)^k \frac{1}{2^{n+1}} \frac{(2n+2-2k)!}{k!\,(n+1-k)!\,(n+1-2k)!}\,.$$

As usual $[k]$ denotes the largest integer $\leq k$.

The functions P_{n+1}^m in (10.7) are called *associated Legendre functions* defined by

$$P_{n+1}^m(t) := (1-t^2)^{m/2} \frac{d^m}{dt^m} P_{n+1}(t), \qquad m = 1, \ldots, n+1\,.$$

For $m = 0$ the associated Legendre function $P_{n+1}^0(t)$ is identical with the corresponding Legendre polynomial $P_{n+1}(t)$. For the calculation of derivatives and norms below we need some properties of the Legendre polynomials and of the associated Legendre functions. The proof of these properties is in part very technical, but it can be done with basic knowledge from analysis. To continue smoothly our considerations we put the necessary results into Appendix 4. We recommend to prove these properties as exercise, see Exercises A.4.6.1, 2, 3, 4.

10.1.4 $\mathbb{H}$-holomorphic spherical functions

We will proceed as follows: The harmonic spherical functions will be continued to harmonic homogeneous polynomials in the unit ball. To these polynomials we apply the operator ∂. The holomorphic homogeneous polynomials obtained in this way will be restricted again to the boundary of the unit ball, and the latter describe the holomorphic spherical functions we are looking for. As an exercise the reader should deal with the construction of applying the adjoint Cauchy–Riemann operator directly to the harmonic spherical functions (without continuation and restriction). The differences should then be discussed, see Exercise 10.4.10.

We consider the continuations of the harmonic spherical functions:

$$\{r^{n+1}\, U_{n+1}^0,\ r^{n+1}\, U_{n+1}^m,\ r^{n+1}\, V_{n+1}^m\,,\ m = 1, \ldots, n+1\}_{n\in\mathbb{N}_0}\,. \tag{10.8}$$

For $n \in \mathbb{N}_0$ we apply the operator ∂ in spherical coordinates (10.3) to (10.8). After restriction to the boundary we get the following holomorphic spherical functions:

$$X_n^0, X_n^m, Y_n^m\,,\ m = 1, \ldots, n+1\,,$$

given by

$$\begin{aligned}
X_n^0 &:= \partial\, (r^{n+1}\, U_{n+1}^0)|_{r=1} \\
&= A^{0,n} + B^{0,n} \cos\varphi\, e_1 + B^{0,n} \sin\varphi\, e_2,
\end{aligned} \tag{10.9}$$

where

$$A^{0,n} := \frac{1}{2} \left(\sin^2\theta\, \frac{d}{dt}[P_{n+1}(t)]_{t=\cos\theta} + (n+1)\cos\theta P_{n+1}(\cos\theta) \right), \tag{10.10}$$

$$B^{0,n} := \frac{1}{2}\left(\sin\theta\cos\theta\frac{d}{dt}[P_{n+1}(t)]_{t=\cos\theta} - (n+1)\sin\theta P_{n+1}(\cos\theta)\right), \quad (10.11)$$

and

$$
\begin{aligned}
X_n^m &:= \partial\left(r^{n+1}\,U_{n+1}^m\right)|_{r=1}\\
&= A^{m,n}\cos m\varphi\\
&\quad + (B^{m,n}\cos\varphi\cos m\varphi - C^{m,n}\sin\varphi\sin m\varphi)\,e_1\\
&\quad + (B^{m,n}\sin\varphi\cos m\varphi + C^{m,n}\cos\varphi\sin m\varphi)\,e_2\;,\quad (10.12)
\end{aligned}
$$

$$
\begin{aligned}
Y_n^m &:= \partial\left(r^{n+1}\,V_{n+1}^m\right)|_{r=1}\\
&= A^{m,n}\sin m\varphi\\
&\quad + (B^{m,n}\cos\varphi\sin m\varphi + C^{m,n}\sin\varphi\cos m\varphi)\,e_1\\
&\quad + (B^{m,n}\sin\varphi\sin m\varphi - C^{m,n}\cos\varphi\cos m\varphi)\,e_2\quad (10.13)
\end{aligned}
$$

with the coefficients

$$A^{m,n} := \frac{1}{2}\left(\sin^2\theta\frac{d}{dt}[P_{n+1}^m(t)]_{t=\cos\theta} + (n+1)\cos\theta\,P_{n+1}^m(\cos\theta)\right) \quad (10.14)$$

$$B^{m,n} := \frac{1}{2}\left(\sin\theta\cos\theta\frac{d}{dt}[P_{n+1}^m(t)]_{t=\cos\theta} - (n+1)\sin\theta\,P_{n+1}^m(\cos\theta)\right) \quad (10.15)$$

$$C^{m,n} := \frac{1}{2}\,m\frac{1}{\sin\theta}\,P_{n+1}^m(\cos\theta)\;, \quad (10.16)$$

$m = 1,\ldots,n+1$.

To proceed safely with the indices in the following calculations we formulate a simple observation as a proposition:

Proposition 10.12. *The holomorphic spherical functions X_n^m and Y_n^m are the zero function for $m \geq n+2$.*

Proof. We have $A^{m,n}$, $B^{m,n}$, $C^{m,n} = 0$ for $m \geq n+2$; as for these m and all $t \in [-1,1]$,

$$P_{n+1}^m(t) = (1-t^2)^{m/2}\,P_{n+1}^{(m)}(t) = 0. \qquad \square$$

A little numerical example may help to motivate the next lengthy calculations. Since the known dimension of $\mathcal{H}_{n,\mathbb{H}}^+$ is $n+1$ we take the real space to have the dimension $4n+4$. This is simple to prove: From the property that the Fueter polynomials $\mathcal{P}_{\nu_1,\nu_2}$, $\nu_1+\nu_2 = n$, constitute a basis in $\mathcal{H}_n^+$, we get at once that every holomorphic homogeneous polynomial of degree n can be represented uniquely as a linear combination of the polynomials $\mathcal{P}_{\nu_1,\nu_2}$, $\mathcal{P}_{\nu_1,\nu_2}e_1$, $\mathcal{P}_{\nu_1,\nu_2}e_2$, and $\mathcal{P}_{\nu_1,\nu_2}e_3$ with real coefficients. Therefore these polynomials constitute a basis in $\mathcal{H}_{n,\mathbb{R}}^+$, and the real dimension indeed is $4n+4$. But by the above construction of the functions X_n^0, X_n^m, Y_n^m, $m = 1,\ldots,n+1$, we get at most $2n+3$ holomorphic polynomials, which may not be linearly independent. In no case is the system complete, as none of the constructed functions has a component of e_3 different from zero. The idea now is easily expressed: we look at

$$\{X_n^0, X_n^m, Y_n^m, X_n^0 e_i, X_n^m e_i, Y_n^m e_i; \, m = 1, \dots, n+1\}, \quad i = 1, \text{ or } i = 2, \text{ or } i = 3.$$

These systems look equivalent but this is not the case. A numerical calculation of the condition number of the Gram matrix shows the following results:

system	degree	condition number
$\{X_n^m\} \cup \{X_n^m e_1\}$	$n = 1$	13,93
	$n = 2$	69,99
	$n = 3$	292,02
	$n = 4$	1158,92
	$n = 7$	70718,72
$\{X_n^m\} \cup \{X_n^m e_3\}$	$n = 1$	3,00
	$n = 2$	5,00
	$n = 3$	7,00
	$n = 4$	9,00
	$n = 7$	15,00

Thus these systems will react quite differently when orthogonalized. Taking into account these numerical results further on we shall deal with the system

$$\{X_n^0, X_n^m, Y_n^m, X_n^0 e_3, X_n^m e_3, Y_n^m e_3; \, m = 1, \dots, n+1\}.$$

We introduce the following notation:

$$
\begin{aligned}
X_{n,0}^m &:= X_n^m, \quad X_{n,3}^m := X_n^m e_3, \quad m = 0, \dots, n+1, \\
Y_{n,0}^m &:= Y_n^m, \quad Y_{n,3}^m := Y_n^m e_3, \quad m = 1, \dots, n+1.
\end{aligned}
$$

Every subspace $\mathcal{H}_n^+$ contains at least two linearly dependent functions which we will delete. This can be seen if we analyze the last two elements of every system. We have

$$Y_{n,3}^{n+1} = X_{n,0}^{n+1} \quad \text{and} \quad X_{n,3}^{n+1} = -Y_{n,0}^{n+1}.$$

This result is obtained from the explicit representations

$$X_n^{n+1} = -C^{n+1,n} \cos n\varphi \, e_1 + C^{n+1,n} \sin n\varphi \, e_2$$

and

$$Y_n^{n+1} = -C^{n+1,n} \sin n\varphi \, e_1 - C^{n+1,n} \cos n\varphi \, e_2.$$

A multiplication by e_3 gives

$$Y_{n,3}^{n+1} = -C^{n+1,n} \cos n\varphi \, e_1 + C^{n+1,n} \sin n\varphi \, e_2 = X_{n,0}^{n+1}$$

and

$$X_{n,3}^{n+1} = -Y_{n,0}^{n+1} .$$

In the following we shall work for all $n \in \mathbb{N}_0$ with the $4n+4$ holomorphic spherical functions

$$\{X_{n,0}^0, X_{n,0}^m, Y_{n,0}^m, X_{n,3}^0, X_{n,3}^l, Y_{n,3}^l; \ m = 1,\dots,n+1, l = 1,\dots,n\} .$$

$$(10.17)$$

Using this system we construct the desired orthonormal basis in $\mathcal{H}_{n,\mathbb{R}}^+$, first relative to the real scalar product $(.,.)_{0,L^2(B_3)}$. The starting point is described by the following theorem:

Theorem 10.13. *For all* $n \in \mathbb{N}_0$ *the subsystems of holomorphic spherical functions* $X_{n,0}^0$, $X_{n,0}^m$, *resp.* $Y_{n,0}^m$ *(*$m = 1,\dots,n+1$*), are orthogonal systems relative to the inner product (10.2) with the norms*

$$||X_{n,0}^0||_{0,L^2(S^2)} = \sqrt{\pi(n+1)}$$

$$(10.18)$$

and

$$||X_{n,0}^m||_{0,L^2(S^2)} = ||Y_{n,0}^m||_{0,L^2(S^2)} = \sqrt{\frac{\pi}{2}(n+1)\frac{(n+1+m)!}{(n+1-m)!}}, \quad m = 1,\dots,n+1.$$

$$(10.19)$$

The calculation of the norms is only possible with great effort. This effort is necessary as we want to explicitly orthogonalize our polynomials. To continue it suffices to know that the norms can be calculated and to know the result. Also this proof has been shifted to Appendix 4.2 to be able to smoothly continue the exposition. An analogous result can be proven for the second subsystem, see Exercise 10.4.11:

Theorem 10.14. *For all* $n \in \mathbb{N}_0$ *the systems* $X_{n,3}^0$, $X_{n,3}^l$ *and* $Y_{n,3}^l$ *(*$l = 1,\dots,n$*) are orthogonal relatively to the inner product (10.2), and the norms are given by*

$$||X_{n,3}^0||_{0,L^2(S^2)} = \sqrt{\pi(n+1)}$$

and

$$||X_{n,3}^l||_{0,L^2(S^2)} = ||Y_{n,3}^l||_{0,L^2(S^2)} = \sqrt{\frac{\pi}{2}(n+1)\frac{(n+1+l)!}{(n+1-l)!}}, \quad l = 1,\dots,n .$$

The second fundamental problem is to find the relations between the subsystems examined so far and to calculate, if possible, the inner products and thus also the angle between the subspaces.

Theorem 10.15. *For all $n \in \mathbb{N}_0$ and $m = 1, \ldots, n+1$; $l = 1, \ldots, n$, we have*

$$
\begin{aligned}
(X_{n,0}^0, X_{n,3}^0)_{0,L^2(S^2)} &= (X_{n,0}^0, X_{n,3}^l)_{0,L^2(S^2)} = (X_{n,0}^0, Y_{n,3}^l)_{0,L^2(S^2)} \\
&= (X_{n,0}^m, X_{n,3}^0)_{0,L^2(S^2)} = (X_{n,0}^m, X_{n,3}^l)_{0,L^2(S^2)} = (Y_{n,0}^m, X_{n,3}^0)_{0,L^2(S^2)} \\
&= (Y_{n,0}^m, Y_{n,3}^l)_{0,L^2(S^2)} = 0
\end{aligned}
$$

and

$$
(X_{n,0}^m, Y_{n,3}^l)_{0,L^2(S^2)} = -(Y_{n,0}^m, X_{n,3}^l)_{0,L^2(S^2)} =
\begin{cases}
0, & m \neq l, \\[2mm]
\dfrac{\pi}{2}\, m\, \dfrac{(n+m+1)!}{(n-m+1)!}, & m = l.
\end{cases}
$$

The proof has again been shifted to the appendix as it is completely technical, no new ideas are necessary. Using the explicit representation of our holomorphic spherical functions all necessary integrals can be easily calculated.

We see that only very few functions from the different systems are not orthogonal. The Gram matrix is only sparse, and a chance remains to have an explicit orthogonalization.

We now denote by

$$
\tilde{X}_{n,0}^0,\ \tilde{X}_{n,0}^m,\ \tilde{Y}_{n,0}^m,\ \tilde{X}_{n,3}^0,\ \tilde{X}_{n,3}^l,\ \tilde{Y}_{n,3}^l;\quad m = 1, \ldots, n+1;\ l = 1, \ldots, n,
$$

the normalized polynomials arising from (10.17). From the above investigations we see that the systems

$$
\left\{ \tilde{X}_{n,0}^0,\ \tilde{X}_{n,0}^m,\ \tilde{Y}_{n,0}^m;\ m = 1, \ldots, n+1 \right\}_{n \in \mathbb{N}_0} \tag{10.20}
$$

and

$$
\left\{ \tilde{X}_{n,3}^0,\ \tilde{X}_{n,3}^l,\ \tilde{Y}_{n,3}^l;\ l = 1, \ldots, n \right\}_{n \in \mathbb{N}_0} \tag{10.21}
$$

are both orthonormal systems. The relations between these subsystems seem to be easier after the normalization. For all $n \in \mathbb{N}_0$ all polynomials of the systems (10.20) and (10.21) are orthogonal apart from

$$
\begin{aligned}
(\tilde{X}_{n,0}^l, \tilde{Y}_{n,3}^l)_{0,L^2(S^2)} &= -(\tilde{Y}_{n,0}^l, \tilde{X}_{n,3}^l)_{0,L^2(S^2)} \\
&= \frac{l}{n+1}, \quad l = 1, \ldots, n\,.
\end{aligned}
$$

We see that for increasing n the angle between the subspaces spanned by the functions (10.20) and (10.21) converges to zero. Thus the parallel projectors onto these subspaces will not be uniformly bounded in L^2, and we cannot avoid the orthogonalization of the complete system. As the result of the orthogonalization we get:

$$X_{n,0}^{0,*} := \tilde{X}_{n,0}^0 , \tag{10.22}$$

$$X_{n,0}^{m,*} := \tilde{X}_{n,0}^m , \tag{10.23}$$

$$Y_{n,0}^{m,*} := \tilde{Y}_{n,0}^m , \tag{10.24}$$

$$Y_{n,3}^{l,*} := \sqrt{s_{n,l}} \left((n+1)\, \tilde{Y}_{n,3}^l - l\, \tilde{X}_{n,0}^l \right) , \tag{10.25}$$

$$X_{n,3}^{0,*} := \tilde{X}_{n,3}^0 , \tag{10.26}$$

$$X_{n,3}^{l,*} := \sqrt{s_{n,l}} \left((n+1)\, \tilde{X}_{n,3}^l + l\, \tilde{Y}_{n,0}^l \right) \tag{10.27}$$

with

$$s_{n,l} = \frac{1}{(n+1)^2 - l^2} , \tag{10.28}$$

$m = 1, \ldots, n+1,\, l = 1, \ldots, n.$

This is easy to see: The first $2n + 3$ functions in (10.22)–(10.24) are already an orthonormal system. For the next $2n + 3 + l$ $(l = 1, \ldots, n)$ functions we calculate stepwise: Let l be fixed and let Y_{nl} be the not yet normalized function which comes from step $(2n + 3 + l)$ of the orthogonalization procedure. This function is calculated explicitly as follows:

$$
\begin{aligned}
Y_{nl} &= \tilde{Y}_{n,3}^l - X_{n,0}^{0,*}(X_{n,0}^{0,*}, \tilde{Y}_{n,3}^l)_{0,L^2(S^2)} \\
&\quad - \sum_{m=0}^{n+1} X_{n,0}^{m,*}(X_{n,0}^{m,*}, \tilde{Y}_{n,3}^l)_{0,L^2(S^2)} - \sum_{m=0}^{n+1} Y_{n,0}^{m,*}(Y_{n,0}^{m,*}, \tilde{Y}_{n,3}^l)_{0,L^2(S^2)} \\
&= \tilde{Y}_{n,3}^l - X_{n,0}^{l,*}(X_{n,0}^{l,*}, \tilde{Y}_{n,3}^l)_{0,L^2(S^2)},
\end{aligned}
$$

here we have used Theorem 10.15. From the construction of the functions in (10.23) it follows that

$$Y_{nl} = \tilde{Y}_{n,3}^l - \tilde{X}_{n,0}^l(\tilde{X}_{n,0}^l, \tilde{Y}_{n,3}^l)_{0,L^2(S^2)},$$

and the substitution of the scalar product's known value gives

$$Y_{nl} = \tilde{Y}_{n,3}^l - \frac{l}{n+1} X_{n,0}^{l,*}.$$

$Y_{n,3}^{l,*}$ has to be calculated by normalization of Y_{nl}. The step $(3n + 4)$ gives (10.26), and the remaining procedure leads us to (10.27) in analogy with the calculations just described.

10.1.5   Completeness in $L^2(B_3) \cap \ker \overline{\partial}$

We will now examine whether the constructed orthonormal system (ONS) is complete in $L^2(B_3)_{\mathbb{R}} \cap \ker \overline{\partial}$. Because of Theorem 10.8 we know that any holomorphic function can be developed into an orthogonal series with functions from the subspaces $\mathcal{H}_n^+$. The spherical functions, constructed in the last subsection, can be extended with the help of the relation (10.4) to an orthogonal system in $\mathcal{H}_{n,\mathbb{R}}^+$, and then can be normalized again. We get the system

$$\frac{1}{2n+3} r^n X_{n,0}^{0,*}, \quad \frac{1}{2n+3} r^n X_{n,0}^{m,*}, \quad \frac{1}{2n+3} r^n Y_{n,0}^{m,*},$$

$$\frac{1}{2n+3} r^n Y_{n,3}^{l,*}, \quad \frac{1}{2n+3} r^n X_{n,3}^{0,*}, \quad \frac{1}{2n+3} r^n X_{n,3}^{l,*}. \qquad (10.29)$$

Every element in $\mathcal{H}_{n,\mathbb{R}}^+$ can be represented by the ONS of the $4n+4$ holomorphic homogeneous polynomials (10.29). So we have proved the result we wanted, and we can formulate the corresponding theorem:

Theorem 10.16. *The system* (10.29) *is complete in* $L^2(B_3)_{\mathbb{R}} \cap \ker \overline{\partial}$.

We will try to construct from the ONS in the real Hilbert space $L^2(B_3)_{\mathbb{R}} \cap \ker \overline{\partial}$ an ONS for the quaternion-valued scalar product in $L^2(B_3)_{\mathbb{H}} \cap \ker \overline{\partial}$. We shall see that it suffices to select suitable functions from the ONS in $L^2(B_3)_{\mathbb{R}} \cap \ker \overline{\partial}$.

Theorem 10.17. *For all $n \in \mathbb{N}_0$, the $n+1$ holomorphic homogeneous polynomials*

$$r^n X_{n,0}^{0,*}, \; r^n X_{n,0}^{2k_1,*}, \; r^n Y_{n,3}^{2k_2,*}, \; k_1 = 1, \ldots, \left[\frac{n+1}{2}\right], \; k_2 = 1, \ldots, \left[\frac{n}{2}\right] \quad (10.30)$$

constitute an orthogonal basis in $\mathcal{H}_{n,\mathbb{H}}^+$.

This result raises the question whether it is possible to define an ONS for $\mathcal{H}_n^+$ using odd indices. This question is answered by the following theorem:

Theorem 10.18. *For all $n \in \mathbb{N}_0$, the $n+1$ holomorphic homogeneous polynomials*

$$r^n X_{n,0}^{2k_1+1,*}, \; r^n Y_{n,3}^{2k_2-1,*}, \; k_1 = 0, \ldots, \left[\frac{n}{2}\right], \; k_2 = 1, \ldots, \left[\frac{n+1}{2}\right],$$

constitute an orthogonal basis in $\mathcal{H}_{n,\mathbb{H}}^+$.

In Appendix A.4.4 the first of these theorems is proved, the second proof is quite analogous.

Corollary 10.19. *Both systems*

$$\left\{ \sqrt{2n+3}\, r^n X_{n,0}^{0,*}, \; \sqrt{2n+3}\, r^n X_{n,0}^{2k_1,*}, \; \sqrt{2n+3}\, r^n Y_{n,3}^{2k_2,*} : \right.$$

$$\left. k_1 = 1, \ldots, \left[\frac{n+1}{2}\right], \; k_2 = 1, \ldots, \left[\frac{n}{2}\right] \right\}_{n \in \mathbb{N}_0} \qquad (10.31)$$

and

$$\left\{ \sqrt{2n+3}\, r^n\, X_{n,0}^{2k_1+1,*}, \ \sqrt{2n+3}\, r^n\, Y_{n,3}^{2k_2-1,*} : \right.$$

$$\left. k_1 = 0, \ldots, \left[\frac{n}{2}\right], \ k_2 = 1, \ldots, \left[\frac{n+1}{2}\right] \right\}_{n\in\mathbb{N}_0} \qquad (10.32)$$

are complete orthonormal systems in $L^2(B_3)_{\mathbb{H}} \cap \ker \overline{\partial}$.

10.2 Fourier expansion in $\mathbb{H}$

After the above extensive preparations the Fourier expansion of a quadratically integrable holomorphic function is merely a formal business. We formulate here only the result for $L^2(B_3)_{\mathbb{R}} \cap \ker \overline{\partial}$ and leave the case $L^2(B_3)_{\mathbb{H}} \cap \ker \overline{\partial}$ to the reader:

Theorem 10.20. *Let* $f \in L^2(B_3)_{\mathbb{R}} \cap \ker \overline{\partial}$. *The function* f *can then be represented with the ONS* (10.22)–(10.27):

$$
\begin{aligned}
f \ =\ & \sum_{n=0}^{\infty} \sqrt{2n+3}\, r^n \left[X_{n,0}^{0,*}\alpha_n + X_{n,3}^{0,*}\beta_n \right. \\
& + \sum_{m=1}^{n} \left(X_{n,0}^{m,*}\gamma_{n,m} + Y_{n,0}^{m,*}\delta_{n,m} + X_{n,3}^{m,*}\varepsilon_{n,m} + Y_{n,3}^{m,*}\varphi_{n,m} \right) \\
& \left. + X_{n,0}^{n+1,*}\gamma_{n,n+1} + Y_{n,0}^{n+1,*}\delta_{n,n+1} \right].
\end{aligned}
$$

Obviously f may be characterized by its coefficients using the Parseval equation:

Theorem 10.21. $f \in L^2(B_3)_{\mathbb{R}} \cap \ker \overline{\partial}$ *is equivalent to*

$$\sum_{n=0}^{\infty} \left[\alpha_n^2 + \beta_n^2 + \sum_{m=1}^{n} \left(\gamma_{n,m}^2 + \delta_{n,m}^2 + \varepsilon_{n,m}^2 + \varphi_{n,m}^2 \right) + \gamma_{n,n+1}^2 + \delta_{n,n+1}^2 \right] < \infty.$$

10.3 Applications

10.3.1 Derivatives of $\mathbb{H}$-holomorphic polynomials

In the introduction to Subsection 10.1.1 it was pointed out that the complex powers z^n have the advantage that their derivatives belong to the same system of functions. We now shall examine what happens when the holomorphic functions of our orthonormal systems will be differentiated. We denote by $\partial^i X_n^0$, $\partial^i X_n^m$ and $\partial^i Y_n^m$ the holomorphic spherical functions which arise by an i-times differentiation of $r^n X_n^0$, $r^n X_n^m$ and $r^n Y_n^m$ by the operator ∂ followed by the restriction to the boundary (here we have to pay attention to distinguish ∂^i and the partial derivative ∂_i). It suffices to examine first the derivatives of the functions X_n^m and Y_n^m. With their help we can easily calculate the derivatives of the functions of the orthogonal systems.

Theorem 10.22. *We have*

$$\partial X_n^m = \partial(r^n X_n^m)|_{r=1}$$

$$= \overset{(1)}{A^{m,n}} \cos m\varphi \qquad\qquad \bullet$$

$$+ (\overset{(1)}{B^{m,n}} \cos\varphi \cos m\varphi - \overset{(1)}{C^{m,n}} \sin\varphi \sin m\varphi)\, e_1$$

$$+ (\overset{(1)}{B^{m,n}} \sin\varphi \cos m\varphi + \overset{(1)}{C^{m,n}} \cos\varphi \sin m\varphi)\, e_2.$$

Proof. Formal differentiation of X_n^m gives the expression

$$\partial X_n^m = \partial(r^n X_n^m)|_{r=1}$$

$$= \overset{(1)}{A^{m,n}} \cos m\varphi$$

$$+ (\overset{(1)}{B^{m,n}} \cos\varphi \cos m\varphi - \overset{(1)}{C^{m,n}} \sin\varphi \sin m\varphi)\, e_1$$

$$+ (\overset{(1)}{B^{m,n}} \sin\varphi \cos m\varphi + \overset{(1)}{C^{m,n}} \cos\varphi \sin m\varphi)\, e_2$$

$$+ \overset{(1)}{E^{m,n}} \sin m\varphi\, e_3$$

with the coefficients

$$\overset{(1)}{A^{m,n}} = \frac{1}{2}\Big(-\sin\theta \frac{d}{d\theta} A^{m,n} + \cos\theta \frac{d}{d\theta} B^{m,n} + \frac{1}{\sin\theta} B^{m,n} + m\frac{1}{\sin\theta} C^{m,n}$$
$$+ n\cos\theta A^{m,n} + n\sin\theta B^{m,n}\Big), \tag{10.33}$$

$$\overset{(1)}{B^{m,n}} = \frac{1}{2}\Big(-\sin\theta \frac{d}{d\theta} B^{m,n} - \cos\theta \frac{d}{d\theta} A^{m,n} + n\cos\theta B^{m,n} - n\sin\theta A^{m,n}\Big), \tag{10.34}$$

$$\overset{(1)}{C^{m,n}} = \frac{1}{2}\Big(-\sin\theta \frac{d}{d\theta} C^{m,n} + m\frac{1}{\sin\theta} A^{m,n} + n\cos\theta C^{m,n}\Big), \tag{10.35}$$

$$\overset{(1)}{E^{m,n}} = \frac{1}{2}\Big(-\cos\theta \frac{d}{d\theta} C^{m,n} - m\frac{1}{\sin\theta} B^{m,n} - \frac{1}{\sin\theta} C^{m,n} - n\sin\theta\, C^{m,n}\Big),$$

$m = 1, \ldots, n+1$. The coefficient functions $\overset{(1)}{E^{m,n}}$ $(m = 1, \ldots, n+1)$ can be calculated with the help of (10.14), (10.15), and (10.16) to show that they vanish.　□

The fact that also the derivatives of the functions X_n^m take only values in span$\{e_0, e_1, e_2\}$ as the functions themselves is remarkable. For the functions Y_n^m we have an analogous result. If we analyze the coefficients of the last theorem more precisely we get a sharper statement:

Theorem 10.23. *Let $n \geq 1$. We then have*

$$\partial X_n^m = (n+m+1)X_{n-1}^m, \quad m = 0, \ldots, n,$$
$$\partial Y_n^m = (n+m+1)Y_{n-1}^m, \quad m = 1, \ldots, n.$$

Proof. Because of the analogous form of the polynomials and their derivatives it suffices to show that

$$\text{(i)} \; \overset{(1)}{A^{m,n}} \; = \; (n+m+1)A^{m,n-1}, \;\; m = 0, \ldots, n,$$

$$\text{(ii)} \; \overset{(1)}{B^{m,n}} \; = \; (n+m+1)B^{m,n-1}, \;\; m = 0, \ldots, n,$$

$$\text{(iii)} \; \overset{(1)}{C^{m,n}} \; = \; (n+m+1)C^{m,n-1}, \;\; m = 1, \ldots, n.$$

This calculation is explicitly carried out in Appendix 4, Theorem A.4.4   □

Indeed the derivatives of our polynomials turn out to be polynomials of lower degree in the same system up to a constant. By iterated application of this result we get corresponding expressions for the derivatives of higher order.

Theorem 10.24. *Let* $n \in \mathbb{N}_0$, $i = 1, 2, \ldots$. *We then have*

$$\partial^i X_n^m \; = \; \left(\prod_{h=1}^{i} (n+m+1-(h-1)) \right) X_{n-i}^m, \;\; m = 0, \ldots, n+1-i,$$

$$\partial^i Y_n^m \; = \; \left(\prod_{h=1}^{i} (n+m+1-(h-1)) \right) Y_{n-i}^m, \;\; m = 1, \ldots, n+1-i.$$

Proof. The proof can be done by induction. The case $i = 1$ is the claim of Theorem A.4.4 (see Exercise 10.4.12).   □

Contrary to the complex case it may happen that the derivative of a polynomial vanishes before the order of the differentiation exceeds the degree of the polynomial:

Corollary 10.25. *For arbitrary* $n \in \mathbb{N}_0$ *we have*

$$\partial^i X_n^m = \partial^i Y_n^m = 0, \; i \geq n - m + 2; \; m = 1, \ldots, n+1.$$

Proof. The proof follows from the last theorem, the representation formula for the derivatives, and Proposition 10.12, see also Exercise 10.4.13.   □

Finally we have to convince ourselves that the derivatives of a basis polynomial do not vanish earlier as described in the last corollary.

Corollary 10.26. *For arbitrary* $n \in \mathbb{N}_0$ *we have*

$$r^n \, X_n^0 \; \in \; (\ker \partial^{n+1} \backslash \ker \partial^n) \cap \ker \overline{\partial},$$
$$r^n \, X_n^m, \; r^n \, Y_n^m \; \in \; (\ker \partial^{n-m+2} \backslash \ker \partial^{n-m+1}) \cap \ker \overline{\partial}, \; m = 1, \ldots, n+1,$$

where ∂^0 *should be identified with the identity.*

Using the above theorems the proof involves only calculations, see Exercise 10.4.14.

We still need to examine the derivatives of the functions in our orthonormal system. As the functions in the orthonormal system are described relatively simply by the functions X_n^m and Y_n^m, the task brings no serious difficulties:

Theorem 10.27. *The derivatives of the functions in the ONS* (10.22)–(10.27) *can be expressed in the same ONS as follows:*

$$\partial X_{n,0}^{0,*} \;=\; \sqrt{n(n+1)}\, X_{n-1,0}^{0,*}, \tag{10.36}$$

$$\partial X_{n,0}^{m,*} \;=\; \sqrt{\frac{n}{n+1}}\, \frac{1}{\sqrt{s_{n,m}}}\, X_{n-1,0}^{m,*}, \tag{10.37}$$

$$\partial Y_{n,0}^{m,*} \;=\; \sqrt{\frac{n}{n+1}}\, \frac{1}{\sqrt{s_{n,m}}}\, Y_{n-1,0}^{m,*}, \tag{10.38}$$

$$\partial Y_{n,3}^{l,*} \;=\; \sqrt{\frac{n+1}{n}}\left(\frac{1}{\sqrt{s_{n-1,l}}} Y_{n-1,3}^{l,*} + \frac{l}{n+1} X_{n-1,0}^{l,*} \right), \tag{10.39}$$

$$\partial X_{n,3}^{0,*} \;=\; \sqrt{n(n+1)}\, X_{n-1,3}^{0,*}, \tag{10.40}$$

$$\partial X_{n,3}^{l,*} \;=\; \sqrt{\frac{n+1}{n}}\left(\frac{1}{\sqrt{s_{n-1,l}}} X_{n-1,3}^{l,*} - \frac{l}{n+1} Y_{n-1,0}^{l,*} \right), \tag{10.41}$$

where $s_{n,m}$ is given by (10.28), *and $m = 1, \dots, n$, $l = 1, \dots, n-1$.*

Proof. The only interesting case is $Y_{n,3}^{l,*}$ and analogously $X_{n,3}^{l,*}$, the remaining steps are formal calculations.

$$
\begin{aligned}
\partial Y_{n,3}^{l,*} &= \sqrt{s_{n,l}}\, \frac{1}{\|X_{n,0}^l\|_{0,L^2(S^2)}} \left((n+1)\partial Y_n^l\, e_3 - l\, \partial X_n^l \right) \\
&= \sqrt{s_{n,l}}\, \frac{\|X_{n-1,0}^l\|_{0,L^2(S^2)}}{\|X_{n,0}^l\|_{0,L^2(S^2)}} (n+l+1) \left((n+1)\tilde Y_{n-1,3}^l - l\, \tilde X_{n-1,0}^l \right) \\
&= \sqrt{\frac{n}{n+1}} \left((n+1)\tilde Y_{n-1,3}^l - l\, \tilde X_{n-1,0}^l \right) . \tag{10.42}
\end{aligned}
$$

Following the construction we get

$$\tilde X_{n-1,0}^l = X_{n-1,0}^{l,*}$$

and

$$\tilde Y_{n-1,3}^l = \frac{1}{n\sqrt{s_{n-1,l}}} Y_{n-1,3}^{l,*} + \frac{l}{n} X_{n-1,0}^{l,*} \;;$$

substitution of these relations into (10.42) gives (10.39). The range of the indices has to be explained yet: The index m runs from 1 to n since

$$\partial X_{n,0}^{n+1,*} = \partial Y_{n,0}^{n+1,*} = 0.$$

The index l runs from 1 to $n-1$ since the linear dependent functions $\partial Y_{n,3}^{n,*}$ and $\partial X_{n,3}^{n,*}$ can be excluded. We indeed get

$$
\begin{aligned}
\partial Y_{n,3}^{n,*} &= \frac{\sqrt{s_{n,n}}}{||X_{n,0}^n||_{0,L^2(S^2)}} \left[(n+1)(2n+1)Y_{n-1,3}^n - n(2n+1)X_{n-1,0}^n\right] \\
&= \frac{\sqrt{s_{n,n}}}{||X_{n,0}^n||_{0,L^2(S^2)}} (2n+1)\left[(n+1)X_{n-1,0}^n - nX_{n-1,0}^n\right] \\
&= \sqrt{2n+1}\frac{||X_{n-1,0}^n||_{0,L^2(S^2)}}{||X_{n,0}^n||_{0,L^2(S^2)}} X_{n-1,0}^{n,*} \\
&= \sqrt{\frac{n}{n+1}} X_{n-1,0}^{n,*}.
\end{aligned}
$$

Using the result (10.37), already proved, we have

$$
\partial Y_{n,3}^{n,*} = \sqrt{s_{n,n}}\,\partial X_{n,0}^{n,*} \tag{10.43}
$$

and analogously

$$
\partial X_{n,3}^{n,*} = -\sqrt{s_{n,n}}\,\partial Y_{n,0}^{n,*}. \tag{10.44}
$$

□

In Exercise 10.4.15 the reader is asked to show that the calculated derivatives (10.36)–(10.41) constitute a basis in $\mathcal{H}_{n-1,\mathbb{R}}^+$. As the number $4n$ is equal to the dimension it suffices to prove the linear independence.

10.3.2 Primitives of $\mathbb{H}$-holomorphic functions

Using the above constructed orthonormal systems of holomorphic functions we now will begin to define holomorphic primitive functions of holomorphic functions. In the complex plane we are able to describe primitives very easily by line integrals. This does not work in higher dimensions as in such case line integrals depend on the curve. We will define primitive functions by using inversion of the differentiation. More precisely, we seek a right inverse operator of the hypercomplex derivative ∂, which maps holomorphic functions into holomorphic functions

Definition 10.28. Every holomorphic function F with the property

$$
\partial F = f \tag{10.45}
$$

for a given holomorphic function f is called a *holomorphic primitive* of f. If for a given $f \in \ker\overline{\partial}$ such a function F exists, we write shortly $Pf := F$.

The idea we shall use is to define the operator P for the functions of our orthonormal system and to extend it to the whole space L^2. It should be remarked that also extensions of this definition are of interest, e.g., an algebraic primitive of holomorphic functions is given by the conjugated Teodorescu operator $\overline{T_G}$ following Definition 7.10. The operator $\overline{T_G}$ is also a right inverse of ∂, but it maps holomorphic functions into harmonic functions. This operator is important for the solution of elliptic boundary value problems, but this will not be pursued here.

Sudbery [152] proved the existence of holomorphic primitives for the class of holomorphic polynomials using vector-analytic methods. Using an ansatz in Fueter polynomials the explicit form of polynomial holomorphic primitives was constructed in [53] for given $C\ell(n)$-valued holomorphic polynomials. But the boundedness of such operator has not been proven. In [14] holomorphic primitives of holomorphic functions have been shown to exist in a suitable domain, but this method was not constructive.

We further remark that holomorphic primitives of one holomorphic function differ by $\mathbb{H}$-*holomorphic constants*, i.e., by functions $f \in \ker \overline{\partial} \cap \ker \partial$ which are only functions of the form $f(x_1, x_2)$ and which satisfy the equation

$$Df = (\partial_1 e_1 + \partial_2 e_2)f = 0,$$

here D is the Dirac operator from Theorem 5.12. As in real and complex analysis we speak sometimes of *the* primitive, meaning that we delete the constants in our calculations.

In a parallel way to the calculation of the derivative of homogeneous holomorphic polynomials we proceed to determine the primitive of a spherical holomorphic function by first extending it into the ball, calculating there a primitive, and then restricting the result if necessary to the surface of the ball. In this manner we can also determine primitives of holomorphic functions.

Definition 10.29. The operator $P : \mathcal{H}_{n,\mathbb{R}}^+ \longrightarrow \mathcal{H}_{n+1,\mathbb{R}}^+$ is defined by

$$P(X_{n,0}^{0,*}) = \frac{1}{\sqrt{(n+1)(n+2)}} X_{n+1,0}^{0,*}, \tag{10.46}$$

$$P(X_{n,0}^{m,*}) = \sqrt{\frac{n+2}{n+1}} \sqrt{s_{n+1,m}} X_{n+1,0}^{m,*}, \tag{10.47}$$

$$P(Y_{n,0}^{m,*}) = \sqrt{\frac{n+2}{n+1}} \sqrt{s_{n+1,m}} Y_{n+1,0}^{m,*}, \tag{10.48}$$

$$P(X_{n,3}^{0,*}) = \frac{1}{\sqrt{(n+1)(n+2)}} X_{n+1,3}^{0,*}, \tag{10.49}$$

$$P(X_{n,3}^{l,*}) = \sqrt{\frac{n+1}{n+2}} \sqrt{s_{n,l}} \left(X_{n+1,3}^{l,*} + \frac{l}{n+1} \sqrt{s_{n+1,l}} Y_{n+1,0}^{l,*} \right), \tag{10.50}$$

$$P(Y_{n,3}^{l,*}) = \sqrt{\frac{n+1}{n+2}} \sqrt{s_{n,l}} \left(Y_{n+1,3}^{l,*} - \frac{l}{n+1} \sqrt{s_{n+1,l}} X_{n+1,0}^{l,*} \right), \tag{10.51}$$

with $n \in \mathbb{N}_0$, $m = 1, \ldots, n+1$ and $l = 1, \ldots, n$.

An analogous definition can be given for the basis (10.30) of $\mathcal{H}_{n,\mathbb{H}}^+$:

Definition 10.30. The operator $P : \mathcal{H}^+_{n,\mathbb{H}} \longrightarrow \mathcal{H}^+_{n+1,\mathbb{H}}$ is defined by

$$P(X^{0,*}_{n,0}) = \frac{1}{\sqrt{(n+1)(n+2)}} X^{0,*}_{n+1,0},$$

$$P(X^{2k_1,*}_{n,0}) = \sqrt{\frac{n+2}{n+1}} \sqrt{s_{n+1,2k_1}} X^{2k_1,*}_{n+1,0},$$

$$P(Y^{2k_2,*}_{n,3}) = \sqrt{\frac{n+1}{n+2}} \sqrt{s_{n,2k_2}} \left(Y^{2k_2,*}_{n+1,3} - \frac{2k_2}{n+1} \sqrt{s_{n+1,2k_2}} X^{2k_2,*}_{n+1,0} \right),$$

with $n \in \mathbb{N}_0$, $k_1 = 1, \ldots, \left[\frac{n+1}{2}\right]$ and $k_2 = 1, \ldots, \left[\frac{n}{2}\right]$.

Our intention is to extend the operator by continuity to the whole space. For this purpose we need the norms of the primitives of our orthonormal basis.

Proposition 10.31. *The norms of the primitives of the holomorphic spherical functions are given by the following formulas:*

$$\|P X^{0,*}_{n,0}\|_{0,L^2(S^2)} = \|P X^{0,*}_{n,3}\|_{0,L^2(S^2)} = \frac{1}{\sqrt{(n+1)(n+2)}},$$

$$\|X^{n+1,*}_{n+1,3} - \sqrt{s_{n+1,n+1}} Y^{n+1,*}_{n+1,0}\|_{0,L^2(S^2)}$$
$$= \|X^{n+1,*}_{n+1,3} + \sqrt{s_{n+1,n+1}} Y^{n+1,*}_{n+1,0}\|_{0,L^2(S^2)} = \sqrt{1 + s_{n+1,n+1}},$$

$$\|X^{n+2,*}_{n+1,0}\|_{0,L^2(S^2)} = \|Y^{n+2,*}_{n+1,0}\|_{0,L^2(S^2)} = 1,$$

$$\|P X^{m,*}_{n,0}\|_{0,L^2(S^2)} = \|P Y^{m,*}_{n,0}\|_{0,L^2(S^2)} = \sqrt{\frac{n+2}{n+1}} \sqrt{s_{n+1,m}}, \quad m = 1, \ldots, n+1,$$

$$\|P X^{l,*}_{n,3}\|_{0,L^2(S^2)} = \|P Y^{l,*}_{n,3}\|_{0,L^2(S^2)} = \sqrt{\frac{n+1}{n+2}} \sqrt{s_{n,l}} \sqrt{1 + \frac{l^2}{(n+1)^2} s_{n+1,l}},$$

$l = 1, \ldots, n.$

Proof. The calculation can be done using the known norms of the elements of our orthonormal system. $\qquad\square$

Theorem 10.32. *The linear operator*

$$P : L^2(B_3)_{\mathbb{R}} \cap \ker \overline{\partial} \longrightarrow L^2(B_3)_{\mathbb{R}} \cap \ker \overline{\partial}$$

is bounded.

Proof. Let $f \in L^2(B_3)_{\mathbb{R}} \cap \ker \overline{\partial}$. We consider the Fourier series relative to the ONS (10.22)–(10.27) in $L^2(B_3)_{\mathbb{R}} \cap \ker \overline{\partial}$:

$$\begin{aligned}
f &= \sum_{n=0}^{\infty} \sqrt{2n+3}\, r^n \left[X^{0,*}_{n,0} \alpha_n + X^{0,*}_{n,3} \beta_n \right.\\
&\quad + \sum_{m=1}^{n} \left(X^{m,*}_{n,0} \gamma_{n,m} + Y^{m,*}_{n,0} \delta_{n,m} + X^{m,*}_{n,3} \varepsilon_{n,m} + Y^{m,*}_{n,3} \varphi_{n,m} \right) \\
&\quad \left. + X^{n+1,*}_{n,0} \gamma_{n,n+1} + Y^{n+1,*}_{n,0} \delta_{n,n+1} \right],
\end{aligned} \tag{10.52}$$

where α_n, β_n, $\gamma_{n,m}$, $\delta_{n,m}$, $\varepsilon_{n,m}$, $\varphi_{n,m}$, $\gamma_{n,n+1}$, $\delta_{n,n+1} \in \mathbb{R}$, $n \in \mathbb{N}_0$, $m = 1, \ldots, n$. The formal application of the operator P leads to the series

$$
Pf = \sum_{n=0}^{\infty} \sqrt{\frac{2n+3}{2n+5}} \sqrt{2n+5}\, r^{n+1} \left\{ X_{n+1,0}^{0,*} \frac{1}{\sqrt{(n+1)(n+2)}}\, \alpha_n \right.
$$

$$
+ X_{n+1,3}^{0,*} \frac{1}{\sqrt{(n+1)(n+2)}}\, \beta_n + \sum_{m=1}^{n} \left[X_{n+1,0}^{m,*} \left(\sqrt{\frac{n+2}{n+1}} \sqrt{s_{n+1,m}}\, \gamma_{n,m} \right. \right.
$$

$$
- \sqrt{\frac{n+1}{n+2}} \sqrt{s_{n,m}} \frac{m}{n+1} \sqrt{s_{n+1,m}}\, \varphi_{n,m} \Bigg) + Y_{n+1,0}^{m,*} \left(\sqrt{\frac{n+2}{n+1}} \sqrt{s_{n+1,m}}\, \delta_{n,m} \right.
$$

$$
+ \sqrt{\frac{n+1}{n+2}} \sqrt{s_{n,m}} \frac{m}{n+1} \sqrt{s_{n+1,m}}\, \varepsilon_{n,m} \Bigg) + X_{n+1,3}^{m,*} \sqrt{\frac{n+1}{n+2}} \sqrt{s_{n,m}}\, \varepsilon_{n,m}
$$

$$
+ Y_{n+1,3}^{m,*} \sqrt{\frac{n+1}{n+2}} \sqrt{s_{n,m}}\, \varphi_{n,m} \Bigg] + X_{n+1,0}^{n+1,*} \sqrt{\frac{n+2}{n+1}} \sqrt{s_{n+1,n+1}}\, \gamma_{n,n+1}
$$

$$
+ Y_{n+1,0}^{n+1,*} \sqrt{\frac{n+2}{n+1}} \sqrt{s_{n+1,n+1}}\, \delta_{n,n+1} \Bigg\}. \tag{10.53}
$$

On the right-hand side we see again a series development with the complete orthonormal system in $L^2(B_3)_{\mathbb{R}} \cap \ker \overline{\partial}$. To be able to apply the Parseval equation we have to prove the convergence of the following series assuming that the series development for f is convergent:

$$
\sum_{n=0}^{\infty} \frac{2n+3}{2n+5} \left\{ \frac{1}{(n+1)(n+2)} |\alpha_n|^2 + \frac{1}{(n+1)(n+2)} |\beta_n|^2 \right.
$$

$$
+ \sum_{m=1}^{n} \left[\left| \sqrt{\frac{n+2}{n+1}} \sqrt{s_{n+1,m}}\, \gamma_{n,m} - \sqrt{\frac{n+1}{n+2}} \sqrt{s_{n,m}} \frac{m}{n+1} \sqrt{s_{n+1,m}}\, \varphi_{n,m} \right|^2 \right.
$$

$$
+ \left| \sqrt{\frac{n+2}{n+1}} \sqrt{s_{n+1,m}}\, \delta_{n,m} + \sqrt{\frac{n+1}{n+2}} \sqrt{s_{n,m}} \frac{m}{n+1} \sqrt{s_{n+1,m}}\, \varepsilon_{n,m} \right|^2
$$

$$
+ \frac{n+1}{n+2} s_{n,m} |\varepsilon_{n,m}|^2 + \frac{n+1}{n+2} s_{n,m} |\varphi_{n,m}|^2 \Bigg]
$$

$$
+ \frac{n+2}{n+1} s_{n+1,n+1} |\gamma_{n,n+1}|^2 + \frac{n+2}{n+1} s_{n+1,n+1} |\delta_{n,n+1}|^2 \Bigg\}. \tag{10.54}
$$

To simplify we estimate: For all $n \in \mathbb{N}_0$ we have

$$
\frac{2n+3}{2n+5} \frac{1}{(n+1)(n+2)} |\alpha_n|^2 < |\alpha_n|^2 \tag{10.55}
$$

and

$$
\frac{2n+3}{2n+5} \frac{1}{(n+1)(n+2)} |\beta_n|^2 < |\beta_n|^2. \tag{10.56}
$$

The inclusion of $s_{n+1,n+1}$ from (10.28) simplifies the coefficients with $\gamma_{n,n+1}$ and $\delta_{n,n+1}$: For all $n \in \mathbb{N}_0$ we have

$$\frac{2n+3}{2n+5}\frac{n+2}{n+1}s_{n+1,n+1}|\gamma_{n,n+1}|^2 = \frac{n+2}{(2n+5)(n+1)}|\gamma_{n,n+1}|^2$$

$$< |\gamma_{n,n+1}|^2, \tag{10.57}$$

$$\frac{2n+3}{2n+5}\frac{n+2}{n+1}s_{n+1,n+1}|\delta_{n,n+1}|^2 < |\delta_{n,n+1}|^2. \tag{10.58}$$

For all other terms we estimate $s_{n,m}$ roughly: For all $n \in \mathbb{N}$ we have

$$\max_{m=1,\dots,n} s_{n+1,m} = \frac{1}{4(n+1)}, \tag{10.59}$$

$$\max_{m=1,\dots,n} s_{n,m} = \frac{1}{2n+1}, \tag{10.60}$$

$$\max_{m=1,\dots,n} \frac{m}{n+1} = \frac{n}{n+1}. \tag{10.61}$$

Thus we find

$$\frac{2n+3}{2n+5}\sum_{m=1}^{n}\left|\sqrt{\frac{n+2}{n+1}}\sqrt{s_{n+1,m}}\,\gamma_{n,m} - \sqrt{\frac{n+1}{n+2}}\sqrt{s_{n,m}}\,\frac{m}{n+1}\sqrt{s_{n+1,m}}\,\varphi_{n,m}\right|^2$$

$$\leq \sum_{m=1}^{n}\left(\frac{n+2}{n+1}s_{n+1,m}|\gamma_{n,m}|^2 + \frac{n+1}{n+2}s_{n,m}\frac{m^2}{(n+1)^2}s_{n+1,m}|\varphi_{n,m}|^2\right.$$

$$\left. + 2\,s_{n+1,m}\sqrt{s_{n,m}}\,\frac{m}{n+1}|\gamma_{n,m}|\,|\varphi_{n,m}|\right)$$

$$\overset{(10.59)-(10.61)}{\leq} \frac{n+2}{4(n+1)^2}\sum_{m=1}^{n}|\gamma_{n,m}|^2 + \frac{n^2}{4(n+2)(2n+1)(n+1)^2}\sum_{m=1}^{n}|\varphi_{n,m}|^2$$

$$+ \frac{n}{4(n+1)^2\sqrt{2n+1}}\sum_{m=1}^{n}|\gamma_{n,m}|\,|\varphi_{n,m}|$$

$$\leq \sum_{m=1}^{n}|\gamma_{n,m}|^2 + \sum_{m=1}^{n}|\varphi_{n,m}|^2 + 2\sum_{m=1}^{n}|\gamma_{n,m}|\,|\varphi_{n,m}| \tag{10.62}$$

$$\leq \sum_{m=1}^{n}|\gamma_{n,m}|^2 + \sum_{m=1}^{n}|\varphi_{n,m}|^2 + 2\left(\sum_{m=1}^{n}|\gamma_{n,m}|^2\right)^{1/2}\left(\sum_{m=1}^{n}|\varphi_{n,m}|^2\right)^{1/2}.$$

To resolve the quadratic terms we used the Cauchy–Schwarz inequality. Since

$$2\left(\sum_{m=1}^{n}|\gamma_{n,m}|^2\right)^{1/2}\left(\sum_{m=1}^{n}|\varphi_{n,m}|^2\right)^{1/2} \leq \sum_{m=1}^{n}|\gamma_{n,m}|^2 + \sum_{m=1}^{n}|\varphi_{n,m}|^2$$

we get from (10.62) for all $n \in \mathbb{N}$,

$$\frac{2n+3}{2n+5}\sum_{m=1}^{n}\left|\sqrt{\frac{n+2}{n+1}}\sqrt{s_{n+1,m}}\,\gamma_{n,m} - \sqrt{\frac{n+1}{n+2}}\sqrt{s_{n,m}}\,\frac{m}{n+1}\sqrt{s_{n+1,m}}\,\varphi_{n,m}\right|^2$$

$$< 2\sum_{m=1}^{n}|\gamma_{n,m}|^2 + 2\sum_{m=1}^{n}|\varphi_{n,m}|^2. \tag{10.63}$$

Analogously for all $n \in \mathbb{N}$ it follows that

$$\frac{2n+3}{2n+5} \sum_{m=1}^{n} \left| \sqrt{\frac{n+2}{n+1}} \sqrt{s_{n+1,m}}\, \delta_{n,m} + \sqrt{\frac{n+1}{n+2}} \sqrt{s_{n,m}} \frac{m}{n+1} \sqrt{s_{n+1,m}}\, \varepsilon_{n,m} \right|^2$$

$$< 2 \sum_{m=1}^{n} |\delta_{n,m}|^2 + 2 \sum_{m=1}^{n} |\varepsilon_{n,m}|^2, \tag{10.64}$$

and

$$\frac{2n+3}{2n+5} \sum_{m=1}^{n} \frac{n+1}{n+2} s_{n,m} |\varepsilon_{n,m}|^2 \stackrel{(10.60)}{\leq} \frac{(2n+3)(n+1)}{(2n+5)(2n+1)(n+2)} \sum_{m=1}^{n} |\varepsilon_{n,m}|^2$$

$$< \sum_{m=1}^{n} |\varepsilon_{n,m}|^2, \tag{10.65}$$

as well as

$$\frac{2n+3}{2n+5} \sum_{m=1}^{n} \frac{n+1}{n+2} s_{n,m} |\varphi_{n,m}|^2 < \sum_{m=1}^{n} |\varphi_{n,m}|^2. \tag{10.66}$$

If we now collect (10.55)–(10.58) and (10.63)–(10.66) we get the estimate

$$|\alpha_n|^2 + |\beta_n|^2 + \sum_{m=1}^{n} \left(2|\gamma_{n,m}|^2 + 3|\varphi_{n,m}|^2 + 2|\delta_{n,m}|^2 + 3|\varepsilon_{n,m}|^2 \right)$$

$$+ |\gamma_{n,n+1}|^2 + |\delta_{n,n+1}|^2$$

$$< 3 \left[|\alpha_n|^2 + |\beta_n|^2 + \sum_{m=1}^{n} \left(|\gamma_{n,m}|^2 + |\varphi_{n,m}|^2 + |\delta_{n,m}|^2 + |\varepsilon_{n,m}|^2 \right) \right.$$

$$\left. + |\gamma_{n,n+1}|^2 + |\delta_{n,n+1}|^2 \right]$$

for all $n \in \mathbb{N}_0$. We know from the Parseval equation for f that

$$3 \sum_{n=0}^{\infty} \left[|\alpha_n|^2 + |\beta_n|^2 + \sum_{m=1}^{n} \left(|\gamma_{n,m}|^2 + |\varphi_{n,m}|^2 + |\delta_{n,m}|^2 + |\varepsilon_{n,m}|^2 \right) \right.$$

$$\left. + |\gamma_{n,n+1}|^2 + |\delta_{n,n+1}|^2 \right] = 3 \|f\|^2_{L^2(B_3)}.$$

Thus the series (10.54) is convergent, it then follows that $Pf \in L^2(B_3)_{\mathbb{R}} \cap \ker \overline{\partial}$. Moreover we proved that

$$\|Pf\|_{L^2(B_3)} \leq \sqrt{3}\, \|f\|_{L^2(B_3)}, \quad f \in L^2(B_3)_{\mathbb{R}} \cap \ker \overline{\partial}, \tag{10.67}$$

so that indeed the operator P is bounded. $\square$

The above estimate can be sharpened easily and leads then to better estimates of the norm of P. But already the rough estimate shows how important the orthogonal system is for us as it allows the application of the Parseval equation. If we tried to give a definition using the more easily defined Fueter polynomials and the Taylor series, we would have had to deal with the eigenvalues of the Gram matrix of the Fueter polynomials, which would have been a quite more difficult task.

10.3.3 Decomposition theorem and Taylor expansion

Let us have another look at the Taylor expansion of a holomorphic function in $\mathbb{C}$. This expansion can also be seen as a splitting or decomposition of the space of quadratically integrable holomorphic functions relative to the kernels of the derivatives $\partial^n / \partial z^n$. This splitting is direct and orthogonal as we use the functions

$$z^n \in \ker \frac{\partial^{n+1}}{\partial z^{n+1}} \setminus \ker \frac{\partial^n}{\partial z^n}.$$

We now seek an analogous approach for holomorphic functions in $C\ell(2)$. With the Corollaries 10.25 and 10.26 we have the appropriate tool.

We start with the investigation in $L^2(B_3)_{\mathbb{R}} \cap \ker \overline{\partial}$. From Corollary 10.25 we know that

$$r^n X_{n,0}^{0,*}, \ r^n X_{n,3}^{0,*} \in \ker \partial^{n+1} \cap \ker \overline{\partial} \tag{10.68}$$

and

$$r^n X_{n,0}^{n+1,*}, \ r^n Y_{n,0}^{n+1,*} \in \ker \partial \cap \ker \overline{\partial}. \tag{10.69}$$

Moreover we have from Corollary 10.26,

$$r^n X_{n,0}^{m,*}, \ r^n Y_{n,0}^{m,*}, \ r^n X_{n,3}^{m,*}, \ r^n Y_{n,3}^{m,*} \in \ker \partial^{n-m+2} \cap \ker \overline{\partial}, \tag{10.70}$$
$$n \geq 1, m = 1, \dots, n.$$

For $n = 0$ obviously $r^n X_{0,0}^{0,*}, \ r^n X_{0,3}^{0,*} \in \ker \partial$ holds. This case is contained in (10.68). With these considerations we are now able to formulate the decomposition theorem:

Theorem 10.33. *The space* $L^2(B_3)_{\mathbb{R}} \cap \ker \overline{\partial}$ *allows the following orthogonal splitting:*

$$L^2(B_3)_{\mathbb{R}} \cap \ker \overline{\partial} = \bigoplus_{n \geq 1} \left((\ker \partial^n \ominus \ker \partial^{n-1}) \cap \ker \overline{\partial} \right). \tag{10.71}$$

Proof. For all $k \in \mathbb{N}$ the intersection $\ker \partial^k \cap \ker \overline{\partial}$ is a closed subspace of $L^2(B_3)_{\mathbb{R}} \cap \ker \overline{\partial}$. With the induced inner product of the Hilbert space $L^2(B_3)_{\mathbb{R}} \cap \ker \overline{\partial}$ the subspace $\ker \partial^k \cap \ker \overline{\partial}$ is also a Hilbert space. On the other hand $\ker \partial^{k-1} \cap \ker \overline{\partial}$ is also a closed subspace of $\ker \partial^k \cap \ker \overline{\partial}$ and in this space possesses an orthogonal complement,

$$\ker \partial^k \cap \ker \overline{\partial} = (\ker \partial^{k-1} \cap \ker \overline{\partial}) \oplus \left((\ker \partial^k \ominus \ker \partial^{k-1}) \cap \ker \overline{\partial} \right), \tag{10.72}$$

where $(\ker \partial^k \ominus \ker \partial^{k-1}) \cap \ker \overline{\partial} = \left(\ker \partial^{k-1} \cap \ker \overline{\partial} \right)^{\perp}$.

The equation (10.72) implies that for all $k \in \mathbb{N}$ the finite decomposition

$$\ker \partial^k \cap \ker \overline{\partial} = \bigoplus_{n=1}^{k} \left((\ker \partial^n \ominus \ker \partial^{n-1}) \cap \ker \overline{\partial} \right) \tag{10.73}$$

can be written. Let us now look at a function $f \in L^2(B_3)_{\mathbb{R}} \cap \ker \overline{\partial}$, which is orthogonal to the subspace $\oplus_{n \geq 1} \left((\ker \partial^n \ominus \ker \partial^{n-1}) \cap \ker \overline{\partial} \right)$. Then f is orthogonal to all terms $(\ker \partial^n \ominus \ker \partial^{n-1}) \cap \ker \overline{\partial}$ $(n \in \mathbb{N})$ and thus also to every finite direct sum of these subspaces, i.e., for all $k \in \mathbb{N}$ we have

$$f \perp \left(\oplus_{n=1}^{k} (\ker \partial^n \ominus \ker \partial^{n-1}) \cap \ker \overline{\partial} \right).$$

From (10.73) it follows again for all $k \in \mathbb{N}$,

$$f \perp \ker \partial^k \cap \ker \overline{\partial}. \tag{10.74}$$

If we analyze our complete orthogonal system

$$\Big\{ \sqrt{2n+3}\, r^n\, X_{n,0}^{0,*},\ \sqrt{2n+3}\, r^n\, X_{n,3}^{0,*},\ \sqrt{2n+3}\, r^n\, X_{n,0}^{m,*},\ \sqrt{2n+3}\, r^n\, Y_{n,0}^{m,*},$$
$$\sqrt{2n+3}\, r^n\, X_{n,3}^{l,*},\ \sqrt{2n+3}\, r^n\, Y_{n,3}^{l,*} : m = 1, \ldots, n+1,\, l = 1, \ldots, n \Big\}_{n \in \mathbb{N}_0}$$

in $L^2(B_3)_{\mathbb{R}} \cap \ker \overline{\partial}$, we then see with (10.68) that

$$\sqrt{2n+3}\, r^n\, X_{n,0}^{0,*},\ \sqrt{2n+3}\, r^n\, X_{n,3}^{0,*} \in \ker \partial^{n+1} \cap \ker \overline{\partial}.$$

The application of (10.74) leads us to the statement

$$(f, \sqrt{2n+3}\, r^n\, X_{n,0}^{0,*})_{L_2(B_3)} = (f, \sqrt{2n+3}\, r^n\, X_{n,3}^{0,*})_{L_2(B_3)} = 0.$$

Analogously it follows from (10.69)–(10.71) and (10.74) that

$$(f, \sqrt{2n+3}\, r^n\, X_{n,0}^{m,*})_{L_2(B_3)} = (f, \sqrt{2n+3}\, r^n\, Y_{n,0}^{m,*})_{L_2(B_3)}$$
$$= (f, \sqrt{2n+3}\, r^n\, X_{n,3}^{m,*})_{L_2(B_3)} = (f, \sqrt{2n+3}\, r^n\, Y_{n,3}^{m,*})_{L_2(B_3)} = 0.$$

We get that f is orthogonal to all basis elements of $L^2(B_3)_{\mathbb{R}} \cap \ker D$ and thus $f = 0$ holds.

We have described the basis of every subspace $(\ker \partial^n \ominus \ker \partial^{n-1}) \cap \ker \overline{\partial}$ $(n \in \mathbb{N})$ by elements of the complete orthonormal system. So it is clear that all these subspaces are pairwise orthogonal. $\qquad \square$

Here we will only state the analogous result for $L^2(B_3)_{\mathbb{H}} \cap \ker \overline{\partial}$. The proof follows the same pattern:

Theorem 10.34. *The space $L^2(B_3)_{\mathbb{H}} \cap \ker \overline{\partial}$ allows the orthogonal splitting*

$$L^2(B_3)_{\mathbb{H}} \cap \ker \overline{\partial} = \bigoplus_{n \geq 1} \left((\ker \partial^n \ominus \ker \partial^{n-1}) \cap \ker \overline{\partial} \right). \tag{10.75}$$

The representations of f following from this theorem are obviously reorderings of the Fourier expansion and inherit the properties of the complex Taylor expansion formulated at the beginning. But one should notice that the subspaces $(\ker \partial^n \ominus \ker \partial^{n-1}) \cap \ker \bar{\partial}$ are of infinite dimension. This follows from the fact that also polynomials of arbitrarily high degree may have the property that their derivative of degree n vanishes. The importance of the theorem consists in substantially simplifying the termwise differentiation of Fourier series. We refer these remarks to the Fourier series solution of partial differential equations.

10.4 Exercises

1. Prove the quaternionic analogy with the theorem of the Riesz brothers: Let F be an $\mathbb{H}$-right-linear functional over the $\mathbb{H}$-valued space $L^2(G)$. Then it allows the representation

$$F(u) = \int\limits_G \overline{f(x)} u(x) d\sigma$$

   with an $\mathbb{H}$-valued function $f \in L^2(G)$.

2. Prove that a function system in the right-linear Banach space is closed if and only if it is complete.

3. Look for conditions for $\{x^{(i)}\}_{i \in \mathbb{N}}$ that the system $\{\phi_{i \in \mathbb{N}} = \mathcal{Q}_0(x - x^{(i)})\}$ is closed in $L^2(G) \cap \ker D$.

4. Examine whether the powers of the variables $z = x_0 + \frac{1}{2}(x_1 e_1 + x_2 e_2)$ and $z = x_0 + \frac{1}{3}(x_1 e_1 + x_2 e_2 + x_3 e_3)$ are holomorphic.

5. Show that the integer powers of the Fueter variables $z_i = x_i - e_i x_0$ are holomorphic, so that they are totally analytic variables.

6. Show that for the variables

$$z(x) = \sum_{k=0}^{3} a_k x_k$$

   the conditions $a_i a_j = a_j a_i$ for all $i, j = 0, 1, 2, 3$ mean that the range of $z(x)$ can be at most a two-dimensional plane.

7. Prove that the mapping

$$\mathbf{I}f(x) = \mathcal{Q}_0(x) f\left(x^{-1}\right)$$

   from $\mathcal{H}^+$ to $\mathcal{H}^-$ is one-to-one. To prove the holomorphy of $\mathbf{I}f$ if f is holomorphic one should prove and use the equation

$$\Delta(xf) = 2\bar{\partial} f + x \Delta f.$$

8. For an orthonormal system of homogeneous holomorphic polynomials H_n^ν, $n = 0, 1, \ldots$; $\nu = 1, \ldots n + 1$ show that

$$
\begin{aligned}
(H_n^\nu, H_k^\mu)_{L^2(B_3)} &= \int_0^1 r^{n+k+2} \left(\int_{S^2} \overline{H_n^\nu} \, H_k^\mu \, |do| \right) dr \\
&= \frac{1}{n+k+3} (H_n^\nu, H_k^\mu)_{L^2(S^2)} .
\end{aligned}
$$

9. With the spherical coordinates $x = r\omega$ and $|x| = r$, $\omega = x/|x|$ show that

$$
\partial_\omega = \sum_{i=0}^2 \omega_i \partial_i = \partial_r,
$$

$$
\begin{aligned}
L &= \sum_{i=0}^2 e_i (r\partial_i - x_i \partial_\omega) \\
&= (-\sin\theta - e_1 \cos\theta \cos\varphi - e_2 \cos\theta \sin\varphi)\partial_\theta \\
&\quad + \frac{1}{\sin\theta} (e_1 \sin\varphi - e_2 \cos\varphi)\partial_\varphi .
\end{aligned}
$$

10. Apply the spherical adjoint Cauchy–Riemann operator ∂ directly to the harmonic spherical functions and discuss the differences with the procedure used in Subsection 10.1.4.

11. Prove Theorem 10.14: *For all $n \in \mathbb{N}_0$ the systems $X_{n,3}^0$, $X_{n,3}^l$ and $Y_{n,3}^l$ ($l = 1, \ldots, n$) are orthogonal relatively to the inner product (10.2), and the norms are given by*

$$
\|X_{n,3}^0\|_{0, L^2(S^2)} = \sqrt{\pi(n+1)}
$$

and by

$$
\|X_{n,3}^l\|_{0, L^2(S^2)} = \|Y_{n,3}^l\|_{0, L^2(S^2)} = \sqrt{\frac{\pi}{2}(n+1)\frac{(n+1+l)!}{(n+1-l)!}}, l = 1, \ldots, n.
$$

12. Prove Theorem 10.24: *Let $n \in \mathbb{N}_0$, $i = 1, 2, \ldots$. We then have*

$$
\partial^i X_n^m = \left(\prod_{h=1}^i (n+m+1-(h-1)) \right) X_{n-i}^m, \quad m = 0, \ldots, n+1-i,
$$

$$
\partial^i Y_n^m = \left(\prod_{h=1}^i (n+m+1-(h-1)) \right) Y_{n-i}^m, \quad m = 1, \ldots, n+1-i.
$$

13. Prove Corollary 10.25: *For arbitrary $n \in \mathbb{N}_0$ we have*

$$
\partial^i X_n^m = \partial^i Y_n^m = 0, \quad i \geq n - m + 2; \quad m = 1, \ldots, n+1.
$$

14. Prove also Corollary 10.26: *For arbitrary* $n \in \mathbb{N}_0$ *we have*

$$r^n X_n^0 \quad \in (\ker \partial^{n+1} \setminus \ker \partial^n) \cap \ker \overline{\partial},$$
$$r^n X_n^m, \quad r^n Y_n^m \in (\ker \partial^{n-m+2} \setminus \ker \partial^{n-m+1}) \cap \ker \overline{\partial}, \quad m = 1, \dots, n+1,$$

where ∂^0 *should be identified with the identity.*

15. Show that the derivatives (10.36)–(10.41) constitute a basis in $\mathcal{H}_{n-1,\mathbb{R}}^+$.

11   Elementary functions

11.1   Elementary functions in $\mathbb{C}$

11.1.1   Exponential function

The functions covered in this section are called *elementary functions*, although they are not really so elementary. We will start by examining the exponential function already known from real analysis, of fundamental importance since it does not only serve as the generator for all the other elementary functions, but it also plays an important role in applications.

Definition 11.1. For all $z \in \mathbb{C}$ the *exponential function* or *e-function* is defined by

$$e^z := \exp(z) := \sum_{n=0}^{\infty} \frac{z^n}{n!}.$$

The radius of convergence $\rho = \infty$ can be evaluated easily (see Exercise 4.4.5), so that e^z is defined in $\mathbb{C}$.

Theorem 11.2. *The exponential function*

(i)   *is holomorphic in $\mathbb{C}$ with*

$$\frac{de^z}{dz} = e^z$$

(ii)   *and for all $z, \zeta \in \mathbb{C}$ the functional equation*

$$e^z e^\zeta = e^{z+\zeta}$$

   holds.

(iii)   *For $x \in \mathbb{R}$ we have $e^x > 0$ with $e^0 = 1$ for all $z \in \mathbb{C}$ the formulae*

$$\overline{e^z} = e^{\overline{z}}, \quad |e^z| = e^x \le e^{|z|} \quad \text{as well as} \quad |e^{ix}| = 1.$$

   are valid. $e^z \ne 0$ for all $z \in \mathbb{C}$.

(iv)   *For all $z \in \mathbb{C}$ we have*

$$e^z = \lim_{n \to \infty} \left(1 + \frac{z}{n}\right)^n.$$

Proof. (i) Theorem 9.5 yields

$$\frac{de^z}{dz} = \sum_{n=1}^{\infty} \frac{z^{n-1}}{(n-1)!} = e^z$$

after shifting the summation index by 1.

(ii) Let ζ be fixed. Define the function

$$g(z) := e^{-z}\, e^{\zeta+z}$$

with the derivative

$$g'(z) = -g(z) + g(z) = 0.$$

Thus g is constant. This leads to $g(0) = g(z) = e^{\zeta}$ for $z = 0$ and consequently

$$e^{\zeta} = e^{-z}\, e^{\zeta+z}.$$

Substituting $\zeta = 0$ the equality $e^z\, e^{-z} = 1$ follows. This gives the functional equation.

(iii) For $x \geq 0$ the power series consists of non-negative terms and at least one positive term, so that $e^x > 0$. The functional equation yields now $e^{-x} > 0$. If we had $e^{z_0} = 0$ for a $z_0 \in \mathbb{C}$ we would obtain

$$1 = e^0 = e^{-z_0}\, e^{z_0} = 0.$$

The equality $\overline{e^z} = e^{\overline{z}}$ is easily obtained by remarking that the coefficients are real. The proof of the remaining equalities is left as Exercise 11.33.

(iv) Considering $\binom{n}{k} = 0$ for $k > n$ we have

$$e^z - \left(1 + \frac{z}{n}\right)^n = \sum_{k=0}^{\infty} \left(\frac{1}{k!} - \binom{n}{k}\frac{1}{n^k}\right) z^k.$$

The coefficients

$$\frac{1}{k!} - \binom{n}{k}\frac{1}{n^k} = \frac{1}{k!}\left(1 - \frac{n}{n}\cdot\frac{n-1}{n}\cdot\ldots\cdot\frac{n-k+1}{n}\right) > 0$$

are positive, leading to

$$\left|e^z - \left(1 + \frac{z}{n}\right)^n\right| \leq \sum_{k=0}^{\infty} \left(\frac{1}{k!} - \binom{n}{k}\frac{1}{n^k}\right)|z|^k = e^{|z|} - \left(1 + \frac{|z|}{n}\right)^n.$$

The term on the right-hand side converges to zero for $n \to \infty$ in agreement with the real case. $\qquad\square$

11.1.2 Trigonometric functions

With the help of the exponential function the so-called *trigonometric functions* can be defined quite easily, but the analogy with the functions of an angle and of the unit circle known from the real case cannot be seen so easily.

Definition 11.3. The *cosine* and *sine* functions are defined by

$$\cos z := \frac{1}{2}(e^{iz} + e^{-iz}), \quad \sin z := \frac{1}{2i}(e^{iz} - e^{-iz}),$$

and belong to the previously mentioned *trigonometric functions*.

Remark 11.4. The term "circle functions" (functions of an angle) usual in the real case can lead to misunderstandings when using a complex argument, since formulae like $\sin ix = -\sinh x$ and $\cos ix = \cosh x$ hold, so that the hyperbolic functions would be a subset of the functions of an angle.

The usual properties certainly continue to hold:

Theorem 11.5.   (i) *For real $z = x$ we have*

$$\cos x = \operatorname{Re} e^{ix}, \quad \sin x = \operatorname{Im} e^{ix}$$

and thus

$$e^{ix} = \cos x + i \sin x \quad (\text{Euler's formula}),$$

as well as $\cos 0 = 1$, $\sin 0 = 0$. Consequently every complex number can be represented in the exponential form $z = re^{i\varphi}$. Recall that $|e^{ix}| = 1$ for all real x (Theorem 11.2 (iii)).

(ii) *For all $z \in \mathbb{C}$ we have the power series representations*

$$\cos z = \sum_{n=0}^{\infty}(-1)^n \frac{z^{2n}}{(2n)!}, \quad \sin z = \sum_{n=0}^{\infty}(-1)^n \frac{z^{2n+1}}{(2n+1)!}.$$

Furthermore cosine is an even *function, sine is an* odd *function, so that*

$$\cos(-z) = \cos z, \quad \sin(-z) = -\sin z.$$

(iii) *The following* sum and difference formulas *hold: for all $z, z_1, z_2 \in \mathbb{C}$ we have*

$$\cos(z_1 + z_2) = \cos z_1 \cos z_2 - \sin z_1 \sin z_2,$$
$$\sin(z_1 + z_2) = \sin z_1 \cos z_2 + \cos z_1 \sin z_2,$$
$$\cos^2 z + \sin^2 z = 1.$$

(iv) *Cosine and sine are holomorphic functions in $\mathbb{C}$ satisfying*

$$\frac{d}{dz}\cos z = -\sin z, \quad \frac{d}{dz}\sin z = \cos z.$$

(v) *According to real analysis the smallest positive real zero of $\cos x$ is $\pi/2$. Then for all $z \in \mathbb{C}$:*

$$\cos(z + 2\pi) = \cos z, \quad \sin(z + 2\pi) = \sin z, \quad e^{z+2\pi i} = e^z.$$

It can be shown that the previous properties are sufficient to define sine and cosine uniquely. According to the theorem the functions known from the real case coincide with the complex ones.

Proof. (i) Using part (iii) of the preceding theorem for real x we get

$$\overline{e^{ix}} = e^{\overline{ix}} = e^{-ix}$$

showing part (i). The values of cos 0 and sin 0 follow from $e^0 = 1$.

(ii) From $i^{2n} = (-1)^n$ and $i^{2n+1} = i(-1)^n$ we obtain

$$e^{iz} = \sum_{k=0}^{\infty} (-1)^k \frac{z^{2k}}{(2k)!} + i \sum_{k=0}^{\infty} (-1)^k \frac{z^{2k+1}}{(2k+1)!}.$$

(iii) The proof is left as an exercise for the interested reader (see Exercise 11.3.1).

(iv) The chain rule yields

$$\frac{d}{dz} \cos z = \frac{1}{2} \frac{d}{dz} (e^{iz} + e^{-iz}) = \frac{1}{2} (ie^{iz} - ie^{-iz}) = -\sin z.$$

The statement for $\sin z$ can be calculated in a similar fashion.

(v) We want to show the existence of a smallest positive zero of the cosine in a short way: It is clear that $\cos 0 = 1$. We use the power series to estimate cos 2:

$$\cos 2 = 1 - \left(2 - \frac{2}{3}\right) - \cdots - \left(\frac{2^{2k}}{(2k)!} - \frac{2^{2k+2}}{(2k+2)!}\right) - \cdots .$$

Notice that the terms in brackets are positive. Indeed for $k > 2$ one gets

$$\frac{2^{2k}}{(2k)!} - \frac{2^{2k+2}}{(2k+2)!} = \frac{2^{2k}}{(2k)!} \left(1 - \frac{2^2}{(2k+1)(2k+2)}\right) > 0.$$

Consequently $\cos 2 < -1/3$. According to the mean value theorem of real analysis cosine must have a zero between 0 and 2. The smallest is called $\pi/2$. Such a smallest zero must exist for continuity reasons. The sum and difference formulas yield $|\sin(\pi/2)| = 1$ and thus $\cos \pi = -1$, $\sin \pi = 0$, and finally we have $\cos(2\pi) = 1$, $\sin(2\pi) = 0$. This leads to

$$
\begin{aligned}
\cos(z + 2\pi) &= \cos z \cos(2\pi) - \sin z \sin(2\pi) = \cos z, \\
\sin(z + 2\pi) &= \sin z \cos(2\pi) + \cos z \sin(2\pi) = \sin z, \\
e^{z+2\pi i} &= e^x e^{i(y+2\pi)} = e^x \left(\cos(y + 2\pi) + i \sin(y + 2\pi)\right) = e^z.
\end{aligned}
$$
$\qquad\square$

11.1.3 Hyperbolic functions

As already mentioned the hyperbolic functions coincide with the trigonometric functions after a rotation by $\pi/2$.

Definition 11.6. The functions *hyperbolic cosine* and *hyperbolic sine* are defined by

$$\cosh z := \frac{1}{2}(e^z + e^{-z}), \quad \sinh z := \frac{1}{2}(e^z - e^{-z})$$

for all $z \in \mathbb{C}$.

In order to complete this section the usual properties are listed:

Theorem 11.7.   (i) *For real* $z = x$ *the hyperbolic functions take real values, with* $\cosh 0 = 1$ *and* $\sinh 0 = 0$.

(ii) *For all* $z \in \mathbb{C}$ *the hyperbolic functions admit the power series expansions*

$$\cosh z = \sum_{n=0}^{\infty} \frac{z^{2n}}{(2n)!}, \quad \sinh z = \sum_{n=0}^{\infty} \frac{z^{2n+1}}{(2n+1)!},$$

$\cosh z$ *is an even function while* $\sinh z$ *is odd.*

(iii) *The sum and difference formulas hold: For all* $z, z_1, z_2 \in \mathbb{C}$ *we have*

$$\begin{aligned}
\cosh(z_1 + z_2) &= \cosh z_1 \cosh z_2 + \sinh z_1 \sinh z_2, \\
\sinh(z_1 + z_2) &= \sinh z_1 \cosh z_2 + \cosh z_1 \sinh z_2, \\
\cosh^2 z - \sinh^2 z &= 1.
\end{aligned}$$

(iv) *The hyperbolic cosine and the hyperbolic sine are holomorphic in* $\mathbb{C}$ *satisfying*

$$\frac{d}{dz} \cosh z = \sinh z, \quad \frac{d}{dz} \sinh z = \cosh z.$$

(v) *The hyperbolic cosine and the hyperbolic sine are periodic with the period* $2\pi i$. *We have the relations*

$$\begin{aligned}
\cosh z &= \cos iz, \quad \sinh z = -i \sin iz, \\
\cos z &= \cosh iz, \quad \sin z = -i \sinh iz, \\
\cos z &= \cos x \cosh y - i \sin x \sinh y, \\
\sin z &= \sin x \cosh y + i \cos x \sinh y,
\end{aligned}$$

with $x = \operatorname{Re} z$, $y = \operatorname{Im} z$.

(vi) *The only zeros of* $\cos z$ *are the real numbers* $z_n = \frac{\pi}{2} + n\pi$, $n \in \mathbb{Z}$; $\sin z$ *becomes zero only at the points* $z_m = m\pi$, $m \in \mathbb{Z}$. *Finally* $\cosh z$ *has the zeros* iz_n *and* $\sinh z$ *vanishes at* iz_m.

The hyperbolic cosine is also known as *catenary*, since the shape of a flexible chain is similar to the graph of the function. This can be obtained by solving second-order ordinary differential equations.

Proof. (i) and (ii) follow immediately from the definition and the power series representation of e^z.

(iii) and (iv): The sum and difference formulas can be computed similarly to the corresponding ones of sine and cosine. The derivatives are a direct consequence of the properties of the exponential function.

(v) The proof is left as Exercise 11.3.2.

(vi) part (v) yields

$$|\cos z|^2 = \cos^2 x \cosh^2 y + \sin^2 x \sinh^2 y = \cos^2 x + \sinh^2 y.$$

This equals zero if and only if $\cos x = 0$, leading to the z_n, and $\sinh y = 0$, occurring only if $y = 0$ according to the definition of hyperbolic sine. The case sine follows using similar considerations. The zeros of the hyperbolic cosine and hyperbolic sine are obtained by rotating by $\pi/2$. $\square$

Finally we want to mention that the well-known functions *tangent* and *cotangent* as well as the *hyperbolic tangent* and the *hyperbolic cotangent* are defined as quotients of sine and cosine, resp. hyperbolic sine and hyperbolic cosine.

11.1.4 Logarithm

Constructing complex inverse functions for the trigonometric functions proves to be a lot more difficult than in the real case. We already encountered this problem when dealing with the integer powers and their inverses in Section 6.1. Now we want to consider the inverse of the exponential function. The formula $w = e^z = e^x e^{iy}$ yields $|w| = e^x$ and $\arg w = y$, so that

$$x = \ln |w|, \quad y = \arg w$$

is the inverse. The problem is that the argument is not uniquely defined, in contrast to the (natural) logarithm of the real line. At first we can state:

Definition 11.8. The *principal value of the logarithm* is defined by

$$\log z := \ln |z| + i \arg z, \quad -\pi < \arg z \le \pi.$$

This definition is not satisfying, because the limitation of the argument was chosen arbitrarily. A different limitation would be possible, too. We shall construct an example of a *Riemann surface* in analogy to Section 6.1.

Obviously the principal value of the logarithm maps the plane denoted by E_0 to the strip

$$|\operatorname{Im} w| < \pi.$$

Taking into account that the exponential function is periodic we will map onto the parallel strips of the w-plane by considering the z-planes

$$E_n : \log z = \ln |z| + i \arg z, \quad -\pi + 2n\pi < \arg z < \pi + 2n\pi, \quad n \in \mathbb{Z}.$$

The plane E_n will be mapped onto the strip

$$-\pi + 2n\pi < \operatorname{Im} w < \pi + 2n\pi.$$

All the planes E_n are cut along the negative real line and connected along the transition of one n to the following or preceding one: the upper edge of E_0 is glued together with the lower edge of E_1 and so on, so that the argument is continuous. E_1 is located above E_0, so that we obtain a helix surface that extends to infinity in both directions.

The logarithm defined on the Riemann surface — called $\mathcal{F}_{\log}$ — is the reasonable inverse of the exponential function:

Definition 11.9. The *logarithm* is defined on $\mathcal{F}_{\log}$ by

$$\log z := \ln|z| + i\arg z, \quad -\pi + 2n\pi < \arg z \leq \pi + 2n\pi, \quad z \in E_n,$$

in a unique way. One is usually unaccustomed to dealing with such domains or Riemann surfaces, but it serves as an appropriate way to deal with the ambiguous argument. There follows:

Proposition 11.10. *On $\mathcal{F}_{\log}$ the logarithm is a holomorphic function satisfying*

$$\frac{d}{dz}\log z = \frac{1}{z}.$$

The proof is straightforward — differentiating the inverse function, $de^z/dz = e^z$, yields:

$$\frac{d}{dz}\log z = \frac{1}{e^{\log z}} = \frac{1}{z}.$$

The dealing with the surface $\mathcal{F}_{\log}$ is still incomplete as in Section 6.1. Since every point on $\mathcal{F}_{\log}$ possesses an ordinary ε-neighborhood on one of the planes — even on the connected edges by combining two semi-circles — the definition for a holomorphic function can be used and the given function can be differentiated. Only the origin common to all the planes has to be left out, such a point is called a *logarithmic singularity* of a Riemann surface.

Finally we will use the logarithm to define the *generalized powers*:

Definition 11.11. For $z, a \in \mathbb{C}$ the power is defined by

$$z^a := \exp(a\log z).$$

This is only defined in a unique way on the Riemann surface of the logarithm; in $\mathbb{C}$ one value of the logarithm has to be decided, usually the principal value. Only the integer powers result in a uniquely defined function in $\mathbb{C}$, because the exponential function is periodic. For rational values of a things are not so complicated. Consider, e.g., $z^{1/2}$, after encircling twice the origin the same values are found again, so that this function can be called "double-valued".

11.2 Elementary functions in $C\ell(n)$

11.2.1 Polar decomposition of the Cauchy–Riemann operator

Defining suitable elementary functions in the paravector space of $\mathbb{R}^{n+1}$ proves to be more difficult than in $\mathbb{C}$. Some aspects coinciding in $\mathbb{C}$ need to be distinguished and taken special care of. In this section we will introduce methods allowing us to define suitable functions.

Therefore it is important to have an appropriate decomposition of the Cauchy–Riemann operator. It should consist of one component along the radial direction and another component representing the tangential derivatives on the n-dimensional sphere.

Definition 11.12. The usual notation is given by

$$L = \sum_{i=0}^{n} e_i L_i(x), \quad L_i(x) = |x|\partial_i - x_i \partial_\omega,$$

$$\partial_\omega = \sum_{i=0}^{n} \omega_i \partial_i = \omega \cdot \nabla, \quad \omega = \sum_{i=0}^{n} \omega_i e_i = \frac{x}{|x|}, \quad \omega_i = \frac{x_i}{|x|}.$$

L and ∂_ω are the operators which allow an appropriate radial decomposition of $\bar{\partial}$. In order to create a better understanding for the representation, study of the effects of the operators L and ∂_ω proves to be useful.

Proposition 11.13. We have:

(i) $\bar{\partial} = \frac{1}{|x|}L + \omega \partial_\omega$ resp. $L = |x|\bar{\partial} - x\partial_\omega$,

(ii) $\partial_i |x| = \frac{x_i}{|x|} = \omega_i$,

(iii) $\partial_\omega x = \omega$,

(iv) $|x|\partial_j \omega_k = \delta_{jk} - \omega_k \omega_j$,

(v) $\omega \cdot L = \sum_{i=0}^{n} \omega_i L_i = 0$.

(i) is the appropriate decomposition, (ii) serves as an auxiliary formula, (iii)–(v) point out the effects of the newly defined operators.

Proof. Showing the parts (i), (ii) and (iii) are left as an exercise (see Exercise 11.3.8). The quotient rule yields part (iv), and we have

$$\partial_j \omega_k = \partial_j \frac{x_k}{|x|} = \frac{\delta_{jk}}{|x|} - \frac{x_j x_k}{|x|^3} = \frac{\delta_{jk} - \omega_j \omega_k}{|x|}.$$

Part (v) can be obtained by using $\sum_{i=0}^{n} \omega_i^2 = 1$ as follows:

$$\sum_{i=0}^{n} \omega_i L_i = \sum_{i=0}^{n} \omega_i(|x|\partial_i - x_i \partial_\omega) = \sum_{i=0}^{n} x_i \partial_i - \sum_{i=0}^{n} \omega_i^2 \sum_{j=0}^{n} x_j \partial_j = 0. \qquad \square$$

The next proposition will cover the effects of our operators on functions depending only on $r = |x|$, such functions are called *radially symmetric*.

Proposition 11.14. *Suppose the function* $f \in C^1(\mathbb{R}^+)$ *depends only on* $|x|$ *and maps to* $C\ell(n)$. *Then*

$$\text{(i)} \ \ Lf = 0 \qquad and \qquad \text{(ii)} \ \ \partial_\omega f = \frac{d}{d|x|} f = f'.$$

We can see clearly that L respects the tangential direction to the sphere, whereas ∂_ω represents the derivative with respect to the radial direction.

Proof. (i) Let $j \in \{0, 1, \ldots, n\}$ be fixed. We have

$$
\begin{aligned}
L_j f(|x|) &= |x|\partial_j f(|x|) - x_j \partial_\omega f(|x|) = |x| f'(|x|)\partial_j |x| - x_j \sum_{i=0}^{n} \omega_i \partial_i f(|x|) \\
&= f'(|x|)\left(x_j - x_j \sum_{i=0}^{n} \omega_i^2\right) = 0.
\end{aligned}
$$

(ii) We deduce

$$\partial_\omega f(|x|) = \sum_{i=0}^{n} \omega_i \partial_i f(|x|) = \sum_{i=0}^{n} \omega_i f'(|x|)\omega_i = f'(|x|). \qquad \square$$

Proposition 11.15. *Suppose that* φ *is a real-valued function defined on the* n-*dimensional unit sphere. Then it satisfies the relations:*

$$\text{(i)} \ \ L\varphi = \text{grad}_\omega \varphi - \omega(\omega \cdot \text{grad}_\omega \varphi) \qquad and \qquad \text{(ii)} \ \ \partial_\omega \varphi = 0.$$

$\text{grad}_\omega \varphi := \sum_{i=0}^{n} e_i \partial_{\omega_i} \varphi$ *denotes the gradient with respect to the variable* ω.

This is the counterpart of the preceding proposition; the function φ does not depend on $|x|$, so that $\partial_\omega \varphi$ equals zero and L takes effect.

Proof. Using the chain rule and 11.13 (iv) we obtain

$$
\begin{aligned}
|x|\partial_i \varphi(\omega) &= |x| \sum_{j=1}^{n} \partial_i \omega_j \partial_{\omega_j} \varphi(\omega) = |x| \sum_{j=1}^{n} \frac{\delta_{ij} - \omega_i \omega_j}{|x|} \partial_{\omega_j} \varphi(\omega) \\
&= \partial_{\omega_i} \varphi(\omega) - \omega_i(\omega \cdot \text{grad}_\omega \varphi)
\end{aligned}
$$

and

$$
\begin{aligned}
\partial_\omega \varphi(\omega) &= \sum_{j=1}^{n} \omega_j \partial_j \varphi(\omega) = \sum_{j,k=0}^{n} \omega_j \frac{\delta_{jk} - \omega_j \omega_k}{|x|} \partial_{\omega_k} \varphi(\omega) \\
&= \frac{1}{|x|} \sum_{k=0}^{n} (\omega_k - \omega_k)\partial_{\omega_k} \varphi(\omega) = 0.
\end{aligned}
$$

This proves part (ii). Thus the second term of L vanishes, so we have then

$$L\varphi(\omega) = \sum_{i=0}^{n} [e_i \partial_{\omega_i} \varphi(\omega) - e_i \omega_i(\omega \cdot \text{grad}_\omega \varphi(\omega))]. \qquad \square$$

Definition 11.16. The operators $E := |x|\partial_\omega$ and $\Gamma := \omega L$ are called the *Euler operator* and the *Dirac operator on the sphere* S^n. The second one is defined as:

$$\Gamma\varphi(\omega) := \omega\left[\text{grad}_\omega\varphi(\omega) - \omega(\omega \cdot \text{grad}_\omega\varphi(\omega))\right].$$

Let us now consider some properties of these operators:

Corollary 11.17. (i) *Suppose that* $\varphi(\omega) = \omega \cdot v$, *where* ω *is an element of the unit sphere* S^n *in the paravector space, and* v *belongs to* $\mathbb{R}^{n+1}$. *Then it satisfies the often-used property of the spherical Dirac operator:*

$$\Gamma(\omega \cdot v) = \omega_0 v + \omega v_0 - 2\omega_0\omega(\omega \cdot v) + \omega \wedge v.$$

The restriction to $\mathbb{R}^n$ $(v_0 = w_0 = 0)$ *yields*

$$\Gamma(\boldsymbol{\omega} \cdot \mathbf{v}) = \boldsymbol{\omega} \wedge \mathbf{v}.$$

For the wedge product $\wedge$ *see Definition A.1.5 (iv) in Appendix 1.*

(ii) *The spherical Dirac operator is of the form*

$$\Gamma = \omega_0 L + \omega L_0 + \sum_{1 \le j < k} e_j e_k (x_j \partial_k - x_k \partial_j).$$

When restricting to $\mathbb{R}^n$ *the formula reduces to the sum.*

(iii) *The following anti-commutator-property holds:*

$$(\omega L + L\omega)f = (2 - n - 2\omega_0\omega)\,f.$$

In the special case of $f(x) = \omega$ *we have:* $(\omega L)\omega = (1-n)\omega.$

Proof. Before the actual proof we want to recall some useful formulas when dealing with paravectors $a = a_0 + \mathbf{a}, b = b_0 + \mathbf{b}$. For vectors we have

$$\mathbf{ab} = -\mathbf{a} \cdot \mathbf{b} + \mathbf{a} \wedge \mathbf{b}.$$

This calculation gets a little more complicated when dealing with paravectors:

$$\begin{aligned}
ab &= a_0 b_0 + a_0 \mathbf{b} + b_0 \mathbf{a} + \mathbf{ab} \\
&= a_0 b + b_0 a - a \cdot b + \mathbf{a} \wedge \mathbf{b} \\
&= a_0 b + b_0 a - a \cdot b + \sum_{1 \le i < j} e_i e_j (a_i b_j - a_j b_i).
\end{aligned}$$

This yields

$$ab + ba = 2(a_0 b + b_0 a - a \cdot b).$$

Now we will use this for our proof:
Define $a := \omega$ and $b := \text{grad}_\omega\varphi - \omega(\omega \cdot \text{grad}_\omega\varphi)$.

(i) Using Definition 11.16 we get

$$\Gamma\varphi(\omega) = \omega L\varphi(\omega) = \omega\left[\text{grad}_\omega\varphi(\omega) - \omega(\omega \cdot \text{grad}_\omega\varphi(\omega))\right].$$

This leads to

$$
\begin{aligned}
\Gamma\varphi(\omega) \;=\;& \omega_0\operatorname{grad}_\omega\varphi(\omega) + \omega\partial_{\omega_0}\varphi(\omega) \\
&+\omega\wedge\operatorname{grad}_\omega\varphi(\omega) - (1+\omega^2)(\omega\cdot\operatorname{grad}_\omega\varphi(\omega)).
\end{aligned}
$$

Using $1+\omega^2 = 2\omega_0\omega$ and $\varphi(\omega)=\omega\cdot v$ it follows that $\operatorname{grad}_\omega(\omega\cdot v)=v$ and consequently

$$
\Gamma(\omega\cdot v) \;=\; \omega_0 v + \omega v_0 + \omega\wedge v - 2\omega_0\omega(\omega\cdot v).
$$

In $\mathbb{R}^n$ ($\omega_0 = v_0 = 0$) the special case holds:

$$
\Gamma(\boldsymbol{\omega}\cdot\mathbf{v}) = \boldsymbol{\omega}\wedge\mathbf{v}.
$$

(ii) Using the paravector formulas we obtain

$$
\Gamma = \omega L = \omega_0 L + \omega L_0 - \omega\cdot L + \sum_{1\le i<j} e_i e_j(\omega_i L_j - \omega_j L_i).
$$

The equality

$$
\omega_i L_j - \omega_j L_i = x_i\partial_j - x_j\partial_i
$$

is easily proven, since $\omega\cdot L = 0$ has been proven earlier. In $\mathbb{R}^n$ only the sum remains.
(iii) We have

$$
(L_i\omega_j) = (|x|\partial_i - x_i\partial_\omega)\omega_j = |x|\partial_i\omega_j - x_i\sum_{k=0}^n \omega_k\partial_k\omega_j.
$$

According to Proposition 11.13 (iv) we get $\partial_i\omega_j = (\delta_{ij} - \omega_i\omega_j)/|x|$, leading to

$$
\begin{aligned}
(L_i\omega_j) \;=\;& \delta_{ij} - \omega_i\omega_j - \frac{x_i}{|x|}\sum_{k=0}^n \omega_k(\delta_{kj} - \omega_k\omega_j) \\
=\;& \delta_{ij} - \omega_i\omega_j - \omega_i\sum_{k=0}^n \omega_k\delta_{kj} + \omega_i\sum_{k=0}^n \omega_k^2\omega_j \\
=\;& \delta_{ij} - \omega_i\omega_j.
\end{aligned}
$$

Consequently,

$$
\begin{aligned}
L(\omega f) \;=\;& \sum_{i,j=0}^n e_i e_j L_i(\omega_j f) = \sum_{i,j=0}^n e_i e_j(L_i\omega_j)f + \sum_{i,j=0}^n e_i e_j\omega_j(L_i f) \\
=\;& \sum_{i,j=0}^n e_i e_j(L_i\omega_j)f - \sum_{i,j=0}^n e_j e_i\omega_j(L_i f) = \sum_{i,j=0}^n e_i e_j(L_i\omega_j)f - (\omega L)f.
\end{aligned}
$$

Finally we have

$$
\begin{aligned}
(L\omega + \omega L)f \;=\;& \sum_{i,j=0}^n e_i e_j(\delta_{ij} - \omega_i\omega_j)f \\
=\;& (1-n)f - \omega^2 f = (2 - n - 2\omega_0\omega)f.
\end{aligned}
$$

$\square$

11.2.2 Elementary radial functions

We will start with defining an exponential function:

Definition 11.18. For paravectors $x = x_0 + \mathbf{x} \in C\ell(n)$ we define

$$e^x := \sum_{k=0}^{\infty} \frac{x^k}{k!}.$$

This series converges absolutely for all x in analogy to the complex case, since according to Section 3.2 we have $|x^n| \le |x|^n$ for a paravector x. Because

$$\exp(|x|)$$

converges, the comparison test yields that e^x converges for all x. The newly defined function is not holomorphic, which is a first disadvantage. This can be seen easily by considering the powers of x which are not holomorphic. Nevertheless at the end of this subsection we shall get around this difficulty.

Theorem 11.19. (i) *For $xy = yx$ we have the functional equation*

$$e^{x+y} = e^x e^y.$$

(ii) *With $\boldsymbol{\omega}(\mathbf{x}) := \mathbf{x}/|\mathbf{x}|$ the exponential function admits the representation*

$$e^x = e^{x_0}(\cos|\mathbf{x}| + \boldsymbol{\omega}(\mathbf{x})\sin|\mathbf{x}|).$$

e^x is paravector-valued.

Since the functions sine and cosine depend on $|\mathbf{x}|$, the term elementary radial function is justified.

Proof. (i) Unfortunately the elegant proof of the complex case, see Theorem 11.2 (ii), does not work here. We have to consider the product (Proposition 9.9) of two power series. The Cauchy product of the series expansions of e^x and e^y in case of commuting x and y yields

$$\sum_{k=0}^{\infty} \frac{(x+y)^k}{k!} = \sum_{k=0}^{\infty} \frac{1}{k!} \sum_{\ell=0}^{k} \binom{k}{\ell} x^\ell y^{k-\ell}$$

$$= \sum_{k=0}^{\infty} \sum_{\ell=0}^{k} \frac{x^\ell y^{k-\ell}}{\ell!(k-\ell)!} = \sum_{\ell=0}^{\infty} \frac{x^\ell}{\ell!} \sum_{m=0}^{\infty} \frac{y^m}{m!} = e^x e^y.$$

(ii) Consequently, $e^x = e^{x_0} e^{\mathbf{x}}$ and $e^{\mathbf{x}}$ remain to be considered. In detail,

$$\sum_{k=0}^{\infty} \frac{\mathbf{x}^k}{k!} = \sum_{\ell=0}^{\infty} \frac{\mathbf{x}^{2\ell}}{(2\ell)!} + \sum_{\ell=0}^{\infty} \frac{\mathbf{x}^{2\ell+1}}{(2\ell+1)!}$$

$$= \sum_{\ell=0}^{\infty} (-1)^\ell \frac{|\mathbf{x}|^{2\ell}}{(2\ell)!} + \boldsymbol{\omega}(\mathbf{x}) \sum_{\ell=0}^{\infty} (-1)^\ell \frac{|\mathbf{x}|^{2\ell+1}}{(2\ell+1)!} = \cos|\mathbf{x}| + \boldsymbol{\omega}(\mathbf{x})\sin|\mathbf{x}|. \qquad \square$$

Example 11.20. Suppose that $e^{x+y} = e^x e^y$. Then the equation $xy = yx$ is not necessarily satisfied. This can be shown via the following counterexample in $\mathbb{H}$:

Suppose that $x = 3\pi e_1$ and $y = 4\pi e_2$. We then obtain $xy = 12\pi^2 e_1 e_2 = -12\pi^2 e_2 e_1 = -yx$. On the one hand we have

$$|x + y| = 5\pi, \quad |x| = 3\pi \quad \text{and} \quad |y| = 4\pi,$$

so that

$$e^{x+y} = \cos 5\pi + \frac{3\pi e_1 + 4\pi e_2}{5\pi} \sin 5\pi = -1.$$

On the other hand we get

$$e^x = \cos 3\pi = -1 \quad \text{and} \quad e^y = \cos 4\pi = 1,$$

leading to $e^{x+y} = e^x e^y$.

Corollary 11.21. (i) *For all* $x \in \mathbb{R}^{n+1}$ *we have* $e^{-x} e^x = 1$, $e^x \neq 0$.

(ii) $e^{kx} = (e^x)^k$ $(k \in \mathbb{Z})$ *(de Moivre's formula)*.

(iii) $e^{\omega(\mathbf{x})\pi} = -1$.

The proof is left as Exercise 11.3.10. Surprisingly the following statement holds:

Corollary 11.22. *The usual limit representation works:*

$$e^x = \lim_{m \to \infty} \left(1 + \frac{x}{m}\right)^m.$$

Proof. The proof is similar to Theorem 11.2 (iv) replacing $|z|$ by $|x|$. □

With the help of the exponential function, trigonometric and hyperbolic functions can be introduced.

Definition 11.23. Define for $x \in \mathbb{R}^{n+1}$,

$$\cos x := \frac{e^{x\omega(\mathbf{x})} + e^{-x\omega(\mathbf{x})}}{2}, \qquad \sin x := -\omega(\mathbf{x}) \frac{e^{x\omega(\mathbf{x})} - e^{-x\omega(\mathbf{x})}}{2},$$

$$\cosh x := \frac{e^x + e^{-x}}{2}, \qquad \sinh x := \frac{e^x - e^{-x}}{2}.$$

Similarly to $\mathbb{C}$ the functions cosine and hyperbolic cosine are even, while sine and hyperbolic sine are odd. The hyperbolic functions satisfy the sum and difference formulas if x and y commute — the proof is similar to Theorem 11.7 (iii), the $\mathbb{C}$ case. Unfortunately the sum and difference formulas for the trigonometric functions do not hold, since $\omega(\mathbf{x} + \mathbf{y})$ is not linear in $\mathbf{x}$ and $\mathbf{y}$, respectively.

Using the definition and Theorem 11.19 (ii) we obtain

$$e^{\pm \mathbf{x}} = \cos |\mathbf{x}| \pm \omega(\mathbf{x}) \sin |\mathbf{x}|$$

and thus

$$\cosh \mathbf{x} = \cos |\mathbf{x}|, \quad \sinh \mathbf{x} = \omega(\mathbf{x}) \sin |\mathbf{x}|.$$

Using the sum and difference formulas, we get

$$\begin{aligned}
\cosh x &= \cosh x_0 \cos |\mathbf{x}| + \sinh x_0 \sinh \mathbf{x}, \\
\sinh x &= \sinh x_0 \cos |\mathbf{x}| + \cosh x_0 \sinh \mathbf{x}.
\end{aligned}$$

Similar formulas can be derived for the trigonometric functions by taking into account that

$$x\omega(\mathbf{x}) = x_0 \omega(\mathbf{x}) + \mathbf{x}\omega(\mathbf{x}) = -|\mathbf{x}| + x_0\omega(\mathbf{x}),$$

$$\omega(x_0\omega(\mathbf{x})) = \frac{x_0\omega(\mathbf{x})}{|x_0|} = (\operatorname{sgn} x_0)\omega(\mathbf{x}),$$

and

$$e^{x\omega(\mathbf{x})} = e^{-|\mathbf{x}|} e^{x_0\omega(\mathbf{x})} = e^{-|\mathbf{x}|}(\cos x_0 + \omega(\mathbf{x}) \sin x_0).$$

Thus we have the representations

$$\begin{aligned}
\cos x &= \cos x_0 \cos \mathbf{x} - \sin x_0 \sin \mathbf{x}, \\
\sin x &= \sin x_0 \cos \mathbf{x} + \cos x_0 \sin \mathbf{x}.
\end{aligned}$$

Consequently the hyperbolic and trigonometric functions are paravector-valued. Considering $\overline{\omega(\mathbf{x})} = -\omega(\mathbf{x})$ one has

$$\begin{aligned}
|\cos x|^2 &= \cos^2 x_0 + \sinh^2 |\mathbf{x}|, \\
|\sin x|^2 &= \sin^2 x_0 + \sinh^2 |\mathbf{x}|.
\end{aligned}$$

Hence the only zeros are already known from the complex case.

Since e^x is paravector-valued, an inverse function comes to mind.

Definition 11.24. For $k \in \mathbb{Z}$ the paravector logarithm in $C\ell(n)$ is defined by

$$\log x := \begin{cases} \ln |x| + \omega(\mathbf{x})(\arccos \frac{x_0}{|x|} + 2k\pi), & |\mathbf{x}| \neq 0 \text{ r } |\mathbf{x}| = 0, \ x_0 > 0 \\ \ln |x| + e_1\pi, & |\mathbf{x}| = 0, \ x_0 < 0. \end{cases}$$

Actually the notation $\log_k x$ would be more pleasing. For $n = 1$ this coincides with the logarithm function introduced in Definition 11.8. One could also replace $e_1\pi$ by $e_j\pi$, so that different definitions are plausible. At least the previously defined function inverts the exponential function:

Corollary 11.25. *The function* $\log x$ *satisfies the properties:*

(i) *The function is the inverse of the exponential function:*

$$e^{\log x} = x, \quad \log e^x = x.$$

(ii) $$\log 1 = 0, \quad \log e_i = \tfrac{\pi}{2} e_i, \quad i = 1, \ldots, n.$$

(iii) *If* $\log x$ *commutes with* $\log y$ *and* x *with* y, *respectively, using an appropriate* k *in the definition, the well-known functional equation is satisfied:*

$$\log (xy) = \log x + \log y.$$

Proof. (i) Obviously,

$$
\begin{aligned}
e^{\log x} &= e^{\ln |x|} e^{\boldsymbol{\omega}(\mathbf{x}) \arccos \frac{x_0}{|x|}} e^{\boldsymbol{\omega}(\mathbf{x}) 2\pi k} \\
&= |x| \left(\cos \arccos \frac{x_0}{|x|} + \boldsymbol{\omega}(\mathbf{x}) \sin \arccos \frac{x_0}{|x|} \right) \\
&= |x| \left(\frac{x_0}{|x|} + \boldsymbol{\omega}(\mathbf{x}) \frac{|\mathbf{x}|}{|x|} \right) = x.
\end{aligned}
$$

Exchanging the functions yields

$$
\begin{aligned}
\log e^x &= \log e^{x_0} (\cos |\mathbf{x}| + \boldsymbol{\omega}(\mathbf{x}) \sin |\mathbf{x}|) \\
&= \ln e^{x_0} + \boldsymbol{\omega}(\mathbf{x}) \arccos \left(\frac{e^{x_0} \cos |\mathbf{x}|}{e^{x_0}} \right) = x_0 + \frac{\mathbf{x}}{|\mathbf{x}|} |\mathbf{x}| = x.
\end{aligned}
$$

(ii) is left as Exercise 11.3.11.

(iii) This proof is similar to the real or complex case.   □

Definition 11.26. Let α be a real number. The generalized power function is defined by the formula

$$x^\alpha := e^{\alpha \log x}.$$

We want to conclude our considerations of elementary radial functions with an example.

Example 11.27. Suppose that $x = \mathbf{x}$ and $\alpha = \tfrac{1}{3}$. Then for the third root of a vector we have:

$$
\begin{aligned}
\mathbf{x}^{\frac{1}{3}} &= \exp \left(\frac{1}{3} \left(\ln |\mathbf{x}| + \boldsymbol{\omega}(\mathbf{x}) \arccos \frac{x_0}{|x|} + \boldsymbol{\omega}(\mathbf{x}) 2k\pi \right) \right) \\
&= \sqrt[3]{|\mathbf{x}|} \left[\cos(\frac{1}{3} \arccos 0 + \frac{2}{3} k\pi) + \boldsymbol{\omega}(\mathbf{x}) \sin(\frac{1}{3} \arccos 0 + \frac{2}{3} k\pi) \right] \\
&= \sqrt[3]{|\mathbf{x}|} \left[\cos \left(\frac{\pi}{6} + \frac{2}{3} k\pi \right) + \frac{\mathbf{x}}{|\mathbf{x}|} \sin \left(\frac{\pi}{6} + \frac{2}{3} k\pi \right) \right] \quad (k = 0, 1, 2).
\end{aligned}
$$

It would be more pleasing if the introduced elementary functions were holomorphic. As already mentioned, this is not the case, so that a huge difference with $\mathbb{C}$ exists. Nevertheless our functions are of the special form

$$f(x) = f_0(x_0, |\mathbf{x}|) + \omega(\mathbf{x}) f_1(x_0, |\mathbf{x}|)$$

with real-valued functions f_0, f_1. Thus it is sound to introduce *radially holomorphic functions*.

Definition 11.28. Abbreviating $|\mathbf{x}| =: r$,

$$\overline{\partial}_{rad} := \frac{1}{2}(\partial_0 + \omega(\mathbf{x})\partial_r), \quad \partial_{rad} = \frac{1}{2}(\partial_0 - \omega(\mathbf{x})\partial_r)$$

are called *radial differential operators*. The notation $\partial_{rad}f =: f'$ will be justified later.

Similarly to the introduction of holomorphic functions in $\mathbb{C}$ we consider the difference $f(x+h) - f(x)$ for a differentiable function f which is given in a domain $G \subset \mathbb{R}^{n+1}$ and has the form

$$f(x) = f_0(x_0, |\mathbf{x}|) + \omega(\mathbf{x}) f_1(x_0, |\mathbf{x}|).$$

Suppose that h has the special form

$$h := h_0 + \omega(\mathbf{x})h_r$$

where h_0, h_r are real. So $\mathbf{x}$ is really only changed in the radial direction. This leads to

$$\omega(\mathbf{x} + \mathbf{h}) = \frac{\mathbf{x} + \omega(\mathbf{x})h_r}{|\mathbf{x} + \omega(\mathbf{x})h_r|} = \frac{\mathbf{x}}{|\mathbf{x}|} \frac{|\mathbf{x}| + h_r}{|\mathbf{x} + \omega(\mathbf{x})h_r|} = \omega(\mathbf{x})$$

for $|h_r| < |\mathbf{x}|$. This condition has to be satisfied anyway, otherwise the denominator could become zero. In addition we have with respect to the variable $|\mathbf{x}|$,

$$|\mathbf{x} + \mathbf{h}| = |\mathbf{x}| \left| 1 + \frac{h_r}{|\mathbf{x}|} \right| = |\mathbf{x}| + h_r.$$

Using

$$\partial_0 = \overline{\partial}_{rad} + \partial_{rad}, \quad \omega(\mathbf{x})\partial_r = \overline{\partial}_{rad} - \partial_{rad},$$

and the ordinary differentiation rules we obtain

$$f(x+h) - f(x) = \partial_0 f_0(x)h_0 + \partial_r f_0(x)h_r + \omega(\mathbf{x})(\partial_0 f_1(x)h_0 + \partial_r f_1(x)h_r) + o(h)$$
$$= (\partial_{rad}f(x))(h_0 + \omega(\mathbf{x})h_r) + (\overline{\partial}_{rad}f(x))(h_0 - \omega(\mathbf{x})h_r) + o(h).$$

One has to pay attention to the fact that $\omega(\mathbf{x})$ does neither depend on x_0 nor on r and has to be treated like a constant with respect to ∂_0 or ∂_r. Finally we get

$$f(x+h) - f(x) = (\partial_{rad}f(x))\, h + (\overline{\partial}_{rad}f(x))\, \overline{h} + o(h).$$

The following definition and the proposition hold in analogy to $\mathbb{C}$:

Definition 11.29. Suppose that a function f, given in a domain $G \subset \mathbb{R}^{n+1}$, has continuous first derivatives and is of the form

$$f(x) = f_0(x_0, |\mathbf{x}|) + \omega(\mathbf{x}) f_1(x_0, |\mathbf{x}|).$$

Then it is referred to as *radially holomorphic* in G if it satisfies

$$f(x + h) - f(x) = (\partial_{rad} f(x)) h + o(h)$$

for all $x \in G$ and $h = h_0 + \omega(\mathbf{x}) h_r \to 0$.
This is equivalent to the existence of the limit

$$\lim_{h \to 0} (f(x + h) - f(x)) h^{-1} = \partial_{rad} f(x) =: f'(x).$$

Corollary 11.30. *A function f is radially holomorphic if and only if it satisfies*

$$\overline{\partial}_{rad} f = 0,$$

which is a Cauchy–Riemann type differential equation (CRD)

$$\partial_0 f_0 = \partial_r f_1, \quad \partial_0 f_1 = -\partial_r f_0.$$

Proof. This proof is similar to the complex case if one considers that for $h \to 0$ the term $\overline{h}/h$ does not possess a limit. $\qquad\square$

The notation $\partial_{rad} f = f'$ is justified by the following statements:

Corollary 11.31. *The previously defined elementary functions are radially holomorphic and satisfy:*

(i) $(e^x)' = e^x$,

(ii) $(\sin x)' = \cos x, \quad (\cos x)' = -\sin x,$

(iii) $(\sinh x)' = \cosh x, \quad (\cosh x)' = \sinh x,$

(iv) $(\log x)' = \frac{1}{x}$,

(v) $(x^\alpha)' = \alpha x^{\alpha-1}$ *for $\alpha \in \mathbb{R}$.*

The proof is left as Exercise 11.3.12 to the interested reader.

11.2.3 Fueter–Sce construction of holomorphic functions

In 1935 R. Fueter [47] developed a concept for creating holomorphic quaternion-valued functions from complex functions which are holomorphic in the upper half-plane. M. Sce [129] and T. Qian [121] extended Fueter's concept to higher dimensions. This is the basic idea:

Let $x = x_0 + \mathbf{x} \in C\ell(n)$ be a paravector and $\omega(\mathbf{x}) = \mathbf{x}/|\mathbf{x}|$. Fueter's concept consists of associating a paravector-valued function h to every complex function f holomorphic in the upper half-plane. In detail:

Definition 11.32. (i) Suppose that a function $f(z) = u(x, y) + iv(x, y)$ is holomorphic in the upper complex half-plane $\mathbb{C}^+$. The associated function is given by

$$h(f)(x) := u(x_0, |\mathbf{x}|) + \boldsymbol{\omega}(\mathbf{x})v(x_0, |\mathbf{x}|).$$

(ii) Furthermore we introduce the *Fueter transform* of f,

$$\tau_n(f) := \kappa_n \, \Delta^{\frac{n-1}{2}} h(f).$$

f is called the primitive of $\tau_n(f)$, κ_n denotes a suitable constant.

For even n it is not a pointwise definition. The theory of pseudo-differential operators takes care of that matter. That is the reason for which we restrict ourselves to odd cases of n. The normalization factor κ_n can be evaluated in the calculation.

Theorem 11.33 (Fueter, Sce, Qian). (i) $h(f) = u + \boldsymbol{\omega}v$ *is a radially holomorphic function.*

(ii) *Suppose that n is odd. Then $\tau_n(f)$ is a right- and left-holomorphic function in $C\ell(n)$. We use the shorthand notation $|\mathbf{x}| =: r$. With $k := \frac{n-1}{2}$ we have*

$$\tau_n(f) = \frac{1}{2^k k!} \Delta^k h(f) = \left(\frac{1}{r} \partial_r \right)^k u + \boldsymbol{\omega}(\mathbf{x}) \left(\partial_r \frac{1}{r} \right)^k v.$$

Part (ii) yields a partial differential equation for h. This formula will be of special use for the case $n = 3$.

Proof. (i) According to Corollary 11.30 a function is radially holomorphic if and only if it satisfies the differential equations corresponding to the CRD in $\mathbb{C}$. Also, a radially holomorphic function will lead to a holomorphic function in $\mathbb{C}^+$.

(ii) We have that $\Delta = \partial_0^2 + \sum_{i=1}^n \partial_i^2$, ∂_0 affects only u and v; instead of $\boldsymbol{\omega}(\mathbf{x})$ we write ω:

$$\partial_0 h = \partial_0 u + \boldsymbol{\omega} \partial_0 v, \quad \partial_0^2 h = \partial_0^2 u + \boldsymbol{\omega} \partial_0 v.$$

For $i = 1, ..., n$ we notice that

$$\partial_i u = \frac{x_i}{r} \partial_r u, \quad \partial_i^2 u = \frac{x_i^2}{r^2} \partial_r^2 u + \frac{r^2 - x_i^2}{r^3} \partial_r u,$$

$$\partial_i \boldsymbol{\omega} = \frac{r^2 e_i - x_i \mathbf{x}}{r^3}, \quad \sum_{i=1}^n \frac{x_i}{r} \partial_i \boldsymbol{\omega} = 0, \quad \sum_{i=1}^n e_i \partial_i \boldsymbol{\omega} = \frac{-n+1}{r},$$

$$\partial_i^2 \boldsymbol{\omega} = \frac{-r^2 \mathbf{x} - 2r^2 x_i e_i + 3x_i^2 \mathbf{x}}{r^5}, \quad \sum_{i=1}^n \partial_i^2 \boldsymbol{\omega} = \frac{-n+1}{r^2} \boldsymbol{\omega}.$$

This leads to

$$\Delta h = \partial_0^2 (u + \boldsymbol{\omega}v) + \partial_r^2 (u + \boldsymbol{\omega}v) + (n-1) \left(\frac{1}{r} \partial_r u + \boldsymbol{\omega} \partial_r \left(\frac{v}{r} \right) \right). \tag{*}$$

In order to show the formula for $\Delta^k h$, we define $u_0 := u$, $v_0 := v$ and for $m = 1, 2, \ldots$

$$u_m = \frac{2m}{r} \partial_r u_{m-1} = 2^m m! \left(\frac{1}{r}\partial_r\right)^m u_0, \quad v_m := 2m\partial_r \frac{v_{m-1}}{r} = 2^m m! \left(\partial_r \frac{1}{r}\right)^m v_0.$$

Using mathematical induction we deduce for $m = 0, 1, 2, \ldots$

$$\partial_0 u_m = \partial_r v_m + 2m\frac{v_m}{r}, \quad \partial_r u_m = -\partial_0 v_m.$$

For $m = 0$ this transforms to $\partial_0 u_0 = \partial_r v_0$, $\partial_r u_0 = -\partial_0 v_0$, i.e., the Cauchy–Riemann equations for radially holomorphic functions. The step from $m - 1$ to m is obtained as follows:

$$\partial_0 u_m = \frac{2m}{r}\partial_r\left(\partial_r v_{m-1} + 2(m-1)\frac{v_{m-1}}{r}\right) = \partial_r v_m + 2m\frac{v_m}{r}$$

and

$$\partial_r u_m = -\partial_r\left(\frac{2m}{r}\partial_0 v_{m-1}\right) = -\partial_0 v_m.$$

We will show the next equality using induction, too,

$$\Delta^m h = \frac{(n-1)(n-3)\cdots(n-2m+1)}{2^m\, m!}(u_m + \omega v_m) =: C_m(u_m + \omega v_m).$$

For $m = 0$ the factor on the right-hand side becomes 1, so that $h = u_0 + \omega v_0$, the induction step is based on the preceding formula $(*)$:

$$
\begin{aligned}
\Delta^m h &= C_{m-1}\Delta(u_{m-1} + \omega v_{m-1})\\
&= C_{m-1}\left[\partial_0^2(u_{m-1} + \omega v_{m-1}) + \partial_r^2(u_{m-1} + \omega v_{m-1})\right.\\
&\quad \left.+(n-1)\left(\frac{1}{r}\partial_r u_{m-1} + \omega\partial_r\left(\frac{v_{m-1}}{r}\right)\right)\right].
\end{aligned}
$$

Since

$$\partial_0^2 u_{m-1} = \partial_r\partial_0 v_{m-1} + \frac{2(m-1)}{r}\partial_0 v_{m-1} = -\partial_r^2 u_{m-1} - \frac{2(m-1)}{r}\partial_r u_{m-1}$$

and

$$\partial_0^2 v_{m-1} = -\partial_r^2 v_{m-1} - 2(m-1)\partial_r\left(\frac{v_{m-1}}{r}\right),$$

we have

$$\Delta^m h = C_{m-1}\frac{n - 2m + 1}{2m}(u_m + \omega v_m) = C_m(u_m + \omega v_m).$$

Substitution of u_m, v_m by u_0, v_0 yields

$$\Delta^m h = (n-1)(n-3)\cdots(n-2m+1)\left(\left(\frac{1}{r}\partial_r\right)^m u_0 + \omega\left(\partial_r\frac{1}{r}\right)^m v_0\right)$$

and with $m = k = \frac{n-1}{2}$,

$$\Delta^k h = 2^k k!\left(\left(\frac{1}{r}\partial_r\right)^k u + \omega\left(\partial_r\frac{1}{r}\right)^k v\right).$$

Showing that $\overline{\partial}\,\tau_n(f) = 0$ is left as Exercise 11.3.13. $\square$

Now we want to take a closer look at the effects of $\tau_n(f)$, restricting ourselves to the case $n = 3$:

Example 11.34 (Exponential function). Let $f(z) = e^z$. The associated function is

$$h(\exp)(x) = e^{x_0}(\cos r + \boldsymbol{\omega}(\mathbf{x})\sin r).$$

This is exactly the radially holomorphic function e^x introduced in the preceding subsection, see Definition 11.18 and Theorem 11.19. The Fueter transform follows by using the previous lemma and a suitable norm factor:

$$\mathrm{EXP}_3(x) := -\frac{1}{2}\tau_3(\exp)(x) = e^{x_0}\left(\frac{\sin r}{r} - \boldsymbol{\omega}(\mathbf{x})\left(\frac{\sin r}{r}\right)_r\right).$$

Denoting

$$\mathrm{sinc}\, r := \frac{\sin r}{r}$$

this can be expressed as

$$\mathrm{EXP}_3(x) = e^{x_0}(\mathrm{sinc}\, r - \boldsymbol{\omega}(\mathbf{x})(\mathrm{sinc}\, r)').$$

This newly-defined exponential function has a different form than the other generalizations of e^x we consider. We gain a holomorphic function, but lose some other properties. At least we have:

(i) $\mathrm{EXP}_3(x) \neq 0$ for all x.

(ii) $\lim_{r \to 0} \mathrm{EXP}_3(x) = e^{x_0}$.

(iii) For real λ the operator ∂ gives

$$\partial \,\mathrm{EXP}_3(\lambda x) = 2\lambda \mathrm{EXP}_3(\lambda x).$$

The factor 2 in part (iii) is based on using ∂, which is equal to $2\partial_z$ in the $\mathbb{C}$ case.

Proof. (i) We have

$$|\mathrm{EXP}_3(x)|^2 = e^{2x_0}\left(\frac{\sin^2 r}{r^2} + \frac{(r\cos r + \sin r)^2}{r^4}\right).$$

This can only become zero if $\sin r = 0$, but then we get $\cos r \neq 0$ in the inner bracket, so that $\mathrm{EXP}_3(x) \neq 0$ for all x.

(ii) It is well known that

$$\frac{\sin r}{r} = 1 - \frac{r^2}{3!} + \cdots,$$

yielding

$$\frac{\sin r}{r} \to 1, \quad \left(\frac{\sin r}{r}\right)' \to 0$$

for $r \to 0$.

(iii) Firstly we conclude that

$$\mathrm{EXP}_3(\lambda x) = e^{\lambda x_0} \left(\frac{\sin \lambda r}{\lambda r} - \omega(\mathbf{x}) \left(\frac{\lambda r \cos \lambda r - \sin \lambda r}{\lambda^2 r^2} \right) \right),$$

because the sign of λ vanishes despite the fact that $|\lambda r| = |\lambda| r$ in some cases. But then we have

$$\partial_0 \mathrm{EXP}_3(\lambda x) = \lambda \, \mathrm{EXP}_3(\lambda x),$$

and since $\mathrm{EXP}_3(\lambda x)$ is holomorphic, using $\overline{\partial} = \partial_0 + D$, we get

$$D \, \mathrm{EXP}_3(\lambda x) = -\partial_0 \mathrm{EXP}_3(\lambda x) = -\lambda \, \mathrm{EXP}_3(\lambda x),$$

and because of $\partial = \partial_0 - D$ we deduce

$$\partial \, \mathrm{EXP}_3(\lambda x) = 2\lambda \, \mathrm{EXP}_3(\lambda x). \qquad \square$$

This helps in solving the so-called ∂-*problem*:

Proposition 11.35. *Suppose that* $L_n = a_n \partial^n + \cdots + a_1 \partial + a_0$ *is a partial differential operator with* $a_k \in \mathbb{R}$. *If* λ_k *is a real root of the algebraic equation*

$$a_n \lambda^n + \cdots + a_1 \lambda + a_0 = 0,$$

the function

$$u_k = \mathrm{EXP}_3 \left(\frac{\lambda_k}{2} x \right)$$

represents a solution of the differential equation $L_n u = 0$. *We can obtain a system of linearly independent solutions with different* λ_k.

The proof follows from part (iii) above.

Example 11.36 (Trigonometric functions). This example will cover the trigonometric functions. The hyperbolic functions are dealt with in the exercises. We shall start with the complex functions

$$\cos z \;=\; \cos x \cosh y - i \sin x \sinh y,$$
$$\sin z \;=\; \sin x \cosh y + i \cos x \sinh y.$$

The associated functions are given by

$$h(\cos)(x) \;=\; \cos x_0 \cosh r - \omega(\mathbf{x}) \sin x_0 \sinh r,$$
$$h(\sin)(x) \;=\; \sin x_0 \cosh r + \omega(\mathbf{x}) \cos x_0 \sinh r,$$

which are, according to Definition 11.23, the radially holomorphic forms of cosine and sine. Theorem 11.33 (ii) yields

$$\mathrm{COS}_3(x) := -\frac{1}{2}\tau_3(\cos)(x) = -\frac{1}{r} \cos x_0 \sinh r + \omega \sin x_0 \left(\frac{\sinh r}{r} \right)_r,$$

$$\mathrm{SIN}_3(x) := -\frac{1}{2}\tau_3(\sin)(x) = -\frac{1}{r} \sin x_0 \sinh r - \omega \cos x_0 \left(\frac{\sinh r}{r} \right)_r.$$

The functions look a little strange, but they are holomorphic. Unfortunately we have to pay for that and lose well-known properties, i.e.,

$$\text{COS}_3^2(x) + \text{SIN}_3^2(x) = \frac{1}{r^2}\sinh^2 r - \left(\frac{1}{r}\cosh r - \frac{1}{r^2}\sinh r\right)^2$$

$$= -\frac{1}{r^2} + \frac{2}{r^3}\cosh r \sinh r - \frac{1}{r^4}\sinh^2 r.$$

This is not very close to the expected result 1. The connection to the exponential function is lost, too. The differentiation properties however hold:

$$\partial_0 \text{COS}_3(x) = -\text{SIN}_3(x), \quad \partial_0 \text{SIN}_3(x) = \text{COS}_3(x).$$

Because the functions are holomorphic, we get $(\partial_0 + D)f = 0$ or $\partial = \partial_0 - D = 2\partial_0$, and finally

$$\partial \text{COS}_3(x) = -2\text{SIN}_3(x), \quad \partial \text{SIN}_3(x) = 2\text{COS}_3(x).$$

The factor 2 comes from using ∂, since $\partial_z = \frac{1}{2}\partial$ in the complex case.

The list of examples could be extended arbitrarily. Tao Qian [121] generalizes Fueters calculations for quaternions by processing the Fueter transform of z^k. F. Sommen [139] computed the Fueter transform of the geometric series

$$\frac{1}{1-z} = 1 + z + z^2 + \cdots,$$

using the Fueter transform of z^k calculated by Qian.

11.2.4 Cauchy–Kovalevsky extension

In this subsection we will deal with a completely different principle for creating holomorphic functions. Real analytic functions defined in a domain $G \subset \mathbb{R}^n$ will be extended to holomorphic functions in $\mathbb{R}^{n+1}$. This is also known as the *Cauchy problem*. Concerning this matter the Russian mathematician Sofya Kovalevsky made important contributions at the end of the 19th century.

SOFYA KOVALEVSKAYA or Sofya Kovalevsky (1850–1891) could study in Germany only after marrying, since studying was impossible for a female in Russia at that time. After attending lectures in Heidelberg she became one of the most well-known students of Weierstraß. In 1874 she graduated with a doctoral degree in Göttingen dealing with the theory of partial differential equations. Ten years later she became the first woman to be offered a professorship at the university of Stockholm. She gained a high reputation due to her award-winning works. She died of pneumonia at a very young age.

Definition 11.37. Let $G \subset \mathbb{R}^n$ be a domain. In addition to G we define the domain $G^* \subset \mathbb{R}^{n+1}$ as follows: G^* consists of all points of the form

$$\mathbf{x} + te_0, \quad (t \in \mathbb{R}, \quad \mathbf{x} \in G \subset \mathbb{R}^n).$$

Given a real analytic function $f(\mathbf{x})$ in $G \subset \mathbb{R}^n$ we are looking for a holomorphic function $f^*(x_0, \mathbf{x})$ in $G^{**} \subset G^*$ with $G \subset G^{**}$ satisfying

$$\overline{\partial} f^*(x_0, \mathbf{x}) = 0 \quad \text{in} \quad G^{**},$$
$$f^*(0, \mathbf{x}) = f(\mathbf{x}), \quad (\textit{initial condition}).$$

This is called the *Cauchy–Kovalevsky extension (CK extension)* of the real function f in the domain G^{**} and is denoted by $CK(f)$.

The idea for solving this problem is based on the relation following from $\overline{\partial} CK(f) = 0$,

$$\partial_0 CK(f) = -D\,CK(f).$$

Formally the equation

$$\partial_0 e^{-x_0 D} = -e^{-x_0 D} D$$

holds, so that we obtain:

Theorem 11.38. $CK(f)(x_0, \mathbf{x}) := e^{-x_0 D} f(\mathbf{x}) = \displaystyle\sum_{m=0}^{\infty} \frac{(-x_0)^m}{m!} D^m f(\mathbf{x})$ *represents the solution of the Cauchy–Kovalevsky problem in an appropriate domain* $G^{**} \subset G^*$ *with* $G \subset G^{**}$.

Proof. Since for a real analytic function all partial derivatives exist, all the $D^m f(\mathbf{x})$ are defined properly. Assuming the convergence of the above series, after shifting the summation index we obtain

$$\partial_0 CK(f)(x_0, \mathbf{x}) = -\sum_{m=0}^{\infty} \frac{(-x_0)^m}{m!} D^{m+1} f(\mathbf{x}) = -D\,CK(f)(x_0, \mathbf{x}),$$

thus the series is holomorphic. The initial condition is satisfied — for $x_0 = 0$ only the $m = 0$ term remains, yielding $CK(f)(0, \mathbf{x}) = f(\mathbf{x})$.

In order to have convergence it is important that real analytic functions possess absolutely converging Taylor series representations at every point $\mathbf{x}$. Denoting $\nabla = (\partial_1, \dots, \partial_n)$ and $\mathbf{k} = (k_1, \dots, k_n)$ the expression

$$f(\mathbf{x} + \mathbf{h}) = \sum_{m=0}^{\infty} \sum_{|\mathbf{k}|=m} \frac{\mathbf{h}^{\mathbf{k}}}{\mathbf{k}!} \nabla^{\mathbf{k}} f(\mathbf{x})$$

converges absolutely for $|\mathbf{h}|$ small enough. Assuming $\mathbf{h} = (h, \dots, h)$ with real h we have $\mathbf{h}^{\mathbf{k}} = h^m$ and

$$|h|^m \sum_{|\mathbf{k}|=m} \frac{1}{\mathbf{k}!} \left| \nabla^{\mathbf{k}} f(\mathbf{x}) \right| =: c_m,$$

with a converging series $\sum c_m$. We want to consider

$$D^m f(\mathbf{x}) = \sum_{i_1,\ldots,i_m=1}^{n} e_{i_1}\ldots e_{i_m}\partial_{i_1}\ldots\partial_{i_m} f(\mathbf{x}) = m! \sum_{|\mathbf{k}|=m} \frac{1}{\mathbf{k}!}e_{\mathbf{k}}\nabla^{\mathbf{k}} f(\mathbf{x}),$$

where $e_{\mathbf{k}} = e_{k_1}\cdots e_{k_m}$. This leads to the estimate

$$|D^m f(\mathbf{x})| \leq \frac{m!}{|h|^m}c_m.$$

Thus we obtain

$$\left|\sum_{m=0}^{\infty}\frac{(-x_0)^m}{m!}D^m f(\mathbf{x})\right| \leq \sum_{m=0}^{\infty}\frac{|x_0|^m}{|h|^m}c_m.$$

For $|x_0| \leq |h|$ our series converges, determining the domain of existence of our solution, $G^{**} \subset G^*$. $\square$

Example 11.39 (Fueter polynomials). Let $P_{\boldsymbol{\varepsilon}_i}(\mathbf{x}) =: x_i$ $(i = 1,\ldots,n)$ denote the special polynomials of a multiindex $\boldsymbol{\varepsilon}_i$ with a 1 in the ith place and zeros elsewhere. Taking into account $D\,P_{\boldsymbol{\varepsilon}_i} = e_i$, we have

$$CK(x_i) = z_i = x_i - x_0 e_i.$$

Here the variables z_i introduced by Fueter appear again.

Suppose that $P_{\mathbf{k}}(\mathbf{x})$ is a homogeneous polynomial of degree k with $k = |\mathbf{k}|$, $a_{\mathbf{k}} \in \mathbb{R}^{n+1}$ and $\mathbf{k} = (k_1,\ldots,k_n)$,

$$P_{\mathbf{k}}(\mathbf{x}) = \sum_{k=|\mathbf{k}|}\mathbf{x}^{\mathbf{k}}a_{\mathbf{k}}.$$

The Cauchy–Kowalewski extension is of the form

$$CK(P_{\mathbf{k}})(x_0,\mathbf{x}) = \sum_{m=0}^{k}\frac{(-x_0)^m}{m!}D^m P_{\mathbf{k}}(\mathbf{x}).$$

The function $CK(P_{\mathbf{k}})$ is constructed to be holomorphic, and it is homogeneous of degree k, because every D decreases the degree of the polynomial by 1. The polynomial $D^m P_{\mathbf{k}}$ is of degree $k - m$, which is balanced out by x_0^m. Additionally we have $D^{k+1}P_{\mathbf{k}} = 0$. The initial conditions yield

$$CK(P_{\mathbf{k}})(0,\mathbf{x}) = P_{\mathbf{k}}(\mathbf{x}).$$

In the special case of the polynomials $\mathbf{x}^{\mathbf{k}}$ we obtain

$$CK(\mathbf{x}^{\mathbf{k}}) = \mathbf{k}!\mathcal{P}_{\mathbf{k}},$$

where $\mathcal{P}_{\mathbf{k}}$ are the Fueter polynomials introduced in Section 6.1. The proof is left as Exercise 11.3.15.

Example 11.40 (Hermite polynomials). Let $\rho_0(x) = e^{-|x|^2/2}$ ($x \in \mathbb{R}^n$) denote the Gaussian distribution function. Because of $|x|^2 = -x^2$ we have $\rho_0(x) = e^{x^2/2}$. Clearly the function ρ_0 is real analytic. The CK extension can be written in the following form:

$$CK(\rho_0)(x) = \sum_{m=0}^{\infty} \frac{(-x_0)^m}{m!} D^m(e^{x^2/2}) = e^{x^2/2} \sum_{m=0}^{\infty} \frac{x_0^m}{m!} H_m(x),$$

with

$$H_m(x) := (-1)^m e^{-x^2/2} D^m(e^{x^2/2}).$$

These functions are called *radial Hermite polynomials* after the French mathematician CHARLES HERMITE (1822–1901). Using a different approach our definition would give the so-called *Rodrigues formula*. The given functions are indeed polynomials, as

$$H_0(x) = 1$$

and the recursion formula

$$H_{m+1}(x) = (x - D)H_m(x)$$

show. The last formula is proved as follows:

$$
\begin{aligned}
H_{m+1}(x) &= (-1)^{m+1} e^{-x^2/2} D^{m+1}(e^{x^2/2}) \\
&= -DH_m(x) - (-1)^{m+1} \left(D e^{-x^2/2} \right) D^m(e^{x^2/2}),
\end{aligned}
$$

since

$$De^{-x^2/2} = e^{-x^2/2} \sum_{i=1}^{n} e_i(-x_i) = -x e^{-x^2/2}.$$

Radial Hermite polynomials are used for constructing higher-dimensional wavelet transforms (see [15]).

We shall now consider the CK extension of the function

$$T(x) := |x|^k e^{x^2/2} P_k(\omega(x)) \quad (\omega \in S^{n-1}, x \in \mathbb{R}^n).$$

Suppose that $P_k(\omega)$ is a (inner) spherical polynomial according to Definition 10.5, a linear combination of Fueter polynomials $\mathcal{P}_k(\omega)$ with $\omega(x) = x/|x|$. Thus the degree of homogeneity is $k = |k|$. We write the CK extension as in the preceding case in the form

$$CK(T)(x) = \sum_{m=0}^{\infty} \frac{(-x_0)^m}{m!} D^m(T(x)) = e^{x^2/2} \sum_{m=0}^{\infty} \frac{x_0^m}{m!} H_{mk}(x) P_k(x)$$

with

$$H_{mk}(x) P_k(x) = (-1)^m e^{-x^2/2} D^m \left(e^{x^2/2} P_k(x) \right).$$

Again the last formula is referred to as *Rodrigues' formula*. The functions $H_{mk}(\mathbf{x})$ are called *generalized radial Hermite polynomials*. Additional information can be found in [139].

Example 11.41 (Plane waves). A complex-valued real analytic function h depending only on the scalar product $\mathbf{x} \cdot \mathbf{y}$ with $(\mathbf{x}, \mathbf{y} \in \mathbb{R}^n)$ is called a *plane wave*. This definition was introduced by F. John in 1955 [67]. Suppose that $\mathbf{y}$ is fixed and h is a given plane wave. The corresponding CK extension in a suitable (open) domain in $\mathbb{R}^{n+1}$ reads as follows:

$$CK(h)(x) =: H(x_0, \mathbf{x}, \mathbf{y}) = \sum_{m=0}^{\infty} \frac{(-x_0)^m}{m!} D^m h(\mathbf{x} \cdot \mathbf{y}).$$

Since h is complex, actually the CK extension would have to be written for each of the two components of h. We have

$$Dh(\mathbf{x} \cdot \mathbf{y}) = \sum_{i=1}^{n} e_i h'(\mathbf{x} \cdot \mathbf{y}) \partial_i (\mathbf{x} \cdot \mathbf{y}) = h'(\mathbf{x} \cdot \mathbf{y}) \mathbf{y}$$

with the derivative h' of h with respect to its variable. Consequently we get the representation

$$H(x, \mathbf{y}) = \sum_{m=0}^{\infty} \frac{(-1)^m}{m!} h^{(m)}(\mathbf{x} \cdot \mathbf{y})(x_0 \mathbf{y})^m,$$

which can be decomposed as follows in view of $\mathbf{y}^2 = -|\mathbf{y}|^2$:

$$\begin{aligned} H(x, \mathbf{y}) &= \sum_{k=0}^{\infty} \frac{(-1)^k}{(2k)!} (x_0|\mathbf{y}|)^{2k} h^{(2k)}(\mathbf{x} \cdot \mathbf{y}) \\ &\quad - \frac{\mathbf{y}}{|\mathbf{y}|} \sum_{k=0}^{\infty} \frac{(-1)^k}{(2k+1)!} (x_0|\mathbf{y}|)^{2k+1} h^{(2k+1)}(\mathbf{x} \cdot \mathbf{y}) \\ &= H_1(\mathbf{x} \cdot \mathbf{y}, x_0|\mathbf{y}|) - \frac{\mathbf{y}}{|\mathbf{y}|} H_2(\mathbf{x} \cdot \mathbf{y}, x_0|\mathbf{y}|). \end{aligned}$$

H_1 and H_2 are real analytic with respect to the new real variables $\mathbf{x} \cdot \mathbf{y}$ and $x_0|\mathbf{y}|$. Extending the variables to complex ones will lead to a function (complex) holomorphic with respect to each variable. Holomorphic functions obtained in this way are called *holomorphic plane waves*.

We will conclude with an interesting special case ([32]). Because of $h^{(k)} = i^k h$ the CK extension $E_{CK}(x_0, \mathbf{x}, \mathbf{y})$ of the complex exponential function

$$h(\mathbf{x} \cdot \mathbf{y}) = e^{i(\mathbf{x} \cdot \mathbf{y})}$$

is defined as

$$E_{CK}(x_0, \mathbf{x}, \mathbf{y}) = e^{i(\mathbf{x} \cdot \mathbf{y})} \left(\sum_{k=0}^{\infty} \frac{(x_0|\mathbf{y}|)^{2k}}{(2k)!} - i \frac{\mathbf{y}}{|\mathbf{y}|} \sum_{k=0}^{\infty} \frac{(x_0|\mathbf{y}|)^{(2k+1)}}{(2k+1)!} \right)$$

$$= e^{i(\mathbf{x} \cdot \mathbf{y})} \left(\cosh(x_0|\mathbf{y}|) - i \frac{\mathbf{y}}{|\mathbf{y}|} \sinh(x_0|\mathbf{y}|) \right).$$

This is a third possible version of an exponential function in $C\ell(n)$. According to the construction this function is holomorphic with respect to the variable $x = x_0 + \mathbf{x}$.

11.2.5   Separation of variables

All holomorphic exponential functions in $C\ell(n)$ yet covered were based on extending a given function in $\mathbb{C}$ or $\mathbb{R}^n$ to $\mathbb{R}^{n+1}$. Considering Theorem 11.2 and the possibility of defining holomorphic functions as solutions of a system of differential equations leads to the idea of taking the properties

$$\frac{de^z}{dz} = e^z, \quad e^0 = 1$$

as the starting point for the generalization. Taking into account the complex case $\frac{1}{2}\partial$ can be considered as the derivative of a holomorphic function $f : \mathbb{R}^{n+1} \to \mathbb{C}\ell(n)$.

We shall restrict ourselves to the case $n = 3$, $f : \mathbb{H} \to \mathbb{H}$, $f = f_0 e_0 + f_1 e_1 + f_2 e_2 + f_3 e_3$. We shall start by studying the role of x_0; let $D = \sum_{i=1}^{3} e_i \partial_i$ denote the Dirac operator as in Theorem 5.12.

Definition 11.42. A function $f : \mathbb{R}^4 \to \mathbb{H}$ satisfying the properties:

1. $\overline{\partial} f = 0$,

2. $\frac{1}{2} \partial f = f$,

3. $f(0) = 1$,

4. $f(x_0, \mathbf{0}) = e^{x_0}$ for $x_0 \in \mathbb{R}$,

5. $f(x) \neq 0$ for all $x \in \mathbb{H}$,

6. $|f(x_0, x_1, 0, 0)| = |f(x_0, 0, x_2, 0)| = |f(x_0, 0, 0, x_3)| = e^{x_0}$ *for* $x_i \in \mathbb{R}$

and being periodic with respect to the variables x_1, x_2, x_3 is called an *exponential function on* $\mathbb{H}$. The last property is required to define a Fourier transform later on.

Theorem 11.43. *Assume that a function* $u(\mathbf{x}) : \mathbb{R}^3 \to \mathbb{H}$ *satisfies the differential equation*

$$Du(\mathbf{x}) = -u(\mathbf{x}).$$

Then

$$f(x) = f(x_0, \mathbf{x}) = e^{x_0} u(\mathbf{x})$$

satisfies the differential equations

$$\bar{\partial} f(x) = 0, \qquad \frac{1}{2}\partial f(x) = f(x).$$

Proof. With $\partial/\partial x_0 = \partial_0$ we get $\partial_0 + D = \bar{\partial}, \partial_0 - D = \partial$ and

$$\frac{1}{2}(\partial_0 \pm D) f(x) = \frac{1}{2}\partial_0 e^{x_0} u(\mathbf{x}) \mp \frac{1}{2}(-D)e^{x_0}u(\mathbf{x}) = \frac{1}{2}f(x) \mp \frac{1}{2}f(x). \qquad \square$$

Every holomorphic solution of $\frac{1}{2}\partial f = f$ must be of the form $f = e^{x_0}u(\mathbf{x})$ with the stated properties of u. This is left as Exercise 11.3.16.

Thus our task is reduced to solving the differential equation

$$Du(\mathbf{x}) = -u(\mathbf{x}).$$

The action of D in $\mathbb{R}^3$ leads to

$$-\Delta u = -Du = u.$$

The Laplacian Δ in $\mathbb{R}^3$ acts as a scalar operator and we have

$$\Delta u_i = -u_i, \quad i = 0, 1, 2, 3.$$

We suppose u_i to be of the form

$$u_i(\mathbf{x}) = r_i(x_1)s_i(x_2)t_i(x_3).$$

The action of Δ gives

$$\begin{aligned}
\Delta u_i(\mathbf{x}) &= r_i''(x_1)s_i(x_2)t_i(x_3) + r_i(x_1)s_i''(x_2)t_i(x_3) + r_i(x_1)s_i(x_2)t_i''(x_3)\\
&= -r_i(x_1)s_i(x_2)t_i(x_3) = -u_i(\mathbf{x}).
\end{aligned}$$

This yields

$$\frac{r_i''(x_1)}{r_i(x_1)} + \frac{s_i''(x_2)}{s_i(x_2)} + \frac{t_i''(x_3)}{t_i(x_3)} = -1.$$

Clearly all quotients need to be constant. Assuming symmetry with respect to x_1, x_2, x_3, we have

$$\frac{r_i''(x_1)}{r_i(x_1)} = \frac{s_i''(x_2)}{s_i(x_2)} = \frac{t_i''(x_3)}{t_i(x_3)} = -\frac{1}{3}.$$

Using the general solutions of the ordinary differential equations for r_i, s_i, t_i,

$$\begin{aligned}
r_i(x_1) &= a_i \cos\frac{x_1}{\sqrt{3}} + b_i \sin\frac{x_1}{\sqrt{3}},\\
s_i(x_2) &= c_i \cos\frac{x_2}{\sqrt{3}} + d_i \sin\frac{x_2}{\sqrt{3}},\\
t_i(x_3) &= e_i \cos\frac{x_3}{\sqrt{3}} + f_i \sin\frac{x_3}{\sqrt{3}},
\end{aligned}$$

and after combining some constants we get

$$
\begin{aligned}
u_i(\mathbf{x}) \;=\; & A_i \cos\frac{x_1}{\sqrt{3}}\cos\frac{x_2}{\sqrt{3}}\cos\frac{x_3}{\sqrt{3}} + B_i \cos\frac{x_1}{\sqrt{3}}\cos\frac{x_2}{\sqrt{3}}\sin\frac{x_3}{\sqrt{3}} \\
&+ C_i \cos\frac{x_1}{\sqrt{3}}\sin\frac{x_2}{\sqrt{3}}\cos\frac{x_3}{\sqrt{3}} + D_i \cos\frac{x_1}{\sqrt{3}}\sin\frac{x_2}{\sqrt{3}}\sin\frac{x_3}{\sqrt{3}} \\
&+ E_i \sin\frac{x_1}{\sqrt{3}}\cos\frac{x_2}{\sqrt{3}}\cos\frac{x_3}{\sqrt{3}} + F_i \sin\frac{x_1}{\sqrt{3}}\cos\frac{x_2}{\sqrt{3}}\sin\frac{x_3}{\sqrt{3}} \\
&+ G_i \sin\frac{x_1}{\sqrt{3}}\sin\frac{x_2}{\sqrt{3}}\cos\frac{x_3}{\sqrt{3}} + H_i \sin\frac{x_1}{\sqrt{3}}\sin\frac{x_2}{\sqrt{3}}\sin\frac{x_3}{\sqrt{3}}.
\end{aligned}
$$

Thus the function we look for has to be periodic and the period can "automatically" be computed from the assumption. This supports the assumption of symmetry — otherwise we would obtain different periods for different directions.

The condition $f(0) = 1$ means $u_0(\mathbf{0}) = 1$, so that $A_0 = 1$, and $u_1(\mathbf{0}) = u_2(\mathbf{0}) = u_3(\mathbf{0}) = 0$ with $A_i = 0$ for $i = 1, 2, 3$.

Since the structure of the function is clear now, we evaluate the differential equation $Du = -u$ in each coordinate. Taking into account that the functions

$$
g_1(\mathbf{x}) := \sin\frac{x_1}{\sqrt{3}}\sin\frac{x_2}{\sqrt{3}}\sin\frac{x_3}{\sqrt{3}}, \quad g_2(\mathbf{x}) := \sin\frac{x_1}{\sqrt{3}}\sin\frac{x_2}{\sqrt{3}}\cos\frac{x_3}{\sqrt{3}},
$$

$$
g_3(\mathbf{x}) := \sin\frac{x_1}{\sqrt{3}}\cos\frac{x_2}{\sqrt{3}}\sin\frac{x_3}{\sqrt{3}}, \quad g_4(\mathbf{x}) := \sin\frac{x_1}{\sqrt{3}}\cos\frac{x_2}{\sqrt{3}}\cos\frac{x_3}{\sqrt{3}},
$$

$$
g_5(\mathbf{x}) := \cos\frac{x_1}{\sqrt{3}}\sin\frac{x_2}{\sqrt{3}}\sin\frac{x_3}{\sqrt{3}}, \quad g_6(\mathbf{x}) := \cos\frac{x_1}{\sqrt{3}}\sin\frac{x_2}{\sqrt{3}}\cos\frac{x_3}{\sqrt{3}},
$$

$$
g_7(\mathbf{x}) := \cos\frac{x_1}{\sqrt{3}}\cos\frac{x_2}{\sqrt{3}}\sin\frac{x_3}{\sqrt{3}}, \quad g_8(\mathbf{x}) := \cos\frac{x_1}{\sqrt{3}}\cos\frac{x_2}{\sqrt{3}}\cos\frac{x_3}{\sqrt{3}}
$$

are linearly independent, comparing coefficients yields 32 linear equations for the remaining 28 free coefficients. We will skip the computation here and refer to the corresponding Exercise 11.3.17. Solving the system of linear equations with parameters B_i, C_i and G_i yields

$$
\begin{aligned}
E_0 &= B_2 - C_3, & E_1 &= \sqrt{3} - B_3 - C_2, \\
E_2 &= -B_0 + C_1, & E_3 &= B_1 + C_0, \\
D_0 &= -\sqrt{3}\,C_3 - G_2, & D_1 &= 1 - \sqrt{3}\,C_2 + G_3, \\
D_2 &= \sqrt{3}\,C_1 + G_0, & D_3 &= \sqrt{3}\,C_0 - G_1, \\
F_0 &= -\sqrt{3}\,B_1 - \sqrt{3}\,C_0 + G_1, & F_1 &= \sqrt{3}\,B_0 - \sqrt{3}\,C_1 - G_0, \\
F_2 &= 2 - \sqrt{3}\,B_3 - \sqrt{3}\,C_2 + G_3, & F_3 &= \sqrt{3}\,B_2 - \sqrt{3}\,C_3 - G_2, \\
H_0 &= -\sqrt{3} + B_3 + 2C_2 - \sqrt{3}\,G_3, & H_1 &= B_2 - 2C_3 - \sqrt{3}\,G_2, \\
H_2 &= -B_1 - 2C_0 + \sqrt{3}\,G_1, & H_3 &= -B_0 + 2C_1 + \sqrt{3}\,G_0.
\end{aligned}
$$

The remaining 12 parameters $B_i, C_i, G_i, i = 0, 1, 2, 3$, need to be determined. We want to use the degrees of freedom left to generalize the property $|e^{iy}| = 1$ of the complex e-function. We require that

$$|u(x_1,0,0)| = |u(0,x_2,0)| = |u(0,0,x_3)| = 1.$$

Using

$$u_i(x_1,0,0) = \delta_{0i} \cos \frac{x_1}{\sqrt 3} + E_i \sin \frac{x_1}{\sqrt 3},$$

with Kronecker's symbol δ_{0i} we have immediately

$$1 = |u(x_1,0,0)|^2 = \cos^2 \frac{x_1}{\sqrt 3} + (E_0^2 + E_1^2 + E_2^2 + E_3^2) \sin^2 \frac{x_1}{\sqrt 3} + 2E_0 \sin \frac{x_1}{\sqrt 3} \cos \frac{x_1}{\sqrt 3}.$$

Consequently we have $E_0 = 0$ and $E_1^2 + E_2^2 + E_3^2 = 1$. The same conditions have to hold for the B_i and the C_i, following from the equations $|u(0,0,x_3)| = 1$ and $|u(0,x_2,0)| = 1$. Since the variables need to be weighted equally we have

$$B_1 = B_2 = B_3 = C_1 = C_2 = C_3 = \frac{1}{\sqrt 3}.$$

Fortunately the conditions for the E_i are satisfied anyway. Consequently the coefficients are

$$B_0 = C_0 = E_0 = 0, \quad B_1 = B_2 = B_3 = C_1 = C_2 = C_3 = E_1 = E_2 = E_3 = \frac{1}{\sqrt 3},$$
$$D_0 = -1 - G_2, \ D_1 = G_3, \qquad\qquad D_2 = 1 + G_0, \qquad D_3 = -G_1,$$
$$F_0 = -1 + G_1, \ F_1 = -1 - G_0, \qquad F_2 = G_3, \qquad\qquad F_3 = -G_2,$$
$$H_0 = -\sqrt 3\, G_3, \ H_1 = -\tfrac{1}{\sqrt 3} - \sqrt 3\, G_2, \ H_2 = -\tfrac{1}{\sqrt 3} + \sqrt 3\, G_1, \ H_3 = \tfrac{2}{\sqrt 3} + \sqrt 3\, G_0.$$

The functions g_i, $i = 1, \ldots, 8$, have to be weighted equally. Thus D_0, F_0, H_0 must equal ± 1, the remaining D_i, F_i, H_i equal $\pm\frac{1}{\sqrt 3}$. This requires

$$G_0 = -1, \ G_1 = G_2 = 0, \ G_3 = \pm\frac{1}{\sqrt 3}.$$

We choose the minus sign and obtain

$$u_0 = g_8 - g_5 - g_3 - g_2 + g_1,$$

$$u_1 = \frac{1}{\sqrt 3}(g_7 + g_6 + g_4 - g_1 - g_5),$$

$$u_2 = \frac{1}{\sqrt 3}(g_7 + g_6 + g_4 - g_1 - g_3),$$

$$u_3 = \frac{1}{\sqrt 3}(g_7 + g_6 + g_4 - g_1 - g_2),$$

with the formerly introduced abbreviations for the functions g_i. Recall that

$$g_8(\mathbf{x}) - g_5(\mathbf{x}) - g_3(\mathbf{x}) - g_2(\mathbf{x}) = \cos \frac{x_1 + x_2 + x_3}{\sqrt 3},$$

$$g_7(\mathbf{x}) + g_6(\mathbf{x}) + g_4(\mathbf{x}) - g_1(\mathbf{x}) = \sin \frac{x_1 + x_2 + x_3}{\sqrt 3},$$

from which finally:

Definition 11.44. If we denote the discovered exponential function with $\mathcal{E}(x)$, we have

$$
\mathcal{E}(x) \;=\; e^{x_0}\left[\left(\cos\frac{x_1+x_2+x_3}{\sqrt{3}} + \sin\frac{x_1}{\sqrt{3}}\sin\frac{x_2}{\sqrt{3}}\sin\frac{x_3}{\sqrt{3}}\right)\right.
$$
$$
+\frac{1}{\sqrt{3}}\left((e_1+e_2+e_3)\sin\frac{x_1+x_2+x_3}{\sqrt{3}} - e_1\cos\frac{x_1}{\sqrt{3}}\sin\frac{x_2}{\sqrt{3}}\sin\frac{x_3}{\sqrt{3}}\right.
$$
$$
\left.\left.-e_2\sin\frac{x_1}{\sqrt{3}}\cos\frac{x_2}{\sqrt{3}}\sin\frac{x_3}{\sqrt{3}} - e_3\sin\frac{x_1}{\sqrt{3}}\sin\frac{x_2}{\sqrt{3}}\cos\frac{x_3}{\sqrt{3}}\right)\right].
$$

In contrast to the other constructions of exponential functions $\mathcal{E}(x)$ is periodic with respect to x_1, x_2, x_3 with the period $2\pi\sqrt{3}$ and has the modulus e^{x_0} along the imaginary line.

We want to examine the behavior of the function $\mathcal{E}$:

Proposition 11.45. *For all $x \in \mathbb{H}$ we have*

$$
\mathcal{E}(x) \neq 0.
$$

Proof. Defining

$$
s = \frac{x_1+x_2+x_3}{\sqrt{3}}, \quad y_1 = \frac{x_1}{\sqrt{3}}, \quad y_2 = \frac{x_2}{\sqrt{3}}, \quad y_3 = \frac{x_3}{\sqrt{3}}
$$

we compute $|\mathcal{E}(x)|^2$ by

$$
|\mathcal{E}(x)|^2 = e^{2x_0}\left\{\cos^2 s + \sin^2 s + \frac{1}{3}\left(\sin^2 y_1 \sin^2 y_2 + \sin^2 y_1 \sin^2 y_3 + \sin^2 y_2 \sin^2 y_3\right.\right.
$$
$$
+6\cos s \sin y_1 \sin y_2 \sin y_3 - 2\sin s\left[\cos y_1 \sin y_2 \sin y_3\right.
$$
$$
\left.\left.\left. + \sin y_1 \cos y_2 \sin y_3 + \sin y_1 \sin y_2 \cos y_3\right]\right)\right\}.
$$

Using the sum and difference formulas for the sine we obtain

$$
|\mathcal{E}(x)|^2 = \frac{1}{3}e^{2x_0}\left\{3 + \sin^2 y_1 \sin^2 y_2 + \sin^2 y_2 \sin^2 y_3 + \sin^2 y_3 \sin^2 y_1\right.
$$
$$
-2\sin(y_1+y_2)\sin y_1 \sin y_2 - 2\sin(y_2+y_3)\sin y_2 \sin y_3
$$
$$
\left.-2\sin(y_3+y_1)\sin y_3 \sin y_1\right\}.
$$

The right-hand side consists of three terms of the form

$$
g_{ij} := 1 + \sin^2 y_i \sin^2 y_j - 2\sin(y_i+y_j)\sin y_i \sin y_j;
$$

estimating from above $|\sin(y_i+y_j)|$ by 1 we obtain

$$
g_{ij} \geq 1 + a^2 b^2 - 2ab = (ab-1)^2.
$$

This is zero only for $a = b = |\sin y_i| = |\sin y_j| = 1$, i.e., y_i and $y_j = \pm\pi/2$. But then we have $y_i + y_j$ equal 0 or $\pm\pi$ and $\sin(y_i+y_j) = 0$. Thus the modulus of $\mathcal{E}(x)$ can never be zero.   □

Now we want to study some other properties of the $\mathcal{E}$-function. For the proof we refer to the exercises, since only computational skills and endurance are required.

Proposition 11.46. *The function $\mathcal{E}(x)$ satisfies the properties stated in Definition 11.42 and:*

(i) $\mathcal{E}(x)$ *is also right-holomorphic, thus it is* biholomorphic.

(ii) $\overline{\partial}\mathcal{E}(x\lambda) = 0$ *for real λ. For $\lambda \in \mathbb{H}$ this is not valid in general.*

(iii) *We get $\mathcal{E}(x\lambda)\overline{\partial} = 0$ for all $\lambda \in \mathbb{H}$, in addition we have*

$$\frac{1}{2}\partial\mathcal{E}(x\lambda) = \lambda\mathcal{E}(x\lambda)$$

for all $\lambda \in \mathbb{H}$.

(iv) $\frac{1}{2}\partial\mathcal{E}(x\lambda)$ *is right-holomorphic, $\lambda\mathcal{E}(x\lambda)$ is also left-holomorphic.*

The function being biholomorphic is pretty surprising. We did not demand it and not every left-holomorphic function is biholomorphic. The behavior of $\mathcal{E}(x\lambda)$ and $\mathcal{E}(\lambda x)$ for $x, \lambda \in \mathbb{H}$ is also very interesting. These functions are used for a solution Ansatz for differential equations and for the Fourier transform. Recall that the formerly considered generalizations of the complex e-function did only allow λx with real λ. These results are in no way self-evident — we only wanted to construct a left-holomorphic exponential function. We stated that $\mathcal{E}(x)$ is also right-holomorphic. This property is invariant under the transformation $x \mapsto x\lambda$. This has some interesting consequences. We have that $\mathcal{E}(x\lambda)$ is not left-holomorphic, but $\lambda\mathcal{E}(x\lambda)$ is biholomorphic. The $\mathcal{E}$-function satisfies more properties than originally intended. Looking back these properties do not appear really strange if one considers functionals of transforms of the type

$$\int_{\mathbb{R}^4} \mathcal{E}(x\lambda)f(x)d\sigma_x.$$

Derivatives pulled out from f to the left-hand side can only be transported in $\mathcal{E}(x\lambda)$ from the right-hand side — due to non-commutativity. We refer our reader to Exercise 11.3.18 dealing with the $\mathcal{E}$-function.

11.3 Exercises

1. Prove Theorem 11.5 (iii): For all $z, z_1, z_2 \in \mathbb{C}$ the following sum and difference formulas hold:

$$\cos(z_1 + z_2) = \cos z_1 \cos z_2 - \sin z_1 \sin z_2,$$
$$\sin(z_1 + z_2) = \sin z_1 \cos z_2 + \cos z_1 \sin z_2,$$
$$\cos^2 z + \sin^2 z = 1.$$

2. Prove Theorem 11.7 (v): The hyperbolic cosine and the hyperbolic sine have the period $2\pi i$. We have the relations

$$
\begin{aligned}
\cosh z &= \cos iz, & \sinh z &= -i\sin iz, \\
\cos z &= \cosh iz, & \sin z &= -i\sinh iz, \\
\cos z &= \cos x \cosh y - i\sin x \sinh y, \\
\sin z &= \sin x \cosh y + i\cos x \sinh y
\end{aligned}
$$

with $z = x + iy$.

3. Evaluate $|e^z|$ and show that $|e^z| \le e^{|z|}$.

4. Compute $|\sin z|^2$. Use this to localize all zeroes of $\sin z$.

5. Evaluate the first five coefficients of the power series of

$$
\tan z = \frac{\sin z}{\cos z}.
$$

6. Localize the zeros and singularities (i.e., the zeros of the denominator) of

$$
\tanh z := \frac{\sinh z}{\cosh z}.
$$

7. Determine the Taylor series expansion of the principal value of the logarithm

$$
\log(1 + z) = \sum_{n=1}^{\infty} \frac{(-1)^{n+1}}{n} z^n.
$$

8. Prove the relations

$$
\overline{\partial} = \frac{1}{|x|} L + \omega \partial_\omega,
$$

$$
\partial_i |x| = \frac{x_i}{|x|} = \omega_i, \quad \partial_\omega x = \omega
$$

in $\mathbb{R}^{n+1}$.

9. Prove that $L\omega_k = e_k - \omega\omega_k$ and $\partial_\omega(\omega_k) = 0$.

10. Prove Corollary 11.21:

a) $e^{-x}e^x = 1$, $e^x \ne 0$,

b) $e^{kx} = (e^x)^k$ $(k \in \mathbb{Z})$,

c) $e^{\omega(x)\pi} = -1$.

11. Show the statements of Corollary 11.25 (ii):

$$\log 1 = 0, \quad \log e_i = \frac{\pi}{2} e_i.$$

12. Prove Corollary 11.31: The elementary functions are radially holomorphic and satisfy

   a) $(e^x)' = e^x$,

   b) $(\sin x)' = \cos x, \quad (\cos x)' = -\sin x$,

   c) $(\sinh x)' = \cosh x, \quad (\cosh x)' = \sinh x$,

   d) $(\log x)' = \frac{1}{x}$,

   e) $(x^\alpha)' = \alpha x^{\alpha-1}$ with $\alpha \in \mathbb{R}$.

13. Prove that $\tau_n(f)$ is holomorphic for n odd according to Theorem 11.33 (ii):

$$\overline{\partial}\tau_n(f) = 0.$$

14. Evaluate the Fueter transform of the hyperbolic functions denoted by $\mathrm{COSH}_3(x)$ and $\mathrm{SINH}_3(x)$, respectively.

15. Establish the relation (see example 11.39):

$$CK(\mathbf{x}^k) = k! \mathcal{P}_k(\mathbf{x}).$$

16. Prove that every solution of $\overline{\partial} f = 0, \partial f = 2f$ has the form (see Theorem 11.43)

$$f(x) = e^{x_0} u(\mathbf{x}).$$

17. Solve the 32 linear equations, appearing in determining the holomorphic exponential function $\mathcal{E}(x)$, for the coefficients $A_i, B_i, C_i, D_i, i = 0, 1, 2, 3$, by evaluating the derivatives of the functions $g_1, \ldots, g_8$ and comparing coefficients.

18. Prove Proposition 11.46:

   a) $\mathcal{E}(x)$ is also right-holomorphic, thus biholomorphic.

   b) $\overline{\partial}\mathcal{E}(x\lambda) = 0$ for real λ. For $\lambda \in \mathbb{H}$ this does not hold in general.

   c) Show that $\mathcal{E}(x\lambda)\overline{\partial} = 0$ for all $\lambda \in \mathbb{H}$, and in addition

$$\frac{1}{2}\partial\mathcal{E}(x\lambda) = \lambda\mathcal{E}(x\lambda)$$

   for all $\lambda \in \mathbb{H}$.

   d) $\frac{1}{2}\partial\mathcal{E}(x\lambda)$ is right-holomorphic, $\lambda\mathcal{E}(x\lambda)$ is also left-holomorphic.

12 Local structure of holomorphic functions

12.1 Behavior at zeros

12.1.1 Zeros in $\mathbb{C}$

In this section we want to discuss mainly local properties of holomorphic functions. First we use Taylor series in $\mathbb{C}$. With their help we can extend to all holomorphic functions the notion of *order of a root*, already known from polynomials.

Definition 12.1 (Order of a root). Let f be a function holomorphic in the domain $G \subset \mathbb{C}$ and $z_0 \in G$. Let in the neighborhood of z_0 the equation

$$f(z) = \sum_{n=k}^{\infty} a_n (z - z_0)^n$$

hold, where $a_k \neq 0$. For $k > 0$ we then say that f has a *zero of order k* in z_0.

The coefficients are uniquely determined in view of Corollary 9.8, and so this definition makes sense. We can read off, that the quotient $f(z)/(z - z_0)^k$ has a finite limit for a zero of order k for $z \to z_0$. This limit is the Taylor coefficient $a_k = f^{(k)}(z_0)/k!$. We can factorize f in a neighborhood of a zero z_0 of order k as follows:

$$f(z) = (z - z_0)^k g(z),$$

where g is a holomorphic function with $g(z_0) \neq 0$. From the last formula it follows then:

Proposition 12.2. *Zeros of holomorphic functions are isolated.*

This proposition is indeed the same as the uniqueness Theorem 9.15 for holomorphic functions, because a non-isolated zero enforces the function to be identically equal to zero.

Example 12.3. For power series we can find the order of a zero only at the point of expansion. We see that the function

$$\sin z = \sum_{n=0}^{\infty} (-1)^n \frac{z^{2n+1}}{(2n+1)!}$$

has a simple zero at $z = 0$. The other zeros are also simple, because of the periodicity, but this is not evident from the power series.

There is no restriction for the zeros of a holomorphic function other than the requirement of being isolated. To show this we want to prove *Weierstraß' product theorem* for an *entire function*, that is a function holomorphic in the whole complex plane. For this we need the notion of an infinite product.

Definition 12.4 (Infinite product). (i) We call the symbol for a sequence of complex numbers (a_n) with $1 + a_n \neq 0$,

$$\prod(1 + a_n),$$

an *infinite product*. It is called *convergent*, if the limit of the *partial products*

$$\lim_{n\to\infty} \prod_{k=1}^{n}(1 + a_k) =: \prod_{k=1}^{\infty}(1 + a_k)$$

exists and is different from 0 and ∞. Using the principal value of the logarithm the product is called *absolutely convergent* if

$$\sum \log(1 + a_n)$$

is absolutely convergent.

(ii) A product with a finite number of factors $1 + a_n = 0$, $n \leq n_0$, is *convergent*, if

$$\prod(1 + a_{n+n_0})$$

is convergent, in the sense of (i). To it we then assign the value 0.

A product is zero, if and only if finitely many factors are zero. We need some properties of infinite products:

Proposition 12.5.

(i) *A necessary condition for the product $\prod(1 + a_n)$ to converge is that $a_n \to 0$.*

(ii) *The product $\prod(1+a_n)$ converges if only if the series $\sum \log(1+a_n)$ converges. Here too we use the principal value of the logarithm.*

(iii) *The product $\prod(1 + a_n)$ converges absolutely if and only if the series $\sum a_n$ converges absolutely.*

Proof. (i) In the case of convergence of the product and in view of

$$1 + a_n = \prod_{k=1}^{n}(1 + a_k) \Big/ \prod_{k=1}^{n-1}(1 + a_k) \to 1$$

for $n \to \infty$ it follows that $a_n \to 0$.

(ii) Using

$$\prod_{k=1}^{n}(1 + a_k) = \exp\left(\sum_{k=1}^{n} \log(1 + a_k)\right),$$

the principal value of the logarithm, and the continuity of the exponential function the convergence of the product follows from the convergence of the series. The inversion is left as Exercise 12.5.2.

(iii) We can assume that $|a_n| \geq 1/2$ only for finitely many factors. For the other n we use the Taylor series of the logarithm (Exercise 11.3.7) if $a_n \neq 0$,

$$\left| \frac{1}{a_n} \log(1 + a_n) - 1 \right| = \left| \sum_{k=2}^{\infty} \frac{a_n^{k-1}}{k} \right| \leq \sum_{k=2}^{\infty} \left(\frac{1}{2} \right)^k = \frac{1}{2}$$

and therewith

$$\frac{1}{2} \leq \left| \frac{\log(1 + a_n)}{a_n} \right| \leq \frac{3}{2}.$$

We can ignore the terms with $a_n = 0$ as $\log(1 + a_n) = 0$, then $\sum |\log(1 + a_n)|$ and $\sum |a_n|$ simultaneously diverge, respectively converge.   □

We are now able to prove the announced theorem:

Theorem 12.6 (Weierstraß' product theorem). *Let (z_k) be a sequence of complex numbers, where $0 < |z_k| \leq |z_{k+1}|$, $k \in \mathbb{N}$, and $z_k \to \infty$. Then natural numbers N_k exist so that*

$$h(z) := \prod_{k=1}^{\infty} E_k(z, z_k)$$

is an entire function with zeros $z_1, z_2, \dots$. The Weierstraß factors $E_k(z, z_k)$ are defined by

$$E_k(z, z_k) := \left(1 - \frac{z}{z_k} \right) \exp\left(\sum_{j=1}^{N_k} \frac{1}{j} \left(\frac{z}{z_k} \right)^j \right).$$

Multiple zeros are expressed by repeated z_k. The factors containing the exponential function are called *generating convergence factors*, because the factors are not necessary if $\sum 1/|z_k|$ already converges.

In conclusion the distribution of the zeros of an entire function has no restriction. So every entire function f can be written in the form

$$f(z) = z^m \, e^{g(z)} \prod_{k=1}^{\infty} E_k(z, z_k),$$

where g is entire and m is the order of the zero of f at $z_0 = 0$. We get the justification of this formula, if we apply Theorem 9.17 to the quotient of f and $z^m \prod E_k(z, z_k)$. This quotient Q is an entire function with no zeros in $\mathbb{C}$. So $\log Q$ can be continued in $\mathbb{C}$ with Theorem 9.17 and is an entire function g, i.e., $Q = e^g$. Now we will turn to the proof of the theorem:

Proof. For natural numbers N_k and $|z| < |z_k|$ the equation

$$E_k(z, z_k) = \exp\left(\log\left(1 - \frac{z}{z_k} \right) + \sum_{j=1}^{N_k} \frac{1}{j} \left(\frac{z}{z_k} \right)^j \right) = \exp\left(- \sum_{j=N_k+1}^{\infty} \frac{1}{j} \left(\frac{z}{z_k} \right)^j \right)$$

holds. If we assume $|z| \leq R$, then for the z_k with $|z_k| \geq 2R$ and $k \geq k_0$ it follows that

$$|\log E_k(z, z_k)| \leq \left|\frac{z}{z_k}\right|^{N_k} \sum_{j=1}^{\infty} \frac{1}{2^j} \leq \left|\frac{z}{z_k}\right|^{N_k}.$$

This estimate provides

$$\sum_{k=k_0}^{\infty} |\log E(z, z_k)| \leq \sum_{k=k_0}^{\infty} \left(\frac{R}{|z_k|}\right)^{N_k} \leq \sum_{k=k_0}^{\infty} \left(\frac{1}{2}\right)^{N_k},$$

and this is convergent for $N_k = k$. In general we can take smaller N_k. The convergence for $|z| \leq R$ is uniform, so that the limit represents a holomorphic function. The product is an entire function since R was arbitrary. $\square$

Example 12.7. The function $\sin \pi z$ has zeros at $z = 0, \pm 1, \ldots$. This results in the product

$$\frac{\sin(\pi z)}{\pi z} = \prod_{k=1}^{\infty} \left(1 - \frac{z^2}{k^2}\right),$$

because

$$\left(1 - \frac{z}{k}\right)\left(1 + \frac{z}{k}\right) = 1 - \frac{z^2}{k^2}.$$

The function

$$f(z) := z \prod_{k=1}^{\infty} \left(1 - \frac{z^2}{k^2}\right)$$

has then the same zeros as $\sin \pi z$. One can prove that this function is really the function $\pi \sin \pi z$.

12.1.2 Zeros in $C\ell(n)$

For a holomorphic function in $C\ell(n)$ the problem concerning zeros is a little bit more difficult than for the zeros in the plane. There are also still open questions. In Section 9.3 we have already noticed that the zeros are not necessarily isolated. In the identity Theorem 9.27 we have shown that the manifold of zeros has at most dimension $n - 1$. But all lower dimensions can occur. This can be seen for example by considering the following function:

$$f(x) := z_1 e_1 + z_2 e_2 + \cdots + z_k e_k = kx_0 + e_1 x_1 + \cdots + e_k x_k \quad (1 \leq k \leq n).$$

This f is biholomorpic and has its zeros at the points with

$$x_0 = x_1 = \cdots = x_k = 0,$$

i.e., on an $(n - k)$-dimensional plane. For $k = n$ we get an isolated point, the origin.

The manifold, on which a holomorpic function is zero in a neighborhood of a point a with $f(a) = 0$, depends among other things on the rank of the Jacobian

$$J_f = \left(\frac{\partial f_i}{\partial x_j} \right).$$

In the case that the rank of J_f is constant in a neighborhood U of the zero, let it equal m with $n + 1 \geq m \geq 2$, Hempfling [61] and in the quaternion case Fueter [49] proved that exactly one manifold of dimension $n + 1 - m$ through the zero exists on which $f(a) = 0$. The case $m = 1$ cannot arise as we have shown already in the identity theorem. We will not prove these results here, because we would need theorems from the theory of ordinary and partial differential equations. Additionally the examples of Zöll [161] show, that the rank of J_f can vary greatly even for isolated zeros.

Nevertheless in the case of $J_f(a)$ with rank $n + 1$ we can conclude from the implicit function theorem that also this rank of J_f is given in a neighborhood of the point a and f denotes a local diffeomorphism. Then the zero at the point a is isolated:

Proposition 12.8. *If the function f in a neighborhood U of the point a is holomorphic, if $f(a) = 0$ and $\det J_f(a) \neq 0$, then the zero of f at a is isolated.*

Propositions about the behavior of a holomorphic function in the neighborhood of manifolds of zeros do not exist. So we want to restrict to isolated zeros at a point a. In this case even the easy notion of the order of a zero is more complicated than in the plane. We know from the previous subsection, that the order of a zero is equal to the index of the first nonvanishing term in the Taylor series. So we can specify in $\mathbb{C}$ the three following characterizations of the order k of a zero:

1. k is the smallest natural number with $f^{(k)}(a) \neq 0$.

2.

$$k := \frac{1}{2\pi i} \int_\gamma \frac{f'(z)}{f(z)} dz$$

holds with γ a simple closed curve around a with no other zeros of f inside. The formula above follows by Cauchy's integral theorem, if we use the Taylor series of f' and f (see also Exercise 12.5.1).

3. The formula

$$k = \frac{1}{2\pi i} \int_{f(\gamma)} \frac{dw}{w}$$

follows from the previous formula using the transformation $w = f(z)$ in the integral.

We cannot use the first definition in $C\ell(n)$, because Zöll [161] provides examples whose Taylor series do not give satisfying results. Also the second definition cannot be applied for lack of a suitable derivative. But the third definition is suitable for $C\ell(n)$ [61], [62]:

Definition 12.9. Let f be left-holomorphic in the domain G and a be an isolated zero of f. Moreover let $B_\varepsilon(a)$ be a sufficiently small ball with radius ε around a and the boundary $S_\varepsilon(a)$, so that no additional zero of f is in $B_\varepsilon(a) \cup S_\varepsilon(a)$. Then

$$\operatorname{ord}(f;a) := \frac{1}{\sigma_n} \int\limits_{F(S_\varepsilon(a))} \mathcal{Q}_0(y) dy^*$$

denotes the *order of the zero* of f at the point a. Here we define

$$F = \sum_{i=0}^{n} e_i f_i$$

to be the restriction of the values of f to $\mathbb{R}^{n+1}$.

The 'cutting' from f to F is an effort, which would have to be analyzed in more detail. We know that $\mathcal{Q}_0$ is the 'simplest' singular holomorphic function having an isolated singularity at the origin. We have introduced

$$\mathcal{Q}_0(y) = \frac{\overline{y}}{|y|^{n+1}}$$

in Definition 7.26, where we used it as kernel for Cauchy's integral formula. It is shown in Proposition 12.49, that the so-defined order is an integer, namely the winding number of the manifold $F(S_\varepsilon)(a)$ around the point a, which must still be defined. So our definition meets the descriptive notion of order for zeros.

As the calculation of the order can technically be difficult, and in addition the manifold is actually defined by $S_\varepsilon(a)$, we can try to transform the integral with the help of the substitution $y = F(x)$ for the variable x. Zöll has realized this: Letting

$$dy_j = \sum_{k=0}^{n} \frac{\partial F_j}{\partial x_k} dx_k$$

it follows (cf. Remark A1.16 c))

$$dy_i^* = (-1)^i \sum_{k_0,\ldots,\hat{k}_i,\ldots k_n=0}^{n} \frac{\partial F_0}{\partial x_{k_0}} \cdots \widehat{\frac{\partial F_i}{\partial x_{k_i}}} \cdots \frac{\partial F_n}{\partial x_{k_n}} dx_{k_0} \wedge \cdots \wedge d\hat{x}_{k_i} \wedge \cdots \wedge dx_{k_n}$$

$$= (-1)^i \sum_{j=0}^{n} \frac{\partial(F_0, \ldots \hat{F}_i \ldots, F_n)}{\partial(x_0, \ldots \hat{x}_j \ldots, x_n)} d\hat{x}_j.$$

Since

$$(-1)^{i+j} \frac{\partial(F_0, \ldots \hat{F}_i \ldots, F_n)}{\partial(x_0, \ldots \hat{x}_j \ldots, x_n)} =: A^{ij}$$

are just the adjoints of J_F, in view of

$$J_F^{-1} = (A^{ij})^\top / \det J_F$$

we have finally the representation

$$dy^* = \sum_{i,j=0}^{n} e_i A^{ij} dx_j^*$$

and for the order the expression:

Corollary 12.10. *With the assumptions of Definition 12.9,*

$$\mathrm{ord}(f;a) = \frac{1}{\sigma_n} \int\limits_{S_\varepsilon(a)} \frac{\overline{F(x)}}{|F(x)|^{n+1}} \sum_{i,j=0}^{n} e_i A^{ij} dx_j^*$$

holds.

Now we can introduce J_F^{-1}, but we encounter the additional restriction $\det J_F \neq 0$. Unfortunately this formula is not promising, because in practical cases the integral is very difficult to calculate. Additional research is needed and hopeful approaches are tackled by Kraußhar. But we will calculate the order of an isolated zero at least in one case:

Example 12.11. We consider the function given at the beginning of this subsection: $f(x) = z_1 e_1 + \cdots + z_n e_n = n x_0 + \mathbf{x}$, where $\mathbf{x}$ is the vector related to x. We do not need to cut f to F, as f is a paravector-valued function. So for $i,j = 1,\ldots,n$ we have:

$$f_0(x) = n x_0, \;\; f_i(x) = x_i$$

and

$$\frac{\partial f_0}{\partial x_0}(x) = n, \;\; \frac{\partial f_0}{\partial x_i}(x) = 0, \;\; \frac{\partial f_i}{\partial x_0}(x) = 0, \;\; \frac{\partial f_i}{\partial x_j} = \delta_{ij}.$$

We get the very simple Jacobian

$$J_f = \begin{pmatrix} n & 0 & \cdots & 0 \\ 0 & 1 & \cdots & 0 \\ \vdots & \vdots & & \vdots \\ 0 & 0 & \cdots & 1 \end{pmatrix}$$

with $\det J_f = n$ and the adjoints $(i,j = 1,\ldots,n)$

$$A^{00} = 1, \;\; A^{0i} = 0, \;\; A^{i0} = 0, \;\; A^{ij} = n \delta_{ij}.$$

From this it follows that

$$\sum_{i,j=0}^{n} e_i A^{ij} dx_j^* = dx_0^* + n d\mathbf{x}^*.$$

and

$$\operatorname{ord}(f;0) = \frac{1}{\sigma_n} \int_{|x|=\varepsilon} \frac{nx_0 - \mathbf{x}}{|nx_0 + \mathbf{x}|^{n+1}} (dx_0^* + nd\mathbf{x}^*).$$

Now we deform the sphere $S_\varepsilon : |x| = \varepsilon$ continuously into the ellipsoid $|nx_0 + \mathbf{x}| = \varepsilon$, not crossing any zeros of f. The order cannot change, as the order is an integer and the transformation is continuous. Now we set

$$x_0(t) = \frac{\varepsilon}{n} y_0(t), \quad x_i(t) = \varepsilon y(t), \quad i = 1, \dots, n,$$

where y is the parametric equation of the unit sphere $y = \sum_{i=0}^{n} y_i e_i$ according to Example A.2.17 c). Furthermore we have

$$|nx_0 + \mathbf{x}| = \varepsilon |y_0 + e_1 y_1 + \cdots + e_n y_n| = \varepsilon, \quad nx_0 - \mathbf{x} = \varepsilon \, \overline{y}.$$

Since x_0 occurs in $dx_1^*, \dots, dx_n^*$, but not in dx_0^* we have

$$dx_0^* + nd\mathbf{x}^* = \varepsilon^n dy^*.$$

Using Example A.2.17 c) we get $dy^* = y|do_1(y)|$, where y is the unit outer normal vector on the unit sphere S^n and $|do_1(y)|$ is the surface element of the unit sphere. Finally the order is

$$\operatorname{ord}(f;0) = \frac{1}{\sigma_n} \int_{S^n} \overline{y}\, y |do_1(y)| = \frac{1}{\sigma_n} \int_{S^n} |do_1(y)| = 1.$$

Of course we expected this value, but the calculation was very complex and for other functions it could be even more involved. As mentioned further research is needed.

Unfortunately in $C\ell(n)$ a theorem equivalent to the Weierstraß' product theorem cannot exist, as the product of holomorphic functions is not holomorphic in general. New considerations are also necessary at this point.

12.2 Isolated singularities of holomorphic functions

12.2.1 Isolated singularities in $\mathbb{C}$

In this subsection we want to examine the behavior of a function f at points in which f is not defined or not holomorphic. First we consider the case of the complex plane. For simplicity we state that $\rho_1 < |z - z_0| < \rho_2$ means the annular domain $\{z : \rho_1 < |z - z_0| < \rho_2\}$.

Definition 12.12 (Singularity). If the function f is holomorphic in the punctured disk $0 < |z - z_0| < R$, but not at z_0, resp. not defined at z_0, we say that f has an *isolated singularity* at z_0. These singularities are classified as follows:

a) The singularity is called *removable*, if f is bounded in $0 < |z - z_0| \le R/2$.

b) The singularity is called a *pole*, if it is not removable and an $n \in \mathbb{N}$ exists, so that $(z - z_0)^n f(z)$ is bounded in $0 < |z - z_0| \le R/2$.

c) Else the singularity is called an *essential singularity*.

d) If a function f is holomorphic in a domain G up to poles, which are not accumulating in G, we call f *meromorphic* in G.

e) We also refer to an essential singularity, if the function f is meromorphic in $0 < |z - z_0| < R$ and has neither a removable singularity nor a pole in x_0.

Now we want to study these singularities, firstly the removable singularities. We prove:

Theorem 12.13 (Riemann's theorem on removable singularities). *Let f be a holomorphic function in $0 < |z - z_0| < R$ and let f have a removable singularity at z_0. Then a number a_0 exists so that the extended function*

$$\tilde{f}(z) = \begin{cases} f(z), & 0 < |z - z_0| < R, \\ a_0, & z = z_0, \end{cases}$$

is holomorphic in $B_R(z_0)$.

Apart from the assumption that f is holomorphic in the punctured disk, it is sufficient to show that $\lim\limits_{r \to 0} r M(r, f) = 0$ with

$$M(r, f) := \max_{|z - z_0| = r} |f(z)|.$$

We recognize that in the case of a removable singularity the problem is only the function value at the point z_0. So we call the function f *holomorphic continuable* at z_0. The assertion of the theorem shows that $M(r, f)$ cannot converge to ∞ for $r \to 0$. But sometimes it is easier to show the weaker assumption in applications.

Proof. The function f is holomorphic in the punctured disk $0 < |z - z_0| < R$. In it f has the convergent Laurent series

$$f(z) = \sum_{n=-\infty}^{\infty} a_n (z - z_0)^n, \quad a_n = \frac{1}{2\pi i} \int_{|\zeta - z_0| = \rho} \frac{f(\zeta)}{(\zeta - z_0)^{n+1}} d\zeta.$$

We consider the coefficients with negative index $n > 0$:

$$\begin{aligned} |a_{-n}| &\le \frac{1}{2\pi} \int_{|\zeta - z_0| = \rho} \frac{M(\rho, f)}{\rho^{-n+1}} |d\zeta| \\ &\le M(\rho, f) \rho^n. \end{aligned}$$

This expression converges with $\rho \to 0$ and all $n \ge 1$ to 0, so that every coefficient of the Laurent series of f with negative index is 0. In conclusion f has a Taylor series and the value at z_0 is defined by

$$a_0 = \frac{1}{2\pi i} \int_{|\zeta - z_0| = \rho} \frac{f(\zeta)}{\zeta - z_0} d\zeta. \qquad \square$$

Example 12.14. Removable singularities appear mostly where a quotient of two holomorphic functions is $0/0$ or ∞/∞, such as

$$\frac{\sin z}{e^z - 1}$$

for $z \to 0$. If we know the power series, by factoring out z in the numerator and denominator we can easily recognize that the quotient for $z \to 0$ converges to 1. The remaining function is obviously holomorphic, because it is the quotient of two different power series different from 0 at $z = 0$. In more complex cases Riemann's theorem on removable singularities ensures that the quotient in $z = 0$ is really a holomorphic function.

Next we turn to the poles:

Theorem 12.15 (Poles). *If f is holomorphic in the punctured disk $0 < |z - z_0| < R$ and has a pole at z_0, then a $k \in \mathbb{N}$ exists so that*

$$f(z) = \frac{g(z)}{(z - z_0)^k}$$

with a holomorphic function g in $B_R(z_0)$ and $g(z_0) \neq 0$. The number k is called order of the pole. The Laurent series of f around z_0 is then given by

$$f(z) = \sum_{n=-k}^{\infty} a_n (z - z_0)^n = \sum_{m=0}^{\infty} a_{m-k}(z - z_0)^{m-k}, \quad a_{-k} \neq 0.$$

This series, which has at least one but only finitely many terms with negative index, characterizes the poles of a function. The part of the Laurent series with negative indices,

$$\sum_{n=-k}^{-1} a_n (z - z_0)^n,$$

has been called the principal part *of f at z_0 in Theorem 9.19.*

The behavior of a function at a pole is clear, because f has the limit ∞ in the sense of the chordal metric and is consequently continuous. We want to emphasize that we always consider the Laurent series of f, which converges in the punctured disk $0 < |z - z_0| < R$, i.e., in a neighborhood of the point z_0. Poles are always isolated according to their definition. In the case of zeros we have proved in Proposition 12.2 that they are isolated. The theorem shows again the result for poles.

Proof. For a suitable n the function $(z - z_0)^n f(z)$ is bounded in the neighborhood of z_0 and so this function has a removable singularity at z_0. Let k be the smallest possible such n, it follows then that

$$(z - z_0)^k f(z) = g(z)$$

with a holomorphic function g in $B_R(z_0)$. Necessarily $g(z_0) \neq 0$, else we could rewrite g in the form $(z - z_0)g_1(z)$ and as a consequence k would not be minimal. From the Taylor series for g we find

$$f(z) = \sum_{n=0}^{\infty} a_n (z - z_0)^{n-k} = \sum_{m=-k}^{\infty} a_{m+k}(z - z_0)^m,$$

i.e., a Laurent series with only finitely many terms with negative indices. There should be at least one term with negative index, else the function would be holomorphic at z_0 and we would have had a removable singularity. Vice versa a function, which is representable by such a Laurent series, has a pole (of order k, if $a_k \neq 0$), because then $(z - z_0)^k f(z)$ is representable by a Taylor series, in particular it is bounded in the neighborhood of z_0. $\qquad\square$

Example 12.16. a) Rational functions have poles at the zeros of the denominator. For example $1/z$ has a simple pole in $z = 0$, $1/z^2$ a double pole and so on; rational functions are meromorphic in $\mathbb{C}$.

b) The cotangent

$$\cot z = \frac{\cos z}{\sin z}$$

has poles of first order at the zeros of the sine, $z_n = n\pi$, $n \in \mathbb{Z}$, as the zeros of the sine are simple. Consequently $\cot z$ is a meromorphic function in $\mathbb{C}$.

Finally we want to turn towards the essential singularities. There are large theories about the behavior of holomorphic functions in a neighborhood of an essential singularity. We want to present only one theorem, which was found by the Italian mathematician FELICE CASORATI (1835–1890), by KARL WEIERSTRASS, and by the Russian mathematician YULIAN V. SOKHOTSKI (1842–1927):

Theorem 12.17 (Theorem of Casorati–Weierstraß–Sokhotski).   (i) *A function f has an essential singularity at z_0 if and only if its Laurent series around z_0 includes infinitely many terms with negative index.*

(ii) *In the neighborhood of an essential singularity z_0, f gets as close as desired to any value in $\hat{\mathbb{C}}$, that is for every $c \in \hat{\mathbb{C}}$ there exists a sequence (z_n) in the punctured disk $0 < |z - z_0| < R$ with $z_n \to z_0$, so that $f(z_n) \to c$ with $n \to \infty$.*

The behavior in the neighborhood of an essential singularity is quite wild and very hard to visualize. Therefore we need complex theories to handle this problem.

Proof. (i) The proposition with the Laurent series follows from the previous theorem: Laurent series with finitely many terms with negative index characterize the poles.

(ii) We now turn towards the harder part of the proposition: If a value $c \neq \infty$ and $\varepsilon > 0, \delta > 0$ existed, so that in the punctured disk $0 < |z - z_0| < \varepsilon$,

$$|f(z) - c| \geq \delta > 0,$$

then the function
$$\frac{1}{f(z) - c}$$
would be bounded by $1/\delta$ and so it would have a removable singularity:
$$\frac{1}{f(z) - c} = g(z) \quad \Rightarrow \quad f(z) = c + \frac{1}{g(z)}$$
with a holomorphic function g, so that f would have at the most a pole or a removable singularity, which would be a contradiction. If the function f were bounded for $c = \infty$ then it would have only a removable singularity. Consequently f has to get as close as desired to every value c. $\qquad\square$

Example 12.18. We take
$$e^{1/z} = \sum_{n=0}^{\infty} \frac{z^{-n}}{n!}$$
as an example of an essential singularity at the point $z = 0$, because the given Laurent series around $z = 0$ has infinitely many terms with negative index.

The example shows a gap in our considerations: How can the behavior of a function f at the point $z = \infty$ be rated? This can be clarified by the following definition:

Definition 12.19 (Behavior at ∞). If the function f is holomorphic in the domain $|z| > R$, then the behavior of f at $z = \infty$ is described by the behavior of
$$f\left(\frac{1}{\zeta}\right)$$
at $\zeta = 0$: f is holomorphic or has a pole or an essential singularity at $z = \infty$, if and only if this is the case for $f(1/\zeta)$ at $\zeta = 0$.

The last example showed already that e^z has an essential singularity at $z = \infty$. We have proved in Theorem 12.6 that the zeros of an entire function can be given at any point with any order, and this holds also for the poles. The corresponding theorem is due to the Swedish mathematician MAGNUS GÖSTA MITTAG-LEFFLER (1846–1927), from whom the mathematical Research Institute that he founded in Stockholm takes the name. This theorem is important in constructing functions:

Theorem 12.20 (Mittag-Leffler). *Let (a_n) be a sequence of complex numbers with $|a_n| \leq |a_{n+1}|$ $(n = 1, 2, \ldots)$ and $a_n \to \infty$. Furthermore let a sequence of principal parts for every a_n be given by*
$$h_n(z) = \sum_{k=1}^{m_n} \frac{A_{nk}}{(z - a_n)^k}$$
with suitable constants A_{nk}. Then a sequence (P_n) of polynomials can be determined so that the series
$$f(z) := \sum_{n=1}^{\infty} (h_n(z) - P_n(z)) \qquad (*)$$

converges uniformly in every compact subset of $\mathbb{C}$, which does not contain any of the a_n. The P_n are called convergence generating terms. *Consequently the function f is meromorphic in $\mathbb{C}$ and has poles at exactly every a_n with the principal part h_n.*

Proof. If $a_1 = 0$, then we set $P_1 = 0$. For every $|a_n| > 0$ h_n is holomorphic in $|z| < |a_n|$ and has a Taylor series

$$h_n(z) = \sum_{k=0}^{\infty} c_{nk} z^k.$$

Now we can determine a partial sum $P_n(z)$ of this Taylor series, so that

$$|h_n(z) - P_n(z)| \leq \frac{1}{2^n}$$

holds in $|z| \leq |a_n|/2$. From this the uniform convergence of the series $(*)$ in every compact subset of $\mathbb{C}$ not containing the point a_n follows. Indeed, if we include the compact set into a disk of radius R, then we have to take only the finitely many $|a_n| \leq 2R$ out of the series, and we have finally a series which is a holomorphic function in view of Theorem 9.3. The added h_n determine the desired poles with the given principal parts.   □

The theorem of Mittag-Leffler can also be proven in any domain G. Finally we can represent every meromorphic function in $\mathbb{C}$ with a series of the form

$$f(z) = f_0(z) + \sum_{n=1}^{\infty} \left(h_n(z) - P_n(z)\right),$$

where $f_0(z)$ is an entire function.

Example 12.21. The Laurent series of the function

$$f(z) := \frac{\pi^2}{\sin^2 \pi z}$$

starts with $(z - n)^{-2}$ at the zeros of the sine. If we construct the (uniformly convergent) series

$$g(z) := \sum_{n=-\infty}^{\infty} \frac{1}{(z - n)^2},$$

then the difference $h(z) := f(z) - g(z)$ is an entire function. The function h is uniformly bounded on the boundary of the squares

$$Q_N : \quad |x| \leq N + \frac{1}{2}, \quad |y| \leq N + \frac{1}{2},$$

where N is a natural number. The periodicity of the function f (period 1) helps on the vertical sides of the squares. The function h is bounded in the whole plane because of the maximum principle Theorem 7.32, consequently h is a constant, see the Theorem of Liouville 7.33. Using the following identity we find that this constant must be zero:

$$h(z) = \frac{1}{2} \left\{ h\left(\frac{z}{2}\right) + h\left(\frac{z+1}{2}\right) \right\} \qquad \text{(Herglotz)}.$$

12.2.2 Isolated singularities in $C\ell(n)$

$C\ell(n)$-holomorphic functions behave considerably differently from $\mathbb{C}$-holomorphic functions at singularities as well as at zeros. A number of questions have to be answered still. We restrict to isolated singularities, which we know at least in the $\mathcal{Q}_\mathbf{k}$ from Definition 7.26. We can thus define analogously to $\mathbb{C}$:

Definition 12.22. If the function f is (left- or right-)holomorphic in the punctured ball $0 < |x - a| < R$, but not at a, resp. is not defined at a, then we say: f has an *isolated singularity* at a. These singularities are classified as follows:

a) The singularity is called *removable*, if f is bounded in $0 < |x - a| < R/2$.

b) The singularity is called a *pole*, if it is not removable and an $m \in \mathbb{N}$ exists with $m \geq n$, so that $|x - a|^m f(x)$ is bounded in $0 < |x - a| < R/2$.

c) Else the singularity is called an *essential singularity*.

d) If a function f is (left- or right-)holomorphic in a domain G except at poles, which do not accumulate in G, then we call f (left- or right-)*meromorphic* in G.

R. Fueter [45] transferred for the first time the concept of complex meromorphic functions to higher dimensions, and he also described the development in a Laurent series in a ballshell (see Theorem 9.28). Now we want to analyze the singularities. For removable singularities we have the same theorem as in $\mathbb{C}$:

Theorem 12.23 (Riemann–Fueter's theorem on removable singularities). *Let f be a holomorphic function in $0 < |x - a| < R$ and let f have an isolated removable singularity at a. Then a value a_0 exists so that the extended function*

$$\tilde{f}(x) = \begin{cases} f(x), & 0 < |x - a| < R, \\ a_0, & x = a, \end{cases}$$

is holomorphic in $B_R(a)$.

The proof is given as Exercise 12.5.7 and is analogous to the complex case. Similar to the complex case we have only to require that $r^n f(x) \to 0$ for $r = |x - a| \to 0$, which sometimes is easier to verify.

The poles cannot be described as simply as in $\mathbb{C}$, but we can prove the following theorem:

Theorem 12.24 (Poles in $C\ell(n)$). *An isolated singularity a of a (left-)holomorphic function in $0 < |x - a| < R$ is a pole if and only if the Laurent series in $0 < |x - a| < R$ has only finitely many terms, but at least one term with a singular function $\mathcal{Q}_\mathbf{k}(x - a)$:*

$$f(x) = \sum_{k=0}^{m} \sum_{|\mathbf{k}|=k} \mathcal{Q}_\mathbf{k}(x - a)b_\mathbf{k} + \sum_{k=0}^{\infty} \sum_{|\mathbf{k}|=k} \mathcal{P}_\mathbf{k}(x - a)a_\mathbf{k}.$$

If in the left sum at least one term with $|\mathbf{k}| = m$ *is different from zero, then we call* $m + n$ *the* order *of the pole and the finite sum with the singular functions* $\mathcal{Q}_{\mathbf{k}}(x - a)$ *is called the* principal part *of* f *at the point* a.

We want to point out explicitly, that poles in $C\ell(n)$ have always at least the order n, as $\mathcal{Q}_0$ is the weakest singularity in $C\ell(n)$.

Proof. We consider only the left-holomorphic case, the right-holomorphic case can be proved in a similar way. First we assume the boundedness of $|x - a|^m f(x)$. According to Theorem 9.28 the coefficients of the Laurent series are

$$b_{\mathbf{k}} = \frac{1}{\sigma_n} \int\limits_{|x-a|=\rho} \mathcal{P}_{\mathbf{k}}(x - a) dx^* f(x).$$

At this point we use the estimate for $\mathcal{P}_{\mathbf{k}}$ according to Corollary 6.5 and the assumptions for f: It follows with $k = |\mathbf{k}|$ and a suitable constant C that

$$|b_{\mathbf{k}}| \leq \frac{1}{\sigma_n} \int\limits_{|x-a|=\rho} C|x - a|^k |dx^*| |x - a|^{-m} \leq C\rho^{k+n-m}.$$

All coefficients $b_{\mathbf{k}}$ must vanish for $k > m - n$, as ρ can be chosen as small as desired.

Now let the Laurent series of f have only finitely many terms with the $\mathcal{Q}_{\mathbf{k}}$, then let m be the largest of the $|\mathbf{k}|$. With the estimate

$$|\mathcal{Q}_{\mathbf{k}}(x - a)| \leq \frac{C_{\mathbf{k}}}{|x - a|^{n+k}} \leq \frac{C_{\mathbf{k}}}{|x - a|^{n+m}}$$

using Proposition 7.27 we get the result for f.   □

Remark 12.25. The behavior of a function at poles in $\mathbb{R}^{n+1}$ is very different from the one in the plane: Of course $\mathcal{Q}_0(x - a) = \overline{(x - a)}/|x - a|^{n+1}$ converges to ∞ for $x \to a$ in the chordal metric as it does in $\mathbb{C}$. But this is not always the case and consequently a meromorphic function need not be continuous at a pole with respect to the chordal metric. For example if we differentiate $\mathcal{Q}_0$ (for simplicity let us take $a = 0$):

$$\partial_i \mathcal{Q}_0(x) = \frac{|x|^2 \overline{e}_i - (n+1)\overline{x} x_i}{|x|^{n+3}},$$

then we get a non-vanishing numerator and the derivative converges to ∞ for $x \to 0$. But this changes if we differentiate again, since in

$$\partial_i \partial_j \mathcal{Q}_0(x) = \frac{n+1}{|x|^{n+5}} \left\{ -(\overline{e}_j x_i + \overline{e}_i x_j)|x|^2 - \delta_{ij} \overline{x} |x|^2 + (n+3)\overline{x} x_i x_j \right\}$$

the bracket vanishes for $x_i = x_j = 0$ if $i \neq j$. This second derivative of $\mathcal{Q}_0$ is even zero in an $(n - 1)$-dimensional plane and converges in this plane to 0 and not to ∞.

Unfortunately there are no available studies about essential singularities in $C\ell(n)$. But the theorem of Mittag-Leffler can be transferred to the case of meromorphic functions in $\mathbb{R}^{n+1}$:

Theorem 12.26 (Mittag-Leffler). *Let a_k be a sequence of numbers of $\mathbb{R}^{n+1}$ and let $|a_k| \leq |a_{k+1}|$ as well as $a_k \to \infty$. In addition let $H_k(x)$ be a sequence of principal parts at the a_k given by*

$$H_k(x) = \sum_{j=0}^{m_k} \sum_{|j|=j} \mathcal{Q}_{\mathbf{j}}(x - a_k) b_{\mathbf{j}k}.$$

Then a sequence (P_k) of holomorphic polynomials can be determined so that the series

$$f(x) := \sum_{k=1}^{\infty} (H_k(x) - P_k(x))$$

converges uniformly in every compact subset of $\mathbb{R}^{n+1}$, which does not contain any of the a_k. The P_k are called convergence generating terms. *The function f is (left)-meromorphic in $\mathbb{R}^{n+1}$ and at the points a_k it has poles precisely with the principal parts H_k.*

The proof is completely analogous to the complex case and is left to the reader as Exercise 12.5.8.

12.3   Residue theorem and the argument principle

12.3.1   Residue theorem in $\mathbb{C}$

The understanding of the local structure of holomorphic or meromorphic functions achieved in the previous section allows us to extend the Cauchy integral theorem and the Cauchy integral formula. Therefore we define at first:

Definition 12.27 (Residue). Let the function f, meromorphic in the domain G, have at $z_0 \in G$ an isolated singularity. Let its Laurent series

$$f(z) = \sum_{n=-\infty}^{\infty} a_n (z - z_0)^n$$

be convergent in $0 < |z - z_0| < R$. We then call

$$a_{-1} = \frac{1}{2\pi i} \int_{|z-z_0|=\rho} f(z) dz =: \operatorname{Res}(f, z_0)$$

the *residue* of f at z_0. We can choose ρ arbitrarily in the interval $(0, R)$, in view of Cauchy's integral theorem.

In the year 1814 the residue was introduced by CAUCHY in his famous work *"Mémoire sur les intégrales définies"*, but basic ideas could be found already in EULER's work.

As a precaution we want to point out that the concept of residues cannot be applied if we consider an essential singularity, which is a limit point of poles. The residue theorem was published by CAUCHY in 1825. The principle was that he reduced quantities given by an integral to differential quantities.

Theorem 12.28 (Residue theorem). *Let $G \subset \mathbb{C}$ be a finitely connected domain with a piecewise smooth boundary. Let the function f be holomorphic in G up to finitely many isolated singularities $a_1, \ldots, a_n$ and continuous in $\overline{G} \setminus \{a_1, \ldots, a_n\}$. Then*

$$\int_{\partial G} f(z)dz = 2\pi i \sum_{k=1}^{n} \mathrm{Res}(f, a_k)$$

holds.

The residue theorem apparently is an extension of the Cauchy integral theorem, because if the function f is holomorphic in G, then the residues are zero and the residue theorem becomes the Cauchy integral theorem.

Proof. We cut out of the domain G little disks $B_\varepsilon(a_k)$, $k = 1, \ldots, n$, which neither intersect themselves nor the boundary of G, each containing only the singularity a_k. Let G_ε be the resulting domain and for this domain we can apply Cauchy's integral theorem, as only finitely many smooth boundaries were added. The function f is holomorphic in G_ε, and consequently

$$\int_{\partial G_\varepsilon} f(z)dz = 0 = \int_{\partial G} f(z)dz - \sum_{k=1}^{n} \int_{|z - a_k| = \varepsilon} f(z)dz.$$

This is already the assertion, because the last integrals are the residues at the points a_k except for the factor $2\pi i$. $\square$

Before we deal with the calculation of residues, we want to consider the point $z = \infty$.

Definition 12.29. If f is holomorphic in $|z| > R$, we call

$$\mathrm{Res}(f, \infty) := -\frac{1}{2\pi i} \int_{|z| = \rho} f(z)dz$$

the *residue* of f at the point $z = \infty$, where we can choose $\rho > R$ arbitrarily.

Also for $z = \infty$ the residue is equal to the coefficient a_{-1} of the Laurent series of f, which converges for $|z| > R$.

Remark 12.30. We want to point out that this residue is not the same as the one we would get if we calculated according to Definition 12.19: In this case we would take the residue of $f(1/\zeta)$ at the point $\zeta = 0$,

$$\frac{1}{2\pi i} \int_{|\zeta|=1/\rho} f\left(\frac{1}{\zeta}\right) d\zeta = -\frac{1}{2\pi i} \int_{|z|=\rho} f(z)\frac{dz}{z^2}.$$

This would just be the coefficient a_1 of the Laurent series around ∞ and in this sense passably, as $a_1 z$ is the first term generating the singularity in ∞. But the definition above is more practical as the following lemma shows:

Proposition 12.31. *Let f be holomorphic in $\mathbb{C}$ up to finitely many points $a_1, \ldots, a_n \in \mathbb{C}$. Then*

$$\sum_{k=1}^{n} \mathrm{Res}(f, a_k) + \mathrm{Res}(f, \infty) = 0$$

holds.

Proof. If $|z| = \rho$ is a circle, which contains in its interior all a_k, $k = 1, \ldots, n$, then

$$\int_{|z|=\rho} f(z)dz = 2\pi i \sum_{k=1}^{n} \mathrm{Res}(f, a_k) = -2\pi i \, \mathrm{Res}(f, \infty)$$

holds. $\qquad\square$

If we now want to calculate integrals by means of residues, it is very helpful to have a method for calculating residues without integrals. In the case of essential singularities this is only possible if the Laurent series is known. Then we simply take the coefficient a_{-1}. Also in the case of poles, knowledge of the Laurent series implies knowledge of the residues. But in this case there exists also another method:

Proposition 12.32 (Calculation of residues at poles). *If the function f has a pole of order k at z_0, we have*

$$(k-1)! \, \mathrm{Res}(f, z_0) = \lim_{z \to z_0} \frac{d^{k-1}}{dz^{k-1}} \left((z - z_0)^k f(z)\right).$$

This formula is particularly easy to handle in the case of simple poles as it then reduces to

$$\mathrm{Res}(f, z_0) = \lim_{z \to z_0} (z - z_0)f(z).$$

Proof. With

$$f(z) = \sum_{n=-k}^{\infty} a_n(z - z_0)^n$$

it immediately follows that

$$(z - z_0)^k f(z) = \sum_{m=0}^{\infty} a_{m-k}(z - z_0)^m.$$

The coefficient a_{-1} is the one with the index $m = k-1$. We get the assertion of the proposition by applying the formula for coefficients of a Taylor series according to Theorem 9.14.

$\square$

Example 12.33. A first example is the calculation of

$$I := \int\limits_{|z-i|=5} \left(\frac{z+2}{z+3} + \frac{z-2}{z+i} \right) dz,$$

which with usual methods would only be possible by considerable effort. But according to the residue theorem we only have to find the poles of the integrand, which here are the points $a_1 = -3$ and $a_2 = -i$. These poles are obviously simple poles and both points are included in the circle of integration, as $|-3-i| < 5$ and $|-2i| < 5$. We calculate the residues at both these points according to the above formula:

$$\text{Res}(f, -3) = \lim_{z \to -3} \left(z + 2 + (z+3)\frac{z-2}{z+i} \right) = -1,$$

and

$$\text{Res}(f, -i) = \lim_{z \to -i} \left((z+i)\frac{z+2}{z+3} + z - 2 \right) = -2 - i.$$

Finally we get the integral we are looking for:

$$I = 2\pi i(-1 - 2 - i) = 2\pi(1 - 3i).$$

12.3.2 Argument principle in $\mathbb{C}$

In this section we shall consider a further generalization of the residue theorem, which is often called the *argument principle*. The most important result is the following:

Theorem 12.34. *Let the function f be meromorphic in the bounded domain $G \subset \mathbb{C}$. Further let $\Gamma \subset G$ be a piecewise smooth curve, which does not contain zeros or poles of f, where z_A is the initial point and z_E the endpoint of Γ. We then have*

$$\exp\left(\int_\Gamma \frac{f'(z)}{f(z)} dz \right) = \frac{f(z_E)}{f(z_A)}.$$

Proof. Let $\gamma : [0,1] \to G$ be a parametrization of Γ. For every point $\gamma(t) \in \Gamma$ there exists a small disk $D(t) := B_{\rho(t)}$ not containing any zero or pole of f. Consequently f'/f is holomorphic in $D(t)$ and has a primitive function

$$F(z) = \int_{z_1}^{z} \frac{f'(\zeta)}{f(\zeta)} d\zeta,$$

where $z_1 \in D(t) \cap \Gamma$. If z_2 is an arbitrary point in $D(t)$ and

$$g(z) := \frac{\exp(F(z) - F(z_2))}{f(z)},$$

then g is holomorphic in $D(t)$ with the derivative

$$g'(z) = -\frac{f'(z)}{f(z)} g(z) + F'(z) g(z) = 0.$$

Hence g is a constant. It follows for $z = z_2$ that

$$g(z) = \frac{1}{f(z_2)} \quad \Rightarrow \quad \exp(F(z) - F(z_2)) = \frac{f(z)}{f(z_2)}.$$

The graph of Γ is a compact set, which can be covered by finitely many disks $D(t)$. These disks $D(t)$ must intersect themselves; we can then find a decomposition $0 = t_0 < t_1 < \cdots < t_{n-1} < t_n = 1$, so that the piece of the curve between $\gamma(t_{j-1})$ and $\gamma(t_j)$ is contained in one of the small disks. We choose the above $z_1 = \gamma(t_{j-1})$, $z_2 = \gamma(t_j)$ and accordingly F_j, and we get

$$\exp\left(\int_\Gamma \frac{f'(\zeta)}{f(\zeta)} d\zeta \right) = \prod_{j=1}^{n} \exp[F_j(\gamma(t_j)) - F_j(\gamma(t_{j-1}))] = \prod_{j=1}^{n} \frac{f(\gamma(t_j))}{f(\gamma(t_{j-1}))} = \frac{f(z_E)}{f(z_A)}. \qquad \square$$

Corollary 12.35. *If the curve Γ is closed, we have*

$$\int_\Gamma \frac{f'(z)}{f(z)} dz = 2n\pi i, \quad n \in \mathbb{Z}.$$

This follows from the fact that for a closed curve the right-hand side in our Theorem 12.34 is 1 and as a consequence the argument of the exponential function can only be an integer multiple of $2\pi i$.

As a first important application we get the definition for the *index* or *winding number* of a curve.

Definition 12.36 (Index, winding number). Let Γ be a closed piecewise smooth curve in $\mathbb{C}$ and $z \notin \Gamma$. We then define

$$I(\Gamma, z) := \frac{1}{2\pi i} \int_\Gamma \frac{d\zeta}{\zeta - z}$$

to be the *index* or the *winding number* of the curve Γ with respect to z.

The previous theorem provides

Corollary 12.37 (Integer index). (i) *The index of a piecewise smooth closed curve Γ with respect to a point z is an integer.*

(ii) *In $\mathbb{C} \setminus \Gamma$ the index is locally constant, thus also in every subdomain of $\mathbb{C} \setminus \Gamma$ it is constant.*

Proof. (i) This follows directly from Corollary 12.35 using the function $f(\zeta) = \zeta - z$.

(ii) On the one side the integral is a continuous function at z and on the other side it is an integer, so it must be locally constant. In addition the graph of Γ is closed and so $\mathbb{C} \setminus \Gamma$ is open. This complement splits into open and connected subsets. As a locally constant function the winding number must be constant in such subdomains. $\square$

This urges us to introduce the following definition:

Definition 12.38 (Interior and exterior of a closed curve). Let Γ be a closed piecewise smooth curve. We call

$$I(\Gamma) := \{z \,:\, z \in \mathbb{C} \setminus \Gamma, \; I(\Gamma, z) \neq 0\},$$

the *interior* of Γ and respectively

$$A(\Gamma) := \{z \,:\, z \in \mathbb{C} \setminus \Gamma, \; I(\Gamma, z) = 0\}.$$

the *exterior* of Γ.

Now we introduce a generalization of Cauchy's integral formula:

Theorem 12.39 (Extended Cauchy's integral formula). *Let $G \subset \mathbb{C}$ be a star-shaped domain and $\Gamma \subset G$ a closed piecewise smooth curve. Let the function f be holomorphic in G. Then for all $z \in G \backslash \Gamma$ we have*

$$I(\Gamma, z)\, f(z) = \frac{1}{2\pi i} \int\limits_{\Gamma} \frac{f(\zeta)}{\zeta - z} d\zeta.$$

The extension consists in the fact that now every arbitrary piecewise smooth curve is allowed and not only the boundary of a domain. This means that Γ can wind around the point z several times or none. Here the common assertion of this theorem is the simple connectivity of G, but we use Theorem 7.4 of Morera in our variant of the proof. According to Morera every holomorphic function in a star-shaped domain possesses a primitive.

Proof. The function

$$g(\zeta) = \begin{cases} \frac{f(\zeta) - f(z)}{\zeta - z}, & \zeta \neq z, \\ f'(z), & \zeta = z \end{cases}$$

is holomorphic in ζ for $\zeta \neq z$ and for $\zeta \to z$ it is continuous. At this point g has a removable singularity and thus g is holomorphic in G. As mentioned above the function possesses a primitive $F(\zeta)$ in the star-shaped domain G in this case. Hence every integral of g along a closed curve Γ is zero, if $z \in G \backslash \Gamma$:

$$\int\limits_{\Gamma} g(\zeta) d\zeta = \int\limits_{\Gamma} \frac{f(\zeta) - f(z)}{\zeta - z} d\zeta = 0$$

so that

$$f(z) \int_\Gamma \frac{d\zeta}{\zeta - z} = \int_\Gamma \frac{f(\zeta)}{\zeta - z} d\zeta,$$

i.e., the assertion, since the factor of $f(z)$ is equal to $2\pi i I(\Gamma, z)$. □

The last point in this section is the already announced *argument principle*.

Theorem 12.40 (Argument principle). *Let G be a finitely connected domain with a piecewise smooth boundary. Let the function f be holomorphic in G up to finitely many poles and continuous in $\overline{G}$. Let f be different from zero on ∂G. Then*

$$\frac{1}{2\pi i} \int_{\partial G} \frac{f'(z)}{f(z)} dz = n(0, f) - n(\infty, f)$$

holds, where $n(0, f)$ is the number of zeros *of f in G (counted according to their order) and $n(\infty, f)$ the* number of poles *(counted correspondingly).*

The argument principle is a possibility to represent the number of zeros and poles of a function by an integral. This is of special importance for the examination of meromorphic functions. The term argument principle is justified, because the integrand is the *logarithmic derivative* of the function f:

$$\frac{f'(z)}{f(z)} dz = d(\log\, f(z)) = d(\ln |f(z)|) + i\, d(\arg f(z)).$$

But as we know, $\ln |f(z)|$ is uniquely defined in $\overline{G}$ and the integral along the contour ∂G of $d \ln |f(z)|$ is zero, so that only the argument of f contributes to the value of the integral. Then the theorem's statement can be rewritten in the form:

$$\frac{1}{2\pi} \int_{\partial G} d \arg\, f(z) = n(0, f) - n(\infty, f).$$

Proof. The logarithmic derivative of the function f is holomorphic in G and continuous in $\overline{G}$ up to the finitely many zeros and poles, which we denote by $a_1, \ldots, a_n$, resp. $b_1, \ldots, b_m$. As usual we exclude these points from G by sufficiently small disks of radius ε. Now we can apply Cauchy's integral theorem to the remaining domain and get

$$\int_{\partial G} \frac{f'(z)}{f(z)} dz = \sum_{j=1}^{n} \int_{|z - a_j| = \varepsilon} \frac{f'(z)}{f(z)} dz + \sum_{k=1}^{m} \int_{|z - b_k| = \varepsilon} \frac{f'(z)}{f(z)} dz.$$

The integrals along the small disks are the residues of f'/f at the corresponding points. We then get:

Let f have a zero at a_j of order p_j and let g be the corresponding holomorphic function with $g(a_j) \neq 0$ and

$$f(z) = (z - a_j)^{p_j} g(z), \quad f'(z) = p_j(z - a_j)^{p_j - 1} g(z) + (z - a_j)^{p_j} g'(z);$$

thus

$$\frac{f'(z)}{f(z)} = \frac{p_j}{z - a_j} + \frac{g'(z)}{g(z)}.$$

As a result f'/f has a simple pole at a_j with the residue p_j. Similarly for a pole in b_k of order q_k we get the residue $-q_k$. Finally we have the desired result

$$\frac{1}{2\pi i} \int_{\partial G} \frac{f'(z)}{f(z)} dz = \sum_{j=1}^{n} p_j - \sum_{k=1}^{m} q_k.$$

$\square$

We conclude this section with a theorem of the French mathematician E. ROUCHÉ (1832–1910), which allows us to compare the number of zeros of two functions:

Theorem 12.41 (Rouché's theorem). *Let the domain G be finitely connected with a piecewise smooth boundary. Let f and g be holomorphic in G and continuous in $\overline{G}$. In addition let f be different from zero on ∂G and let us assume on ∂G,*

$$|g(z)| < |f(z)|.$$

Then the functions f and $f + g$ have the same number of zeros in G.

Proof. The function $f + \lambda g$ is different from zero on ∂G for $0 \le \lambda \le 1$. Then

$$n(0, f + \lambda g) = \frac{1}{2\pi i} \int_{\partial G} \frac{f'(z) + \lambda g'(z)}{f(z) + \lambda g(z)} dz$$

gives the number of zeros of $f + \lambda g$. The integral is continuous in λ and is an integer, thus it must be constant for all λ, $0 \le \lambda \le 1$. The integral values for $\lambda = 0$ und $\lambda = 1$ give the assertion

$$n(0, f) = n(0, f + g).$$

$\square$

12.3.3 Residue theorem in $C\ell(n)$

In this section we can proceed in a parallel way as done in $\mathbb{C}$, but only in the case of isolated singularities. The theory referring to higher dimensional singularities needs to be refined, maybe with the general theories of J. LERAY [89] and F. NORGUET [112].

First we define:

Definition 12.42 (Residue). Let the left-holomorphic function f have a pole or an essential singularity at the point a and let it have the Laurent series

$$f(x) = \sum_{k=0}^{\infty} \sum_{|\mathbf{k}|=k} \mathcal{P}_{\mathbf{k}}(x - a) a_{\mathbf{k}} + \sum_{k=0}^{\infty} \sum_{|\mathbf{k}|=k} \mathcal{Q}_{\mathbf{k}}(x - a) b_{\mathbf{k}}$$

in $0 < |x - a| < R$. We then call the coefficient

$$b_{\mathbf{0}} = \frac{1}{\sigma_n} \int_{|x-a|=\rho} dx^* f(x) =: \operatorname{Res}(f, a)$$

the *residue* of f at the point a. The radius ρ can be chosen arbitrarily between 0 and R.

In the case of a right-holomorphic function f, the residue is defined by

$$b_0 = \frac{1}{\sigma_n} \int\limits_{|x-a|=\rho} f(x)dx^* =: \operatorname{Res}(f,a).$$

In the following we restrict ourselves to the left-holomorphic case, as the right-holomorphic case presents no serious differences. As in the plane we have the

Theorem 12.43 (Residue theorem). *Let G be a domain in $\mathbb{R}^{n+1}$ with a sufficiently smooth boundary manifold. Let the function f be continuous in $\overline{G}$ and left-holomorphic in G apart from finitely many isolated singularities $a_1, \ldots, a_m$. We then have*

$$\int\limits_{\partial G} dx^* f(x) = \sigma_n \sum_{k=1}^{m} \operatorname{Res}(f, a_k).$$

We omit the proof, as it is similar to the proof in $\mathbb{C}$.

We can also define the residue at $x = \infty$ as in $\mathbb{C}$:

Definition 12.44. If f is left-holomorphic for $|x| > R$, we define

$$\operatorname{Res}(f, \infty) := -\frac{1}{\sigma_n} \int\limits_{|x|=\rho} dx^* f(x)$$

to be the *residue* of f at the point $x = \infty$. In the formula we can choose $\rho > R$ arbitrarily.

The resulting lemma is also the same as in $\mathbb{C}$:

Proposition 12.45. *Let f be left-holomorphic in $\mathbb{R}^{n+1}$ apart from finitely many points $a_1, \ldots, a_m \in \mathbb{R}^{n+1}$. We then have*

$$\sum_{k=1}^{m} \operatorname{Res}(f, a_k) + \operatorname{Res}(f, \infty) = 0.$$

We also omit this proof, as it is nearly literally the same as in $\mathbb{C}$. Interestingly enough is the fact that there is also a parallel in calculating the residue at poles according to Proposition 12.32:

Proposition 12.46 (Calculation of residues at poles). *If the left-holomorphic function f has a pole of order $m + n$ at a, then*

$$\operatorname{Res}(f, a) = \frac{1}{m!} \lim_{r \to 0} \partial_r^m r^{m+n} \omega f(x)$$

holds, where $r := |x - a|$ and $\omega := (x - a)/r$.

Proof. For simplicity of notation we set $a = 0$. The Laurent series of f in its Taylor part is composed of terms like

$$\mathcal{P}_{\mathbf{k}}(x)a_{\mathbf{k}} = r^k \mathcal{P}_{\mathbf{k}}(\omega)a_{\mathbf{k}},$$

where $k := |\mathbf{k}| \geq 0$. As terms depending on ω will not be affected by differentiation with respect to r, it follows that

$$\frac{1}{m!}\partial_r^m r^{m+n} \omega \mathcal{P}_{\mathbf{k}}(x)a_{\mathbf{k}} = \frac{1}{m!}\left(\partial_r^m r^{m+n+k}\right)\omega \mathcal{P}_{\mathbf{k}}(\omega)a_{\mathbf{k}} \to 0$$

with $r \to 0$, because a factor $r^{n+k} \to 0$ remains after differentiation.

The first singular term $\mathcal{Q}_0(x)b_0$ becomes

$$\omega \mathcal{Q}_{\mathbf{k}}(x)b_0 = \frac{\omega\overline{\omega}}{r^n}b_0 = \frac{b_0}{r^n}$$

in view of $\omega\overline{\omega} = 1$, and thus we find as requested

$$\frac{1}{m!}\partial_r^m r^{m+n} \omega \mathcal{Q}_0(x)b_0 = \frac{1}{m!}\partial_r^m r^m b_0 = b_0.$$

We apply Proposition 7.27 to the further singular terms $\mathcal{Q}_{\mathbf{k}}(x)b_{\mathbf{k}}$ and get

$$\mathcal{Q}_{\mathbf{k}}(x) = \frac{q_{\mathbf{k}}(x)}{r^{n+2k+1}} = \frac{q_{\mathbf{k}}(\omega)}{r^{n+k}}.$$

We obtain the expression (we have $m \geq k \geq 1$)

$$\frac{1}{m!}\partial_r^m r^{m-k}\omega q_{\mathbf{k}}(\omega)b_{\mathbf{k}}.$$

The differentiation of this term gives zero since $m - k < m$; thus the limit is 0.   □

12.3.4 Argument principle in $C\ell(n)$

Also in this section we can proceed for many steps in a parallel way as in $\mathbb{C}$, but some questions are still left open. In particular we consider only the dimensions 0 and n, although there are theories for the dimensions in between. In this case we would have to deal with differential forms of corresponding degree. We do not want to consider this case and refer the reader to [33] and [138]. At first we define

Definition 12.47 (Index, winding number). Let M be an n-dimensional closed sufficiently smooth manifold in $\mathbb{R}^{n+1}$ with $x \notin M$. Then we call

$$I(M,x) := \frac{1}{\sigma_n}\int_M dy^* \mathcal{Q}_0(y - x) = \frac{1}{\sigma_n}\int_M \mathcal{Q}_0(y - x)dy^*$$

the *(Kronecker) index* or the *winding number* of the manifold M with respect to the point x.

First we formulate the following proposition:

Proposition 12.48. *We have*

$$d\{g(dx_i \wedge dx)^* f\} = -\{(\partial_i g)dx^* f + gdx^*(\partial_i f)\} + \{(g\overline{\partial})dx_i^* f + gdx_i^*(\overline{\partial}f)\}.$$

The proof is a simple calculation and is left as Exercise 12.5.9. In addition we have:

Proposition 12.49 (Integer index). (i) *The index of a sufficiently smooth n-dimensional manifold M with respect to a point $x \notin M$ is an integer.*

(ii) *The index is locally constant, i.e., in every subdomain of $\mathbb{R}^{n+1} \setminus M$ it is a constant.*

First some remarks: We have shown in Example A.2.17 c), that

$$dy^* = \nu |do|$$

with the normal vector ν. As we shall prove that the index is an integer, it follows that

$$I(M, x) = \frac{1}{\sigma_n} \int_M \frac{\nu \cdot (y - x)}{|y - x|^{n+1}} |do|$$

with the dot-product between the quantities ν and $y - x$ understood as vectors. We recognize that this is the Kronecker index in topology, which is the number of windings of the manifold M around the point x.

Proof. (i) The correlation of both specified forms of the index results from changing to a parametric representation as in the previous remark because the dot-product is commutative.

If $x \notin M$ then $I(M, x)$ is a right- and left-holomorphic function, because the integrand is differentiable. We use at the moment the right holomorphic form

$$I(M, x) = \frac{1}{\sigma_n} \int_M dy^* \mathcal{Q}_0(y - x).$$

This form can be expanded in a Laurent series for $|x| > R$ with a large enough R:

$$I(M, x) = \sum_{k=0}^{\infty} \sum_{|\mathbf{k}|=k} a_{\mathbf{k}} \mathcal{P}_{\mathbf{k}}(x) + \sum_{k=0}^{\infty} \sum_{|\mathbf{k}|=k} b_{\mathbf{k}} \mathcal{Q}_{\mathbf{k}}(x),$$

where

$$a_{\mathbf{k}} = \frac{1}{\sigma_n} \int_{|u|=\rho} I(M, u)du^* \mathcal{Q}_{\mathbf{k}}(u), \quad b_{\mathbf{k}} = \frac{1}{\sigma_n} \int_{|u|=\rho} I(M, u)du^* \mathcal{P}_{\mathbf{k}}(u).$$

According to Example A.2.17 c) we have $du^* = \rho^n \nu |do_1|$, where ν is the outer unit normal vector and $|do_1|$ is the surface element of the unit ball. Furthermore let $u = \rho \omega$, so that with $k = |\mathbf{k}|$,

$$\mathcal{P}_{\mathbf{k}}(u) = \rho^k \mathcal{P}_{\mathbf{k}}(\omega), \quad \mathcal{Q}_{\mathbf{k}}(u) = \frac{1}{\rho^{k+n}} q_{\mathbf{k}}(\omega)$$

holds. As a consequence with a suitable constant C we have

$$|a_{\mathbf{k}}| \leq \frac{C}{\rho^n} \rho^n \frac{1}{\rho^{k+n}} \to 0$$

as $\rho \to \infty$. Thus we get $a_{\mathbf{k}} = 0$ for all $\mathbf{k}$, as the coefficients do not depend on ρ. We use the Fubini theorem to prove that also the $b_{\mathbf{k}}$ vanish:

$$b_{\mathbf{k}} = \frac{1}{\sigma_n^2} \int\limits_{M} dy^* \int\limits_{|u|=\rho} \mathcal{Q}_0(y-u) du^* \mathcal{P}_{\mathbf{k}}(u).$$

Using Cauchy's integral formula we get

$$b_{\mathbf{k}} = -\frac{1}{\sigma_n} \int\limits_{M} dy^* \mathcal{P}_{\mathbf{k}}(y).$$

From the properties of the Fueter polynomials of Theorem 6.2 (ii) we have

$$\partial_n \mathcal{P}_{\mathbf{k}+\varepsilon_n}(y) = (k_n + 1)\mathcal{P}_{\mathbf{k}}(y),$$

where ε_n is a multiindex, which has only a 1 in the n-th place and else zeros. We apply Proposition 12.48 and Stokes' Theorem A.2.18 to the formula

$$b_{\mathbf{k}} = -\frac{1}{(k_n + 1)\sigma_n} \int\limits_{M} dy^* \partial_n \mathcal{P}_{\mathbf{k}+\varepsilon_n}(y),$$

where we have to set $g = 1$, $f = \mathcal{P}_{\mathbf{k}+\varepsilon_n}$ and $i = n$ in Proposition 12.48. So we get

$$0 = \int\limits_{\partial M} (dy_n \wedge dy)^* f(y) = -\int\limits_{M} dy^* \partial_n f(y) = b_{\mathbf{k}},$$

where the 0 arises from $\partial M = \emptyset$. Hence $I(M, x)$ vanishes outside M. $I(M, x)$ is an integer also in the other subdomains of $\mathbb{R}^{n+1} \setminus M$, because the value changes only by an integer while crossing M.

The latter is obtained as follows: Let K be a sufficiently small ball with the center at a regular point of M, then K is divided into two domains K_1 and K_2. The boundaries of K_1 and K_2 are given by $\partial K_1 = \partial K_{01} + M_0$ and $\partial K_2 = \partial K_{02} - M_0$, where M_0 is the intersection between K and M and the numbering is properly chosen. According to Cauchy's integral formula we have

$$\frac{1}{\sigma_n} \int\limits_{\partial K_1} dy^* \mathcal{Q}_0(y-x) = 1,$$

where $x \in K_1$. So we get

$$I(M, x) = \frac{1}{\sigma_n} \int\limits_{M} dy^* \mathcal{Q}_0(y-x) = 1 + \left(\int\limits_{M \setminus M_0} - \int\limits_{\partial K_{01}} \right) dy^* \mathcal{Q}_0(y-x).$$

The right-hand side of the formula is right-holomorphic in K; if the index in K_1 is constant, then it must be constant in K according to the identity Theorem 9.27. The index changes by the calculated value 1 while crossing M or by another integer number, if M_0 has to be counted several times.

(ii) It was proved just now, that the index is an integer, so it has to be constant in each subdomain of the complement of M. □

Now we could define the interior and exterior of a manifold in a way similar to the plane, but we will not do this. We could also develop the n-th homology group with this index, but this is also not done here.

We point out further parallels to the plane, which are partly conjectures for which a proof is missing. We conjecture the theorem.

Theorem 12.50 (Extended Cauchy's integral theorem). *Let the function f be left holomorphic in a domain $G \subset \mathbb{R}^{n+1}$. In addition let $M \subset G$ be an n-dimensional closed sufficiently smooth manifold, i.e., with no boundary. Then both*

(i) *the Cauchy integral theorem,*

$$\int_M dy^* f(y) = 0,$$

(ii) *and the Cauchy integral formula ($x \in G \setminus M$),*

$$I(M, x) f(x) = \frac{1}{\sigma_n} \int_M \mathcal{Q}_0(y - x) dy^* f(y),$$

hold.

An argument principle as in $\mathbb{C}$ is not known in $C\ell(n)$. An approach is given in the definition of the order of a zero in Corollary 12.10. But to prove the step from this formula to an argument principle will be a hard job. The inclusion of poles is not possible at the present state of the theory, as no integral formula is available for calculating the order of a pole.

12.4 Calculation of real integrals

Real integrals of the form

$$\int_a^b f(x) dx$$

can often be calculated with the help of the residue theorem. Therefore we want to give some hints and examples. The interval $[a, b]$ can be finite or infinite, at first we assume that it is finite. Now a Jordan curve Γ' in the upper half-plane is added to the interval, so that a closed curve $\Gamma = [a, b] + \Gamma'$ arises, which bounds the domain G. Let the function f be meromorphically continuable in $G \cup \Gamma'$, then

$$\int_\Gamma f(z)dz = \int_a^b f(x)dx + \int_{\Gamma'} f(z)dz = 2\pi i \sum_k \mathrm{Res}(f, a_k)$$

holds, where the sum includes all residues of f in G. The basis of all the following calculation is the next proposition, named after the french mathematician CAMILLE JORDAN (1838–1922).

Proposition 12.51 (Jordan's lemma). *Let a be fixed and let (C_{R_n}) be a sequence of arcs $C_{R_n} := \{z : |z| = R_n, \mathrm{Im}\, z > -a\}$ with $R_n \to \infty$. Let the function f be defined on all arcs C_{R_n} and let it converge for $n \to \infty$ uniformly to zero. Then for an arbitrary $\lambda > 0$,*

$$\lim_{n \to \infty} \int_{C_{R_n}} f(z)e^{i\lambda z}dz = 0$$

holds.

Proof. Let $M(R_n) := \max\limits_{z \in C_{R_n}} |f(z)|$ and assume at first $a > 0$. According to Figure 12.1 we see that $\alpha_n = \arcsin(a/R_n)$ is the angle between the segment connecting the origin with A and the positive real axis. As $(1/x)\arcsin x \to 1$ for $x \to 0$, the equation $R_n \alpha_n = a(R_n/a)\arcsin(a/R_n) \to a$ for $n \to \infty$ holds. According to the figure we get the estimate

$$|e^{i\lambda z}| = e^{-\lambda y} \le e^{a\lambda}$$

for the arcs Γ_1 from A to B and Γ_4 from D to E. From the latter it follows that

$$\left| \int_{\Gamma_j} f(z)e^{i\lambda z}dz \right| \le M(R_n)e^{a\lambda}\alpha_n R_n, \quad j = 1, 4,$$

since $\alpha_n R_n$ is the length of the arcs. This term converges to zero, as $M(R_n)$ converges to zero and $\alpha_n R_n$ is bounded by assumption.

We use the known estimate

$$\sin \varphi \ge \frac{2}{\pi}\varphi$$

on the arc Γ_2 from B to C and get

$$|e^{i\lambda z}| = e^{-\lambda R_n \sin \varphi} \le \exp\left(-\frac{2\lambda R_n}{\pi}\varphi\right).$$

We then have

$$\left| \int_{\Gamma_2} f(z)e^{i\lambda z}dz \right| \le M(R_n)R_n \int_0^{\pi/2} \exp\left(-\frac{2\lambda R_n}{\pi}\varphi\right) d\varphi = M(R_n)\frac{\pi}{2\lambda}(1 - e^{-\lambda R_n}).$$

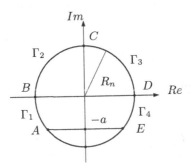

Figure 12.1

This integral converges to zero as $M(R_n)$ converges to zero. The integral along the arc Γ_3 from C to D is treated similarly. So the lemma is proven in the case of positive a. In the case of negative a the proof simplifies and only the arcs Γ_2 and Γ_3 have to be considered, but this has already been proven. □

The sequence of arcs C_{R_n} can be replaced by a system of arcs C_R with a real parameter $R \to \infty$ without changing the proof. We then have

$$\lim_{R \to \infty} \int_{C_R} f(z)e^{i\lambda z}dz = 0.$$

Let us consider some examples:

Example 12.52. This example is due to the French mathematician PIERRE SIMON LAPLACE (1749–1827), whom we shall meet many times again. We will calculate the following integral:

$$I := \int_0^\infty \frac{\cos x}{x^2 + c^2}dx, \quad c \in \mathbb{R}.$$

We continue the integrand into the complex plane by

$$F(z) := \frac{e^{iz}}{z^2 + c^2}.$$

We have to choose $\lambda = 1$ and $f(z) = 1/(z^2 + c^2)$ to apply Jordan's lemma. In particular we have for $|z| = R$,

$$|f(z)| \leq \frac{1}{R^2 - c^2},$$

and this term converges to zero independently of the argument of z for $R \to \infty$. If we choose $a = 0$ in the Jordan lemma, then C_R is the arc of the half-circle around zero with radius R in the upper half-plane. Jordan's lemma gives

$$\int_{C_R} f(z)dz = \int_{C_R} \frac{1}{z^2 + c^2} e^{iz} dz \to 0$$

for $R \to \infty$. We now apply the residue theorem to the upper half-circle:

$$\int_{-R}^{R} \frac{e^{ix}}{x^2 + c^2} dx + \int_{C_R} \frac{e^{iz}}{z^2 + c^2} dz = 2\pi i \operatorname{Res}(F, ci) = 2\pi i \frac{e^{-c}}{2ci} = \frac{\pi}{c} e^{-c}.$$

We have to consider that obviously the only pole is ic in the upper half-plane and $z^2 + c^2 = (z - ic)(z + ic)$. Finally for $R \to \infty$ we get

$$\int_{-\infty}^{\infty} \frac{e^{ix}}{x^2 + c^2} dx = \frac{\pi}{c} e^{-c}.$$

Then we have to split the integral into real and imaginary parts

$$\int_{-\infty}^{\infty} \frac{\cos x}{x^2 + c^2} dx = \frac{\pi}{c} e^{-c}, \quad \int_{-\infty}^{\infty} \frac{\sin x}{x^2 + c^2} dx = 0,$$

where the last integral follows noticing that sine is an odd function. As cosine is an even function, we get finally for the required integral

$$I = \frac{\pi}{2c} e^{-c}.$$

Example 12.53. This example dates back to EULER in 1781. We need to evaluate the integral

$$I := \int_{0}^{\infty} \frac{\sin x}{x} dx.$$

We use as continuation into the complex plane

$$F(z) = \frac{e^{iz}}{z}.$$

This function has a simple pole at the origin. To apply Jordan's lemma we have to take $f(z) = 1/z$ with $a = 0$ and $\lambda = 1$. We go around the pole at the origin, as it is located on the path of integration, by integrating along a small half circle C_ε with radius ε in the upper half-plane:

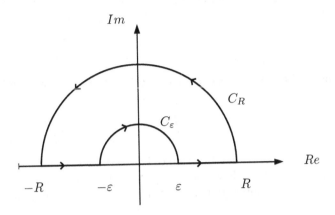

Figure 12.2

We apply Cauchy's integral theorem to the two arcs C_R, C_ε and the two intervals $[-R, -\varepsilon]$, $[\varepsilon, R]$,

$$\left(\int_{-R}^{-\varepsilon} - \int_{C_\varepsilon} + \int_{\varepsilon}^{R} + \int_{C_R} \right) \frac{e^{iz}}{z} dz = 0.$$

Because of $|f(z)| = 1/R \to 0$ on C_R, Jordan's lemma yields

$$\lim_{R \to \infty} \int_{C_R} \frac{e^{iz}}{z} dz = 0.$$

We still have to find an estimate for the integral along the arc C_ε. In the neighborhood of the origin the equation

$$F(z) = \frac{1}{z} + P(z), \quad P(z) = \frac{e^{iz} - 1}{z}$$

holds, where $|P(z)| \le K$ with a suitable constant K. The arc C_ε is parameterized by $z = \varepsilon e^{it}$, $0 \le t \le \pi$, so that we finally get (cf. Theorem 7.13)

$$\int_{C_\varepsilon} f(z) dz = \int_{C_\varepsilon} \frac{dz}{z} + \int_{C_\varepsilon} P(z) dz = -\pi i + O(\varepsilon)$$

and that is the same as

$$\lim_{\varepsilon \to 0} \left(\int_{-\infty}^{-\varepsilon} + \int_{\varepsilon}^{\infty} \right) \frac{e^{ix}}{x} dx = \pi i.$$

If in the first integral of the above formula we substitute $x' := -x$, it follows that

$$\int_0^\infty \frac{e^{ix} - e^{-ix}}{x} dx = \pi i,$$

where the integral is a proper integral, but it does not converge absolutely. Finally
we get

$$\int\limits_0^\infty \frac{\sin x}{x} dx = \frac{\pi}{2}.$$

Example 12.54. We consider a whole class of real integrals which have the form

$$\int\limits_0^{2\pi} R(\cos t, \sin t) dt.$$

Here $R(x, y)$ is a rational function of two variables, i.e., a ratio of two polynomials
in x and y with the additional assumption that $R(\cos t, \sin t)$ is finite for all t. If t
runs through the interval $[0, 2\pi]$, then $z = e^{it}$ moves along the unit circle. In view
of

$$\cos t = \frac{1}{2}(e^{it} + e^{-it}), \qquad \sin t = \frac{1}{2i}(e^{it} - e^{-it})$$

we get

$$\int\limits_0^{2\pi} R(\cos t, \sin t) dt = \frac{1}{i} \int\limits_{|z|=1} R\left(\frac{1}{2}(z + z^{-1}), \frac{1}{2i}(z - z^{-1})\right) \frac{dz}{z}.$$

Now the integrand is a rational function $F(z)$ and by the residue theorem we have

$$\int\limits_0^{2\pi} R(\cos t, \sin t) dt = 2\pi \sum_k \text{Res}(F, a_k),$$

where the sum is taken over all residues of F in $|z| < 1$. As a specific example let
us consider the integrand

$$R(\cos t, \sin t) = \frac{1}{1 + 3\sin^2 t}.$$

Here we get

$$F(z) = \frac{1}{z}\left[\frac{1}{1 + \frac{-3}{4}(z - z^{-1})^2}\right] = \frac{-4z}{3(z^2 - 3)(z + \frac{1}{\sqrt{3}})(z - \frac{1}{\sqrt{3}})}.$$

In the unit circle the integrand has only the two poles $1/\sqrt{3}$ and $-1/\sqrt{3}$ with the
residues

$$\text{Res}(F, \frac{1}{\sqrt{3}}) = \frac{1}{4}, \quad \text{Res}(F, \frac{-1}{\sqrt{3}}) = \frac{1}{4}.$$

We thus find the required integral

$$\int\limits_0^{2\pi} \frac{dt}{1 + 3\sin^2 t} = \pi.$$

12.5 Exercises

1. Show that the order k of a root of a holomorphic function f at a point a is given by the formula
$$k = \frac{1}{2\pi i} \int_\gamma \frac{f'(z)}{f(z)} dz,$$
where γ is a simple closed curve, in which no other root of f lies. For γ a circle centered at a with a small enough radius can be chosen.

2. Show that the convergence of the series $\sum \log(1+a_n)$ follows from the convergence of the infinite product $\prod(1+a_n)$. The principal value of the logarithm has to be taken.

3. Show that
$$\prod \left(1 - \frac{z}{c+n}\right) e^{z/n}$$
converges for all $c \in \mathbb{C}$ and $n = 1, 2, \ldots$, if the constant c is a non-negative integer.

4. Find the isolated singularities of
$$\cot z = \frac{\cos z}{\sin z},$$
and calculate the residues. Determine the first five coefficients of the Laurent series around the point $z = 0$.

5. Describe the missing steps in Example 12.21, i.e., show the equation
$$\frac{\pi^2}{\sin^2 \pi z} = \sum_{n=-\infty}^{\infty} \frac{1}{(z-n)^2}.$$

6. Prove the equation
$$\pi \cot \pi z = \frac{1}{z} + 2 \sum_{n=1}^{\infty} \frac{z}{z^2 - n^2}.$$

7. Prove Riemann–Fueter's Theorem 12.23 on removable singularities: *Let f be a holomorphic function in $0 < |x - a| < R$ and let f have a removable singularity at a. Then a value a_0 exists so that the extended function with*
$$\tilde{f}(x) = \begin{cases} f(x), & 0 < |x-a| < R, \\ a_0, & x = a \end{cases}$$
is holomorphic in $B_R(a)$.

8. Prove Mittag-Leffler's Theorem 12.26 in $C\ell(n)$: *For given a_k with $|a_k| \le |a_{k+1}| \to \infty$ and given principal parts $H_k(x)$ at the a_k holomorphic polyno-mials $P_k(x)$ exist, so that the series*

$$f(x) := \sum_{k=1}^{\infty}(H_k(x) - P_k(x))$$

is a meromorphic function in $\mathbb{R}^{n+1}$ with poles a_k and principal parts $H_k(x)$.

9. Prove

$$d\{g(dx_i \wedge dx)^* f\} = -\{(\partial_i g)dx^* f + gdx^*(\partial_i f)\} + \{(g\overline{\partial})dx_i^* f + gdx_i^*(\overline{\partial}f)$$

(cf. Proposition 12.48).

10. Let $R(x)$ be a rational function with no poles on the real axis. In addition let $R(x)$ vanish of order 2 for $x \to \infty$, i.e., let $x^2|R(x)|$ be bounded. Map the complex z-plane with

$$w = \frac{z - i}{z + i}$$

to the w-plane. Determine the inverse mapping and the integral

$$I = \int_{-\infty}^{\infty} R(x)dx$$

by applying the mapping. By which residues can the integral be described? Apply this procedure to the example

$$R(x) = \frac{1}{1 + x^2}.$$

11. Determine the value of the integral

$$I = \int_{|z+2|=4} \frac{dz}{(z - 1)(z + 1)(z + i)}.$$

12. Calculate

$$\int_{0}^{\infty} \frac{dx}{1 + x^4}.$$

13. Calculate *Fresnel's integrals*:

$$\int_{0}^{\infty} \cos(x^2)dx = \int_{0}^{\infty} \sin(x^2)dx = \sqrt{\frac{\pi}{8}}.$$

13 Special functions

13.1 Euler's Gamma function

13.1.1 Definition and functional equation

In this section we present and examine the first non-elementary function in $\mathbb{C}$, namely *Euler's Gamma function* also called *Euler's integral of second kind*. This function was studied by Euler for the first time in 1764. His goal was to find an interpolation of the factorial function.

Definition 13.1 (Gamma function). The function given by the integral

$$\Gamma(z) = \int_0^\infty t^{z-1} e^{-t} dt,$$

where $\operatorname{Re} z > 0$, is called the *Gamma function*.

In the integral the power with complex exponent is defined by

$$t^{z-1} := e^{(z-1)\ln t},$$

since the basis is real. This power is a uniquely defined holomorphic function with respect to z. We keep this definition in mind for the following similar integrals. Of course we have to think about the convergence of the integral: Owing to $|t^{z-1}| = t^{x-1}$ and the e-function e^{-t}, the convergence is obtained for $t \to \infty$ for all x. The lower boundary causes bigger difficulties: As in this case $e^{-t} \to 1$ for $t \to 0$ only the factor t^{x-1} is important in the integrand. So the integral converges for $x > 0$ and the definition is usable in the case $x = \operatorname{Re} z > 0$.

Now we want to prove the first properties of Γ:

Theorem 13.2.   (i) *The integral in the definition of* $\Gamma(z)$ *converges absolutely and uniformly for* $0 < \rho \leq \operatorname{Re} z \leq R$. *As a consequence* Γ *is a holomorphic function for* $\operatorname{Re} z > 0$ *and in particular it is positive for* $z = x > 0$.

(ii) *If* $\operatorname{Re} z > 0$, *then the functional equation*

$$\Gamma(z+1) = z\,\Gamma(z)$$

holds. More generally for all $n \in \mathbb{N}$ *we have*

$$\Gamma(z+n) = z(z+1)(z+2)\cdots(z+n-1)\Gamma(z).$$

(iii) *For all* $n \in \mathbb{N}$ *we have*
$$\Gamma(n) = (n-1)!.$$

Γ is defined as a holomorphic function in the right half-plane; the definition in the left half-plane will be dealt with after the proof. Indeed the function Γ interpolates the factorial function, as we will prove in (iii).

Proof. (i) The equation

$$\int_0^\infty |t^{z-1}|e^{-t}dt \le \int_0^1 t^{\rho-1}dt + \int_1^\infty t^{R-1}e^{-t}dt,$$

holds for $\rho \le \operatorname{Re} z \le R$. In addition both integrals on the right-hand side converge uniformly with respect to x since they are independent of x. According to Section 9.1 we can differentiate under the integral and so Γ is a holomorphic function if $\operatorname{Re} z > 0$. As the integrand is positive for $x > 0$, $\Gamma(x)$ is positive as well.

(ii) We integrate by parts

$$\Gamma(z+1) = \int_0^\infty t^z e^{-t}dt = \lim_{\beta \to \infty} -t^z e^{-t}\big|_{t=0}^\beta + z\int_0^\infty t^{z-1}e^{-t}dt = z\Gamma(z),$$

where $\operatorname{Re} z > 0$. The general formula is obtained by induction on n.

(iii) The assertion results from (ii) by induction: $\Gamma(n+1) = n\Gamma(n)$ and considering

$$\Gamma(1) = \int_0^\infty e^{-t}dt = \lim_{\beta \to \infty} -e^{-t}\big|_0^\beta = 1.$$

$\square$

Now we can continue Γ into the left half-plane:

Definition 13.3 (Gamma function). Let $\Gamma(z)$ be defined by

$$\Gamma(z) := \frac{1}{z(z+1)(z+2)\cdots(z+n-1)}\Gamma(z+n),$$

where $0 \ge \operatorname{Re} z > -n$. As $n \in \mathbb{N}$ can be chosen arbitrarily this definition holds for every point in $\mathbb{C}$.

According to Definition 9.16 we have continued Γ holomorphically to every point of the left half-plane except the negative integers. The identity theorem 9.15 implies that such a continuation is uniquely determined, so that the above formula gives the same value for different n.

Corollary 13.4. *The function Γ is meromorphic in the whole complex plane $\mathbb{C}$ and at the points $z_n = -n$, $n = 0, 1, 2, \ldots$, it has simple poles with the residues*

$$\operatorname{Res}(\Gamma(z), z_n) = \frac{(-1)^n}{n!}.$$

Γ has an essential singularity at the point $z = \infty$.

Proof. As $\Gamma(z+n)$ is holomorphic if $x > -n$, so $\Gamma(z)$ is also holomorphic except for the zeros in the denominator at $0, -1, -2, \ldots, -(n-1)$. Since n is taken arbitrarily, $\Gamma(z)$ is holomorphic for all $z \in \mathbb{C}$ apart from the poles at the non-positive integers. The poles are simple according to the definition and so the residues can be calculated by

$$\operatorname{Res}(\Gamma(z), -n) = \lim_{z \to -n} (z+n)\Gamma(z)$$

$$= \lim_{z \to -n} \frac{1}{z(z+1)...(z+n-1)}\Gamma(z+n+1) = \frac{1}{(-1)^n n!}\Gamma(1).$$

As $z = \infty$ is an accumulation point of poles it is an essential singularity.

$\square$

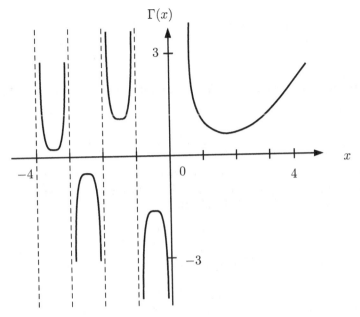

Figure 13.1

We want to prove some — partly amazing — properties of Γ. A comprehensive presentation is given in the "Bible" of classical analysis: WHITTAKER–WATSON [159]. We mostly proceed like E.C. TITCHMARSH in his also classical book [155]. Before we prove functional equations a proposition follows:

Proposition 13.5 (Beta function). *If* $\operatorname{Re} z > 0$, $\operatorname{Re} \zeta > 0$ *the equation*

$$B(z, \zeta) := \frac{\Gamma(z)\Gamma(\zeta)}{\Gamma(z+\zeta)} = \int_0^\infty \frac{u^{\zeta-1}}{(1+u)^{z+\zeta}}\, du = \int_0^1 \lambda^{z-1}(1-\lambda)^{\zeta-1}\, d\lambda$$

holds. The Beta function *is also called* Euler's integral of first kind.

Proof. We can interchange the order of integration in the product

$$\Gamma(z)\Gamma(\zeta) = \int_0^\infty t^{z-1}e^{-t}\, dt \int_0^\infty \tau^{\zeta-1}e^{-\tau}\, d\tau,$$

for $\operatorname{Re} z > 0$ and $\operatorname{Re} \zeta > 0$ in view of the absolute and local uniform convergence. First we substitute $\tau = tu$, $d\tau = t\, du$, then $v = t(1+u)$, $dv = (1+u)dt$. This results in

$$
\begin{aligned}
\Gamma(z)\Gamma(\zeta) &= \int_0^\infty t^{z-1}e^{-t}\, dt \int_0^\infty t^\zeta u^{\zeta-1}e^{-tu}\, du \\
&= \int_0^\infty \frac{u^{\zeta-1}}{(1+u)^{z+\zeta}}\, du \int_0^\infty v^{z+\zeta-1}e^{-v}\, dv \\
&= \Gamma(z+\zeta) \int_0^\infty \frac{u^{\zeta-1}}{(1+u)^{z+\zeta}}\, du.
\end{aligned}
$$

The second integral in the lemma can be easily transformed by the substitution

$$\lambda = \frac{1}{1+u}, \quad u = \frac{1-\lambda}{\lambda}, \quad du = -\frac{d\lambda}{\lambda^2}. \qquad \Box$$

Theorem 13.6 (Legendre's duplication formula).   (i) *We have* $\Gamma(\tfrac{1}{2}) = \sqrt{\pi}$.

(ii) *For all* $z \in \mathbb{C}$ *the Legendre's duplication formula holds:*

$$\Gamma(2z)\Gamma(\tfrac{1}{2}) = 2^{2z-1}\Gamma(z)\Gamma(z + \tfrac{1}{2}).$$

The value of $\Gamma(\tfrac{1}{2})$ is already amazing, because nothing in the defining integral indicates the connection to π. The formula was discovered by the French mathematician ADRIEN-MARIE LEGENDRE (1751–1833).

Proof. (i) We set $z = \zeta = \tfrac{1}{2}$ in Proposition 13.5 and get

$$\Gamma^2(\tfrac{1}{2}) = \Gamma(1) \int_0^\infty \frac{du}{\sqrt{u}(1+u)} = 2 \arctan \sqrt{u}\Big|_0^\infty = \pi.$$

This gives the assertion since $\Gamma(\tfrac{1}{2}) > 0$.

(ii) Now we set $z = \zeta$ in Proposition 13.5 and get

$$\frac{\Gamma^2(z)}{\Gamma(2z)} = \int_0^1 \lambda^{z-1}(1-\lambda)^{z-1}d\lambda = 2 \int_0^{1/2} \lambda^{z-1}(1-\lambda)^{z-1}d\lambda,$$

where the last equality holds because of the symmetry around $\lambda = 1/2$. If we substitute

$$\lambda = \frac{1}{2} - \frac{1}{2}\sqrt{t}, \quad \lambda(1-\lambda) = \frac{1}{4}(1-t), \quad d\lambda = -\frac{1}{4}\frac{dt}{\sqrt{t}},$$

then according to Proposition 13.5 we get

$$\frac{\Gamma^2(z)}{\Gamma(2z)} = 2^{1-2z} \int_0^1 (1-t)^{z-1}\frac{dt}{\sqrt{t}} = 2^{1-2z}\frac{\Gamma(z)\Gamma(\tfrac{1}{2})}{\Gamma(z + \tfrac{1}{2})},$$

which is what we wanted. But we remark that the proof makes the assumption $\operatorname{Re} z > 0$. However, in view of the identity theorem the functional equation holds in the whole domain of the function Γ, in our case $\mathbb{C} \setminus \{0, -1, -2, \dots\}$. $\Box$

Another amazing functional equation is the following, whose proof is a good application of the residue theorem:

Theorem 13.7. *For all* $z \in \mathbb{C}$ *we have the functional equation*

$$\Gamma(z)\Gamma(1 - z) = \frac{\pi}{\sin \pi z}.$$

Proof. We set $\zeta = 1 - z$ in Proposition 13.5 and obtain

$$\Gamma(z)\Gamma(1 - z) = \int_0^\infty \frac{u^{-z}}{1 + u}du.$$

First we restrict our considerations to real $z = x$ where $0 < x < 1$. We then calculate the integral along the following path γ: We go along the real axis from $u = \rho$ to $u = R$, then along the circle $|u| = R$ in the positive direction to $u = R$, from there to $u = \rho$ and finally back along the circle $|u| = \rho$ in negative direction to $u = \rho$. The problem here is the argument of the complex u: If we start on the real axis with $\arg u = 0$, we then reach $\arg u = 2\pi$ after going around $|u| = R$. We have to pay attention to this while calculating the second integral. The closed integral is evaluated by the residue theorem, where we have to take into account the simple pole at -1 with the residue $(-1)^{-x} = e^{-i\pi x}$:

$$
\begin{aligned}
2\pi i\, e^{-i\pi x} &= \int_\gamma \frac{u^{-x}\, du}{1+u} \\
&= \int_\rho^R \frac{e^{-x\ln t}\, dt}{1+t} + \int_R^\rho \frac{e^{-x(\ln t + 2\pi i)}\, dt}{1+t} \\
&\quad + i\int_0^{2\pi} \frac{R^{1-x} e^{-i(x-1)\varphi}\, d\varphi}{1+Re^{i\varphi}} + i\int_{2\pi}^0 \frac{\rho^{1-x} e^{-i(x-1)\varphi}\, d\varphi}{1+\rho e^{i\varphi}} \\
&= (1 - e^{-2\pi i x}) \int_\rho^R \frac{t^{-x}\, dt}{1+t} + I_3 + I_4 .
\end{aligned}
$$

I_3 is the integral along the circle with radius R and I_4 the integral along the circle with radius ρ. Both integrals converge to 0 for $R \to \infty$, respectively $\rho \to 0$, and we have for I_3,

$$
|I_3| \leq \int_0^{2\pi} \frac{R^{1-x}\, d\varphi}{R-1} \leq 2\pi \frac{R^{1-x}}{R-1} .
$$

The last quotient converges to zero since $0 < x < 1$ and $R \to \infty$. With a similar estimation we can prove $I_4 \to 0$ for $\rho \to 0$. The remaining formula

$$
2\pi i\, e^{-i\pi x} = (1 - e^{-2\pi i x}) \int_0^\infty \frac{t^{-x}\, dt}{1+t} = (1 - e^{-2\pi i x})\Gamma(x)\Gamma(1-x)
$$

is equivalent to

$$
\Gamma(x)\Gamma(1-x) = \frac{\pi}{\sin \pi x} .
$$

This equation is now proven on the line segment $0 < \operatorname{Re} z < 1$. As both sides of the equation are holomorphic functions (apart from the poles) and the line segment is a set with accumulation points, the equation holds then in the whole plane. $\qquad \square$

13.1.2 Stirling's theorem

We want to prove another important property of the Gamma function, which describes the behavior for $\operatorname{Re} z \to \infty$. The Scottish mathematician JAMES STIRLING (1692–1770) discovered this property. We want to avoid too long a proof and so first we prove some propositions:

Proposition 13.8. *For* $\operatorname{Re} z > 0$ *the formula*

$$
\log \Gamma(z) = -\gamma z - \log z + \sum_{n=1}^{\infty} \left(\frac{z}{n} - \log\left(1 + \frac{z}{n}\right) \right),
$$

holds, where γ is the so called Euler–Mascheroni constant

$$\gamma := \lim_{N \to \infty} \left(1 + \frac{1}{2} + \cdots + \frac{1}{N} - \log N \right) = 0,5772157\ldots \; .$$

Proof. We apply again Proposition 13.5, but this time to $z - h$ and h, where we assume $\mathrm{Re}\, z > 1$. We have

$$\frac{\Gamma(z - h)\Gamma(h)}{\Gamma(z)} = \int_0^1 t^{h-1}(1 - t)^{z-h-1} dt.$$

We expand both sides with respect to h and compare the coefficients of $h^0 = 1$, where h is real and $0 < h < x$. On the left side we obtain

$$\frac{1}{\Gamma(z)} \left(\Gamma(z) - h\Gamma'(z) + \cdots \right) \left(\frac{1}{h} - a_0 + \cdots \right) = \frac{1}{h} - \frac{\Gamma'(z)}{\Gamma(z)} - a_0 + \cdots,$$

where $-a_0$ is the corresponding coefficient of the Laurent series of $\Gamma(h)$ around $h = 0$. On the right-hand side we get

$$\left. \frac{t^h}{h} \right|_{t=0}^1 + \int_0^1 t^{h-1}[(1 - t)^{z-h-1} - 1] dt = \frac{1}{h} + \int_0^1 [(1 - t)^{z-1} - 1] \frac{dt}{t} + o(1),$$

where the -1 in the square brackets is a term providing convergence for $t = 0$. Furthermore the right integral equals just the first integral above if $h = 0$. So it follows

$$\frac{\Gamma'(z)}{\Gamma(z)} = \int_0^1 [1 - (1 - t)^{z-1}] \frac{dt}{t} - a_0,$$

where we substitue

$$\frac{1}{t} = \frac{1}{1 - (1 - t)} = \sum_{n=0}^{\infty} (1 - t)^n$$

and integrate termwise along the interval $[\varepsilon, 1]$. We then calculate the limit for $\varepsilon \to 0$ and obtain

$$\frac{\Gamma'(z)}{\Gamma(z)} = \sum_{n=0}^{\infty} \left(\frac{1}{n + 1} - \frac{1}{n + z} \right) - a_0.$$

If we combine $1/(n + 1)$ with the term $1/(n + 1 + z)$, we obtain

$$\frac{\Gamma'(z)}{\Gamma(z)} = -\frac{1}{z} + \sum_{n=1}^{\infty} \left(\frac{1}{n} - \frac{1}{n + z} \right) - a_0.$$

We integrate this equation along the line segment from $c > 0$ to z:

$$\log \Gamma(z) - \log \Gamma(c) - \log c = -\log z + \sum_{n=1}^{\infty} \left(\frac{z - c}{n} - \log \frac{n + z}{n + c} \right) - a_0(z - c),$$

where we use the principal values of the logarithms. As $\Gamma(c)\, c = \Gamma(c + 1) \to 1$ for $c \to 0$, we obtain almost the final assertion:

$$\log \Gamma(z) = -a_0 z - \log z + \sum_{n=1}^{\infty} \left(\frac{z}{n} - \log \left(1 + \frac{z}{n} \right) \right).$$

Finally we need only to determine a_0 by setting $z = 1$ into the above equation,

$$a_0 = \sum_{n=1}^{\infty} \left(\frac{1}{n} - \ln\left(1 + \frac{1}{n}\right) \right)$$

$$= \lim_{N \to \infty} \sum_{n=1}^{N} \left(\frac{1}{n} - \ln(n+1) + \ln n \right) = \lim_{N \to \infty} \left(\sum_{n=1}^{N} \frac{1}{n} - \ln(N+1) \right) =: \gamma,$$

where we can use $\ln N$ instead of $\ln(N+1)$ in view of $\ln(N+1) - \ln N = \ln(1+1/N) \to 0$ for $N \to \infty$. $\qquad \square$

Now the claim of this section:

Theorem 13.9 (Stirling's theorem). *If* $\operatorname{Re} z > 0$ *and* $|\arg z| \le \frac{\pi}{2} - \delta$, $\delta > 0$, *uniformly in* z *we have*

$$\log \Gamma(z) = (z - \tfrac{1}{2}) \log z - z + \tfrac{1}{2} \ln 2\pi + R(z),$$

where the remainder is given by

$$R(z) = \int_0^{\infty} \frac{[t] - t + \frac{1}{2}}{t + z} dt = O\left(\frac{1}{|z|}\right)$$

for $z \to \infty$ *and* $[t]$ *denotes the largest integer* $\le t$.

We still need an additional proposition for better structuring the proof. The first part of the proposition is a kind of interpolation for Γ and the second is already Stirling's theorem, but in the case of integers.

Proposition 13.10. (i) *For real* $c > 0$ *and real* $z = x > 0$ *we have*

$$\Gamma(x + c) = \Gamma(x) x^c \left(1 + O\left(\frac{1}{x^{c+1}}\right) \right).$$

(ii) *For all* $n \in \mathbb{N}$ *and* $n \to \infty$ *we have*

$$\ln(n!) = \ln \Gamma(n + 1) = (n + \tfrac{1}{2}) \ln n - n + \ln \sqrt{2\pi} + O\left(\frac{1}{n^2}\right).$$

Proof. (i) First we assume $c > 1$. According to Proposition 13.5, which is a fundamental result for the Gamma function, the equation

$$\frac{\Gamma(x)\Gamma(c)}{\Gamma(x+c)} = \int_0^1 \lambda^{x-1}(1 - \lambda)^{c-1} d\lambda,$$

holds. If we now substitute $\lambda = e^{-t}$, $d\lambda = -e^{-t}dt$, we then obtain

$$\frac{\Gamma(x)\Gamma(c)}{\Gamma(x+c)} = \int_0^{\infty} e^{-tx}(1 - e^{-t})^{c-1} dt$$

$$= \int_0^{\infty} t^{c-1} e^{-tx} dt - \int_0^{\infty} \{t^{c-1} - (1 - e^{-t})^{c-1}\} e^{-xt} dt$$

$$= x^{-c}\Gamma(c) - R(c, x),$$

where $R(c, x)$ represents the last of both integrals. In view of $1 - e^{-t} < t$ for $t > 0$ the last integral is positive, and since $1 - e^{-t} > t - t^2/2$ for $0 < t < 1$ we have the estimate

$$R(c, x) \leq \int_0^1 \{1 - (1 - t/2)^{c-1}\} t^{c-1} e^{-xt} dt + \int_1^\infty t^{c-1} e^{-xt} dt$$

$$\leq K \int_0^1 t^c e^{-xt} dt + \int_1^\infty t^c e^{-xt} dt \leq \frac{K+1}{x^{c+1}} \Gamma(c+1),$$

where we used $(1 - t/2)^{c-1} \geq 1 - Kt$ for a suitable K. With a little more transformation, assuming $c \leq 1$, the assertion follows by the functional equation in Theorem 13.2 (ii).

(ii) The following lines can be interpreted by the Riemann–Stieltjes integral or by direct calculation ($[t]$ is again the biggest integer not greater than t):

$$\ln(n!) = \sum_{k=1}^n \ln k = \int_{1-}^{n+} \ln t\, d[t] = [t] \ln t\Big|_{1-}^{n+} - \int_1^n \frac{[t]}{t} dt$$

$$= (n + \tfrac{1}{2}) \ln n - n + 1 - \int_1^n \frac{[t] - t + \tfrac{1}{2}}{t} dt.$$

The numerator of the remainder is written in such a special form in view of the following property: With

$$\Phi(t) := \int_1^t ([\tau] - \tau + \tfrac{1}{2}) d\tau,$$

we obtain

$$\int_m^{m+1} ([\tau] - \tau + \tfrac{1}{2}) d\tau = 0$$

and so $\Phi(n) = 0$ for all n. For the remainder we thus get

$$\int_1^n \frac{\Phi'(t)}{t} dt = \frac{\Phi(t)}{t}\Big|_1^n + \int_1^n \frac{\Phi(t)}{t^2} dt = \int_1^n \frac{\Phi(t)}{t^2} dt.$$

Hence the remainder is monotonically decreasing with respect to n and bounded. It converges to $-C + 1$, where C is a constant and we obtain

$$\ln(n!) = (n + \tfrac{1}{2}) \ln n - n + C + O\left(\frac{1}{n^2}\right).$$

For determining the constant C, we set $z = n$ in Legendre's duplication formula:

$$\ln \Gamma(2n) + \ln \Gamma(\tfrac{1}{2}) = (2n - 1) \ln 2 + \ln \Gamma(n) = \ln \Gamma(n + \tfrac{1}{2})$$

or with the equation $\ln \Gamma(n + 1) = \ln(n!)$ just proved

$$(2n - 1 + \tfrac{1}{2}) \ln(2n - 1) - (2n - 1) + C + \ln \sqrt{\pi}$$
$$= (2n - 1) \ln 2 + 2(n - \tfrac{1}{2}) \ln(n - 1) - 2(n - 1) + 2C + \ln \sqrt{n - 1} + o(1).$$

Comparing the constant terms results in

$$-\frac{1}{2} \ln 2 + \ln \sqrt{\pi} + C = -\ln 2 + 2C$$

and as a consequence

$$C = \ln \sqrt{2\pi}. \qquad \qquad \square$$

Proof of the Stirling theorem. We obtain from the previous lemma

$$\log \Gamma(z) = -\gamma z - \log z + \sum_{n=1}^{\infty} \left(\frac{z}{n} - \log \left(1 + \frac{z}{n} \right) \right).$$

Now we substitute the sum from 1 to ∞ by a sum from 1 to N and let N diverge to ∞. If we add and subtract in the sum $z \ln N$, then the sum vanishes partly with γz and we are left with

$$\log \Gamma(z) = - \log z + \lim_{N \to \infty} \left\{ z \ln N - \sum_{n=1}^{N} \log \left(1 + \frac{z}{n} \right) \right\}.$$

Indicating by S_N the term in braces we obtain

$$S_N = z \ln N + \ln(N!) - \sum_{n=1}^{N} \log(n + z). \qquad (*)$$

We now evaluate the remainder given in the theorem between the limits 0 and N.

$$
\begin{aligned}
R_N : \quad &= \int_0^N \frac{[t] - t + \frac{1}{2}}{t + z} dt = -N + \sum_{n=0}^{N-1} \int_n^{n+1} \frac{n + z + \frac{1}{2}}{t + z} dt \\
&= -N + \sum_{n=0}^{N-1} \left[\left(n + z + \frac{1}{2} \right) \log(n + 1 + z) - \left(n + z + \frac{1}{2} \right) \log(n + z) \right].
\end{aligned}
$$

By adding and subtracting a 1 in the first bracket of the sum a telescoping sum is obtained and we find

$$R_N = -N - \left(z + \frac{1}{2} \right) \log z + \left(N + z + \frac{1}{2} \right) \log(N + z) - \sum_{n=1}^{N} \log(n + z).$$

Substituting the latter into the formula for S_N we then use Proposition 13.10 (ii) to eliminate $\ln(N!)$:

$$
\begin{aligned}
S_N &= z \ln N + \ln(N!) + N + \left(z + \frac{1}{2} \right) \log z - \left(N + z + \frac{1}{2} \right) \log(N + z) + R_N \\
&= \left(z + \frac{1}{2} \right) \log z - z \log \left(1 + \frac{z}{N} \right) - \left(N + \frac{1}{2} \right) \log \left(1 + \frac{z}{N} \right) + \ln \sqrt{2\pi} + R_N + o(1).
\end{aligned}
$$

We now let $N \to \infty$ and obtain

$$\lim_{N \to \infty} S_N = (z + \tfrac{1}{2}) \log z - z + \ln \sqrt{2\pi} + R(z),$$

where $R(z)$ is the remainder from the theorem. So the assertion is proven except for the estimation of the remainder. In the proof of the previous proposition we have introduced the function

$$\Phi(t) = \int_0^t ([\tau] - \tau + \tfrac{1}{2}) d\tau,$$

which vanishes at the natural numbers and oscillates between 0 and $-1/8$. We set

$$R(z) = \int_0^{\infty} \frac{\Phi'(t)}{t + z} dt = \int_0^{\infty} \frac{\Phi(t)}{(t + z)^2} dt,$$

and with $z = re^{i\varphi}$ it follows that

$$
\begin{aligned}
|R(z)| &\leq \frac{1}{8} \int_0^\infty \frac{dt}{t^2 + r^2 + 2rt \cos \varphi} \\
&\leq \frac{1}{8} \int_0^\infty \frac{dt}{t^2 + r^2 - 2tr \cos \delta} \\
&= \frac{\pi}{8r \sin \delta},
\end{aligned}
$$

where the last step is obtained by calculating the integral.   □

13.2 Riemann's Zeta function

13.2.1 Dirichlet series

We discuss the function class of *Dirichlet series*, before introducing another non-elementary function in $\mathbb{C}$, *the Riemann Zeta function*, which also belongs to this class. The class of Dirichlet series contains some interesting functions.

JOHANN PETER GUSTAV LEJEUNE DIRICHLET (1805–1859) had Wallonian ancestors who came from Verviers in Belgium. He studied in Paris in the years 1822–1826, received in 1827 an honorary doctorate by the University of Bonn and, supported by ALEXANDER VON HUMBOLDT, became privatdozent and professor in Breslau. In 1829 he went to Berlin, where he became full professor in 1839. In 1855 he was appointed as the successor of GAUSS in Göttingen. Not only his pioneering scientific work in mathematics and mathematical physics had essential influence in mathematics, but also his exemplary lectures were of utmost importance. Among his students we find the well known mathematicians B. RIEMANN, E.E. KUMMER, L. KRONECKER, G. EISENSTEIN and R. DEDEKIND.

Definition 13.11 (Dirichlet series). A series of the form

$$
\sum \frac{a_n}{n^z}
$$

is called a *Dirichlet series*, where $a_n, z \in \mathbb{C}$ and $n \in \mathbb{N}$.

For Dirichlet series the notation $s = \sigma + it$ is often used instead of z, but we do not want to have too much notation. The slightly more general type of series,

$$
\sum a_n e^{-\lambda_n z}, \quad \lambda_n \in \mathbb{C},
$$

called *generalized Dirichlet series*, is also sometimes considered. We have to study the convergence behavior:

Theorem 13.12 (Convergence abscissa). *If the Dirichlet series converges at z_0, then it is uniformly convergent in the angular sector*

$$
|\arg(z - z_0)| \leq \frac{\pi}{2} - \delta,
$$

where δ can be arbitrarily chosen in $0 < \delta < \pi/2$. Thus a Dirichlet series possesses a convergence abscissa

$$x_c := \inf\{x : \sum a_n n^{-z} \text{ convergent in } z = x + iy\},$$

so that the series converges and becomes a holomorphic function if $\operatorname{Re} z > x_c$.

Proof. With the transformations $a'_n := a_n n^{-z_0}$ and $z' := z - z_0$ we can transform the convergence point into the origin. Then the series $\sum a_n$ converges; let us set

$$R_n := \sum_{k=n+1}^{\infty} a_k.$$

We have $R_n \to 0$ for $n \to \infty$, and for $x > 0$ we obtain

$$\sum_{n=M}^{N} \frac{a_n}{n^z} = \sum_{n=M}^{N} \frac{R_{n-1} - R_n}{n^z} = \sum_{n=M}^{N} R_n \left(\frac{1}{(n+1)^z} - \frac{1}{n^z} \right) + \frac{R_{M-1}}{M^z} - \frac{R_N}{(N+1)^z}.$$

In view of

$$\left| \frac{1}{(n+1)^z} - \frac{1}{n^z} \right| = \left| z \int_n^{n+1} \frac{dt}{t^{z+1}} \right| \leq |z| \int_n^{n+1} \frac{dt}{t^{x+1}} = \frac{|z|}{x} \left(\frac{1}{n^x} - \frac{1}{(n+1)^x} \right)$$

and as $|R_n| \leq \varepsilon$ for $n > M - 1$ is independent of z, it follows that

$$\left| \sum_{n=M}^{N} \frac{a_n}{n^z} \right| \leq \frac{\varepsilon |z|}{x} \sum_{n=M}^{N} \left(\frac{1}{n^x} - \frac{1}{(n+1)^x} \right) + \frac{\varepsilon}{M^x} + \frac{\varepsilon}{(N+1)^x} \leq \frac{\varepsilon |z|}{x} \frac{2}{M^x} \leq \frac{2\varepsilon |z|}{x}.$$

Finally we have

$$\frac{|z|}{x} = \sqrt{1 + \frac{y^2}{x^2}} = \frac{1}{|\cos \varphi|} \leq \frac{1}{\sin \delta}$$

since $|\varphi| \leq \frac{\pi}{2} - \delta$, and from this the statement of the theorem

$$\left| \sum_{n=M}^{N} \frac{a_n}{n^z} \right| \leq \frac{2\varepsilon}{\sin \delta}.$$

If x_c is the convergence abscissa defined in the theorem, then there is a point z_0 at which the series converges and $x_0 = \operatorname{Re} z_0$ is arbitrarily close to x_c. According to the first part of the proof, the series is a holomorphic function in the half-plane $x > x_0$ (as δ can be chosen arbitrarily small), and so the series is also a holomorphic function in the half-plane $x > x_c$. $\qquad \square$

Things are similar but easier in the case of absolute convergence.

Proposition 13.13 (Absolute convergence). *The domain in which the Dirichlet series converges absolutely is also a half-plane (which can degenerate into the whole plane or the empty set), given by*

$$x > x_a := \inf\{x : \sum |a_n| n^{-x} \text{ is convergent}\}.$$

Proof. Since n^{-x} decreases as x increases, the convergence at a point x_0 results in the convergence at all points $x > x_0$. As x_a (as well as x_c) can be $+\infty$ or $-\infty$, the half-plane in which convergence occurs can degenerate into the empty set or into the whole plane.   □

The following example shows that x_a and x_c need not be equal: From analysis it is known that

$$\sum \frac{(-1)^{n+1}}{n^x}$$

is absolutely convergent, if $x > 1$, but according to the Leibniz criterion (the terms have alternating signs and the absolute value converges monotonically to 0) the series converges for $x > 0$, so $x_c = 0 < 1 = x_a$ really holds.

13.2.2   Riemann's Zeta function

The most known Dirichlet series is the *Riemann Zeta function*:

Definition 13.14 (Riemann's Zeta function). The series

$$\zeta(z) = \sum_{n=1}^{\infty} \frac{1}{n^z}.$$

is called *Riemann's Zeta function.*

The series has the convergence abscissas $x_c = x_a = 1$, because it converges absolutely for $x > 1$ and at $z = 1$ it diverges as it becomes the *harmonic series*. There is an extensive theory about the Zeta function and also many books on this topic. We give a short presentation.

Theorem 13.15 (Euler's product formula). *In the half-space* $\operatorname{Re} z > 1$,

$$\zeta(z) = \prod_{p} \left(1 - \frac{1}{p^z}\right)^{-1}$$

holds, where the product ranges over all primes p.

Infinite products are defined and described explicitly in Definition 12.4 and Proposition 12.5. The convergence is given by the existence of

$$\lim_{N \to \infty} \prod_{p \le N} \left(1 - \frac{1}{p^z}\right)^{-1},$$

where the limit should not be zero.

Proof. Let $N > 0$ be a natural number and define the product

$$\Pi_N := \prod_{p \le N} \left(1 - \frac{1}{p^z}\right)^{-1} = \prod_{p \le N} \left(1 + \frac{1}{p^z} + \frac{1}{p^{2z}} + \cdots\right).$$

Expanding these finitely many terms leads to

$$\prod_N = 1 + \sum{}' \frac{1}{n^z},$$

where the sum ranges over all natural numbers, which can be represented as the product of powers of primes $p \leq N$. The latter are at least all natural numbers $\leq N$ and so we obtain

$$\left| \prod_N - \sum_{n=1}^{N} \frac{1}{n^z} \right| \leq \sum_{n=N+1}^{\infty} \frac{1}{n^x}.$$

If $N \to \infty$, then the right side converges to 0 and the sum in the absolute value to $\zeta(z)$. $\qquad\square$

This theorem points out the connection of $\zeta(z)$ with the primes. Unfortunately we can not explore this relation in this book. But we want to hint at the prime number theorem, which gives the number of primes $\leq N$ and can be shown with the help of the statement that $\zeta(z)$ has no zero for $\operatorname{Re} z \geq 1$. Already Euler's product formula shows that $\zeta(z)$ has no zero for $x > 1$. The following functional equation is also an important statement, and shows in addition that $\zeta(z)$ can holomorphically be continued into the whole plane except for one simple pole at $z = 1$.

Theorem 13.16 (Functional equation). *The Riemann Zeta function $\zeta(z)$ can meromorphically be expanded into the whole plane and has a simple pole at $z = 1$ with residue 1. Furthermore $\zeta(z)$ satisfies the functional equation*

$$\zeta(1 - z) = 2^{1-z} \pi^{-z} \cos \frac{\pi z}{2} \, \Gamma(z) \zeta(z).$$

Proof. If $x = \operatorname{Re} z > 1$ from

$$\Gamma(z) \frac{1}{n^z} = \int_0^{\infty} \left(\frac{\tau}{n} \right)^{z-1} e^{-\tau} \frac{d\tau}{n} = \int_0^{\infty} t^{z-1} e^{-nt} dt,$$

we have

$$\Gamma(z)\zeta(z) = \int_0^{\infty} t^{z-1} \sum_{n=1}^{\infty} e^{-nt} \, dt = \int_0^{\infty} \frac{t^{z-1}}{e^t - 1} dt. \qquad (*)$$

We consider the integral

$$J(z) := \int_C \frac{(-u)^{z-1}}{e^u - 1} du,$$

where C is the curve along the real axis from ∞ to δ with $0 < \delta < 1$ and then along the positive circle with radius δ around 0 and back along the real axis from δ to ∞. This integral converges locally uniformly in the z-plane, because the convergence at ∞ is ensured by the e-function. So J is a holomorphic function.

The problem of $(-u)^{z-1} = e^{(z-1)\log(-u)}$ being a multivalued function is solved by stating that $\log(-u)$ for $u = -\delta$ should be a real number. If $u = te^{i\varphi}$, then $-\pi$ has to be chosen as argument of $\log(-u)$ on the integral from ∞ to δ, because after a half positive circle around zero to $u = -\delta$ we find the argument to be 0. As a result the argument on the integral from δ to ∞ is π. Combining these integrals along the real axis and remarking that $\sin \pi(z - 1) = -\sin \pi z$ results in

$$\int_\delta^\infty \left(-e^{(z-1)(\ln t - i\pi)} + e^{(z-1)(\ln t + i\pi)} \right) \frac{dt}{e^t - 1} = -2i \sin \pi z \int_\delta^\infty \frac{t^{z-1}}{e^t - 1} dt.$$

On the circle with radius δ the denominator can be represented by $t(1 + O(t))$ ($\delta \to 0$), for the numerator we get

$$(-u)^{z-1} = \delta^{x-1} e^{y(\varphi - \pi)} = O(\delta^{x-1}).$$

The integrand in the integral along the small circle is $O(\delta^{x-2})$ ($\delta \to 0$), this integral converges for $x > 2$ and δ going to zero. Setting this in $(*)$ gives

$$-2i \sin \pi z \Gamma(z) \zeta(z) = \lim_{\delta \to 0} J(z).$$

As the integrand of $J(z)$ does not cross any poles for $0 < \delta < 1$, the integral $J(z)$ is independent of δ and we can ignore the limit: So for $x > 2$ we obtain

$$-2i \sin(\pi z) \Gamma(z) \zeta(z) = J(z).$$

We have thus expanded $\zeta(z)$ into the whole plane, because all other appearing functions are defined in $\mathbb{C}$. The simple poles of $\Gamma(z)$ at the points $z = 0, -1, -2, \ldots$ cancel with the simple zeros of $\sin(\pi z)$ at these points, so that $\zeta(z)$ is holomorphic for $\operatorname{Re} z \le 0$. Only at $z = 1$ is the zero of $\sin(\pi z)$ not balanced, and so there we find a simple pole. Since we do not really know $J(z)$, we calculate the residue with the help of $(*)$:

$$\Gamma(z) \zeta(z) = \int_0^1 \left(\frac{1}{e^t - 1} - \frac{1}{t} \right) t^{z-1} dt + \frac{1}{z - 1} + \int_1^\infty \frac{t^{z-1}}{e^t - 1} dt$$

holds if $x > 0$. The integrals converge uniformly in the neighborhood of $z = 1$, so that $\operatorname{Res}(\zeta(z), 1) = 1$.

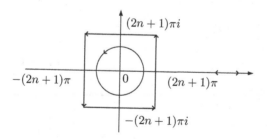

Figure 13.2

Finally we turn towards the functional equation: We transform the circle in our above curve C into a square centered at zero, with the sides of length $2(2n + 1)\pi$ parallel to the axes. The integrals along the real axis change to integrals from ∞ to $(2n + 1)\pi$ and back to ∞. We call this new curve C_n. By doing so we cross the poles of $(e^u - 1)^{-1}$ at the points $\pm 2k\pi i$, $k = 1,\ldots,n$. Therefore we have to apply the residue theorem. The residue of the integrand at $2k\pi$ is

$$\lim_{u \to 2k\pi i} (u - 2k\pi i)\frac{e^{(z-1)\log(-u)}}{e^u - 1} = e^{(z-1)(\log 2k\pi - i\pi/2)} = i(2k\pi)^{z-1}e^{-i\pi z/2},$$

since we have to choose the argument of the logarithm as before. The residue of the integrand at $-2k\pi i$ is similar:

$$\lim_{u \to -2k\pi i} (u + 2k\pi i)\frac{e^{(z-1)\log(-u)}}{e^u - 1} = e^{(z-1)(\log 2k\pi + i\pi/2)} = -i(2k\pi)^{z-1}e^{i\pi z/2}.$$

The sum of both residues is

$$-2i^2(2k\pi)^{z-1}\sin\frac{\pi z}{2},$$

and thus it follows that

$$J(z) = \int_{C_n} \frac{(-u)^{z-1}}{e^u - 1}\,du = 4\pi i(2\pi)^{z-1}\sin\frac{\pi z}{2}\sum_{k=1}^{n} k^{z-1}.$$

If we now let $n \to \infty$, then the integrals along the the real axis converge to zero. On the sides of the square $|u| = O(n)$ holds and for the denominator $|e^u - 1| \geq A > 0$, because on the vertical sides of the square e^u diverges to ∞ or converges to 0. On the horizontal sides the values of e^u are equal for all n and do not approach 1 due to the periodicity. So the integral has the order $O(n^x)$ on the square, as the sides have the length $O(n)$. In the case $x < 0$ the integral converges to 0, and so we obtain

$$-2i\sin(\pi z)\Gamma(z)\zeta(z) = -2i(2\pi)^z\sin\frac{\pi z}{2}\sum_{k=1}^{\infty} k^{z-1}.$$

The assumption $x < 0$ is also essential for the convergence of the last sum, this sum is then just $\zeta(1 - z)$ and we have proved

$$\zeta(1 - z) = (2\pi)^{-z}\frac{\sin \pi z}{\sin\frac{\pi z}{2}}\Gamma(z)\zeta(z) = 2^{1-z}\pi^{-z}\cos\frac{\pi z}{2}\Gamma(z)\zeta(z).$$

By the identity Theorem 9.15 this equation holds for all z, in which both sides are holomorphic, i.e., in $\mathbb{C}$ except the poles. $\qquad\square$

We further obtain from the functional equation: Since $\Gamma(z)$ has no zero for $x > 0$, due to Stirling's theorem, and also $\zeta(z)$ has no zero for $x > 1$, as we saw above, $\zeta(1 - z)$ has the same zeros $z = 2n+1$ as $\cos(\pi z/2)$. So the Riemann Zeta function has zeros at $z = -2n$, $n = 1, 2,\ldots$, these are the so-called trivial zeros. At $z = 0$ there is no zero, because in the functional equation the pole of $\zeta(z)$ at $z = 1$ is compensated by the zero of $\cos(\pi z/2)$. One can prove that Riemann's Zeta function has infinitely many zeros in the open strip $0 < x < 1$, and these are the non-trivial or essential zeros. Riemann has conjectured that all essential zeros are located on the *critical line* $x = \frac{1}{2}$, the famous *Riemann hypothesis*, a proof of which is still missing. But numerical computations have shown that up to large values of y all essential zeros are indeed located on the critical line.

13.3   Automorphic forms and functions

13.3.1   Automorphic forms and functions in $\mathbb{C}$

In this section we will handle the basic theory of automorphic functions and forms. They have an important significance in many areas of mathematics and its applications. Since the last century the theory of holomorphic automorphic forms is a major item of classical complex analysis as well as of analytic number theory.

For defining automorphic functions we first have to define *discrete groups* of Möbius transformations:

Definition 13.17 (Discrete group). We understand the Möbius transformations as the group $SL(2, \mathbb{C})$ made of all (2×2)-matrices with determinant 1. The norm of an element

$$T = \left(\begin{array}{cc} a & b \\ c & d \end{array} \right)$$

is defined by

$$\| T \| := \left(|a|^2 + |b|^2 + |c|^2 + |d|^2 \right)^{1/2}.$$

We call a subgroup $H \subset SL(2, \mathbb{C})$ *discrete*, if for all positive C there are only finitely many $T \in H$ with

$$\| T \| \leq C.$$

The group $SL(2, \mathbb{Z})$ is called the *special linear group*.

From the definition we have that all discrete groups are countable (or finite). At this point we give two important examples:

Example 13.18. a) The group $\mathbb{Z}$ is a discrete group to which belong the translations by an integer $z + m$, thus $a = d = 1$, $c = 0$, and $b = m \in \mathbb{Z}$.

b) Let ω_1, ω_2 be two $\mathbb{R}$-linear independent complex numbers, i.e., $\operatorname{Im}(\overline{\omega}_1 \omega_2) \neq 0$. They span a parallelogram in $\mathbb{C}$. The set $\Omega := \mathbb{Z}\omega_1 + \mathbb{Z}\omega_2$ is again a discrete group, its elements are the translations $z + m\omega_1 + n\omega_2$ with $m, n \in \mathbb{Z}$. This corresponds to a Möbius transformation with $a = d = 1$, $c = 0$ and $b = \omega := m\omega_1 + n\omega_2$. The group Ω spans a two-dimensional lattice in $\mathbb{C}$. The parallelogram spanned by ω_1, ω_2 is also called the *periodic parallelogram*.

These are of course some of the simplest examples of discrete groups. Any other parallelogram in the lattice can be chosen as a periodic parallelogram. We now define automorphic functions with the help of discrete groups:

Definition 13.19.   (i) A function f, holomorphic or meromorphic in a domain G, is called *automorphic* with respect to the discrete group H, if for all $T \in H$,

a) $T(G) \subset G$,

b) $f(T(z)) = f(z)$.

(ii) Functions which are automorphic with respect to the group $\mathbb{Z}$ are called *simple periodic functions.*

(iii) Functions which are automorphic with respect to the group Ω are called *double periodic functions* or *elliptic functions.*

The simple periodic functions include of course all elementary trigonometric functions, but we will not have a closer look at them, rather we examine the elliptic or double periodic functions more precisely.

Examples of simple or double periodic functions in one complex variable were first announced by G. EISENSTEIN (1823–1852) in the year 1847 [38] as well as in lectures from K. WEIERSTRASS in 1863. Eisenstein introduced the following function series:

Definition 13.20 (Meromorphic translative Eisenstein series). The series

$$
\epsilon_m^{(1)}(z;\mathbb{Z}) := \begin{cases} \frac{1}{z} + \sum\limits_{k\in\mathbb{Z}\backslash\{0\}} \left(\frac{1}{z+k} - \frac{1}{k}\right), & m = 1, \\ \sum\limits_{k\in\mathbb{Z}} \frac{1}{(z+k)^m}, & m \geq 2 \end{cases}
$$

and

$$
\epsilon_m^{(2)}(z;\Omega) := \begin{cases} \frac{1}{z^2} + \sum\limits_{\omega\in\Omega\backslash\{0\}} \left(\frac{1}{(z+\omega)^2} - \frac{1}{\omega^2}\right), & m = 2, \\ \sum\limits_{\omega\in\Omega} \frac{1}{(z+\omega)^m}, & m \geq 3 \end{cases}
$$

are called *meromorphic translative Eisenstein series,* where Ω and $\mathbb{Z}$ are the above described discrete groups.

These series are meromorphic functions for the corresponding translation group. At the points $z = k$, resp. $z = \omega$, are located poles of according order; similarly to the theorem of Mittag-Leffler the uniform convergence can be shown in $|z| < R$ apart from neighborhoods of the poles (we refer to the Exercises 13.4.7 and 13.4.8). If the uniform convergence is shown, then the simple, resp. double, periodicity follows since the translations $z + 1$, resp. $z + \omega_i$, $i = 1, 2$, give rise only to admissible reorderings of the series.

The elementary trigonometric functions can be constructed with the help of $\epsilon_m^{(1)}$, while the second type of series generates the double periodic Weierstraßelliptic functions with respect to one complex variable.

The elliptic functions are particularly characterized by their special value distribution. This distribution is described completely by the following three theorems of Liouville:

Theorem 13.21 (First Liouville theorem). *Every double periodic function f which is holomorphic in the whole complex plane is a constant function.*

Proof. Let f be an entire elliptic function and P an arbitrary periodic parallelogram of f. We remark that P is compact. Since f is holomorphic everywhere in the complex plane, f is bounded on P: There is $M \in \mathbb{R}$ with $|f(z)| \leq M$ for all $z \in P$. Let z be an arbitrary point in $\mathbb{C}$, then there is an $\omega \in \Omega$, so that $z + \omega \in P$. Hence f is bounded in the whole $\mathbb{C}$. According to Liouville's Theorem 7.33, f is a constant. $\qquad\square$

Theorem 13.22 (Second Liouville theorem). *The sum of the residues of an elliptic function f in a periodic parallelogram is zero.*

This theorem shows that an elliptic function cannot have one simple pole in its periodic parallelogram, then the sum of the residues would not be zero.

Proof. Without loss of generality we assume that f has no poles located on the boundary of the periodic parallelogram P. Otherwise in the following calculation we can consider a periodic parallelogram displaced at $\alpha \in \mathbb{C}$ having no pole of f on its boundary. Such a P always exists since the set of poles of f in P is discrete and finite. We obtain

$$2\pi i \sum_{c \in P} \mathrm{Res}(f, c) = \sum_{c \in P} \int_{|z-c|=\varepsilon} f(z)dz = \int_{\partial P} f(z)dz$$

$$= \int_0^{\omega_1} f(z)dz + \int_{\omega_1}^{\omega_1+\omega_2} f(z)dz + \int_{\omega_1+\omega_2}^{\omega_2} f(z)dz + \int_{\omega_2}^0 f(z)dz$$

$$= \int_0^{\omega_1} f(z)dz + \int_0^{\omega_2} f(z+\omega_1)dz + \int_{\omega_1}^0 f(z+\omega_2)dz + \int_{\omega_2}^0 f(z)dz = 0,$$

where we used the periodicity of f in the last step. $\qquad\square$

Theorem 13.23 (Third Liouville theorem). *Let P_0 be the union of the interior of the periodic parallelogram with the half-open lines $[0, \omega_1)$ and $[0, \omega_2)$, P_0 is called* cell *or a* fundamental area. *An elliptic function is defined completely by its values in P_0. Then the sum of the order of all a-points of f in P_0 is independent of $a \in \hat{\mathbb{C}}$ and is called the* order of f:

$$\sum_{c \in P_0} \mathrm{ord}(f - a; c) =: \mathrm{ord}\, f.$$

Proof. The order of an a-point of f, i.e., a point z with $f(z) = a$, is defined by the order of the zero of $f(z) - a$. Without loss of generality we assume again, that f has no a-points nor poles located on ∂P. They are thus located in the interior of P_0. Since f is meromorphic and double periodic, the function $g(z) := f'(z)/(f(z) - a)$ is also meromorphic and double periodic. If we now apply the previous theorem and the argument principle to the function g, we then obtain the assertion. $\qquad\square$

Example 13.24. Among the elliptic functions the function series $\epsilon_m^{(2)}(z; \Omega)$ are characterized by the fact that every elliptic function can be constructed through these series by a *finite* sum up to a constant.

We want to point out that the function $\epsilon_2^{(2)}(z;\Omega)$ is called the *Weierstraß ℘-function*, where the stylized $\wp$ is pronounced like a normal p. Since the transformation $z = -z$ is only a reordering of the series, $\wp(z)$ is an even function,

$$\wp(-z) = \wp(z),$$

and the derivative

$$\wp'(z) = -2 \sum_{\omega \in \Omega} \frac{1}{(z+\omega)^3}$$

is then odd. We want to derive the differential equation of the $\wp$-function as a first step into the extensive theory of elliptic functions: The difference $\wp(z) - z^{-2}$ is holomorphic in a neighborhood of zero and it has, for sufficiently small $|z|$, the expansion

$$\wp(z) - \frac{1}{z^2} = \frac{1}{20}g_2 z^2 + \frac{1}{28}g_3 z^4 + O(z^6).$$

It is not hard to see (cf. Exercise 13.4.9) that

$$g_2 = 60 \sum_{\omega}' \frac{1}{\omega^4}, \qquad g_3 = 140 \sum_{\omega}' \frac{1}{\omega^6},$$

where $\sum'$ stand for the summation over all $\omega \neq 0$. After differentiation we obtain

$$\wp'(z) = \frac{-2}{z^3} + \frac{1}{10}g_2 z + \frac{1}{7}g_3 z^3 + O(z^5).$$

Calculating $\wp^3$ and $\wp'^2$ we find

$$\wp^3(z) = \frac{1}{z^6} + \frac{3}{20}g_2\frac{1}{z^2} + \frac{3}{28}g_3 + O(z^2),$$

$$\wp'^2(z) = \frac{4}{z^6} - \frac{2}{5}g_2\frac{1}{z^2} - \frac{4}{7}g_3 + O(z^2)$$

and thus

$$\wp'^2(z) - 4\wp^3(z) + g_2\wp(z) + g_3 = O(z^2).$$

The function on the left-hand side is holomorphic in the neighborhood of the origin and it is also an elliptic function with poles at most at $\omega \in \Omega$. Hence it cannot have poles and is constant according to Theorem 13.21. If we let $z \to 0$ the right hand side of the last equation is zero, thus we have obtained:

Proposition 13.25. *The Weierstraß℘-function satisfies the differential equation*

$$\wp'^2(z) = 4\wp^3(z) - g_2\wp(z) - g_3.$$

Example 13.26. A further family of classical Eisenstein series is

$$G_m(z) = \sum_{(c,d) \in \mathbb{Z} \times \mathbb{Z} \setminus \{(0,0)\}} (cz+d)^{-m}, \qquad \mathrm{Im}\, z > 0, \quad m \geq 4, \; m \text{ even.}$$

These series are holomorphic functions in the upper half-plane:

$$H^+(\mathbb{C}) := \{z \in \mathbb{C} \mid \operatorname{Im} z > 0\}.$$

Obviously they have singularities located at the rational points on the real axis. We can transform these functions $G_m(z)$ at every $z \in H^+(\mathbb{C})$ as follows:

$$f(T(z)) = (cz + d)^{-m} f\left(\frac{az + b}{cz + d}\right),$$

where T is the Möbius transformation associated with the matrix

$$M = \begin{pmatrix} a & b \\ c & d \end{pmatrix} \in SL(2, \mathbb{Z}).$$

We call the functions f with such a transformation behavior an *automorphic form of weight m* (here) with respect to the complete linear group $SL(2, \mathbb{Z})$. They are indeed classical examples for *modular forms* and play an important role first of all in number theory. This is justified by the fact that their Fourier expansion contains sums of powers of divisors and the Riemann Zeta function:

Theorem 13.27. *Let $m \geq 4$ be an even number; the Eisenstein series $G_m(z)$ has the following Fourier representation in the upper half-plane:*

$$G_m(z) = 2\zeta(m) + 2\frac{(2\pi i)^m}{(m-1)!} \sum_{n=1}^{\infty} \tau_{m-1}(n) e^{2\pi i n z},$$

where $\zeta(m)$ is the Riemann Zeta function and τ_m are sums of the powers of divisors,

$$\tau_m(n) = \sum_{r \geq 1, r \mid n} r^m.$$

Proof. We can reorder the series $G_m(z)$ as follows:

$$G_m(z) = 2\zeta(m) + 2\sum_{j=1}^{\infty} \epsilon_m^{(1)}(jz; \mathbb{Z}),$$

where $\epsilon_m^{(1)}(z; \mathbb{Z})$ are the series according to definition 13.20. The next step is the calculation of the Fourier representation of $\epsilon_m^{(1)}$. In the case $m = 2$ we obtain, according to Example 12.21,

$$\sum_{n \in \mathbb{Z}} (z + n)^{-2} = \left(\frac{\pi}{\sin(\pi z)}\right)^2 = \left(\frac{2\pi i}{e^{\pi i z} - e^{-\pi i z}}\right)^2 = (2\pi i)^2 e^{2\pi i z} \left[\frac{1}{1 - e^{2\pi i z}}\right]^2,$$

and if $\operatorname{Im} z > 0$,

$$= (2\pi i)^2 e^{2\pi i z} \sum_{r=1}^{\infty} r e^{2\pi i z(r-1)} = (2\pi i)^2 \sum_{r=1}^{\infty} r e^{2\pi i r z},$$

where we applied the derivative of the geometric series. By termwise differentiation and by induction we obtain for all integers $m \geq 2$ the equation

$$\epsilon_m^{(1)}(z; \mathbb{Z}) = \frac{(2\pi i)^m}{(m-1)!} \sum_{r=1}^{\infty} r^{m-1} e^{2\pi i r z}.$$

We plug this formula into the equation for $G_m(z)$ and finally obtain

$$
\begin{aligned}
G_m(z) &= 2\zeta(m) + 2\frac{(2\pi i)^m}{(m-1)!} \sum_{j=1}^{\infty} \sum_{r=1}^{\infty} r^{m-1} e^{2\pi i r j z} \\
&= 2\zeta(m) + 2\frac{(2\pi i)^m}{(m-1)!} \sum_{n=1}^{\infty} \left(\sum_{r \geq 1, r|n} r^{m-1} \right) e^{2\pi i n z}.
\end{aligned}
$$

$\square$

Remark 13.28. A further systematic method for constructing holomorphic modular forms with respect to the specialized linear group is to use suitable and in the upper half-plane bounded holomorphic functions $f : H^+(\mathbb{C}) \to \mathbb{C}$. These functions must belong to the class of automorphic forms with weight k of the translation group $\mathbb{Z}$. Then we can sum the expressions $f(T(z))$ over a complete representative system of right cosets of $SL(2, \mathbb{Z})$ modulo the translation invariant group $\mathbb{Z}$, i.e.,

$$\sum_{T \in SL(2,\mathbb{Z})/\mathbb{Z}} (cz + d)^{-m} f(T(z)).$$

If k is sufficiently large this series converges and delivers then modular forms with respect to the complete specialized linear group $SL(2, \mathbb{Z})$. Function series of this type are often called *Poincaré series* in an extended meaning. The simplest non-trivial example results from the setting $f = 1$. Then Poincaré's series are equal to the classical Eisenstein series $G_m(z)$ up to a normalization factor. On the other hand all modular forms with positive even integer weight $k \geq 4$ with respect to the group $SL(2, \mathbb{Z})$ can be constructed from the Eisenstein series G_m.

The method just exposed has the advantage that it can be applied in a simple way to other discrete groups, like important congruence groups in number theory, and that it thereby provides further examples of automorphic forms.

13.3.2 Automorphic functions and forms in $C\ell(n)$

The systematic development of the general theory of holomorphic modular forms with respect to one complex variable was mainly developed by H. Poincaré, F. Klein and R. Fricke [116, 71].

O. Blumenthal (1904), C.L. Siegel (in the 1930s) and their students began to study holomorphic functions of several complex variables. In 1949 H. Maaß [99] introduced a higher dimensional type of non-analytic automorphic forms (Maaß' waveforms), that are eigenfunctions of the hyperbolic Laplacian. These two higher dimensional generalizations raised high interest and are still in the focus of current research.

Variations of holomorphic Siegel modular forms of quaternionic symplectic groups, and in general of orthogonal groups, are still regarded with increasing interest. We refer therefore to A. Krieg [84] as well as E. Freitag and C.F. Hermann [41].

In another direction E. Kähler [68], J. Elstrodt, F. Grunewald and J. Mennicke [39, 40] as well as A. Krieg [85, 86], V. Gritsenko [51] and other authors study complex-valued generalizations of the non-analytic forms of discrete subgroups of the Vahlen group (group of the Vahlen matrices according to Definitions 6.16 and 6.18) in higher dimensional half-spaces of the Euclidean space $\mathbb{R}^n$. Automorphic forms on m-times Cartesian products of quaternionic half-spaces are of great interest, such as in works of O. Richter and H. Skogman [123, 124].

None of the above listed higher dimensional variations of automorphic forms are solutions of the Cauchy–Riemann differential equations.

In works of A.C. Dixon [36], R. Fueter [46, 48, 49, 50] and J. Ryan [125] we find the first contributions to holomorphic generalizations of the special double periodic Weierstraßelliptic functions.

A systematic theory of holomorphic automorphic forms for the general arithmetic subgroup of the Vahlen group, including generalizations of the modular group and their congruence groups, is developed in [81].

This subsection is based on the accomplishment of R.S. Kraußhar (cf. [81]), who in his habilitation thesis and the following works established the basis of such a generalization. We only want to give a short introduction into the central aspects of this theory. First we study the simplest type of discrete subgroups of the Vahlen group in $\mathbb{R}^{n+1}$, being the translation groups, which operate on $\mathbb{R}^{n+1}$: For each arbitrary set of p $\mathbb{R}$-linear independent vectors $\omega_1, \ldots, \omega_p \in \mathbb{R}^{n+1}$, where $p \in \{1, \ldots, n+1\}$, $\Omega_p = \mathbb{Z}\omega_1 + \cdots + \mathbb{Z}\omega_p$ is a p-dimensional lattice in $\mathbb{R}^{n+1}$. Since the associated translation group $\mathcal{T}(\Omega_p)$ is generated by the matrices

$$\begin{pmatrix} 1 & \omega_1 \\ 0 & 1 \end{pmatrix}, \ldots, \begin{pmatrix} 1 & \omega_p \\ 0 & 1 \end{pmatrix},$$

it operates discontinuously on $\mathbb{R}^{n+1}$ according to its Möbius transformation $T(x) = x + \omega_j$, $j = 1, \ldots, p$. This group is discrete since only finitely many T with limited norm exist. So we can say that we obtain meromorphic automorphic functions with respect to a general discrete translation group $\mathcal{T}(\Omega_p)$ by adding up all fundamental solutions of the $\bar{\partial}$-operator $\mathcal{Q}_0(x) = \overline{x}/|x|^{n+1}$, resp. its derivatives $\mathcal{Q}_{\mathbf{m}}(x) = (-1)^{|\mathbf{m}|} \nabla^{\mathbf{m}}(\overline{x}/|x|^{n+1})/\mathbf{m}!$, according to Definition 7.26 at the points $x + \omega_i$ over the whole translation group. The following definition provides an exact description:

Definition 13.29 (Meromorphic Eisenstein series with respect to translation subgroups). Let $p \in \{1, 2, \ldots, n+1\}$ and let $\omega_1, \ldots, \omega_p$ be $\mathbb{R}$-linear independent vectors in $\mathbb{R}^{n+1}$, furthermore let $\Omega_p = \mathbb{Z}\omega_1 + \cdots + \mathbb{Z}\omega_p$ be the corresponding lattice.

We define the associated meromorphic translative Eisenstein series by

$$\epsilon_{\mathbf{m}}^{(p)}(x;\Omega_p) = \sum_{\omega \in \Omega_p} \mathcal{Q}_{\mathbf{m}}(x+\omega),$$

where $\mathbf{m}$ is a multiindex and $|\mathbf{m}| \geq \max\{0, p - n + 1\}$. If $p = n$, $\mathbf{m} = \mathbf{0}$ as well as $p = n + 1$, $|\mathbf{m}| = 1$ we set

$$\epsilon_{\mathbf{m}}^{(p)}(x;\Omega_p) = \mathcal{Q}_{\mathbf{m}}(x) + \sum_{\omega \in \Omega_p \setminus \{0\}} \left[\mathcal{Q}_{\mathbf{m}}(x+\omega) - \mathcal{Q}_{\mathbf{m}}(\omega) \right].$$

Also in the remaining case $p = n+1$ and $\mathbf{m} = \mathbf{0}$, holomorphic Eisenstein series can be constructed [74], but only those which have at least two different singularities in every periodic box.

The convergence of this series is proved in ([81], Chapter 2). We use on the one hand the estimate in Proposition 7.27,

$$|\mathcal{Q}_{\mathbf{m}}(x)| \leq \frac{C_{\mathbf{m}}}{|x|^{n+|\mathbf{m}|}},$$

and on the other one a lemma of Eisenstein, [38], stating that the series

$$\sum_{(m_1,\ldots,m_p)\in\mathbb{Z}^p\setminus\{0\}} |m_1\omega_1 + \cdots + m_p\omega_p|^{-(p+\alpha)}$$

converges if and only if $\alpha > 0$. Hence the series in the definition converge absolutely except at the poles. The limit function is a meromorphic function with poles located at the lattice points. The functions are p-times periodic, because a translation of the argument by ω_i is only a renumbering of the series, which is allowed in view of the absolute convergence.

The series $\epsilon_{\mathbf{m}}^{(p)}$ with $p < n$ generalize the series from Definition 13.20 to the Clifford analysis. They provide an elementary foundation for constructing meromorphic generalizations of various classical trigonometric function, as described in [81], Chapter 2. The special series $\epsilon_0^{(p)}$ is a p-times periodic meromorphic generalization of the classical cotangent. Thus the p-times periodic meromorphic generalizations of the classical tangent, cosecant, and secant as well as the squared cosecant and secant are constructed from the cotangent by superposition:

Example 13.30. Let $\mathcal{V}_p(2)$ denote the canonical representative system of the factor module $\Omega_p/2\Omega_p$. We obtain the meromorphic generalization of the tangent, cosecant, and secant in the following way:

$$\tan^{(p)}(x) := -\sum_{v\in\mathcal{V}_p\setminus\{0\}} \epsilon_0^{(p)}\left(x + \frac{v}{2}\right),$$

$$\csc^{(p)}(z) := \frac{1}{2^{n-2}}\epsilon_0^{(p)}\left(\frac{x}{2}\right) - \epsilon_0^{(p)}(x),$$

$$\sec^{(p)}(z) := \sum_{v\in\mathcal{V}_p\setminus\{0\}} \csc^{(p)}\left(x + \frac{v}{2}\right).$$

From the partial derivatives of $\csc^{(p)}$ and $\sec^{(p)}$ we get the meromorphic generalizations of the squared cosecant and secant.

Every p-times periodic meromorphic function with $p < n + 1$ can be represented by a finite sum of p-times periodic cotangents $\epsilon_0^{(p)}$ and (resp. or) their partial derivatives apart from an entire function. This can be proven as a consequence of Theorem 12.26 of Mittag-Leffler.

In the special case $p = n + 1$ the series $\epsilon_\mathbf{m}^{(n+1)}$ are meromorphic generalizations of Weierstraß' elliptic functions. Especially $\epsilon_\mathbf{m}^{(n+1)}$, where $|\mathbf{m}| = 1$, defines the *generalized Weierstraß$\wp$-function*. This was examined first by A.C. Dixon, R. Fueter, and J. Ryan.

Every $(n + 1)$-times periodic meromorphic function in $\mathbb{R}^{n+1}$ can be represented by finitely many series $\epsilon_\mathbf{m}^{(n+1)}$ apart from a constant. The study of the generalized elliptic functions was also pursued by Krausshar in [81]. In particular this was done from number theoretical and complex analytic viewpoints, including explicit applications to the theory of special L^2-spaces.

One of the most fundamental properties is that all functions $\epsilon_\mathbf{m}^{(p)}$ can be characterized by special functional equations. These equations generalize the known cotangent duplication formula. All function series $\epsilon_\mathbf{m}^{(p)}$ (except in the case $p = n$, where the index is $\mathbf{m} = \mathbf{0}$) satisfy the following multiplication formula:

Theorem 13.31. *Let Ω_p be a p-dimensional lattice in $\mathbb{R}^{n+1}$ and let $\mathbf{m}$ be a multi-index. In addition if $p = n + 1$, we assume $|\mathbf{m}| \geq 1$. In all cases*

$$r^{n+|\mathbf{m}|} \epsilon_\mathbf{m}^{(p)}(rx) = \sum_{v \in \mathcal{V}_p(r)} \epsilon_\mathbf{m}^{(p)}\left(x + \frac{v}{r}\right),$$

holds, where $\mathcal{V}_p(r)$ is the canonical representative system of $\Omega_p/r\Omega_p$ and $r \geq 2$ is a natural number.

We remark that the canonical representative system $\mathcal{V}_p(r)$ is composed of the quantities $m_1\omega_1 + \cdots + m_p\omega_p$, where m_i, $0 \leq m_i < r$, are integers. And so $\mathcal{V}_p(r)$ has exactly r^p elements.

Proof (outline). In the cases $|\mathbf{m}| \geq \max\{0, p - n + 1\}$ the formula can be shown by a direct reordering argument. In the case $p = n$ and $\mathbf{m} = \mathbf{0}$ we use the following identity of arrangement of antipodal lattice points:

$$\sum_{v \in \mathcal{V}_p(r) \backslash \{0\}} \sum_{w \in \Omega_p \backslash \{0\}} [\mathcal{Q}_0(rw + v) - \mathcal{Q}_0(rw)] = - \sum_{v \in \mathcal{V}_p(r) \backslash \{0\}} \mathcal{Q}_0(v).$$

This formula and some clever arguments for reordering give the claim.

An argument of integration shows that the theorem holds also in the remaining cases apart from a paravector-valued constant $C \in \mathbb{R}^{n+1}$. Finally the *Legendre relation* for generalized elliptic functions leads to $C = 0$. So the theorem is shown in all cases. We refer the reader to [81], Chapter 2.5, for an explicit proof.   □

Also the converse holds:

Theorem 13.32. *Let Ω_p be a p-dimensional lattice in $\mathbb{R}^{n+1}$ and $\mathbf{m}$ a multiindex with $|\mathbf{m}| \geq 1$ in the case $p = n + 1$. Moreover let $g : \mathbb{R}^{n+1} \to C\ell(n)$ be a meromorphic function with the principal parts $\mathcal{Q}_{\mathbf{m}}(x - \omega)$ at every lattice point $\omega \in \Omega_p$. If g satisfies the functional equation*

$$r^{n+|\mathbf{m}|} g(rx) = \sum_{v \in \mathcal{V}_p(r)} g\left(x + \frac{v}{r}\right),$$

where $2 \leq r \in \mathbb{N}$, then there is a Clifford number $C \in C\ell(n)$ and $g(x) = \epsilon_{\mathbf{m}}^{(p)}(x) + C$ for all $x \in \mathbb{R}^{n+1} \backslash \Omega_p$.

Proof. First we consider the function $s(x) := g(x) - \epsilon_{\mathbf{m}}^{(p)}(x)$, which is holomorphic in the whole space $\mathbb{R}^{n+1}$. From the previous theorem it follows that

$$r^{n+|\mathbf{m}|} s(rx) = \sum_{v \in \mathcal{V}_p(r)} s\left(x + \frac{v}{r}\right), \quad s(0) = s_0, \tag{$*$}$$

where $s_0 \in C\ell(n)$. We assume that s is not a constant and define $\beta := |\omega_1| + \cdots + |\omega_p|$. According to the maximum principle in Theorem 7.32 there is a point $c \in \partial B_{r\beta}(0)$, so that

$$|s(x)| < |s(c)|$$

for all $x \in B_{r\beta}(0)$. Moreover

$$\left| \frac{c + v}{r} \right| = \left| \frac{1}{r}\left(c + \sum_{i=1}^{p} \alpha_i \omega_i\right) \right| < \beta + \beta \leq r\beta$$

holds for all $0 \leq \alpha_i < r$. Hence from $(*)$ by setting $rx = c$ we find

$$r^{n+|\mathbf{m}|} |s(c)| = \left| \sum_{v \in \mathcal{V}_p(r)} s\left(\frac{c+v}{r}\right) \right| \leq \sum_{v \in \mathcal{V}_p(r)} \left| s\left(\frac{c+v}{r}\right) \right| < r^p |s(c)| \leq r^{n+|\mathbf{m}|} |s(c)|.$$

But this is a contradiction. The assumption that s is not a constant is false, thus proving the theorem. $\qquad\square$

A certain analogue to the complex case is that the generalized elliptic functions play a major role in the value distribution. Both next theorems are a direct generalization of Liouville's first two theorems.

Theorem 13.33. *Every entire $(n + 1)$-times periodic function is constant.*

Proof. The theorem can be proved in strict analogy with the complex case in Theorem 13.21 using the higher dimensional Liouville's theorem, Proposition 7.33. $\qquad\square$

Theorem 13.34. *The sum of the residues of an $(n+1)$-times periodic meromorphic function f vanishes in a periodic box.*

Proof. We perform the proof for the special case that f has only isolated singularities. However the theorem holds also in the general case, where the residues need to be applied in the sense of Leray–Norguet, resulting then in a markedly more technical effort. So we assume that f has only isolated singularities. Without loss of generality we furthermore assume that the fundamental box P spanned by the vectors $\omega_1, \ldots, \omega_{n+1}$ has no poles located on its boundary. An $(n+1)$-dimensional box is composed of exactly $2(n+1)$ different n-dimensional boundary planes B_j. In addition we define B_j', $j = 1, \ldots, n+1$, where $B_j' = B_j + \omega_j$ is that boundary plane which arises by a translation of B_j by the vector ω_j. The normal vectors of these planes are outwards oriented. Hence B_j and B_j' have opposite orientation. According to the residue Theorem 12.43 for a left-holomorphic function we have

$$
\sigma_n \sum_{c \in P} \operatorname{Res}(f, c) = \int_{\partial P} dx^* f(x) = \sum_{j=1}^{n+1} \left(\int_{B_j} dx^* f(x) + \int_{B_j'} dx^* f(x) \right)
$$

$$
= \sum_{j=1}^{n+1} \left(\int_{B_j} dx^* f(x) - \int_{B_j} dx^* f(x + \omega_j) \right) = 0,
$$

where we used the periodicity of f and the opposite orientation of B_j and B_j'.   □

In a special case an analogy to the third Liouville's theorem could be proved recently [62]. The difficulty while transferring the theorem into higher dimension is the fact that no quotients of meromorphic functions are allowed:

Theorem 13.35. *Let f be an $(n+1)$-times periodic paravector-valued meromorphic function having only isolated poles. Moreover we assume for $a \in \mathbb{R}^{n+1}$ that f has only isolated a-points. Let P be a periodic box with the property that neither poles nor a-points of f are located on its boundary planes. We denote the poles inside P by $b_1, \ldots, b_\nu$. Let $\delta > 0$ be sufficiently small, so that the open punctured ball $B_\delta(b_i) \backslash \{b_i\}$ contains neither poles nor a-points. Then with the order of a zero given by Definition 12.9,*

$$
\sum_{c \in P \backslash \{b_1, \ldots, b_\nu\}} \operatorname{ord}(f - a; c) = -\sum_{i=1}^{\nu} p(f - a; b_i)
$$

holds, where

$$
p(f - a; b_i) := \frac{1}{\sigma_n} \int_{F_i} \mathcal{Q}_0(y) dy^*,
$$

and $F_i := (f - a)(S_\delta(b_i))$ is the image of $S_j(b_i)$ under $f - a$.

Proof. In contrast to the complex case we cannot deduce this assertion from the previous one since the quotient of holomorphic functions need not be holomorphic in $\mathbb{R}^{n+1}$. For proving the theorem we have to calculate explicitly the appearing integrals. There are only finitely many a-points inside of P in view of the isolation of the a-points. We denote them by $t^1, \ldots, t^\mu$. Then an $\varepsilon > 0$ exists, so that for every $j \in \{1, \ldots, \mu\}$ the balls $B_\varepsilon(t^j)$ are pairwise disjoint and moreover ∂P and the $B_\delta(b_i)$ have no common points. As a result in $P \backslash \bigcup\limits_{j=1}^{\mu} B_\varepsilon(t^j)$ no a-points are left. Henceforth we label the boundary planes of P again as $B_1, B'_1, \ldots, B_n, B'_n$. By setting $H_j := (f - a)(S_\varepsilon(t^j))$ we calculate

$$\sum_{c \in P \backslash \{b_1, \ldots, b_\nu\}} \mathrm{ord}(f - a; c) + \sum_{i=1}^{\nu} p(f - a; b_i)$$

$$= \frac{1}{\sigma_n} \sum_{j=1}^{\mu} \int\limits_{H_j} \mathcal{Q}_0(y) dy^* + \sum_{i=1}^{\nu} p(f - a; b_i) = \frac{1}{\sigma_n} \int\limits_{(f-a)(\partial P)} \mathcal{Q}_0(y) dy^*$$

$$= \frac{1}{\sigma_n} \sum_{k=1}^{n+1} \left(\int\limits_{(f-a)(B_k)} \mathcal{Q}_0(y) dy^* + \int\limits_{(f-a)(B'_k)} \mathcal{Q}_0(y) dy^* \right)$$

$$= \frac{1}{\sigma_n} \sum_{k=1}^{n+1} \left(- \int\limits_{(f-a)(B'_k)} \mathcal{Q}_0(y) dy^* + \int\limits_{(f-a)(B'_k)} \mathcal{Q}_0(y) dy^* \right) = 0.$$

In the last step of the calculation we used the periodicity of f since $f(x) - a$ and $f(x - \omega_k) - a$ have the same range, but the orientation of B_k is opposite to the one of B'_k. $\qquad \square$

The generalized Weierstraß elliptic functions in Example 13.30 are moreover used as generating functions for higher dimensional meromorphic generalizations of the classical Eisenstein series G_m:

Definition 13.36. Let $\mathbf{m}$ be a multiindex, then for all odd $|\mathbf{m}| \geq 3$ the meromorphic Eisenstein series associated to the generalized Weierstraß $\wp$-function according to Definition 13.29 is defined by

$$G_{\mathbf{m}}(x) := \sum_{(\alpha, \omega) \in \mathbb{Z} \times \Omega_n \backslash \{(0,0)\}} \mathcal{Q}_{\mathbf{m}}(\alpha x + \omega), \quad x \in H^+(\mathbb{R}^{n+1}) := \{x \in \mathbb{R}^{n+1} : x_n > 0\}.$$

Let now Ω_n be an n-dimensional lattice in $\mathrm{span}_{\mathbb{R}}\{e_0, \ldots, e_{n-1}\}$. The lattice should be non-degenerate, i.e., the ω_i should be $\mathbb{R}$-linear independent.

To show the convergence of the series we apply the estimate in Proposition 7.27 to the partial derivatives of $\mathcal{Q}_0$,

$$|\mathcal{Q}_{\mathbf{m}}(x)| \leq \frac{C_{\mathbf{m}}}{|x|^{n+|\mathbf{m}|}}.$$

If we decompose x into $x = \mathbf{y} + x_n e_n$, where $\mathbf{y} = x_0 + x_1 e_1 + \cdots + x_{n-1} e_{n-1}$, then for $x \in H^+(\mathbb{R}^{n+1})$ we always find $x_n > 0$. In the next step with the help of a classical argument of compactification we can show that for every $\varepsilon > 0$ there is a real $\rho > 0$, so that all $(\alpha, \omega) \in \mathbb{Z} \times \Omega_n$ uniformly satisfy the equation

$$|\alpha x + \omega| \geq \rho |\alpha e_n + \omega|.$$

Here x is an element of the corresponding vertical stripe

$$V_\varepsilon(H^+(\mathbb{R}^{n+1})) := \left\{ x = \mathbf{y} + x_n e_n \in H^+(\mathbb{R}^{n+1}) \ : \ |\mathbf{y}| \leq \frac{1}{\varepsilon}, x_n \geq \varepsilon \right\}.$$

Finally we obtain the estimate

$$\sum_{(\alpha,\omega)\in\mathbb{Z}\times\Omega_n\backslash\{(0,0)\}} |\mathcal{Q}_{\mathbf{m}}(\alpha x + \omega)|$$

$$\leq \qquad \rho^{-(n+|\mathbf{m}|)} C_{\mathbf{m}} \sum_{(\alpha,\omega)\in\mathbb{Z}\times\Omega_n\backslash\{(0,0)\}} |\alpha e_n + \omega|^{-(|\mathbf{m}|+n)}.$$

The series occurring in the last row is an *Epstein Zeta function*, from which we obtain the convergence of our series in the case $|\mathbf{m}| \geq 2$.

In case of a multiindex $\mathbf{m}$ with even $|\mathbf{m}|$ the series vanishes identically. But this is not the case for all indices $\mathbf{m}$ with odd length $|\mathbf{m}| \geq 3$, a fact which we now investigate.

As in the complex case these function series have interesting Fourier expansions from the number theoretical point of view, since in the Fourier expansion occur representation numbers of multiple sums of powers of divisors and a vector-valued generalization of Riemann's Zeta function.

Definition 13.37 (Generalized Riemann's Zeta function in $C\ell(n)$). If $|\mathbf{m}|$ is odd we then define the series

$$\zeta_M^{\Omega_n}(\mathbf{m}) = \frac{1}{2} \sum_{\omega\in\Omega_n\backslash\{0\}} \mathcal{Q}_{\mathbf{m}}(\omega),$$

which is called *Riemann's Zeta function in $C\ell(n)$*. The subscript M should evoke a connection to 'meromorphic' to distinguish this one from the many other generalized Zeta functions.

Further information can be found in [81], Chapter 2.4. From Proposition 7.27 it follows that the above defined Zeta function is paravector-valued. We want to point out explicitly that the variable $\mathbf{m}$, corresponding to the z in the classical Riemann Zeta function, is not continuous since the function is only defined in the case of multiindices $\mathbf{m}$. Hence further research is needed. We now turn to the Fourier expansion:

Theorem 13.38 (Fourier expansion). *The Eisenstein series $G_{\mathbf{m}}$ belonging to the orthonormal lattice, where $\mathbf{m} = (0, m_1, \ldots, m_n, 0)$ and $|\mathbf{m}| \geq 3$ is odd, have the following Fourier expansion in the upper half-space*

$$G_{\mathbf{m}}(x) = 2\zeta_M^{\Omega_n}(\mathbf{m}) + \sigma_n(2\pi i)^{|\mathbf{m}|} \sum_{\mathbf{s}\in\mathbb{Z}^n\backslash\{0\}} \tau_{\mathbf{m}}(\mathbf{s})(ie_n + \frac{\mathbf{s}}{|\mathbf{s}|})e^{2\pi i\langle\mathbf{s},\mathbf{x}\rangle}e^{-2\pi|\mathbf{s}|x_n}.$$

As usual σ_n *is the area of the unit sphere* S^n *in* $\mathbb{R}^{n+1}$. *Furthermore* $\tau_{\mathbf{m}}(\mathbf{s})$ *is defined by*

$$\tau_{\mathbf{m}}(\mathbf{s}) = \sum_{\mathbf{r}|\mathbf{s}} \mathbf{r}^{\mathbf{m}},$$

where $\mathbf{r}|\mathbf{s}$ *means that a natural number* a *exists with* $a\mathbf{r} = \mathbf{s}$.

At this point we see a nice correspondence between the Fourier expansion of the classical Eisenstein series (Definition 13.20) and the structure of the Fourier expansion of the higher dimensional ones in Definition 13.29. Instead of the ordinary Riemann Zeta function we have the vector valued Zeta function from Definition 13.37. We want to emphasize that the generalized Zeta function is compatible downwardly to the ordinary one, because in case of two dimensions if we rewrite Definition 13.37 and use paravector formalism we obtain the classical Zeta function.

Furthermore the Zeta function occurring here has a close relation to Epstein's Zeta function shown in [75]. In the closely related work [27] it was more precisely shown that every vector component of the generalized Zeta function can be represented by a finite sum of scalar-valued Dirichlet series of the form

$$\delta(P(\cdot), s) = \sum_{\mathbf{g} \in \mathbb{Z}^n \setminus \{0\}} P(\mathbf{g})(g_1^2 + \cdots + g_n^2)^{-s}.$$

Here P is a real-valued polynomial with respect to $g_1, \ldots, g_n$, and s is a complex number which satisfies the condition $\mathrm{Re}(s) - \deg(P) > (n-1)/2$. The polynomial P was calculated explicitly in [27].

The sums of divisors occurring in the Fourier expansion of the complex Eisenstein series according to Definition 12.38 are generalized by the term $\tau_{\mathbf{m}}$ of the theorem. And again this term $\tau_{\mathbf{m}}$ can be expressed by sums of divisors:

$$\sigma_{\mathbf{m}}(\mathbf{s}) = \mathbf{s}^{\mathbf{m}} \sigma_{-|\mathbf{m}|}(\gcd(s_1, \ldots, s_n)).$$

The holomorphic plane wave function comes in as a consequence of holomorphy as it is the natural generalization of the classical exponential function.

Moreover we notice that infinitely many Fourier coefficients do not vanish. Hence the series $G_{\mathbf{m}}(z)$ represent non-trivial functions for all multiindices $\mathbf{m}$, where $|\mathbf{m}| \geq 3$ is odd.

Proof (outline). First we expand the partial series $\epsilon_{\mathbf{m}}^{(n)}(x; \Omega_n)$ ($|\mathbf{m}| \geq 2$) into a Fourier series on $H^+(\mathbb{R}^{n+1})$,

$$\sum_{\mathbf{r} \in \mathbb{Z}^n} \alpha_f(\mathbf{r}, x_n) e^{2\pi i \langle \mathbf{r}, \mathbf{x} \rangle}.$$

By a direct calculation we find

$$\alpha_f(0, x_n) = \int_{[0,1]^n} \left(\sum_{\mathbf{m} \in \mathbb{Z}^n} \mathcal{Q}_{\mathbf{m}}(x + \mathbf{m}) \right) d\sigma = 0.$$

In the case $\mathbf{r} \neq \mathbf{0}$ we use iterated partial integration; we integrate $Q_{\mathbf{m}}$ until it becomes Q_0 and then differentiate the exponential terms. After finitely many iterations we obtain the following result:

$$\alpha_f(\mathbf{r}, x_n) = (2\pi i)^{|\mathbf{m}|} \mathbf{r}^{\mathbf{m}} \int_{\mathbb{R}^n} Q_0(x) e^{-2\pi i \langle \mathbf{r}, \mathbf{x} \rangle} \, d\sigma.$$

The value of the remaining integral is known and can be found in [136, 92, 27] or calculated with the residue theorem. So we get

$$\int_{\mathbb{R}^n} Q_0(z) e^{-2\pi i \langle \mathbf{r}, \mathbf{x} \rangle} \, d\sigma = \frac{\sigma_n}{2} \left(i e_n + \frac{\mathbf{r}}{|\mathbf{r}|} \right) e^{-2\pi |\mathbf{r}| x_n}.$$

From $x_n > 0$ we find the formula

$$\epsilon_{\mathbf{m}}^{(n)}(x; \Omega_n) = \frac{\sigma_n}{2} \sum_{\mathbf{r} \in \mathbb{Z}^n \setminus \{0\}} (2\pi i)^{|\mathbf{m}|} \mathbf{r}^{\mathbf{m}} \left(i e_n + \frac{\mathbf{r}}{|\mathbf{r}|} \right) e^{-2\pi |\mathbf{r}| x_n} e^{2\pi i \langle \mathbf{r}, \mathbf{x} \rangle}.$$

We can now reorder the series $G_{\mathbf{m}}(x)$ as follows:

$$G_{\mathbf{m}}(x) = 2\zeta_M^{\Omega_n}(\mathbf{m}) + 2 \sum_{a=1}^{\infty} \sum_{\mathbf{m} \in \mathbf{Z}_n} \epsilon_{\mathbf{m}}(ax; \Omega_n),$$

where we use the notation introduced above. Finally we put $\epsilon_{\mathbf{m}}^{(n)}(ax; \Omega_n)$ in the reordered series,

$$G_{\mathbf{m}}(x) = 2\zeta_M^{\Omega_n}(\mathbf{m}) + 2(2\pi i)^{|\mathbf{m}|} \left(\frac{\sigma_n}{2} \right) \sum_{a=1}^{\infty} \sum_{\mathbf{r} \in \mathbb{Z}^n} \mathbf{r}^{\mathbf{m}} \left(i e_n + \frac{\mathbf{r}}{|\mathbf{r}|} \right) e^{2\pi i \langle a\mathbf{r}, \mathbf{x} \rangle} e^{-2\pi |a\mathbf{r}| x_n},$$

which can then be reordered to give the needed form.   □

In the next part of this section we will recognize that the series $G_{\mathbf{m}}$ are fundamental elements to construct classes of holomorphic modular forms with respect to bigger discrete groups:

Definition 13.39.   (i) The *$C\ell(n)$-valued modular groups* Γ_p are generated by the following matrices for $p < n + 1$:

$$\begin{pmatrix} 0 & 1 \\ -1 & 0 \end{pmatrix}, \begin{pmatrix} 1 & e_0 \\ 0 & 1 \end{pmatrix}, \ldots, \begin{pmatrix} 1 & e_p \\ 0 & 1 \end{pmatrix}.$$

(ii) The *main congruence groups* of order M, $N \geq 1$, of Γ_p are defined by

$$\Gamma_p[N] := \left\{ M = \begin{pmatrix} a & b \\ c & d \end{pmatrix} \in \Gamma_p, a - 1, b, c, d - 1 \in N\mathcal{O}_p \right\},$$

where

$$\mathcal{O}_p := \sum_A \mathbb{Z} e_A$$

is the integer additive subgroup in $C\ell(n)$.

(iii) Let $\mathcal{T}_p$ be the group generated by the matrices

$$\begin{pmatrix} 1 & e_0 \\ 0 & 1 \end{pmatrix}, \ldots, \begin{pmatrix} 1 & e_p \\ 0 & 1 \end{pmatrix},$$

let $\mathcal{T}_p[N]$ be the group of matrices generated by the matrices in $\Gamma_p[N]$ where $c = 0$ and $a = d = 1$. Finally let $\mathcal{R}_p[N]$ be a representative system of the right cosets of $\mathcal{T}_p[N]\backslash\Gamma_p[N]$.

In contrast to the classical complex case the series $G_{\mathbf{m}}(x)$ are not modular forms of the complete modular group Γ_n. The singular points of $G_{\mathbf{m}}(x)$ are the rational points $\mathbb{Q}e_0 + \cdots + \mathbb{Q}e_{n-1}$, which are at least invariant under Γ_n.

Now we will specify two construction theorems which generalize the classical definition of Poincaré series, see Remark 13.28, to Clifford analysis. These theorems provide a systematic example of non-trivial holomorphic automorphic forms of the group $\Gamma_p[N]$, where $N \geq 1$ and $\Gamma_p = \Gamma_p[1]$.

We abbreviate

$$(f|M)(x) := \overline{\mathcal{Q}_0(cx + d)} f(T(x))$$

where $M \in \Gamma_p[N]$. First we show a construction theorem giving examples of holomorphic automorphic forms of the group $\Gamma_p[N]$ where $p < n$ and $N \geq 3$.

Theorem 13.40. *Let $1 \leq p < n$ and $N \geq 3$ be a natural number. Moreover let $f : H^+(\mathbb{R}^{n+1}) \to C\ell(n)$ be a bounded holomorphic function invariant under the translation group $\mathcal{T}_p$. Then*

$$g(z) := \sum_{M \in \mathcal{R}_p[N]} (f|M)(x), \qquad x \in H^+(\mathbb{R}^{n+1}),$$

is a $C\ell(n)$-valued holomorphic function, which is uniformly bounded on every compact subset of $H^+(\mathbb{R}^{n+1})$. In addition for all $M \in \Gamma_p[N]$ the transformation behavior is $g(x) = (g|M)(x)$.

Proof (outline). For showing the absolute convergence of the series, we just have to prove that the series

$$\sum_{M \in \mathcal{R}_p[N]} |\mathcal{Q}_0(cx + d)| \leq C \sum_{M \in \mathcal{R}_p[N]} |ce_n + d|^n$$

converges for $n > p + 1$ on $H^+(\mathbb{R}^{n+1})$ in view of the boundedness of f. From classical arguments of compactification we have that for every $\varepsilon > 0$ there is a real number $\rho > 0$ with

$$|cx + d| \geq \rho |ce_n + d| \quad \text{for all } x \in V_\varepsilon(H^+(\mathbb{R}^{n+1})) \text{ and } \begin{pmatrix} * & * \\ c & d \end{pmatrix} \in \Gamma_p[N].$$

The series

$$\sum_{M \in \mathcal{R}_p[N]} (|c| + |d|)^{-\alpha}$$

converges if $\alpha > p + 1$.

Since f is a holomorphic function from Weierstraß' convergence Theorem 9.3 we have that the limit function g is holomorphic on the half-space.

To show that g is an automorphic form of $\Gamma_p[N]$, we choose an arbitrary matrix $A \in \Gamma_p[N]$. For the associated $T \in SL(2, \mathbb{Z})/\mathbb{Z}$ and due to the homogeneity of the weight factors $\mathcal{Q}_0(ab) = \mathcal{Q}_0(b)\mathcal{Q}_0(a)$, we then obtain

$$g(T(x)) = \sum_{M \in \mathcal{R}_p[N]} \overline{\mathcal{Q}_0(c_M T(x) + d_M)} f(T(x))$$

$$= \sum_{M \in \mathcal{R}_p[N]} \overline{\mathcal{Q}_0 \left(\frac{c_M(a_A x + b_A)\overline{(c_A x + d_A)} + d_M(c_A x + d_A)\overline{(c_A x + d_A)}}{|c_A x + d_A|^2} \right)} f(T(x))$$

$$= \sum_{M \in \mathcal{R}_p[N]} \overline{\mathcal{Q}_0 \left(\frac{c_A x + d_A}{|c_A x + d_A|^2} \right)} \, \overline{\mathcal{Q}_0 \left((c_M a_A + d_M c_A)x + c_M b_A + d_M d_A \right)} f(T(x))$$

$$= \overline{[\mathcal{Q}_0(c_A x + d_A)]^{-1}} \sum_{M \in \mathcal{R}_p[N]} \overline{\mathcal{Q}_0(c_{MA} x + d_{MA})} f(T(x))$$

$$= \overline{[\mathcal{Q}_0(c_A x + d_A)]^{-1}} g(x),$$

where we have set $c_{MA} := c_M a_A + d_M c_A$ and $d_{MA} := c_M b_A + d_M d_A$. The last step is a consequence of reordering, which is allowed due to the invariance of f under $\mathcal{T}_p[N]$.

In the case $N = 1, 2$ this construction delivers only the zero function, then the negative identity matrix $-I$ is an element of $\Gamma_p[N]$. Exactly in these cases $g(z) = (g|I)(z) = -g(z)$ holds for all $x \in H^+(\mathbb{R}^{n+1})$, as the automorphic factor $\mathcal{Q}_0$ is odd. $\qquad \square$

Example 13.41. a) The simplest non-trivial example of a holomorphic automorphic form of $\Gamma_p[N]$ with $p < k - 2$ and $N \geq 3$ provides the following $\Gamma_p[N]$-*Eisenstein series*:

$$\mathcal{G}^{(p,N)}(x) = \sum_{M \in \mathcal{R}_p[N]} \overline{\mathcal{Q}_0(cx + d)}.$$

The convergence and regularity properties follow from the previous Theorem 13.40 by setting $f = 1$. To prove that $\mathcal{G}^{(p,N)}$ is not vanishing identically in the case $N \geq 3$, we consider the following limit, where $x = x_0 + \mathbf{x}$:

$$\lim_{x_0 \to \infty} \mathcal{G}^{(p,N)}(x) = \sum_{M \in \mathcal{R}_p[N]} \lim_{x_0 \to \infty} \overline{\mathcal{Q}_0(cx + d)}$$

$$= \sum_{M \in \mathcal{R}_p[N], c_M = 0} \overline{\mathcal{Q}_0(d)} = 1.$$

b) If we substitute f in the theorem by the holomorphic Eisenstein series $G_\mathbf{m}$, which we have considered before, we obtain further examples of non-trivial $C\ell(n)$-valued automorphic functions of the group $\Gamma_p[N]$ with $p < n - 1$ and $N \geq 3$:

$$E_\mathbf{m}^p(x) = \sum_{M \in \mathcal{R}_p[N]} (\tilde{G}_\mathbf{m}|M)(x).$$

Here we have $\tilde{G}_{\mathbf{m}}(x) := G_{\mathbf{m}}(\mathbf{y} + x_n e_n; N\mathbb{Z}^n)$. From a limit argument similar to the previous one we have

$$\lim_{x_n \to \infty} E_{\mathbf{m}}^p(x) = 2\zeta_M^{\Omega_n}(\mathbf{m}).$$

Since the terms $2\zeta_M^{\Omega_n}(\mathbf{m})$ are exactly the Laurent coefficients of the series $\epsilon_{\mathbf{m}}^n(x)$, multiindices $\mathbf{m}$ have to exist with $2\zeta_M^{\Omega_n}(\mathbf{m}) \neq 0$ or else we would obtain $\epsilon_{\mathbf{m}}^n(x) = Q_{\mathbf{m}}(x)$, which would be a contradiction with the periodicity of $\epsilon_{\mathbf{m}}^n(x)$.

At the end of this section we want to present a second construction theorem, which also delivers non-trivial holomorphic modular forms in the case of the groups $\Gamma_p[N]$ with $N = 1, 2$ and even $p = n-1$. This theorem was proven in [78]. The underlying idea is to make the construction with the help of two automorphic factors and a further auxiliary variable. We restrict to $\mathbb{R}^n$ and to $x = \mathbf{x}$. The following theorem provides functions with the transformation behavior

$$f(x, y) = (f\|M)(x, y) := \mathcal{Q}_0\widetilde{(cx + d)}f(M\langle x \rangle, M\langle y \rangle)\mathcal{Q}_0(x\hat{c} + \hat{d}) \qquad (13.1)$$

under the operation of the whole group $\Gamma_p[N]$. Here $\hat{x}$ is the reversion according to Proposition 3.10 and $\tilde{x}$ is the main involution of $C\ell(n)$ according to Definition 3.5.

To avoid trivial examples we denote such a function as a *non-trivial holomorphic automorphic form*, only if its restriction to the diagonal $x = y$ is a non-constant automorphic C^∞-form of $\Gamma_p[N]$.

Theorem 13.42. *Let $p \in \{1, \ldots, n-1\}$ and $N \geq 3$ be a natural number. Moreover let $H_2^+(\mathbb{R}^n) := H^+(\mathbb{R}^n) \oplus H^+(\mathbb{R}^n)$. If $f : H_2^+(\mathbb{R}^n) \to C\ell(n)$ is a bounded function satisfying the equation $\overline{\partial}_x f(x, y) = f(x, y)\overline{\partial}_y = 0$ for all $(x, y) \in H_2^+(\mathbb{R}^n)$, and if $f(T(x), T(y)) = f(x, y)$ holds for all $T \in \mathcal{T}_p$, then*

$$g(x, y) := \sum_{M \in \mathcal{R}_p[N]} (f\|M)(x, y)$$

is a left-holomorphic function with respect to x and a right-holomorphic one with respect to y. Furthermore $g(x, y)$ satisfies

$$g(x, y) = (g\|M)(x, y)$$

for all $(x, y) \in H_2^+(\mathbb{R}^n)$ and all $M \in \Gamma_p[N]$.

The proof is similar to the proof of the previous theorem. In every vertical stripe $\mathcal{V}_\varepsilon(\mathbb{R}^n) \times \mathcal{V}_\varepsilon(\mathbb{R}^n)$ the estimate

$$\sum_{M \in \mathcal{R}_p[N]} \left| \mathcal{Q}_0\widetilde{(cx + d)}f(M\langle x \rangle, M\langle y \rangle)\mathcal{Q}_0(y\hat{c} + \hat{d}) \right|$$

$$\leq L \sum_{M \in \mathcal{R}_p[N]} \frac{1}{|ce_n + d|^{2n}}$$

holds, where L is a positive constant. The abscissa of the absolute convergence of the series in the last line is $p < 2n - 3$. Hence in spaces of dimension $n \geq 3$ we obtain the absolute convergence for all $p \leq n-1$. The function g is biholomorphic in view of Weierstraß' convergence theorem. The automorphic property under simultaneous operations of Γ_p, resp. $\Gamma_p[N]$, can be shown by using the homogeneity of $\mathcal{Q}_0$ by means of reordering arguments.

Example 13.43. a) The simplest non-trivial examples of holomorphic automorphic forms of $\Gamma_p[1]$ and $\Gamma_p[2]$ (for all $p \leq n - 1$) are the following biholomorphic *Eisenstein series*, which were first discovered in [78]:

$$\mathcal{F}(x,y) \;=\; \sum_{M \in \mathcal{R}_p[N]} (1 \| M)(x,y)$$

$$= \; \mathcal{Q}_0(\widetilde{c_M x + d_M}) \mathcal{Q}_0(y \hat{c}_M + \hat{d}_M), \qquad (x,y) \in H_2^+(\mathbb{R}^n).$$

To prove that the series are indeed non-trivial examples for $N = 1, 2$, we use again the limit arguments. Thus in the case of the group $\Gamma_p[1] = \Gamma_p$ with $p \leq n$ we obtain

$$\lim_{x_n \to \infty} \mathcal{F}(x_n, x_n) = \sum_{M \in \mathcal{R}_p, c_M \neq 0} \underbrace{\lim_{x_n \to \infty} \mathcal{Q}_0(\widetilde{cx_0 + d}) \mathcal{Q}_0(x_0 \hat{c} + \hat{d})}_{=0}$$

$$+ \sum_{M \in \mathcal{R}_p, c_M = 0} \widetilde{\mathcal{Q}_0(d)} \mathcal{Q}_0(\hat{d}) = 2 \sum_{A \subseteq \{1, \ldots, p\}} e_A \, \overline{e_A} = 2^{p+1}.$$

Hence the restriction to the diagonal $\mathcal{F}(x, x)$ is indeed a non-constant function. Moreover the function is in the class C^∞ with respect to the single paravector variable x and has the transformation behavior as in formula (13.1) (just before Theorem 13.42). A similar argument can be used in the case $N = 2$, where the limit is $2 \neq 0, \infty$.

b) We obtain further non-trivial examples while substituting the following product of holomorphic Eisenstein series for f,

$$f(x, y) = G_{\mathbf{m}}(x; \mathbb{Z}^n) G_{\mathbf{m}}(y; \mathbb{Z}^n).$$

The verification — again limit arguments can be used — that these series are further examples is left as Exercise 13.4.10.

Final remark: We obtain similar results for polyholomorphic functions in general real and complex Minkowski spaces, which are holomorphic with respect to several Clifford variables. This function class allows the handling of several fundamental problems of analytic number theory, of the theory of Bergman and Hardy spaces on hyperbolic polyhedral domains, and of harmonic analysis on conformal plane spin-manifolds. For a further deepening and more extensive description of this theory we refer the reader to the new book [81] and the current articles [26, 25, 82, 83].

13.4 Exercises

1. Show that

$$\int_0^\infty e^{-zt}\,dt = \frac{1}{z}$$

   is a holomorphic function in $\operatorname{Re} z > 0$ and separate the real and imaginary part.

2. Show that

$$\int_0^\infty e^{-zt^2}\,dt = \frac{\sqrt{\pi}}{2\sqrt{z}},$$

   $\operatorname{Re} z > 0$.

3. Show that in $\mathbb{C}$,

$$\Gamma(z) = \int_1^\infty t^{z-1}e^{-t}\,dt + \sum_{n=0}^\infty \frac{(-1)^n}{n!(z+n)}$$

   holds except at the poles.

4. Show that for $\alpha > 0$, $\beta > 0$ and real x, y,

$$\int_x^y (x - t)^{\alpha-1}(t - y)^{\beta-1}\,dt = \frac{\Gamma(\alpha)\Gamma(\beta)}{\Gamma(\alpha+\beta)}(x - y)^{\alpha+\beta-1}.$$

5. Show that, if $0 < \operatorname{Re} z < 1$,

$$\Gamma(z)\zeta(z) = \int_0^\infty t^{z-1}\left(\frac{1}{e^t - 1} - \frac{1}{t}\right)dt.$$

6. Show that, if $-1 < \operatorname{Re} z < 0$,

$$\Gamma(z)\zeta(z) = \int_0^\infty t^{z-1}\left(\frac{1}{e^t - 1} - \frac{1}{t} + \frac{1}{2}\right)dt.$$

7. Show that the series

$$\epsilon_m^{(1)}(z; \mathbb{Z}) = \begin{cases} \frac{1}{z} + \sum\limits_{k\in\mathbb{Z}\setminus\{0\}}\left(\frac{1}{z+k} - \frac{1}{k}\right), & m = 1, \\ \sum\limits_{k\in\mathbb{Z}}\frac{1}{(z+k)^m}, & m \geq 2 \end{cases}$$

   converges uniformly for $|z| \leq R$ except at the poles. Thus these series are meromorphic functions in $\mathbb{C}$.

8. Show that also the Eisenstein series

$$
\epsilon_m^{(2)}(z;\Omega) =
\begin{cases}
\frac{1}{z^2} + \displaystyle\sum_{\omega\in\Omega\setminus\{0\}} \left(\frac{1}{(z+\omega)^2} - \frac{1}{\omega^2}\right), & m = 2, \\[2mm]
\displaystyle\sum_{\omega\in\Omega} \frac{1}{(z+\omega)^m}, & m \geq 3
\end{cases}
$$

converges uniformly except at the poles. Thus these series are also meromorphic functions in $\mathbb{C}$.

9. Prove that the Taylor series

$$
\wp(z) - \frac{1}{z^2} = \frac{1}{20}g_2 z^2 + \frac{1}{28}g_3 z^4 + O(z^6)
$$

in a neighborhood of the origin satisfies

$$
g_2 = 60 \sum_\omega {}' \frac{1}{\omega^4}, \quad g_3 = 140 \sum_\omega {}' \frac{1}{\omega^6}.
$$

10. Show that the product of Eisenstein series

$$
f(x,y) = G_{\mathbf{m}}(x; \mathbb{Z}^n) G_{\mathbf{m}}(y; \mathbb{Z}^n)
$$

is a holomorphic and automorphic function (cf. Example 13.43 b).

Appendix

A.1 Differential forms in $\mathbb{R}^n$

A.1.1 Alternating linear mappings

As we have to use differential forms we give here a short introduction.

Definition A.1.1. Let V and W be real vector spaces.

(i) A mapping $\Phi : V^q \to W$ is called q-times *multilinear* or *q-linear* if it is $\mathbb{R}$-linear in each of its q variables. The set of q-linear mappings from V to W is denoted by $L^q(V, W)$. The number q is called the *degree* of Φ. Such multilinear mappings are also called *multilinear forms* if $W = \mathbb{R}$.

(ii) $\Phi \in L^q(V, W)$ is called *alternating* or *skew symmetric* if for every permutation $\sigma \in \text{perm}(q)$,

$$\sigma\Phi(x_1, \ldots, x_q) := \Phi(x_{\sigma(1)}, \ldots, x_{\sigma(q)}) = (\text{sgn}\,\sigma)\Phi(x_1, \ldots, x_q).$$

Here $\text{perm}(q)$ is the permutation group of q elements. The set of alternating mappings in $L^q(V, W)$ is denoted by $A^q(V, W)$.

(iii) $\Phi \in L^q(V, W)$ is called *symmetric* if $\sigma\Phi = \Phi$ holds for every permutation σ. The set of symmetric mappings is denoted by $S^q(V, W)$.

We have $A^q \subset L^q, S^q \subset L^q$ and $A^q \oplus S^q = L^q$.

Instead of real vector spaces one may use complex vector spaces without problems, then $\mathbb{R}$ has to be substituted by $\mathbb{C}$. Scalar products in $V = \mathbb{R}^n$ are such multilinear mappings with values in $W = \mathbb{R}$.

Proposition A.1.2.   (i) *If* $\{e_1, \ldots, e_n\}$ *is a basis of* V, *then a q-linear mapping is uniquely determined by its values on the sets of q basis elements* $\{e_{j_1}, \ldots, e_{j_q}\}$, *values which can be chosen arbitrarily in* W.

(ii) $L^q(V, W), A^q(V, W)$, *and* $S^q(V, W)$ *are real vector spaces with operations given by the addition in* W *and the multiplication by real numbers.*

(iii) *Introducing norms we have for each pair of norms,* $|.|_1$ *in* V *and* $|.|_2$ *in* W *that*

$$|\Phi|_{12} := \sup\{|\Phi(x_1, \ldots, x_q)|_2 : |x_1|_1 \leq 1, \ldots, |x_q|_1 \leq 1\}$$

is finite and a norm of the vector space $L^q(V, W)$ *with*

$$|\Phi(x_1, \ldots, x_q)|_2 \leq |\Phi|_{12}|x_1|_1 \cdots |x_q|_1.$$

(iv) *If q vectors* $x_1, \ldots, x_q$ *are linearly dependent if and only if* $\Phi(x_1, \ldots, x_q) = 0$, *then* Φ *is alternating.*

(v) *The vector spaces* A^q *and* S^q *are closed relative to norm convergence.*

Proof. (i) $\Phi \in V^q$, then

$$\Phi(x^1, ..., x^k) = \sum_{i_q=0^n} x^1_{i_1} ... x^q_{i_q} \Phi(e_{i_1}, ..., e_{i_q})$$

which proves the assertion.

(ii) follows from the (multi)-linearity of L^q, A^q and S^q.

(iii) is obtained by using operator norms in finite spaces and matrix calculus.

(iv) It remains only to prove one direction. Let $q = 2$, then

$$\Phi(x_i, x_j) = -\Phi(x_j, x_i) \quad (i \neq j)$$

and

$$\Phi(x_i, x_j) + \Phi(x_j, x_i) = \Phi(x_i + x_j, x_j) + \Phi(x_i + x_j, , x_i) = \Phi(x_i + x_j, x_i + x_j) = 0. \quad \square$$

We explicitly remark that an alternating mapping changes sign if two variables are interchanged. Therefore an alternating mapping has the value zero if two of its variables are equal.

We shall study now the alternating mappings in more detail, we start with a

Definition A.1.3. (i) For $\Phi \in L^q$ the mapping $\alpha_q : L^q(V, W) \to A^q(V, W)$ with

$$\alpha_q(\Phi) := \sum_{\sigma \in \mathrm{perm}(q)} (\mathrm{sgn}\, \sigma) \sigma \Phi$$

is called *anti-symmetrization.*

(ii) For $\Phi \in L^{p+q}(V, W)$ the mapping $(\alpha_p, \alpha_q) : L^{p+q}(V, W) \to A^{p,q}(V, W)$ is defined by

$$(\alpha_p, \alpha_q)\Phi(x_1, \ldots, x_p, x_{p+1}, \ldots, x_{p+q})$$
$$:= \sum_{\sigma \in \mathrm{perm}(p)} (\mathrm{sgn}\, \sigma) \sum_{\tau \in \mathrm{perm}(q)} (\mathrm{sgn}\, \tau) \Phi(x_{\sigma(1)}, \ldots, x_{\sigma(p)}, x_{p+\tau(1)}, \ldots, x_{p+\tau(q)}).$$

(iii) The *symmetrization* is analogously defined by

$$\beta_q(\Phi) := \sum_{\sigma \in \mathrm{perm}(q)} \sigma \Phi,$$

(β_p, β_q) has to be introduced correspondingly.

Clearly these mappings are linear. The mappings in the set $A^{p,q}(V, W)$ are alternating separately the first p and the last q variables. Every $\Phi_q \in A^q$ can be represented as an image of a $\Phi'_q \in L^q$, in every case $\alpha_q \Phi'_q = \Phi_q$ if Φ_q is already alternating. Now we define a mapping which is important for us:

Proposition A.1.4. *The mapping $\alpha_{p,q} : A^{p,q} \to A^{p+q}$, defined by*

$$\alpha_{p,q}((\alpha_p, \alpha_q)\Phi) := \alpha_{p+q}(\Phi),$$

is well defined and linear. An analogous mapping $\beta_{p,q}$ is defined for the symmetrization, and a corresponding assertion holds.

The proof may be found, e.g., in [22] or [65].

From now on we assume W to be an algebra $\mathcal{A}$ such that we can multiply in it. For us $\mathcal{A}$ will generally be $\mathbb{R}$ but $\mathbb{C}$, $\mathbb{H}$, or the Clifford algebra $C\ell(n)$ are equally possible. Now we are able to define:

Definition A.1.5. (i) The direct sum of the $L^q(V, \mathcal{A})$, i.e., is the set of all formal finite sums of elements of all $L^q(V, \mathcal{A})$, is denoted by $L^\infty(V, \mathcal{A})$:

$$L^\infty(V, \mathcal{A}) := \bigoplus_{q=0}^{\infty} L^q(V, \mathcal{A}).$$

(ii) $L^\infty(V, \mathcal{A})$ becomes an algebra with the following tensor product: For $\Phi_p \in L^p(V, \mathcal{A})$ and $\Phi_{p'} \in L^{p'}(V, \mathcal{A})$ one defines $\Phi_p \otimes \Phi_{p'} \in L^{p+p'}(V, \mathcal{A})$ by

$$(\Phi_p \otimes \Phi_{p'})(x_1, \dots, x_{p+p'}) := \Phi_p(x_1, \dots, x_p)\Phi_{p'}(x_{p+1}, \dots, x_{p+p'}).$$

$\mathcal{A}$ is a subalgebra of $L^\infty(V, \mathcal{A})$ with $a \otimes \Phi_p = a\Phi_p$, $a \in \mathcal{A}$.

(iii) $A^\infty(V, \mathcal{A})$ and $S^\infty(V, \mathcal{A})$ are analogously defined.

(iv) In A^∞ the following *exterior product* or *alternating product* or *wedge product* is defined:

$$\wedge : A^p(V, \mathcal{A}) \times A^q(V, \mathcal{A}) \to A^{p+q}(V, \mathcal{A})$$

with

$$\Phi_p \wedge \Phi_q := \alpha_{p,q}(\Phi_p \otimes \Phi_q).$$

This makes A^∞ a (graduated) algebra, the *exterior algebra* or *Graßmann algebra*.

(v) Quite analogously the *symmetric product*

$$\vee : S^p(V, \mathcal{A}) \times S^q(V, \mathcal{A}) \to S^{p+q}(V, \mathcal{A})$$

is defined using $\beta_{p,q}$.

We remark explicitly that A^∞ and S^∞ are not subalgebras of L^∞, but only linear subspaces, as the exterior and the symmetric product are defined only in A^∞, resp. S^∞. The following rules hold:

Proposition A.1.6. (i) *The multiplications $\wedge$ and $\vee$ are distributive and associative, the latter if the algebra $\mathcal{A}$ is associative.*

(ii) *For commutative $\mathcal{A}$ (especially $\mathcal{A} = \mathbb{R}, \mathbb{C}$) we have*

$$\Phi_p \wedge \Phi_q = (-1)^{pq} \Phi_q \wedge \Phi_p$$
$$\Phi_p \vee \Phi_q = \Phi_q \vee \Phi_p.$$

For the proof we refer to the books cited above and recommend it as an exercise (see Exercise A.1.3.3). The distributivity follows easily from the distributivity in L^∞; for the associativity and the commutation rule one has to represent Φ_p, Φ_q as images of elements of $L^\infty(V, \mathcal{A})$ and then to use their properties.

Example A.1.7. a) The $\Phi_0 \in L^0(V, \mathcal{A})$ are the elements of $\mathcal{A}$. As no variables exist they cannot be exchanged and we have $A^0 = S^0 = L^0$, moreover α_0 and β_0 equal the identity. The exterior multiplication by a Φ_0 runs as follows:

$$\Phi_0 \wedge \Phi_p = \alpha_{0,p}(\alpha_0 \Phi_0 \otimes \alpha_p \tilde{\Phi}_p)$$
$$= \alpha_{0+p} \Phi_0 \otimes \tilde{\Phi}_p = \Phi_0(\alpha_p \tilde{\Phi}_p) = \Phi_0 \Phi_p.$$

b) The elements in L^1 resp. A^1 or S^1 are functions of only one variable. Therefore one cannot exchange variables also for the Φ_1, we have $L^1 = A^1 = S^1$ and $\alpha_1 = \beta_1 = id$. The exterior product of elements of A^1 has the form

$$\Phi_1 \wedge \tilde{\Phi}_1 = \alpha_{1,1}(\alpha_1 \Phi_1 \otimes \alpha_1 \tilde{\Phi}_1) = \alpha_2(\Phi_1 \otimes \tilde{\Phi}_1).$$

We get

$$(\Phi_1 \wedge \tilde{\Phi}_1)(x_1, x_2) = \Phi_1(x_1)\tilde{\Phi}_1(x_2) - \Phi_1(x_2)\tilde{\Phi}_1(x_1).$$

c) The product xy of two vectors in $\mathbb{R}^n$, interpreted as 1-vectors in the Clifford algebra $C\ell(n)$, is a 2-form with values in $\mathcal{A} = C\ell(n)$. At the end of Subsection 3.2.2 we found that the symmetrized product $(xy + yx)/2$ is the usual scalar product $x \cdot y$; the anti-symmetrized product

$$\frac{xy - yx}{2} = x \wedge y$$

appears to be the exterior product, also called the *Graßmann product*. The Graßmann product of two 1-vectors is a 2-vector or *bivector*. It describes the oriented volume of the parallelogramm, spanned by the vectors x and y as we have seen in $\mathbb{R}^3$ in Section 2 following Proposition 2.42.

Obviously also products with more factors are defined, so a product of k 1-vectors gives a so-called k-form and we have also the exterior product, which gives a k-vector . The exterior multiplication of a k-vector and an ℓ-vector gives a $(k+\ell)$-vector. The commutation rules of the last proposition hold obviously also here. For the special case of $\mathbb{R}^n$ the A^k are denoted by $\Lambda^k(\mathbb{R}^n)$ and A^∞ by $\Lambda(\mathbb{R}^n)$.

d) Especially the literature in physics represents the scalar and the exterior product by two operators. On the one hand this is the *annihilation operator* J_x of the scalar or interior multiplication and on the other hand the *creation operator* E_x of the exterior multiplication. Both operators are defined by the following recursions: Let x be a 1-vector and $x^{(k)} = x^1 \wedge x^2 \wedge \cdots \wedge x^k$ with 1-vectors $x^1, \ldots, x^k$. One defines then

$$J_x(1) := 0, \quad J_x(x^{(k)}) := \sum_{i=1}^{k} (-1)^{i-1}(x \cdot x^i) x^1 \wedge \cdots \wedge x^{i-1} \wedge x^{i+1} \wedge \cdots \wedge x^k$$

and

$$E_x(1) := x, \quad E_x(x^{(k)}) := x \wedge x^{(k)}.$$

Both operators have to be continued linearly to $\Lambda(\mathbb{R}^n)$. From the definition we can at once conclude

Proposition A.1.8. (i) *For $y \in \Lambda^k(\mathbb{R}^n)$ we have $E_x(y) \in \Lambda^{k+1}(\mathbb{R}^n)$ and $J_x(y) \in \Lambda^{k-1}(\mathbb{R}^n)$. In particular for $y \in \Lambda^1(\mathbb{R}^n)$ we have $J_x(y) = x \cdot y$.*

(ii) *For $x \in \Lambda^1(\mathbb{R}^n)$ and $y, z \in \Lambda(\mathbb{R}^n)$ the relation*

$$E_x(y) \cdot z = y \cdot J_x(z)$$

holds showing J_x to be the conjugate operator to E_x relative to the interior product.

(iii) *In $\Lambda(\mathbb{R}^n)$ we have always $E_x^2 = J_x^2 = 0$ and so $x \wedge x = 0$.*

The proof is given as Exercise A.1.3.4.

e) If we look for alternating forms for $V = \mathbb{R}^n$ and $\mathcal{A} = \mathbb{R}$ we see that non-trivial alternating forms exist only up to degree n: A Φ_p can be split up into expressions of the form

$$\Phi_p(e_{i_1}, \ldots, e_{i_p})$$

because of the distributivity; here the e_j are the basis elements of V. If $p > n$ at least two basis elements have to be equal, therefore the alternating form is zero on a set with $p > n$ basis elements (an exchange of the two equal variables does not change the value of the form but its sign).

Besides $A^0(\mathbb{R}^n, \mathbb{R}) = \mathbb{R}$ it is easy to describe $A^1(\mathbb{R}^n, \mathbb{R}) = L^1(\mathbb{R}^n, \mathbb{R})$: $L^1(\mathbb{R}^n, \mathbb{R})$ contains precisely the $\Phi_1(x) = a \cdot x$ with $a, x \in \mathbb{R}^n$. Also the n-forms are easy to compute: with

$$x_k = \sum_{j=1}^{n} x_{kj} e_j$$

we have $\Phi_n(e_{j_1}, \ldots, e_{j_n}) = 0$ if two of the e_{j_i} are equal. This gives

$$
\begin{aligned}
\Phi_n(x_1, \ldots, x_n) &= \sum_{j_1,\ldots,j_n=1}^{n} x_{1j_1} \cdots x_{nj_n} \Phi_n(e_{j_1}, \ldots, e_{j_n}) \\
&= \sum_{\sigma \in \text{perm}(n)} x_{1\sigma(1)} \cdots x_{n\sigma(n)} \Phi_n(e_{\sigma(1)}, \ldots, e_{\sigma(n)}) \\
&= \left(\sum_{\sigma \in \text{perm}(n)} (\text{sgn}\,\sigma) x_{1\sigma(1)} \cdots x_{n\sigma(n)} \right) \Phi_n(e_1, \ldots, e_n) \\
&= \Phi_n(e_1, \ldots, e_n) \det(x_1, \ldots, x_n).
\end{aligned}
$$

Consequently only one n-linear form exists in $\mathbb{R}^n$ up to a factor.

A.1.2 Differential forms

We have now provided the necessary notions for our real purpose and we are able to define differential forms:

Definition A.1.9. Let G be a domain in V.

(i) A mapping $\omega_p : G \to A^p(V, \mathcal{A})$ is called a *differential form of degree p* with values in $\mathcal{A}$. For $x \in G$ we have therefore $\omega(x) \in A^p(V, \mathcal{A})$. If variables from V^p are needed we add them in brackets:

$$ \omega(x)[h_1, \ldots, h_p]. $$

If ω_p is m-times continuously differentiable (in the sense of mappings between vector spaces) we write
$$ \omega_p \in C^m(G). $$

(ii) For every $x \in G$ and using Definition A.1.5 (iv) we define $\omega_p \wedge \omega_q$ by

$$ (\omega_p \wedge \omega_q)(x) := \omega_p(x) \wedge \omega_q(x). $$

(iii) For $\omega_p \in C^1(G)$,

$$ d\omega_p := \alpha_{1,p}(\omega_p') : G \to A^{p+1}(V, \mathcal{A}) $$

is called the *exterior differential* or *total differential* of ω_p. If $d\omega_p = 0$ then ω_p is called *closed* or *exact*. If an ω_{p-1} exists with $\omega_p = d\omega_{p-1}$, then ω_p is called a *total* differential form. The operator d is also called a *Cartan operator*.

Remark A.1.10. a) As $A^0(V, \mathcal{A}) = \mathcal{A}$ a differential form of degree 0 is a function defined in G with values in $\mathcal{A}$. As the Examples A.1.7 show we have

$$ \omega_0 \wedge \omega_p = \omega_0 \omega_p, \quad \omega_p \wedge \omega_0 = \omega_p \omega_0 $$

and

$$\begin{aligned}
(\omega_1 \wedge \tilde{\omega}_1)(x)[h_1, h_2] &= \omega_1(x) \wedge \tilde{\omega}(x)[h_1, h_2] \\
&= (\omega_1 \otimes \tilde{\omega}_1 - \tilde{\omega}_1 \otimes \omega_1)(x)[h_1 h_2] \\
&= \omega_1(x)[h_1]\tilde{\omega}_1(x)[h_2] - \omega_1(x)[h_2]\tilde{\omega}_1(x)[h_1].
\end{aligned}$$

Luckily one has very seldom to go into such details.

b) In definition (iii) a derivative ω'_p is used. This is always meant in the sense of mappings between vector spaces: If $f : G \to \mathcal{A}$ is a function defined in a domain $G \subset V$, it is differentiable at a point $x_0 \in G$ if and only if an $f'(x_0) \in L^1(V, \mathcal{A})$ exists such that

$$f(x) = f(x_0) + f'(x_0)[x - x_0] + |x - x_0|o(1).$$

Here we have used a notation for functions which have the character of a remainder term, as introduced by the German mathematicians PAUL BACHMANN (1837–1920) and EDMUND LANDAU (1877–1938):

Definition A.1.11 (Bachmann–Landau symbols). A function $g(h)$ with values in a vector space and depending on a variable h in a neighborhood of the origin of another vector space V is denoted by $o(1)$ — read '*little*-o *of 1*' —, if $g(h)$ converges to 0 for $h \to 0$.

If the function g is bounded for $h \to 0$ it is denoted by $O(1)$ – read '*big*-O *of 1*' –. If the product $ko(1)$, resp. $kO(1)$, with another function k is defined one writes simply $o(k)$, resp. $O(k)$.

A sum and a product of such functions are a function of the same type, similarly the multiplication with a bounded function does not change the type. That is the advantage of this notation; we shall use it often with profit.

So, for a differential form $\omega_p : G \to A^p(V, \mathcal{A})$ differentiation means the existence of an equation

$$\omega_p(x) = \omega_p(x_0) + \omega'_p(x_0)[x - x_0] + |x - x_0|o(1)$$

for $x \to x_0$, here $\omega'_p(x_0) \in L(V, A^p(V, \mathcal{A}))$.

It is important for the first and the higher derivatives that we always can put canonically

$$L(V, L^p(V, \mathcal{A})) \cong L^{p+1}(V, \mathcal{A});$$

this is done as follows: $\Phi \in L(V, L^p(V, \mathcal{A}))$ assigns to an $x_0 \in V$ a $\Phi(x_0) \in L^p(V, \mathcal{A})$, which is a linear operation relative to x_0. Therefore

$$\Phi(x_0)[x_1, \dots, x_p]$$

is a linear function relative to all variables; it is a multilinear function of grade $p + 1$, and so it is an element in $L^{p+1}(V, \mathcal{A})$.

In particular for the differentiation of our differential form we have

$$\omega_p'(x_0)[x_1,\ldots,x_p] \in L^{p+1}(V,\mathcal{A}),$$

here the derivative is already assumed to be alternating relative to the p variables. That shows

$$\alpha_{1,p}(x_0)\omega_p'$$

to be reasonably defined and alternating with degree $p+1$.

c) Following the theorem of Schwarz the second and therefore also the higher derivatives of a function $f : G \to \mathcal{A}$ are symmetric. For $f'' \in L^2(V,\mathcal{A})$ this means

$$f''(x_0)[h,k] = f''(x_0)[k,h] \quad (h,k \in V).$$

We need rules for calculating with differential forms:

Proposition A.1.12. *Let the algebra* $\mathcal{A}$ *be associative. We then have:*

(i) *The exterior product* $\wedge$ *is associative.*

(ii) *The exterior differential d is additive, i.e.,* $d(\omega + \tilde{\omega}) = d\omega + d\tilde{\omega}$.

(iii) *In a commutative algebra* $\mathcal{A}$ *we have*

$$\omega_p \wedge \omega_q = (-1)^{pq}\omega_q \wedge \omega_p.$$

(iv) *From* $\omega_p = \alpha_p \Phi_p$ *it follows that* $d\omega_p = \alpha_{p+1}\Phi_p'$.

(v) *For d the following Leibniz rule holds:*

$$d(\omega_p \wedge \omega_q) = (d\omega_p) \wedge \omega_q + (-1)^p \omega_p \wedge (d\omega_q).$$

(vi) *If* $\omega_p \in C^2(G)$ *we have* $d(d\omega_p) = 0$: *The exterior differential of a differential form is also a total differential form.*

Proof. (i), (ii) and (iii) follow directly from the rules in Proposition A.1.6 for alternating forms.

(iv) We have

$$
\begin{aligned}
\omega_p' &= (\alpha_p \Phi_p)' = \sum_{\sigma \in \mathrm{perm}(p)} (\mathrm{sgn}\,\sigma)(\sigma \Phi_p)' \\
&= \sum_{\sigma \in \mathrm{perm}(p)} (\mathrm{sgn}\,\sigma)\sigma \Phi_p'[\cdot,\cdots] = (\alpha_1,\alpha_p)\Phi_p' = \alpha_{p+1}\Phi_p' \\
d\omega_p &= \alpha_{1,p}(\omega_p') = \alpha_{1,p}[(\alpha_p \Phi_p)'] = \alpha_{1,p}[(\alpha_1,\alpha_p)\Phi_p'] = \alpha_{p+1}(\Phi_p').
\end{aligned}
$$

Here σ in the second line applies only to the last variables of Φ', that is just the same as the application of (α_1,α_p) to Φ_p'.

(v) Again with $\omega_p = \alpha_p \Phi_p$ and $\omega_q = \alpha_q \Phi_q$ it follows that

$$
\begin{aligned}
d(\omega_p \wedge \omega_q) &= d(\alpha_p \Phi_p \wedge \alpha_q \Phi_q) = d[\alpha_{p,q}(\alpha_p \Phi_p \otimes \alpha_q \Phi_q)] \\
&= d(\alpha_{p+q}(\Phi_p \otimes \Phi_q)) = \alpha_{p+q+1}(\Phi_p \otimes \Phi_q)' \\
&= \alpha_{p+q+1}(\Phi_p' \otimes \Phi_q + \Phi_p \otimes \Phi_q') \\
&= \alpha_{p+1,q}(\alpha_{p+1}, \alpha_q)\Phi_p' \otimes \Phi_q + \alpha_{p+q+1}\Phi_p \otimes \Phi_q'.
\end{aligned}
$$

Within the last summand the additional variable, coming up with the differentiation of Φ_q, has to be shifted to the first place, where this additional variable has to be, following the definition of the derivative. If we do that with a permutation τ, we get with $p + q + 1 =: r$,

$$
\alpha_r \tau \Phi = \sum_{\sigma \in \mathrm{perm}(r)} (\mathrm{sgn}\,\sigma)\sigma\tau\Phi = (\mathrm{sgn}\,\tau) \sum_{\sigma \in \mathrm{perm}(r)} (\mathrm{sgn}\,\sigma\tau)\sigma\tau\Phi = (\mathrm{sgn}\,\tau)\alpha_r \Phi.
$$

If we apply this in our differentiation formula because of $\mathrm{sgn}\,\tau = (-1)^p$ (the additional variable in Φ_q' has to be shifted over the p variables of Φ_p) we get

$$
\begin{aligned}
d(\omega_p \wedge \omega_q) &= \alpha_{p+1,q}\omega_p' \otimes \omega_q + (\mathrm{sgn}\,\tau)\alpha_{p,q+1}(\alpha_p, \alpha_{q+1})\Phi_p \otimes \Phi_q' \\
&= (d\omega_p) \wedge \omega_q + (-1)^p \omega_p \wedge (d\omega_q),
\end{aligned}
$$

and that is the assertion.

(vi) From $\omega_p = \alpha_p \Phi_p$ it follows that $d\omega_p = \alpha_{p+1}\Phi_p'$ and

$$
d(d\omega_p) = \alpha_{p+2}\Phi_p'' = \sum_{\sigma \in \mathrm{perm}(p+2)} (\mathrm{sgn}\,\sigma)\sigma\Phi_p'' = \sum_{\sigma \in \mathrm{perm}(p+2)} \sigma\Phi_p''.
$$

Now, following the theorem of Schwarz, Φ_p'' is symmetric in the first two variables, arising from the differentiations. Therefore we split up σ in the following way: $\sigma = (\rho, \tau)$, where ρ is related only to the first two variables. Then for every τ there appears once ρ and once ρ', relative to the first two exchanged variables. ρ and ρ' give for σ just different signs, while Φ_p'' is unchanged because of the symmetry of the second derivatives. Therefore the sum is zero indeed.

$\square$

We need a representation of differential forms by a basis, which is the usual form of working with differential forms. A further operation will follow.

Theorem A.1.13. (i) *A differential form* $\omega_q : G \to A^q(V, \mathcal{A})$ *possesses the following* canonical representation *depending on the basis (if $\mathcal{A}$ is of dimension greater than 1, the representation holds for each component separately):*

Let $\{e_1, \ldots, e_n\}$ be a basis of V, $x = \sum_{j=1}^n x_j e_j$. Let

$$
dx_j : G \to A^1(V, \mathbb{R}) \quad (= L(V, \mathbb{R})) \quad \text{with} \quad dx_j[x] := x_j,
$$

be the projection of x onto the j-th coordinate. The dx_j do not depend on the points in G, they are constant differential forms of degree 1. For ω_q we have then the representation

$$
\omega_q(x) = \sum_{j_1 < \cdots < j_q} a_{j_1 \ldots j_q}(x)\, dx_{j_1} \wedge \cdots \wedge dx_{j_q},
$$

here j_i runs through $1, \ldots, n$ and the $a_{j_1 \ldots j_q}(x)$ are functions which are differentiable as often as the differential form itself.

(ii) *A differential form of degree 0, $\omega_0(x)$, is a function $a(x)$; a differential form of degree 1 has the form*

$$\omega_1(x) = \sum_{j=1}^{n} a_j(x) dx_j, \quad a_j(x) = \omega_j(x)[e_j].$$

Moreover $\omega_q(x) = 0$ for $q > n$ and $\omega_n(x) = a(x) dx_1 \wedge \cdots \wedge dx_n$.

Proof. (i) The differential form $\omega_q(x)[h_1, \ldots, h_q]$ is uniquely defined by the values on the sets of q basis elements $e_{i_1}, \ldots, e_{i_q}$, where it is useful to assume $i_1 < \cdots < i_q$. On this set of basis elements $dx_{i_1} \wedge \cdots \wedge dx_{i_q}$ has the value 1, on all other sets of basis elements the value 0. The proof uses mathematical induction relative to q: From $dx_i[e_j] = \delta_{ij}$ and the induction starts $dx_i = \alpha_1(dx_i)$; it then follows that

$$dx_{i_1} \wedge \cdots \wedge dx_{i_{q-1}} \wedge dx_{i_q} = \alpha_{q-1}(dx_{i_1} \wedge \cdots \wedge dx_{i_{q-1}}) \wedge dx_{i_q}$$
$$= \alpha_{q-1,1}(\alpha_{q-1}(dx_{i_1} \wedge \cdots \wedge dx_{i_{q-1}}) \otimes dx_{i_q}$$
$$= \alpha_{q-1,1}(\alpha_{q-1} \otimes_{j=1}^{q-1} dx_{i_j} \otimes \alpha_1 dx_{i_q}$$
$$= \alpha_{q-1,1}(\alpha_{q-1}, \alpha_1) dx_{i_1} \otimes \cdots \otimes dx_{i_q}$$
$$= \alpha_q(dx_{i_1} \otimes \cdots \otimes dx_{i_q}).$$

If one substitutes here the values $e_{j_1}, \ldots, e_{j_q}$ one can prove the equation

$$(dx_{i_1} \wedge \cdots \wedge dx_{i_q}[e_{j_1}, \ldots, e_{j_q}] = \delta_{i_1, j_1} \cdots \delta_{i_q, j_q}, \quad j_1 < \cdots < j_q,$$

(see Exercise A.1.3.6). This gives

$$\omega_q(x) = \sum_{i_1 < \cdots < i_q} \omega_q(x)[e_{i_1}, \ldots, e_{i_q}] dx_{i_1} \wedge \cdots \wedge dx_{i_q};$$

the property to be a basis and the uniqueness of the representation are now easy to see. (ii) follows without difficulties from Examples A.1.7 a, b, e. $\square$

Corollary A.1.14. (i) *For a function f we have*

$$df(x) = \sum_{j=1}^{n} \frac{\partial f}{\partial x_j} dx_j.$$

(ii) *The local increasing of a function f is described by $(h \to 0)$*

$$f(x + h) - f(x) = df[h] + |h| o(1).$$

Proof. (i) We have $df = \alpha_1 f' = f'$, therefore $df[e_j] = f'[e_j] = \partial f / \partial x_j$ and

$$df = \sum_{j=1}^{n} \frac{\partial f}{\partial x_j} dx_j.$$

(ii) We know from real analysis that

$$f(x+h) - f(x) = \sum_{i=1}^{n} \frac{\partial f}{\partial x_i} h_i + |h| o(1)$$

with the components h_i of h. Then from

$$df[h] = \sum_{i=1}^{n} \frac{\partial f}{\partial x_i} dx_i[h] = \sum_{i=1}^{n} \frac{\partial f}{\partial x_i} h_i$$

the desired equation follows:

$$f(x+h) - f(x) = df[h] + |h| o(1). \qquad \square$$

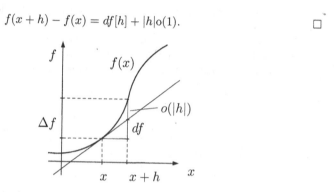

Figure A.1

At the end of this appendix we are going to define a useful operation with differential forms without saying anything about the background. The notion is helpful for the definition of holomorphic functions and also for the integration.

Definition A.1.15 (Hodge operator). For $\omega_q(x) := \sum\limits_{i_1 < \cdots < i_q} a_{i_1 \ldots i_q}(x) dx_{i_1} \wedge \cdots \wedge dx_{i_q}$
the *Hodge operator* is defined as follows:

$$\omega_q^*(x) := \sum_{i_1 < \cdots < i_q} a_{i_1 \ldots i_q}(x)(dx_{i_1} \wedge \cdots \wedge dx_{i_q})^*$$

and

$$(dx_{i_1} \wedge \cdots \wedge dx_{i_q})^* := (\operatorname{sgn} \sigma) dx_{i_{q+1}} \wedge \cdots \wedge dx_{i_n}$$

if σ is the permutation with $\sigma(j) = i_j$.

Remark A.1.16. a) We have in particular

$$\omega_0^* = f^* = f dx_1 \wedge \cdots \wedge dx_n.$$

For the integration we shall identify $dx_1 \wedge \cdots \wedge dx_n$ with the volume element $d\sigma$.
b) Obviously we have $\omega_q^{**} = \omega_q$.

c) The form dx^* is of special interest for us, firstly we have

$$dx_i^* = (-1)^{i-1} dx_1 \wedge \cdots \wedge dx_{i-1} \wedge dx_{i+1} \wedge \cdots \wedge dx_n =: (-1)^{i-1} d\hat{x}_i,$$

the sign comes from the shifting of the index i to the first place. The abbreviation $d\hat{x}_i$ saves much typing. So dx^* is a differential form of degree $(n-1)$, the evaluation of its value gives:

$$dx^*[h_1, \ldots, h_{n-1}] = \sum_{i=1}^n e_i \, dx_i^*[h_1, \ldots, h_{n-1}]$$

$$= \sum_{i=1}^n (-1)^{i-1} e_i \, \alpha_{n-1} (dx_1 \otimes \cdots \otimes dx_{i-1} \otimes dx_{i+1} \otimes \cdots \otimes dx_n)[h_1, \ldots, h_{n-1}]$$

$$= \sum_{i=1}^n (-1)^{i-1} e_i \cdot$$

$$\sum_{\sigma \in \mathrm{perm}(n-1)} (\mathrm{sgn}\,\sigma) dx_1[h_{\sigma(1)}] \cdots dx_{i-1}[h_{\sigma(i-1)}] dx_{i+1}[h_{\sigma(i)}] \cdots dx_n[h_{\sigma(n-1)}]$$

$$= \sum_{i=1}^n (-1)^{i-1} e_i \sum_{\sigma \in \mathrm{perm}(n-1)} (\mathrm{sgn}\,\sigma) h_{\sigma(1),1} \cdots h_{\sigma(i-1),i-1} h_{\sigma(i),i+1} \cdots h_{\sigma(n-1),n}$$

$$= \det \begin{pmatrix} e_1 & \cdots & e_n \\ h_{11} & \cdots & h_{1n} \\ \vdots & & \vdots \\ h_{n-1,1} & \cdots & h_{n-1,n} \end{pmatrix}.$$

Here the determinant has to be developed formally with respect to the first line as we know from the cross product in $\mathbb{R}^3$, moreover we put $h_i = (h_{i,1}, \ldots, h_{i,n})$. One can call this expression the *vector product* or *cross product* in $\mathbb{R}^n$:

$$dx^*[h_1, \ldots, h_{n-1}] =: \bigtimes_{i=1}^{n-1} h_i.$$

Obviously this vector is orthogonal to all $h_1, \ldots, h_{n-1}$, as the scalar product with another vector a gives the determinant which as first line has the components of the vector a:

$$a \cdot \left(\bigtimes_{i=1}^{n-1} h_i \right) = \det \begin{pmatrix} a \\ h_1 \\ \vdots \\ h_{n-1} \end{pmatrix}.$$

We shall use this fact while integrating over manifolds. We can look upon the product in the last formula as a *mixed n-product* of the n vectors $a, h_1, \ldots, h_{n-1}$.

This is an operation with n incoming vectors, therefore no multiplication in the usual sense. This product gives the volume of the box spanned by the n vectors, we know this for $\mathbb{R}^3$ and here we have a generalization. As the determinant is anti-symmetric relative to the h_i one can ask for the relation to $h_1 \wedge \cdots \wedge h_{n-1}$, which is also an anti-symmetric product. To look for this relation we define

$$e_j^* := (-1)^{j-1} e_1 \wedge \cdots \wedge e_{j-1} \wedge e_{j+1} \wedge e_n,$$

and we get with quite similar calculations as in Example A.1.7 d

$$h_1 \wedge \cdots \wedge h_{n-1} = \det \begin{pmatrix} e_1^* & \cdots & e_n^* \\ h_{11} & \cdots & h_{1n} \\ \vdots & & \vdots \\ h_{n-1,1} & \cdots & h_{n-1,n} \end{pmatrix}$$

(see Exercise A.1.3.5). The mapping $e_j \leftrightarrow e_j^*, j = 0, \ldots n$, defines an isomorphism between the vector spaces $C\ell(n)^1$ and $C\ell(n)^{n-1}$, which gives a duality in $C\ell(n)$ if we continue it to the other degrees. Expressions such as $dx^*[h_1, \ldots, h_{n-1}]$ and $h_1 \wedge \cdots \wedge h_{n-1}$ are dual.

A.1.3   Exercises

1. Prove Proposition A.1.2.

2. Prove that the mapping $\alpha_{p,q} : A^{p,q} \to A^{p+q}$, defined by

$$\alpha_{p,q}((\alpha_p, \alpha_q)\Phi) := \alpha_{p+q}(\Phi),$$

   is well defined and linear (see Proposition A.1.4).

3. For commutative algebras $\mathcal{A}$ prove the rule, see Proposition A.1.6 (ii),

$$\Phi_p \wedge \Phi_q = (-1)^{pq} \Phi_q \wedge \Phi_p.$$

4. Prove Proposition A.1.8

   (i) *For $y \in \Lambda^k(\mathbb{R}^n)$ we have $E_x(y) \in \Lambda^{k+1}(\mathbb{R}^n)$ and $J_x(y) \in \Lambda^{k-1}(\mathbb{R}^n)$. In particular for $y \in \Lambda^1(\mathbb{R}^n)$ we have $J_x(y) = x \cdot y$.*

   (ii) *For $x \in \Lambda^1(\mathbb{R}^n)$ and $y, z \in \Lambda(\mathbb{R}^n)$ the relation*

   $$E_x(y) \cdot z = y \cdot J_x(z)$$

   holds. That shows J_x to be the conjugate operator to E_x relative to the interior product.

   (iii) *In $\Lambda(\mathbb{R}^n)$ we have always $E_x^2 = J_x^2 = 0$.*

5. Using the definitions in Remark A.1.16 c), prove

$$e_j \wedge e_j^* = e_1 \wedge \cdots \wedge e_n$$

and for vectors $h_1, \ldots, h_{n-1}$,

$$h_1 \wedge \cdots \wedge h_{n-1} = \det \begin{pmatrix} e_1^* & \cdots & e_n^* \\ h_{11} & \cdots & h_{1n} \\ \vdots & & \vdots \\ h_{n-1,1} & \cdots & h_{n-1,n} \end{pmatrix}.$$

6. Prove

$$(dx_{i_1} \wedge \cdots \wedge dx_{i_q})[e_{j_1}, \ldots, e_{j_q}] = \delta_{i_1 j_1} \cdots \delta_{i_q j_q}.$$

A.2 Integration and manifolds

A.2.1 Integration

A.2.1.1 Integration in $\mathbb{R}^{n+1}$

The integral theorems proved in Section 7 are a basic and important tool in function theory in the plane as well as in space. For this purpose we give here the necessary definitions and theorems within the theory of integration. The functions to be integrated are differential forms and the domains of integration will be manifolds. For this integration the degree of the differential form and the dimension of the manifold have to be the same.

The simplest manifolds, although they are not usually referred to in this way, are the domains in $\mathbb{R}^{n+1}$ or in $\mathbb{H}$, where our functions are defined. We remind the reader that we have $\mathbb{R}^3$ as domain of definition in $\mathbb{H}$ if we look at $\mathbb{H}$ as $C\ell(2)$, otherwise we deal with mappings from $\mathbb{H}$ into $\mathbb{H}$, which are defined in $\mathbb{R}^4$ and where we have to choose $n = 3$.

We have seen in Appendix 1, Theorem A.1.13 (ii), that in $\mathbb{R}^{n+1}$ up to a factor only one differential form exists:

$$f(x)dx_0 \wedge dx_1 \wedge \cdots \wedge dx_n =: f(x)d\sigma.$$

We identify now $dx_0 \wedge dx_1 \wedge \cdots \wedge dx_n$ with the volume element $d\sigma$ of integration in the sense of Riemann or Lebesgue. The reason is the identical behavior if the variables are transformed, a fact we shall see later. For a function with real values we apply the usual notion of integration from real analysis:

Definition A.2.1 (Integral over a domain). If a function $f = \sum_A f_A e_A$ is given in a domain in $\mathbb{H}$ or $\mathbb{R}^{n+1}$ with values in $\mathbb{H}$ or $C\ell(n)$, then the integral over f is defined by integrating the components of f separately:

$$\int_G f(x)d\sigma := \sum_A e_A \int_G f_A(x)d\sigma.$$

If necessary the variable of integration is added as an index to $d\sigma$: $d\sigma_x$.

We give some rules for integrals:

Proposition A.2.2.   (i) *If two domains or open sets G_1 and G_2 are disjoint, then we have*

$$\int_{G_1 \cup G_2} f(x)d\sigma = \int_{G_1} f(x)d\sigma + \int_{G_2} f(x)d\sigma.$$

(ii) *In $C\ell(n)$*

$$\left| \int_G f(x)d\sigma \right| \leq \int_G |f(x)|d\sigma$$

holds.

(iii) *We have always*

$$\overline{\int_G f(x)d\sigma} = \int_G \overline{f(x)}d\sigma.$$

Improper integrals are included here, e.g., for G being an unbounded domain of integration.

Proof. (i) is simply the additivity of the integral relative to its domain.

(ii) We denote the integral to be estimated by J, for $J = 0$ nothing has to be proved. Let now $J \neq 0$, as the algebraic structure does not matter for the estimate. We look at f as a function with values in $C\ell(2^n)$, then J is a paravector and has an inverse. We have with $a := J/|J|$ obviously $|a| = 1$ and

$$
\begin{aligned}
|J| &= a^{-1}J = \int_G a^{-1}f(x)d\sigma = \int_G \mathrm{Sc}(a^{-1}f(x))d\sigma \\
&\leq \int_G |a^{-1}f(x)|d\sigma = \int_G |f(x)|d\sigma.
\end{aligned}
$$

Here Sc denotes also the real part in $\mathbb{C}$.

(iii) follows from the definition. $\square$

A.2.1.2 Transformation of variables

The transformation of variables is important for integration over manifolds, therefore we have to study how a differential form behaves if the variables are transformed.

Definition A.2.3. Let $G \subset \mathbb{R}^{n+1}$ and $H \subset \mathbb{R}^p$ ($1 \leq p \leq n+1$) be domains and let $\varphi : H \to G$ with $x = \varphi(t)$ be an injective mapping from H to G with $\varphi \in C^{m+1}(H)$. The derivative φ' is assumed to have rank p at all points of G. Let $\omega_p \in C^m(G)$ be a differential form of degree p in G, then the differential form $\omega_p \circ \varphi$ in H is defined by ($h_i \in \mathbb{R}^p$)

$$(\omega_p \circ \varphi)(t)[h_1,\ldots,h_p] := \omega_p(\varphi(t))[\varphi'(t)[h_1],\ldots,\varphi'(t)[h_p]].$$

We remark that φ' is a $((n+1) \times p)$-matrix and it is not always quadratic.

Proposition A.2.4. *The form $\omega_p \circ \varphi$ is a differential form of degree p in H that is m-times continuously differentiable.*

Proof. Linearity and anti-symmetry follow from these properties of ω_p and the linearity of φ', the differentiability follows from the chain rule. For the chain rule we need $\varphi \in C^{m+1}(H)$ to have $\varphi' \in C^m(H)$. $\square$

Two differential forms ω_p and ω_q satisfy the rules:

Proposition A.2.5 (Rules for variable transformation).

$$
\begin{array}{lll}
\text{a)} & (\omega_p \wedge \omega_q) \circ \varphi & = (\omega_p \circ \varphi) \wedge (\omega_q \circ \varphi), \\
\text{b)} & (d\omega_p) \circ \varphi & = d(\omega_p \circ \varphi), \\
\text{c)} & (\omega_p \circ \varphi) \circ \psi & = \omega_p \circ (\psi \circ \varphi).
\end{array}
$$

Proof. We only sketch the proof (see Exercise A.2.3.1): One has to use the representation $\omega_p = \alpha_p \Phi$ with the anti-symmetrization α_p and a multilinear mapping Φ as in Appendix 1. Then φ and α_p commute as one has only to substitute into sums. Together with the definition of the exterior product assertion a) follows. The assertion in b) can be proved using the definition of the exterior derivative and the symmetry of φ'' according to the theorem of Schwarz. Finally the assertion of c) follows similarly from the chain rule for $\psi \circ \varphi$. □

The behavior of the canonical representation is also important when the variables are transformed:

Proposition A.2.6. *Let the differential form $\omega_p(x)$ be given in canonical representation:*

$$
\omega_p(x) = \sum_{i_1 < \cdots < i_p} a_{i_1 \ldots i_p}(x)\, dx_{i_1} \wedge \cdots \wedge dx_{i_p}.
$$

For a transformation of variables $x = \varphi(t)$ we then have

$$
(\omega_p \circ \varphi)(t) = \left(\sum_{i_1 < \cdots < i_p} (a_{i_1 \ldots i_p} \circ \varphi)(t)\, \frac{\partial(\varphi_{i_1}, \ldots, \varphi_{i_p})}{\partial(t_1, \ldots, t_p)} \right) dt_1 \wedge \cdots \wedge dt_p.
$$

Proof. The proof is again only sketched. Similarly as in Remark A.1.16 c) the substitution of $\varphi'(t)[h_1], \ldots, \varphi'(t)[h_p]$ in $dx_{i_1} \wedge \cdots \wedge dx_{i_p}$ gives the Jacobians written in the assertion and the expression for $dt_1 \wedge \cdots \wedge dt_p[h_1, \ldots, h_p]$. □

Remark A.2.7. a) The case $p = n + 1$, dealt with in the last subsection, is of special interest for us. If we transform the variables in $f(x)dx_0 \wedge \cdots \wedge dx_n$ we get, following the last proposition,

$$
f(\varphi(t))\frac{\partial(\varphi_0, \ldots, \varphi_n)}{\partial(t_0, \ldots, t_n)}dt_0 \wedge \cdots \wedge dt_n = f(\varphi(t))\det\varphi'(t)dt_0 \wedge \cdots \wedge dt_n.
$$

This is precisely the transformation rule for integrals, known from real analysis, with the so-called *Jacobian* $J(\varphi) = \det \varphi'$. Here we see the reason to identify $dx_0 \wedge \cdots \wedge dx_n$ with the volume element of integration $d\sigma$.

b) We are now going to formulate the so-called Poincaré lemma:

Proposition A.2.8 (Poincaré's lemma). *Let G be a star-shaped domain relative to $x_0 \in G$, i.e., for all $x \in G$ the line from x_0 to x lies completely in G. If for a differential form $\omega_p \in C^1(G)$ we have $d\omega_p = 0$ in G, then ω_p is constant for $p = 0$ and for $p > 0$ a differential form ω_{p-1} exists with*

$$
\omega_p = d\omega_{p-1}.
$$

One approaches the proof by using the integral

$$\omega_{p-1}(x)[h_1, \ldots, h_{p-1}] := \int_0^1 t^{p-1} \omega_p(tx)[x, h_1, \ldots, h_{p-1}] dt.$$

The reader is asked to prove the statement (see Exercise A.2.3.2).

A.2.1.3 Manifolds and integration

We are looking now at the domains of integration in $\mathbb{R}^{n+1}$ which have at most dimension n. We start with a definition:

Definition A.2.9. (i) A *p-dimensional manifold* M_p in $\mathbb{R}^{n+1}$ is a set with the following properties:

For every $x_0 \in M_p$ a relatively open neighborhood $U(x_0)$ exists together with a homeomorphic mapping $\varphi : H \to U(x_0)$ of a domain $H \subset \mathbb{R}^p$ onto $U(x_0)$, which is bijective and continuous in both directions. (U, ψ) with $\psi = \varphi^{-1}$ is called a *chart*, also a *(local) coordinate system*, a set of charts is called an *atlas*.

(ii) The manifold M_p is defined to be k-times continuously differentiable ($\in C^k$), if for every pair of charts (U_1, ψ_1) and (U_2, ψ_2) with $U_1 \cap U_2 \neq \emptyset$ the mapping

$$\chi = \psi_2 \circ \varphi_1 : \psi_1(U_1 \cap U_2) \to \psi_2(U_1 \cap U_2)$$

is k-times continuously differentiable. Moreover the $\varphi_i \in C^k$ are assumed to have rank $\varphi_i' = p$ at all points of H. A manifold, which is at least in C^1, is called *smooth manifold* (sometimes for this notion C^∞ is assumed).

(iii) Moreover M_p is called *orientable* if $\det \chi' > 0$ always.

(iv) The manifold M_p has a *boundary*, if charts exist which map onto a half-ball such that the plane boundary component of the half-ball corresponds to a part of the boundary of M_p.

We deal only with smooth manifolds, at most with structures which are composed by finitely many smooth manifolds, thus we speak of *piecewise smooth manifolds*. We remark that

$$\varphi'(t)[e_i] = \frac{\partial \varphi(t)}{\partial t_i};$$

these vectors span the *tangential space* $T_x(M_p)$ at the point $x = \varphi(t)$. The tangential space is independent of the special chart as the passage to another chart only changes the basis of the tangential space.

Unfortunately our intuition fails in higher dimensions and we can use only analogies with dimensions 2 and 3. Here some examples follow:

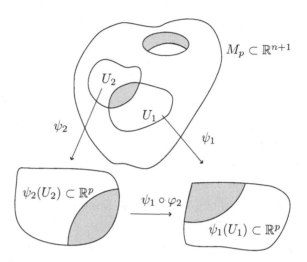

Figure A.2

Example A.2.10. a) **Curves.** In every dimension we have 1-dimensional manifolds, called *curves*; for each curve, the corresponding mapping φ is called a *parametric representation* of the curve. Generally for a curve, one mapping $\varphi : [a, b] \to \mathbb{R}^{n+1}$ is sufficient. The point $\varphi(a)$ is called the *initial point* of the curve, $\varphi(b)$ the *endpoint*. If the initial and endpoint coincide the curve is called *closed*. If φ is injective, i.e., it meets no point twice — or $\varphi(t_1) \neq \varphi(t_2)$ holds for $t_1 \neq t_2$ — (with the possible exception of initial and endpoint), the curve is called a *Jordan curve*. A strictly monotone and surjective function $\tau : [a, b] \to [\alpha, \beta]$, which is continuously differentiable and has a positive derivative in the whole interval including the boundary points, is called an *admissible parameter transformation*.

A curve is called *smooth*, if at all points $\varphi'(t) \neq 0$, *piecewise smooth*, if it can be split into finitely many smooth curves.

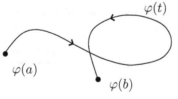

Figure A.3

If Γ is a curve represented by $\varphi(t)$ with $t \in [a, b]$ then $-\Gamma$ denotes the curve run in the opposite direction. A possible parametrization is $\tilde{\varphi}(\tau) := \varphi(-\tau)$ with $\tau \in [-b, -a]$.

If Γ_1 and Γ_2 are two smooth or piecewise smooth curves, where the endpoint of Γ_1 coincides with the initial point of Γ_2, one can join both parametric representations and gets a piecewise smooth sum $\Gamma_1 + \Gamma_2$. But we allow also formal sums of finitely many curves, which may occur as boundaries of plane domains.

b) Curves may occur as boundaries of two-dimensional manifolds. One sees this simply in the plane, e.g., a circle is the boundary of a disc. If we choose for a circle the parametric representation

$$\varphi(t) = z_0 + R(\cos t + i \sin t), \quad 0 \leq t \leq 2\pi,$$

the interpretation as a manifold needs to remove the special situation of the point $z_1 := z_0 + R$. The corresponding $\psi = \varphi^{-1}$ constitutes together with the circle and without the point z_1 a chart; a second chart has to include the point z_1, e.g., by $t \in [-\pi, \pi]$. These two charts would constitute an atlas of the circle.

c) A quadrangle in the plane has four lines as boundary; the whole boundary is a piecewise smooth curve, but not a manifold. One may call it a piecewise smooth manifold.

d) In $\mathbb{R}^3$ besides curves we have two-dimensional manifolds, e.g., the sphere with center in the origin, which may be described by spherical coordinates: For $-\pi < t_1 < \pi$, $-\frac{\pi}{2} < t_2 < \frac{\pi}{2}$ one has

$$x_0 = R\cos t_1 \cos t_2, \quad x_1 = R\sin t_1 \cos t_2, \quad x_2 = R\sin t_2$$

(see Example A.2.17 a). Similarly to the circle in the plane this chart represents the sphere; to get an atlas at least one further chart has to be added.

e) Similarly to the quadrangle the boundary of a cuboid in $\mathbb{R}^3$ is formed by six quadrangles; it is not a smooth manifold, but it is composed of finitely many smooth manifolds. So we may also call this boundary a piecewise smooth manifold. The edges have to be looked at as singular sets.

f) Another example is the *torus* (cf. Figure A.4). The core of such a ring may be described by a circle of radius R in the plane $x_0 = 0$; the torus is then parameterized by circles of radius r around the points of the core.

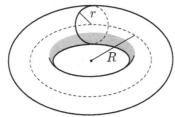

Figure A.4

We get with $0 < t_1, t_2 < 2\pi$,

$$x_0 = r \sin t_1,$$
$$x_1 = (R + r \cos t_1) \cos t_2,$$
$$x_2 = (R + r \cos t_1) \sin t_2.$$

One of the circles around points of the core is not covered; for an atlas of the torus at least a second chart is necessary, where one has to define t_1 and t_2 suitably.

g) Further examples are the projective plane or the Möbius strip. The last one arises if a paper strip is glued together twisted, a fact that may be described also mathematically. It is an example of a non-orientable manifold.

Remark A.2.11. The *orientation of a manifold* M_p is generally chosen such that for a chart (U, ψ) with $\psi^{-1} = \varphi : H \to U$ in H the usual orientation of $\mathbb{R}^p$ is given.

For the *orientation of the boundary* ∂M_p we proceed as follows: Let (U, ψ) be a chart mapping a part of the boundary into a (hyper)plane. We choose this plane to be $t_1 = 0$ so that the image of U lies in $\{t_1 < 0\}$. Then we choose at each point t of the boundary a coordinate system $e_1, \ldots, e_p$ oriented correspondingly to $\mathbb{R}^p$ such that e_1 points to the positive t_1-direction. The other $e_2, \ldots, e_p$ correspond to tangential vectors of the boundary of M_p; they span the tangential space of ∂M_p. This definition is independent from the chosen chart, if we pass to another chart the Jacobian is positive (see Definition A.2.9 (iii)) so that the orientation is not influenced. The other orientation is then denoted by $-\partial M_p$.

We are now going to integrate over such manifolds:

Definition A.2.12. Let $M_p \in C^1$ be an orientable manifold of dimension p and let one of its charts (U, ψ) be given with $\psi^{-1} = \varphi : H \to U$. Let ω_p be a differential form of degree p in a domain $G \supset M_p$. Then we define

$$\int_U \omega_p := \int_H \omega_p \circ \varphi := \int_H \omega_p(\varphi(t))[\varphi'(t)[e_1], \ldots, \varphi'(t)[e_p]] d\sigma_t$$

with the canonical basis $e_1, \ldots, e_p$ of $\mathbb{R}^p \supset H \ni t$.

We have to confirm that this is well defined, i.e., that another chart leads to the same result:

So, let $\tilde{\psi}$ belong to another chart $(\tilde{U}, \tilde{\psi})$, for simplicity we assume that the sets U and $\tilde{U}$ in M_p are the same. Moreover let $t = \chi(\tilde{t}) = \psi \circ \tilde{\varphi}(\tilde{t})$ be the mapping from $\tilde{H}$ to H with $\det \chi' > 0$ and $\varphi \circ \chi = \tilde{\varphi}$. Then we get

$$\int_H \omega_p(\varphi(t))[\varphi'(t)[e_1], \ldots, \varphi'(t)[e_p]] d\sigma_t$$

$$= \int_{\tilde{H}} \omega_p(\varphi(\chi(\tilde{t})))[\varphi'(\chi(\tilde{t}))[e_1], \ldots, \varphi'(\chi(\tilde{t}))[e_p]] \det \chi'(\tilde{t}) d\sigma_{\tilde{t}}$$

$$= \int_{\tilde{H}} \omega_p(\tilde{\varphi}(\tilde{t}))[\tilde{\varphi}'(\tilde{t}) \circ \chi'^{-1}(\chi(\tilde{t}))[e_1], \ldots, \tilde{\varphi}'(\tilde{t}) \circ \chi'^{-1}(\chi(\tilde{t}))[e_p]] \det \chi'(\tilde{t}) d\sigma_{\tilde{t}}.$$

With $\chi'^{-1}[e_j] =: h_j =: \sum_{i=1}^{p} \alpha_{ij} \tilde{e}_i$ and $\omega_p(\tilde{\varphi}(\tilde{t})) =: \Phi(\tilde{t})$,

$$
\begin{aligned}
\Phi[\tilde{\varphi}'[h_1], \ldots, \tilde{\varphi}'[h_p]] &= \sum_{i_1,\ldots,i_p=1}^{p} \Phi[\tilde{\varphi}'[\tilde{e}_{i_1}], \ldots, \tilde{\varphi}'[\tilde{e}_{i_p}]] \alpha_{i_1 1} \cdots \alpha_{i_p p} \\
&= \sum_{\sigma \in \mathrm{perm}(p)} \Phi[\tilde{\varphi}'[\tilde{e}_1], \ldots, \tilde{\varphi}'[\tilde{e}_p]] (\mathrm{sgn}\,\sigma) \alpha_{1\sigma(1)} \cdots \alpha_{p\sigma(p)} \\
&= \Phi[\tilde{\varphi}'[\tilde{e}_1], \ldots, \tilde{\varphi}'[\tilde{e}_p]] \det(\alpha_{ij})
\end{aligned}
$$

follows. We know that $h_j = \chi'^{-1}[e_j] = \partial \chi^{-1}/\partial t_j$, thus $\alpha_{ij} = \partial \chi_i^{-1}/\partial t_j$ and

$$
\det(\alpha_{ij}) = \det\left(\frac{\partial \chi_i^{-1}}{\partial t_j}\right) = \det \chi'^{-1} = \frac{1}{\det \chi'}.
$$

This gives at last as desired

$$
\int_H \omega_p(\varphi(t))[\varphi'(t)[e_1], \ldots, \varphi'(t)[e_p]] d\sigma_t = \int_{\tilde{H}} \omega_p(\tilde{\varphi}(\tilde{t}))[\tilde{\varphi}'(\tilde{t})[\tilde{e}_1], \ldots, \tilde{\varphi}'(\tilde{t}[\tilde{e}_p]] d\sigma_{\tilde{t}}.
$$

So indeed $\int_U \omega_p$ is well defined. The next step is to define the integral over the whole manifold. We need here an important theorem:

Theorem A.2.13 (Partition of unity). *Let $K \subset \mathbb{R}^p$ be compact and $(U_i)_{i \in I}$ be an open covering of K. Then functions $f_i : \mathbb{R}^p \to [0,1]$ exist with $f_i \in C^\infty$, $i \in I$, such that with*

$$
\mathrm{supp}\, f_i := \overline{\{x : x \in \mathbb{R}^p, f_i(x) \neq 0\}}
$$

we have:

a) $\mathrm{supp}\, f_i \subset U_i$.

b) *For all $x \in K$ exists an index $i \in I$ with $0 \le f_i(x) \le 1$. $\sum_i f_i(x) \le 1$ for all $x \in \mathbb{R}^p$.*

c) $\sum_i f_i(x) = 1$ *for all $x \in K$.*

d) *For all $x \in K$ there exists a neighborhood U such that only finitely many functions $f_i (i \in I)$ are not vanishing on U.*

Here $\mathrm{supp}\, f$ is called the *support* of the function f. For the proof we refer to [22]. Functions of the form

$$
f(x) = \begin{cases} c \exp\left(\frac{1}{|x|^2 - r^2}\right), & |x| < r,\ c \in \mathbb{R}, \\ 0, & |x| \ge r \end{cases}
$$

are the essential device for the proof. Now we are able to define:

Definition A.2.14. Let $M_p \in C^1$ be an orientable manifold and ω_p be a differential form in a neighborhood of M_p. We assume either ω_p to have compact support or M_p to be contained in a compact set. Then one uses the partition of unity so that $\omega_p = \sum_i f_i \omega_p$ and $f_i \omega_p$ has its support in a chart U_i of an atlas of M_p. Now one is able to define

$$\int_{M_p} \omega_p := \sum_i \int_{U_i} f_i \omega_p.$$

The independence of the chosen atlas, resp. the given covering, follows if we look at the intersections $U_i \cap V_j$, the charts $(V_j, \tilde{\psi}_j)$ being from the second partition.

Now we shall look at some special cases; interesting for us are the degrees $p = 1$ and $p = n$:

Example A.2.15. a) Case $p = 1$: Our manifold is a curve $M_1 = \Gamma$ given by $x = \varphi(t)$, $t \in [t_0, t_1]$. A differential form of degree 1 is of the form

$$\omega_1(x) = \sum_{i=0}^{n} a_i(x) dx_i,$$

the $a_i(x)$ being functions with values in the considered algebra. Because of

$$\frac{d\varphi(t)}{dt} = (x'_0(t), \ldots, x'_n(t))$$

we get for the integral

$$\int_\gamma \omega_1 = \int_{t_0}^{t_1} \sum_{i=0}^{n} a_i(\varphi(t)) x'_i(t) dt.$$

For $\omega_1 = f(x) dx g(x)$ the integral simplifies to

$$\int_\gamma f(x) dx g(x) = \int_{t_0}^{t_1} f(\varphi(t)) \varphi'(t) g(\varphi(t)) dt.$$

For these line integrals the following rules hold:

$$
\begin{array}{rcll}
\int_{-\gamma} f(x) dx & = & -\int_\gamma f(x) dx & \text{(orientation)}, \\
\int_{\gamma_1} f(x) dx + \int_{\gamma_2} f(x) dx & = & \int_{\gamma_1 + \gamma_2} f(x) dx & \text{(additivity)}, \\
\int_\gamma (\alpha f(x) + \beta g(x)) dx & = & \alpha \int_\gamma f(x) dx + \beta \int_\gamma g(x) dx & \text{(linearity)}, \\
\left| \int_\gamma f(x) dx \right| & \leq & \int_{t_0}^{t_1} |f(x) x'(t)| dt & \text{(modulus inequality)}.
\end{array}
$$

For the proof see Exercise A.2.3.3. Analogous formulas hold if the form $f(x) dx g(x)$ is integrated, which may happen if the algebra is not commutative.

b) Case $n = 2$ and $p = 1$: We are in the plane and the line integrals setting $x_0 =: x$ and $x_1 =: y$ have the form

$$\int_\gamma a dx + b dy = \int_{t_0}^{t_1} (a(\varphi(t))x'(t) + b(\varphi(t))y'(t)) dt.$$

One can view $a dx + b dy$ as a scalar product, thus $(a, b) \cdot (x', y') dt$. Here (x', y') is a tangential vector of the curve; the integral consequently sums up the components of the vector field (a, b) which point to the tangential direction. If we view the integrand as a product $(-b, a) \cdot (-y', x')$ the components of $(-b, a)$ in normal direction are summed up. This measures, e.g., a stream flow through the curve. For the theory of curves it is important that one can use a parameter called the *arc length*:

$$s(t) = \int_{t_0}^{t} |(x'(\tau), y'(\tau))| d\tau.$$

The real number $s(b)$ gives the *length* of the curve. This parameter simplifies many formulas but we shall not go into detail here.

c) Case $p = n$: We have to integrate over n-dimensional surfaces in $\mathbb{R}^{n+1}$, where the differential form $\omega_n = f(x) dx^* g(x)$ is of interest to us. In Remark A.1.16 c) we have already studied this form obtaining

$$dx^*[h_1, \ldots, h_n] = \underset{i=1}{\overset{n}{\times}} h_i.$$

This is a vector product of the n vectors h_i. While integrating we have to use

$$h_i = \varphi'(t)[e_i] = \frac{\partial \varphi(t)}{\partial t_i},$$

thus

$$\int_{M_n} f(x) dx^* g(x) = \int_H f(\varphi(t)) \left(\underset{i=1}{\overset{n}{\times}} \frac{\partial \varphi(t))}{\partial t_i} \right) g(\varphi(t)) d\sigma_t.$$

Here we assume for simplicity that only one chart is sufficient, otherwise we have to modify things correspondingly. The vector product has the following meaning: One multiplies the n tangential vectors, the vector product is orthogonal to all of them, thus it is the normal to our manifold or surface. Therefore dx^* is the oriented surface element for the integration over the manifold with its value in the corresponding algebra, the notation do or do_x is used here:

$$dx^* = \left(\underset{i=1}{\overset{n}{\times}} \frac{\partial \varphi(t)}{\partial t_i} \right) do_t =: do.$$

In any case one has to test the correct orientation.

Similarly to curves, the integral

$$\int_{M_n} |dx^*| = \int_H \left| \underset{i=1}{\overset{n}{\times}} \frac{\partial \varphi}{\partial t_i} \right| d\sigma_t = \int_H |do|$$

gives the volume or area of the considered manifold, thus for a two-dimensional surface in $\mathbb{R}^3$ the usual area.

The next example requires a proposition we already found for the quaternions (Proposition 2.41):

Proposition A.2.16 (Lagrange's identity). *Let h_i, $i = 1, \ldots, n$, be vectors in $\mathbb{R}^{n+1}$ and let*

$$h := \underset{i=1}{\overset{n}{\times}} h_i.$$

Then the identity of Lagrange holds:

$$|h|^2 = \det (h_i \cdot h_j)_{i,j=1,\ldots,n}.$$

Proof. Following Remark A.1.16 c) we have for an arbitrary paravector a,

$$a \cdot h = \det \begin{pmatrix} a \\ h_1 \\ \vdots \\ h_n \end{pmatrix}, \quad \text{thus}$$

$$(h \cdot h)^2 = \det \begin{pmatrix} h \\ h_1 \\ \vdots \\ h_n \end{pmatrix} \det(h\, h_1 \ldots h_n) = \det \begin{pmatrix} h \cdot h & h \cdot h_1 & \cdots & h \cdot h_n \\ h_1 \cdot h & h_1 \cdot h_1 & \cdots & h_1 \cdot h_n \\ \vdots & \vdots & & \vdots \\ h_n \cdot h & h_n \cdot h_1 & \cdots & h_n \cdot h_n \end{pmatrix}.$$

Because of $h \cdot h_i = 0$, $i = 1, \ldots, n$, we develop using the first row and get as desired

$$|h|^4 = (h \cdot h)^2 = |h|^2 \det (h_i \cdot h_j). \qquad \square$$

Example A.2.17. a) As an example of integration over a manifold we shall compute the surface area of the unit sphere $S^n \subset \mathbb{R}^{n+1}$, which we need for our integral theorems. On the unit sphere

$$x_0^2 + x_1^2 + \cdots + x_n^2 = 1,$$

x_n runs at most between -1 and 1, therefore t_n is uniquely determined by

$$x_n = \sin t_n$$

in the interval $I := [-\pi/2, \pi/2]$. From $x_0^2 + \cdots + x_{n-1}^2 = 1 - x_n^2 = \cos^2 t_n$ follows that x_{n-1} runs between $-\cos t_n$ and $\cos t_n$, thus a t_{n-1} is uniquely determined in I by

$$x_{n-1} = \cos t_n \sin t_{n-1}.$$

The procedure is to be continued up to

$$x_2 = \cos t_n \cdots \cos t_3 \sin t_2,$$
$$x_0^2 + x_1^2 = \cos^2 t_n \cdots \cos^2 t_2.$$

Then we have to define

$$x_0 = \cos t_n \cdots \cos t_2 \cos t_1,$$
$$x_1 = \cos t_n \cdots \cos t_2 \sin t_1,$$

with $t_1 \in [-\pi, \pi]$, as it is known from the plane. The procedure may be summarized as

$$x_i = \cos t_n \cdots \cos t_{i+1} \sin t_i, \quad i = 0, \ldots, n, \quad t_0 := \frac{\pi}{2}.$$

We get further

$$\frac{\partial x_i}{\partial t_j} = \begin{cases} 0 & \text{for} \quad j < i, \\ \cos t_n \cdots \cos t_i & \text{for} \quad j = i, \\ -\cos t_n \cdots \cos t_{j+1} \sin t_j \cos t_{j-1} \cdots \cos t_{i+1} \sin t_i & \text{for} \quad j > i. \end{cases}$$

The reader may prove in Exercise A.2.3.6 that

$$\frac{\partial x}{\partial t_j} \cdot \frac{\partial x}{\partial t_j} = \prod_{k=j+1}^{n} \cos^2 t_k$$

and for $i \neq j$,

$$\frac{\partial x}{\partial t_i} \cdot \frac{\partial x}{\partial t_j} = 0.$$

Using the identity of Lagrange, Proposition A.2.16 for the surface element of the sphere we get

$$|do| = \left(\det \left(\frac{\partial x}{\partial t_i} \cdot \frac{\partial x}{\partial t_j} \right) \right)^{1/2} d\sigma_t$$

and with some simple calculations

$$|do| = \prod_{i=2}^{n} (\cos t_n \cdots \cos t_{i+1}) d\sigma_t$$
$$= \cos t_2 \cos^2 t_3 \cdots \cos^{n-1} t_n d\sigma_t.$$

This gives for the surface area σ_n of the sphere S^n,

$$\sigma_n = \int_{t_1=-\pi}^{\pi} \int_{t_2=-\pi/2}^{\pi/2} \cdots \int_{t_n=-\pi/2}^{\pi/2} \cos t_2 \cdots \cos^{n-1} t_n \, dt_n \cdots dt_1$$
$$= 2\pi \prod_{k=1}^{n-1} \int_{t=-\pi/2}^{\pi/2} \cos^k t \, dt =: 2\pi \prod_{k=1}^{n-1} I_k.$$

Using partial integration we get a recursion formula for the I_k with $k > 1$,

$$\int_{-\pi/2}^{\pi/2} \cos^k t\, dt = 2 \int_0^{\pi/2} (\cos^{k-2} t - \cos^{k-2} t \sin^2 t) dt = I_{k-2} - \frac{1}{k-1} I_k$$

or

$$I_k = \frac{k-1}{k} I_{k-2}.$$

Together with $I_0 = \pi$, $I_1 = 2$ one can compute

$$I_{2k} = \frac{(2k)!\pi}{2^{2k}(k!)^2}, \quad I_{2k+1} = \frac{2^{2k+1}(k!)^2}{(2k+1)!}.$$

We know from Section 13.1 that the last expression may be written using the Gamma function

$$I_k = \frac{\Gamma\left(\frac{1}{2}\right) \Gamma\left(\frac{n+1}{2}\right)}{\Gamma\left(\frac{n+2}{2}\right)}.$$

From $\sigma_n = I_{n-1}\sigma_{n-1}$ and $\sigma_0 = 2$, resp. $\sigma_1 = 2\pi$, it follows at last that

$$\sigma_n = 2 \frac{\Gamma^{n+1}\left(\frac{1}{2}\right)}{\Gamma\left(\frac{n+1}{2}\right)}.$$

This looks quite symmetric and short but the effort for the calculation is large. To find the known value for σ_1 we have to refer to $\Gamma(\frac{1}{2}) = \sqrt{\pi}$ and $\Gamma(1) = 1$ (Theorems 13.2 (iii) and 13.6 (i)).

b) The calculations in the last example will be used now to deal with the transformation to spherical coordinates for the ball. There we may have coordinates

$$y = rx$$

with x from the example a). If we move to the coordinates $r, t_1, \ldots, t_n$ we are interested in the Jacobian

$$J := \frac{\partial(y_0, y_1, \ldots, y_n)}{\partial(r, t_1, \ldots, t_n)}.$$

For it we have

$$\frac{\partial y}{\partial r} = x, \quad \frac{\partial y}{\partial t_i} = r \frac{\partial x}{\partial t_i}$$

with the derivatives from example a). On account of $x \cdot x = 1$ it follows that

$$x \cdot \frac{\partial x}{\partial t_i} = 0,$$

so that for the Jacobian with $h_i := \partial x / \partial t_i$ we get

$$J^2 = r^{2n} \det \begin{pmatrix} x \\ h_1 \\ \vdots \\ h_n \end{pmatrix} \det(x\, h_1 \ldots h_n) = r^{2n} \det \begin{pmatrix} 1 & 0 & \cdots & 0 \\ 0 & h_1 \cdot h_1 & \cdots & 0 \\ \vdots & \vdots & & \vdots \\ 0 & 0 & \cdots & h_n \cdot h_n \end{pmatrix}.$$

To determine the sign of J one may insert $t_1 = t_2 = \cdots = t_n = 0$. This gives just the unit matrix, so J is positive and

$$d\sigma_y = r^n dr |do_x|.$$

For an integral over a ball $B_R(0)$ of radius R with center at the origin we obtain

$$\int_{B_R(0)} f(y) d\sigma_y = \int_0^R r^n \int_{\partial B_1(0)} f(rx) |do_x| dr = \frac{\sigma_n}{n+1} R^{n+1}.$$

We know this formula very well from the plane and the space $\mathbb{R}^3$.

c) We can see yet from the considerations in the example b) that $x, h_1, \ldots, h_n$ has a positive determinant. Thus it is a system of vectors as we assumed for the orientation of a boundary manifold. The paravector x is the outer normal of the sphere (with radius R around the origin), as x equals the vector product of the tangential vectors $h_1, \ldots, h_n$. Therefore for such a ball we have

$$dy^* = do_y = x|do_y| = R^n x |do_x|.$$

A.2.2   Theorems of Stokes, Gauß, and Green

A.2.2.1   Theorem of Stokes

We deal now with a fundamental theorem of analysis which is also an important tool for us. It generalizes the fundamental theorem of differential and integral calculus in $\mathbb{R}$,

$$f(b) - f(a) = \int_a^b f'(x) dx,$$

to manifolds and differential forms. The fundamental theorem shows the difference of the values of a function f at the boundary points to be equal to the integral of the derivative $f'(x)$ over the whole interval. Quite similarly the theorem of the Irish mathematician GEORGE GABRIEL STOKES (1819–1903) states the following:

Theorem A.2.18 (Theorem of Stokes). *Let $M_{p+1} \in C^1$ be an orientable, bounded, and smooth manifold of dimension $p+1$ with sufficiently smooth boundary ∂M_{p+1}*

*which may be oriented according to Remark A.2.11. Let ω_p be a differential form
of degree p, continuously differentiable in a neighborhood of M_{p+1}. We then have*

$$\int\limits_{\partial M_{p+1}} \omega_p = \int\limits_{M_{p+1}} d\omega_p.$$

This corresponds precisely to the fundamental theorem of differential and integral
calculus: the integral of the differential form over the boundary of the manifold
equals the integral of the derivative of the differential form over the manifold itself.
We are not able to go into the deep proof of this theorem, thus we have to refer
to the literature, e.g., Amann and Escher [8].

One of the difficulties of Stoke's theorem is the precise formulation of the assumptions regarding the manifold's boundary. It has to be made by finitely many
smooth manifolds whose boundaries must have in turn p-dimensional measure 0.
We skip the precise definition as we shall deal in this book only with manifolds and
boundaries which fulfil the 'sufficient' from the theorem. We shall deal with discs,
balls, or cuboids, very rarely with more complicated manifolds. The cuboid is a
good example for the boundary to have wedges and vertices, but not too many.

A.2.2.2 Theorem of Gauß

We shall look at the theorem of Stokes in the plane, where it is named mostly
after Gauß and the Russian mathematician MIKHAIL VASILEVICH OSTROGRADSKI
(1801–1862). Here we are able to say a bit more about the assumptions of the
theorem of Stokes. We start with a definition:

Definition A.2.19 (Connectivity of a domain).   (i) A compact and connected set
   is called a *continuum.*

(ii) A domain G is called k-*times connected* if its boundary ∂G consists of pre-
    cisely n disjoint continua. For the *1-time* or *simple connectivity* $\partial G = \emptyset$ is
    possible. If k is not specified one speaks also of *finite connectivity.*

(iii) If no k exists as assumed in (ii) the domain is of *infinite connectivity.*

This definition holds independently of the dimension. But in the plane we are able
to define simple sufficient assumptions for the boundary, if the domain is of finite
connectivity.

Theorem A.2.20 (Theorem of Gauß–Ostrogradski). *Let G be a plane domain of finite connectivity; the boundary ∂G has to consist of finitely many, piecewise smooth
curves which are oriented such that G lies on their left side. Let the functions u
and v be in $C^1(\overline{G})$; we then have*

$$\int\limits_{\partial G} u(x,y)dx + v(x,y)dy = \int\limits_{G} (v_x(x,y) - u_y(x,y))d\sigma_{(x,y)}.$$

Proof. The proof is very simple: The manifold M_2 is here the domain G, the differential form is $\omega_1 = udx + vdy$. Then the assertion follows at once from Stoke's theorem if one considers

$$d\omega_1 = du \wedge dx + dv \wedge dy = (-u_y + v_x)dx \wedge dy. \qquad \square$$

We shall write the theorem in complex form which is more convenient for us:

Theorem A.2.21 (Theorem of Gauß in $\mathbb{C}$). *Making the same assumptions as in the last theorem with a function $f \in C^1(\overline{G})$ we have*

$$\int_{\partial G} f(z)dz = 2i \int_G \partial_{\overline{z}} f(z)d\sigma.$$

Since $dz^* = dy - idx = -idz$ we may formulate the assertion of the theorem in the form

$$\int_{\partial G} f(z)dz^* = \int_G \overline{\partial} f(z)d\sigma,$$

as we shall do in higher dimensions. Here we have used the notation $\overline{\partial} = 2\partial_{\overline{z}}$ from the beginning of Section 5.1. The proof is again a simple conclusion from Stoke's theorem, clearly also from the above theorem of Gauß–Ostrogradski:

$$d(fdz) = df \wedge dz = (\partial_{\overline{z}} f)d\overline{z} \wedge dz$$

and

$$d\overline{z} \wedge dz = (dx - idy) \wedge (dx + idy) = 2idx \wedge dy = 2id\sigma.$$

In higher dimensions we give the theorem of Gauß directly in an algebraic form which is more convenient for us:

Theorem A.2.22 (Theorem of Gauß in $C\ell(n)$). *Let $G \subset \mathbb{R}^{n+1}$ be a domain of finite connectivity with sufficiently smooth boundary ∂G, let the boundary be oriented correspondingly to Remark A.2.11, i.e., with an outward pointing normal. Let $f, g \in C^1(\overline{G})$. We then have*

$$\int_{\partial G} f dx^* g = \int_G \left((f\overline{\partial})g + f\overline{\partial}g \right) d\sigma.$$

Proof. Following the product rule for the total differential in Proposition A.1.12 (v) we have

$$d(fdx^*g) = df \wedge dx^*g + (-1)^n fdx^* \wedge dg,$$

and similarly to the proof of Theorem 5.11,

$$df \wedge dx^* = \sum_{i=0}^{n} \frac{\partial f}{\partial x_i} e_i dx_i \wedge dx_i^*,$$

and analogously for $dx^* \wedge dg$. Finally the assertion follows from

$$dx_i \wedge dx_i^* = d\sigma \quad \text{resp.} \quad dx_i^* \wedge dx_i = (-1)^n d\sigma. \qquad \square$$

Remark A.2.23. For $G \subset \mathbb{R}^n$ with analogous assumptions and the Dirac operator D according to Theorems 5.12, resp. 5.20, we have

$$\int_{\partial G} f d\mathbf{x}^* g = \int_G ((fD)g + fDg) d\sigma.$$

A.2.2.3   Theorem of Green

Unfortunately the names for the last and the next theorems are not fixed. We name the next theorem after the English mathematician GEORGE GREEN (1793–1841), although it follows directly from the theorem of Gauß when choosing $g = 1$ and substituting f by $f\partial$.

Theorem A.2.24 (Green's formula). *Let G be a domain of finite connectivity in $\mathbb{R}^{n+1}$ and with sufficiently smooth boundary ∂G appropriately oriented. For $f \in C^2(\overline{G})$ we have*

$$\int_{\partial G} (f\partial) dx^* = \int_G \Delta f d\sigma.$$

Here

$$\Delta = \partial \overline{\partial} = \overline{\partial} \partial = \sum_{i=0}^{n} \frac{\partial^2}{\partial x_i^2}$$

is the Laplace operator.

The last theorem in this appendix will be a theorem of Green in $\mathbb{R}^n$, in which $\overline{\partial} = D$, the Dirac operator defined in Theorems 5.12 and 5.20. For the Dirac operator we have $\overline{D} = -D$ and $D^2 = -\Delta_n$, where for clarity the index of Δ shows the dimension.

Theorem A.2.25 (Theorem of Green). *Let G be a domain in $\mathbb{R}^n$ of finite connectivity whose boundary ∂G is sufficiently smooth and appropriately oriented, we assume f and g in $C^2(\overline{G})$. We then have*

$$\int_{\partial G} ((fD) d\mathbf{x}^* g - f d\mathbf{x}^* Dg) = \int_G (f\Delta_n g - (\Delta_n f)g) d\sigma.$$

Proof. We write down the theorem of Gauß in $\mathbb{R}^n$ once with fD and once with Dg:

$$\int_{\partial G} (fD) d\mathbf{x}^* g = \int_G ((fD)(Dg) - (\Delta_n f)g) d\sigma,$$

$$\int_{\partial G} f d\mathbf{x}^* (Dg) = \int_G (-f\Delta_n g + (fD)(Dg)) d\sigma.$$

Subtraction of both the equations gives the assertion. $\qquad\square$

A.2.3 Exercises

1. Prove the rules in Proposition A.2.5:

   a) $(\omega_p \wedge \omega_q) \circ \varphi = (\omega_p \circ \varphi) \wedge (\omega_q \circ \varphi)$,

   b) $(d\omega_p) \circ \varphi = d(\omega_p \circ \varphi)$,

   c) $(\omega_p \circ \varphi) \circ \psi = \omega_p \circ (\varphi \circ \psi)$.

2. Prove Poincaré's Lemma A.2.8:
   Let G be a star-shaped domain relative to the point $x_0 \in G$, i.e., for all $x \in G$ the line from x_0 to x should be contained completely in G. Then if for a differential form $\omega_p \in C^1(G)$ the statement $d\omega_p = 0$ in G holds, one has for $p = 0$ a constant form ω_p, and for $p > 0$ a differential form ω_{p-1} exists with

$$\omega_p = d\omega_{p-1}.$$

3. Prove the rules for line integrals in Example A.2.15 a:

$$\int_{-\gamma} f(x)dx = -\int_\gamma f(x)dx,$$
$$\int_{\gamma_1} f(x)dx + \int_{\gamma_2} f(x)dx = \int_{\gamma_1+\gamma_2} f(x)dx,$$
$$\int_\gamma (af(x) + bg(x))dx = a\int_\gamma f(x)dx + b\int_\gamma g(x)dx,$$
$$\left| \int_\gamma f(x)dx \right| \le \int_a^b |f(x)x'(t)|dt.$$

For the last rule we refer to Proposition A.2.2 (ii).

4. Let γ be a smooth curve parameterized by $x(t)$, $t \in [a, b]$, let $f(x)$ be continuous on γ. We then define

$$\int_\gamma f(x)\overline{dx}g(x) := \int_a^b f(x(t))\overline{x'(t)}g(x(t))dt.$$

Prove that

$$\overline{\int_\gamma f(x)dxg(x)} = \int_\gamma \overline{g(x)}\,\overline{dx}\,\overline{f(x)}$$

and

$$\left| \int_\gamma f(x)\overline{dx}g(x) \right| \le \int_a^b |f(x(t))\overline{x'(t)}g(x(t))|dt.$$

5. Let $G \subset \mathbb{C}$ be a domain and let $g, h : G \to \mathbb{C}$ be continuous. Let γ be a piecewise smooth curve in G with initial point z_A and endpoint z_E. Moreover we define

$$\int_\gamma [g(z)dz + h(z)d\bar{z}] := \int_\gamma g(z)dz + \int_\gamma h(z)d\bar{z}.$$

Prove: If a differentiable function $f : G \to \mathbb{C}$ exists with $\partial_z f = g$ and $\partial_{\bar{z}} f = h$ one has

$$\int_\gamma [g(z)dz + h(z)d\bar{z}] = f(z_E) - f(z_A).$$

What does this mean in the case $h = 0$?

6. Prove the orthogonality of the tangential vectors to S^n in Example A.2.17 a and the equations

$$\frac{\partial x}{\partial t_j} \cdot \frac{\partial x}{\partial t_j} = \prod_{k=j+1}^{n} \cos^2 t_k.$$

7. Compute the integral

$$\int_{|x|<1} \frac{d\sigma}{|x|}.$$

8. Let G be a domain of finite connectivity in $\mathbb{C}$ with piecewise smooth boundary. Prove

$$\int_{\partial G} \bar{z}dz = 2i\sigma(G),$$

here $\sigma(G)$ is the area of G.

9. Prove the theorem of Gauß in the following form:

$$\int_{\partial G} f(x)d\bar{x}^* = \int_G (\partial f)(x)d\sigma_x.$$

A.3 Some function spaces

We need some function spaces, particularly for working with the Teodorescu transformation, and we shall present them in this appendix. We deal always with functions $f : E \to Y$, $E \subset X$, where we may have

X identified with $\mathbb{R}, \mathbb{C}, \mathbb{H}$ or $\mathbb{R}^{n+1}$, and Y identified with $\mathbb{R}, \mathbb{C}, \mathbb{H}, \mathbb{R}^{n+1}$ or $C\ell(n)$;

in every case we assume the Euclidean metric.

A.3.1 Spaces of Hölder continuous functions

Ludwig Otto Hölder (1859–1937) introduced the following definition:

Definition A.3.1. Let be $0 < \mu \leq 1$. A function $f : E \to Y$ defined on a set $E \subset X$ is called *Hölder continuous on* E with the *Hölder exponent* μ, if the quantity

$$|f|_{\mu,E} := \sup_{x,y \in E; x \neq y} \frac{|f(x) - f(y)|}{|x - y|^\mu}$$

is finite, i.e., if a constant $A > 0$ exists such that for $x, y \in E$,

$$|f(x) - f(y)| \leq A|x - y|^\mu$$

always holds. The constant A is called a *Hölder constant*. For the exponent $\mu = 1$ we speak of *Lipschitz continuity*. One uses also the expressions: f satisfies a *Hölder condition* or a *Lipschitz condition* .

Below, some properties of such functions are listed:

Proposition A.3.2. *Let $f : E \to Y$, $E \subset X$ be given. We then have:*

(i) *Hölder continuous functions are continuous.*

(ii) *For compact E and $|f(x) - f(y)| \leq Ar^\mu$ for all $r = |x - y| < \delta$, $\delta > 0$ the function f satisfies a Hölder condition in the whole E.*

(iii) *If the real components of f possess partial derivatives in a convex and compact set E and these derivatives are bounded there, then f satisfies a Lipschitz condition in E.*

(iv) *If f satisfies a Hölder condition in a convex and compact set E with exponent $\mu > 1$, then f is constant.*

(v) *Not every continuous function is Hölder continuous.*

Proof. (i) is contained in the definition.

(ii) We assume $r > \delta$. From (i) and the compactness of E follows $|f(x)| \leq M$ in E, therefore we have for $r \geq \delta$,

$$|f(x) - f(y)| \leq \frac{2M}{\delta^\mu} \delta^\mu \leq \frac{2M}{\delta^\mu} r^\mu.$$

Thus for arbitrary $x, y \in E$ we have the inequality $|f(x) - f(y)| \leq A'|x - y|^\mu$ if we choose $A' = \max\{A, 2M/\delta^\mu\}$.

(iii) On the line from x to y we can apply the mean value theorem of differential calculus to the real components f_i of f. As the components' partial derivatives are assumed to be bounded, maybe by M_i, we get at once inequalities of the form

$$|f_i(x) - f_i(y)| \leq M_i|x - y|.$$

These estimates can be put together for the corresponding estimate of $|f(x) - f(y)|$.

(iv) The Hölder condition is also satisfied for the components f_i of f. From $\mu > 1$ it follows for all $x, x_0 \in E$

$$\lim_{x \to x_0} \frac{|f_i(x) - f_i(x_0)|}{|x - x_0|} \leq \lim_{x \to x_0} A|x - x_0|^{\mu - 1} = 0.$$

Therefore the partial derivatives are zero and the f_i are constant.

To prove (v) it suffices to produce one such function. E.g., we may choose $f : [0, \frac{1}{2}] \to \mathbb{R}$ and

$$f(t) = \begin{cases} \frac{1}{\ln t} & (0 < t \leq \frac{1}{2}), \\ 0 & (t = 0). \end{cases}$$

Obviously f is continuous in $t = 0$ but not Hölder continuous:

$$|f(t) - f(0)| = \frac{1}{|\ln t|} \leq At^\mu$$

will not be satisfied for any A since $|\ln t|t^\mu \to 0$ for $t \to 0$ for all $\mu > 0$. $\qquad \square$

Definition A.3.3. The set of all functions which satisfy on a compact set E a Hölder condition with exponent μ is denoted by $H^\mu(E)$.

Introducing the metric

$$\|f\|_{\mu, E} = \max_{z \in E} |f(x)| + \sup_{x, y \in E} \frac{|f(x) - f(y)|}{|x - y|^\mu}$$

one can prove that $H^\mu(E)$ is a *Banach space*.

A.3.2 Spaces of differentiable functions

Let $G \subset X$ be a domain and m be a positive integer. By $C^m(G)$ we denote the space of all functions $f : G \to Y$ which are m-times continuously differentiable in G. If Y is of dimension at least 2 the components f_i of f are assumed to fulfil this condition. We equip this space with the topology of uniform convergence of functions f and their derivatives $\nabla^\alpha f$ for all $|\alpha| \leq m$ on all compact subsets $K \subset G$. Here $\nabla = (\partial_0, \partial_1, \ldots, \partial_n)$ is the vector operator of differentiation which we introduced in Definition 7.24, and $\alpha = (\alpha_0, \alpha_1, \ldots, \alpha_n)$ is a multiindex; we define

$$\nabla^\alpha f := \partial_0^{\alpha_0} \partial_1^{\alpha_1} \ldots \partial_n^{\alpha_n} f.$$

The subspace $B^m(G)$ may contain all functions f from $C^m(G)$, which are bounded on G together with all derivatives $\nabla^\alpha f$ for $|\alpha| \leq m$. For $B^m(G)$ we introduce the norm with the real components f_i of f,

$$\|f\|_{m,G} := \max_i \max_{0 \leq j \leq m} \max_{|\alpha|=j} \sup_{x \in G} |\nabla^\alpha f_i(x)|,$$

which will be a Banach space. In particular we have

$$\|f\|_{0,G} := \max_i \sup_{x \in G} |f_i(x)|.$$

By $C^\infty(G)$ one denotes the space of all functions $f : G \to Y$ possessing all derivatives ∇^α of arbitrary degree. And $C_0^\infty(G)$ denotes the subspace of all functions f in $C^\infty(G)$ which have compact support $\text{supp}\, f$ in G; we had defined in Theorem A.2.13,

$$\text{supp}\, f = \overline{\{x \in G : f(x) \neq 0\}}.$$

The space $C_0^\infty(B_r(a))$ contains, e.g., the important function

$$\varphi(x) = \begin{cases} \exp\left(\frac{1}{|x-a|^2-r^2}\right), & \text{if } |x-a| < r, \\ 0, & \text{otherwise.} \end{cases}$$

For $s = m + \mu$ the space $B^s(G) \subset B^m(G)$ denotes the space of all functions, which have in G Hölder continuous derivatives of degree m with the Hölder exponent μ. In $B^s(G)$ we introduce the norm

$$\|f\|_{s,G} = \max_i \max_{0 \leq j \leq m} \max_{|\alpha|=j} \sup_{x \in G} |\nabla^\alpha f_i(x)| + \max_i \max_{|\alpha|=m} \|\nabla^\alpha f_i\|_{\mu,G}$$

and get a Banach space. Also the notation $C^{m,\mu}(G)$ is used.

A.3.3 Spaces of integrable functions

Let $G \subset X$ be a domain and p a positive real number. Then $L^p(G)$ denotes the space of all equivalence classes of Lebesgue measurable functions $f : G \to Y$ for which $|f|^p$ is integrable over G. With the norm

$$\|f\|_{p,G} := \left(\int_G |f(x)|^p d\sigma\right)^{1/p}$$

$L^p(G)$ for $p \geq 1$ becomes a Banach space. Moreover for $p = 2$ we have a Hilbert space with the scalar product

$$(f,g)_2 := (f,g)_{2,G} := \int_G \overline{f(x)} g(x) d\sigma.$$

But in this case we have to restrict Y to $\mathbb{R}, \mathbb{C}, \mathbb{H}$, or $\mathbb{R}^{n+1} \subset C\ell(n)$. The space of all Lebesgue measurable functions $f : G \to Y$ with

$$\text{vrai max}_G |f(x)| < \infty$$

is denoted by $L^\infty(G)$. Here vrai max is the *essential maximum*, which is defined as the infimum of all real a such that $|f(x)| < a$ up to a set of measure 0. With the norm $\|f\|_{\infty, G} := \text{vrai max}_G |f(x)|$ the space $L^\infty(G)$ becomes a Banach space.

The space L^1_{loc} contains all functions f defined almost everywhere in G which are locally integrable, i.e., for every measurable compact set K one has $f \in L^1(K)$.

We remark explicitly that the Hilbert space $L^2(G)$ together with the above defined scalar product possesses the norm

$$\|f\|_2 := (f, f)^{1/2}.$$

This is quite analogous to the usual norm for functions which have vector values. The following properties hold:

(i) For all $f \in L^2(G)$ one has $(f, f)_2 > 0$. Moreover $(f, f)_2 = 0$ if and only if $f = 0$.

(ii) For $f, g \in L^2(G)$ one has $(f, g)_2 = \overline{(g, f)}_2$, where the bar denotes the conjugation in $Y = \mathbb{C}, \mathbb{H}, \mathbb{R}^{n+1}$.

(iii) For $f, g, h \in L^2(G)$ one has the distributive rule $(f + g, h)_2 = (f, h)_2 + (g, h)_2$.

(iv) For $\lambda \in X = \mathbb{R}, \mathbb{C}, \mathbb{H}, \mathbb{R}^{n+1}$ one has $\lambda(f, g)_2 = (\overline{\lambda}f, g)_2$ and $(f, g)_2\lambda = (f, g\lambda)_2$.

The proof is recommended as an exercise.

A.3.4 Distributions

The sequence $(\varphi_m) \subset C_0^\infty(G)$ is said to converge to a function φ in the sense of the space $\mathcal{D}(G)$ if

(i) a compact set $K \subset G$ exists such that $\text{supp}(\varphi_m) \subset K$ holds for all m, and

(ii) $\nabla^\alpha \varphi_m \to \nabla^\alpha \varphi$ converges uniformly on K for all indices α.

We denote then the space $C_0^\infty(G)$ equipped with this topology by $\mathcal{D}(G)$. Its dual space $\mathcal{D}'(G)$ relative to $Z := \mathbb{R}, \mathbb{C}$ or $\mathbb{H}$ is called the *space of Schwartz's distributions*. The space $\mathcal{D}'(G)$ is thus the vector space of all continuous linear functionals $T : \mathcal{D}(G) \to Z$ having the following properties:

Let there be $T, S \in \mathcal{D}'(G)$ and $\lambda \in Z$; we then have

$$\begin{aligned}
(T + S)(\varphi) &= T(\varphi) + S(\varphi), \\
(\lambda T)(\varphi) &= \lambda(T(\varphi)), \\
T(\varphi + \lambda\psi) &= T(\varphi) + \lambda T(\psi)
\end{aligned}$$

for $\varphi, \psi \in \mathcal{D}(G)$. In $Z = \mathbb{H}$ one has to take care of the non-commutativity. In $\mathcal{D}'(G)$ the topology is defined by the following convergence criterion:

$T_m \to T$ is convergent in $\mathcal{D}'(G)$ if and only if $T_m(\varphi) \to T(\varphi)$ in Z for all $\varphi \in \mathcal{D}(G)$. One defines the derivative $S = \nabla^\alpha T \in \mathcal{D}'(G)$ of a distribution $T \in \mathcal{D}'(G)$ in the distributional sense by

$$S(\varphi) := (-1)^{|\alpha|} T(\nabla^\alpha \varphi) \quad \text{for all } \varphi \in \mathcal{D}(G).$$

We remark that the partial differential operator $\nabla^\alpha : \mathcal{D}'(G) \to \mathcal{D}'(G)$ is always continuous, i.e., from $T_m \to T$ in $\mathcal{D}'(G)$ it follows necessarily that $\nabla^\alpha T_m \to \nabla^\alpha T$ in $\mathcal{D}'(G)$.

A.3.5 Hardy spaces

Let there be $0 < p < \infty$. A holomorphic function f in $B_1(0) \subset \mathbb{C}, \mathbb{H}$ or $\mathbb{R}^{n+1}$ belongs to the *Hardy space* $H^p(B_1(0))$ if the condition

$$\sup_{0<r<1} \int_{|x|=1} |f(rx)|^p do_1(x) < \infty$$

is satisfied. Analogously the *harmonic Hardy space* is defined for harmonic functions u by $h^p(B_1(0))$: A harmonic function u belongs to $h^p(B_1(0))$ if it satisfies the condition

$$\sup_{0<r<1} \int_{|x|=1} |u(rx)|^p do_1(x) < \infty.$$

A.3.6 Sobolev spaces

Let $G \subset X$ be a domain and m be a positive integer. The space, denoted by $W^{m,p}(G)$, containing all equivalence classes of functions $f \in L^p(G)$ whose derivatives (in the distributional sense) $\nabla^\alpha f$ for $|\alpha| \leq m$ belong to $L^p(G)$ is called *Sobolev space*.

With the norms

$$\|f\|_{m,p,G} := \int_G \left(\sum_{|\alpha| \leq m} |\nabla^\alpha f(x)|^p \, do \right) \quad (1 \leq p < \infty),$$

$$\|f\|_{m,\infty,G} := \max_{|\alpha| \leq m} \mathrm{vraimax}_{x \in G} |\nabla^\alpha f(x)| \quad (p = \infty)$$

the Sobolev space becomes a Banach space. For $m = 0$ we have

$$W^{0,p}(G) = L^p(G).$$

The space $W_0^{m,p}(G)$ is just the closure of $\mathcal{D}(G)$ using the topology of the space $W^{m,p}(G)$.

Moreover let Γ be the sufficiently smooth boundary of the domain G and $B^{m-\frac{1}{p},p}(\Gamma)$ the space of boundary values g of all functions $f \in W^{m,p}(G)$. A norm in $B^{m-\frac{1}{p},p}(\Gamma)$ is given by

$$\|g\|_{m-\frac{1}{p},p} = \inf \|f\|_{m,p},$$

here the infimum has to be taken over all functions in $W^{m,p}(G)$ generating on Γ the same trace g. Such spaces are also called *spaces of Besov type*.

A.4 Properties of holomorphic spherical functions

This appendix is intended to present some calculations and proofs in detail, which have been overlooked in Section 10 to facilitate the general understanding. Principally what is to be done to calculate norms and angles is clear. But to be correct and to make our constructions verifiable, we have to compute the values used. We need only tools from real analysis.

A.4.1 Properties of Legendre polynomials

Legendre polynomials and Legendre functions are solutions of a second-order ordinary differential equation:

$$(1 - t^2)(P_{n+1}^m(t))'' - 2t(P_{n+1}^m(t))' + \left((n+1)(n+2) - m^2 \frac{1}{1 - t^2} \right) P_{n+1}^m(t) = 0,$$

$$m = 0, \ldots, n + 1. \tag{A.4.1}$$

They satisfy also the recursion formulas

$$(1 - t^2)(P_{n+1}^m(t))' = (n + m + 1)P_n^m(t) - (n+1)\, t\, P_{n+1}^m(t)\,, \tag{A.4.2}$$

$$(1 - t^2)^{1/2}(P_{n+1}^m(t))' = P_{n+1}^{m+1}(t) - m\,(1 - t^2)^{-1/2}\, t\, P_{n+1}^m(t)\,, \tag{A.4.3}$$

$$(1 - t^2)^{1/2} P_{n+1}^m(t) = \frac{1}{2n + 3}\left(P_{n+2}^{m+1}(t) - P_n^{m+1}(t) \right)\,, \tag{A.4.4}$$

and formulas, which are called three-terms recurrence relations,

$$(n + 1 - m)P_{n+1}^m(t) - (2n + 1)\, t\, P_n^m(t) + (n + m)P_{n-1}^m(t) = 0\,, \tag{A.4.5}$$

$m = 0, \ldots, n + 1$. We have for $m = n \geq 1$,

$$P_m^m(t) = (2m - 1)!!\,(1 - t^2)^{m/2}\,.$$

These functions are orthogonal in pairs in $L_2([-1, 1])$,

$$\int_{-1}^{1} P_{n+1}^m(t) P_{n+1}^l(t)\, dt = 0,\ m \neq l,$$

and the norms may be computed to give

$$\int_{-1}^{1} \left(P_{n+1}^m(t) \right)^2 dt = \frac{2}{2n + 3} \frac{(n + 1 + m)!}{(n + 1 - m)!},\ m = 0, \ldots, n + 1. \tag{A.4.6}$$

The Legendre functions (not the Legendre polynomials) are also orthogonal in the following weighted L^2-space:

$$\int_{-1}^{1} P_{n+1}^m(t) \, P_{n+1}^k(t) \, (1 - t^2)^{-1} \, dt = \begin{cases} 0, & m \neq k, \\[2mm] \dfrac{(n + 1 + m)!}{m \, (n + 1 - m)!}, & m = k, \end{cases} \qquad (A.4.7)$$

$m, k = 1, \ldots, n + 1$.

The reader interested in more details on Legendre polynomials and functions is referred to [7] and [128].

A.4.2 Norm of holomorphic spherical functions

Theorem A.4.1. *For all $n \in \mathbb{N}_0$ the subsystems of holomorphic spherical functions $X_{n,0}^0$, $X_{n,0}^m$ resp. $Y_{n,0}^m$ ($m = 1, \ldots, n + 1$) are orthogonal systems with the norms*

$$\|X_{n,0}^0\|_{0,L^2(S^2)} = \sqrt{\pi(n + 1)} \qquad (A.4.8)$$

and

$$\|X_{n,0}^m\|_{0,L^2(S^2)} = \|Y_{n,0}^m\|_{0,L^2(S^2)} = \sqrt{\frac{\pi}{2}(n + 1)\frac{(n + 1 + m)!}{(n + 1 - m)!}},$$

$$m = 1, \ldots, n + 1. \qquad (A.4.9)$$

Proof. A simple calculation shows the orthogonality:

$$(X_{n,0}^0, X_{n,0}^m)_{0,L^2(S^2)} = \int_0^{\pi} (A^{0,n} A^{m,n} + B^{0,n} B^{m,n}) \sin\theta \, d\theta \int_0^{2\pi} \cos m\varphi \, d\varphi$$
$$= 0, \quad m = 1, \ldots, n + 1,$$

$$(X_{n,0}^0, Y_{n,0}^m)_{0,L^2(S^2)} = \int_0^{\pi} (A^{0,n} A^{m,n} + B^{0,n} B^{m,n}) \sin\theta \, d\theta \int_0^{2\pi} \sin m\varphi \, d\varphi$$
$$= 0, \quad m = 1, \ldots, n + 1,$$

$$(X_{n,0}^{m_1}, Y_{n,0}^{m_2})_{0,L^2(S^2)} = \int_0^{\pi} (A^{m_1,n} A^{m_2,n} + B^{m_1,n} B^{m_2,n}) \sin\theta \, d\theta$$
$$\cdot \int_0^{2\pi} \cos m_1\varphi \sin m_2\varphi \, d\varphi - \int_0^{\pi} C^{m_1,n} C^{m_2,n} \sin\theta \, d\theta \int_0^{2\pi} \sin m_1\varphi \cos m_2\varphi \, d\varphi$$
$$= 0, \quad m_1, m_2 = 1, \ldots, n + 1,$$

$$(X_{n,0}^{m_1}, X_{n,0}^{m_2})_{0,L^2(S^2)} = \int_0^{\pi} (A^{m_1,n} A^{m_2,n} + B^{m_1,n} B^{m_2,n}) \sin\theta \, d\theta$$
$$\cdot \int_0^{2\pi} \cos m_1\varphi \cos m_2\varphi \, d\varphi + \int_0^{\pi} C^{m_1,n} C^{m_2,n} \sin\theta \, d\theta \int_0^{2\pi} \sin m_1\varphi \sin m_2\varphi \, d\varphi$$
$$= 0, \quad m_1 \neq m_2,$$

$$(Y_{n,0}^{m_1}, Y_{n,0}^{m_2})_{0,L^2(S^2)} = \int_0^\pi (A^{m_1,n} A^{m_2,n} + B^{m_1,n} B^{m_2,n}) \sin\theta d\theta$$

$$\cdot \int_0^{2\pi} \sin m_1\varphi \sin m_2\varphi d\varphi + \int_0^\pi C^{m_1,n} C^{m_2,n} \sin\theta d\theta \int_0^{2\pi} \cos m_1\varphi \cos m_2\varphi d\varphi$$

$$= 0, \quad m_1 \neq m_2 .$$

To compute the norms (A.4.8) and (A.4.9) we use the formulas (A.4.6)–(A.4.7) and define

$$N_{n+1}^m := \int_{-1}^1 (P_{n+1}^m(t))^2 \, dt = \frac{2}{2n+3} \frac{(n+1+m)!}{(n+1-m)!}, \qquad (A.4.10)$$

$$M_{n+1}^m := \int_{-1}^1 (1-t^2)^{-1} (P_{n+1}^m(t))^2 \, dt = \frac{(n+1+m)!}{m(n+1-m)!} . \qquad (A.4.11)$$

From (10.9)–(10.11) we get

$$\|X_{n,0}^0\|_{0,L^2(S^2)}^2 = \int_{S^2} \overline{X_{n,0}^0} X_{n,0}^m \, |do| = 2\pi \int_0^\pi \left[(A^{0,n})^2 + (B^{0,n})^2 \right] \sin\theta d\theta$$

$$= \frac{\pi}{2} \int_0^\pi \left[\sin^2\theta \left[\frac{d}{dt}[P_{n+1}(t)]_{t=\cos\theta} \right]^2 d\theta + (n+1)^2 (P_{n+1}(\cos\theta))^2 \right] \sin\theta d\theta .$$

We substitute $t = \cos\theta$, use the definition of the associated Legendre functions, and get

$$\|X_{n,0}^0\|_{0,L^2(S^2)}^2 \;=\; \frac{\pi}{2} \int_{-1}^1 \left[(1-t^2)((P_{n+1}(t))')^2 + (n+1)^2 (P_{n+1}(t))^2 \right] dt .$$

$$=\; \frac{\pi}{2} \int_{-1}^1 (P_{n+1}^1(t))^2 \, dt + \frac{\pi}{2}(n+1)^2 \int_{-1}^1 (P_{n+1}(t))^2 \, dt$$

$$=\; \frac{\pi}{2} N_{n+1}^1 + \frac{\pi}{2}(n+1)^2 N_{n+1}^0$$

$$=\; \pi(n+1) .$$

To prove (A.4.9) we deal only with the functions $X_{n,0}^m$ ($m = 1, \ldots, n+1$). For $m = 1, \ldots, n+1$ the equation (10.12) gives

$$\|X_{n,0}^m\|_{0,L^2(S^2)}^2 \;=\; \int_{S^2} \overline{X_{n,0}^m} X_{n,0}^m \, |do|$$

$$=\; \int_0^\pi \int_0^{2\pi} \left[((A^{m,n})^2 + (B^{m,n})^2) \cos^2 m\varphi + (C^{m,n})^2 \sin^2 m\varphi \right] \sin\theta d\varphi d\theta$$

$$=\; \pi \left[\int_0^\pi ((A^{m,n})^2 + (B^{m,n})^2) \sin\theta d\theta + \int_0^\pi (C^{m,n})^2 \sin\theta d\theta \right] . \qquad (A.4.12)$$

We compute the particular integrals: Together with (10.14) and (10.15) we get

$$(A^{m,n})^2 + (B^{m,n})^2 = \frac{1}{4} \left\{ \sin^2\theta \left[\frac{d}{dt}[P_{n+1}^m(t)]_{t=\cos\theta} \right]^2 + (n+1)^2 (P_{n+1}^m(\cos\theta))^2 \right\},$$

and transforming the variables by $t = \cos\theta$ it follows that

$$\int_0^\pi \left((A^{m,n})^2 + (B^{m,n})^2 \right) \sin\theta d\theta = \int_{-1}^1 \left((A^{m,n})^2 + (B^{m,n})^2 \right) dt$$

$$= \frac{1}{4} \int_{-1}^1 (1 - t^2) \left[(P_{n+1}^m(t))' \right]^2 dt + \frac{1}{4}(n+1)^2 N_{n+1}^m . \qquad \text{(A.4.13)}$$

To compute the right-hand side of (A.4.13) we use the recursion formulas (A.4.3) and get

$$\int_{-1}^1 (1 - t^2) \left[(P_{n+1}^m(t))' \right]^2 dt = N_{n+1}^{m+1} + m^2 \int_{-1}^1 (1 - t^2)^{-1} t^2 (P_{n+1}^m(t))^2 dt$$

$$-2m \int_{-1}^1 (1 - t^2)^{-1/2} t P_{n+1}^m(t) P_{n+1}^{m+1}(t) dt .$$

$$\text{(A.4.14)}$$

The first integral on the right-hand side can now be evaluated completely:

$$\int_{-1}^1 (1 - t^2)^{-1} t^2 (P_{n+1}^m(t))^2 dt \;=\; -\int_{-1}^1 (1 - t^2)^{-1} (1 - t^2 - 1) (P_{n+1}^m(t))^2 dt$$

$$=\; -\int_{-1}^1 (P_{n+1}^m(t))^2 dt + \int_{-1}^1 (1 - t^2)^{-1} (P_{n+1}^m(t))^2 dt$$

$$=\; -N_{n+1}^m + M_{n+1}^m . \qquad \text{(A.4.15)}$$

Therefore (A.4.14) reads

$$\int_{-1}^1 (1 - t^2) \left[(P_{n+1}^m(t))' \right]^2 dt = N_{n+1}^{m+1} - m^2 N_{n+1}^m + m^2 M_{n+1}^m$$

$$-2m \int_{-1}^1 (1 - t^2)^{-1/2} t P_{n+1}^m(t) P_{n+1}^{m+1}(t) dt , \qquad \text{(A.4.16)}$$

but we still have to evaluate the integral on the right-hand side of (A.4.16). We use again the recursion formula (A.4.4) in the form,

$$(1 - t^2)^{-1/2} P_{n+1}^m(t) = \frac{1}{2n+3} \left[(1 - t^2)^{-1} P_{n+2}^{m+1}(t) - (1 - t^2)^{-1} P_n^{m+1}(t) \right] ,$$

$$\text{(A.4.17)}$$

and the two-step-formula (A.4.5),

$$t P_{n+1}^{m+1}(t) = \frac{1}{2n+3} \left[(n+1-m) P_{n+2}^{m+1}(t) + (n+m+2) P_n^{m+1}(t) \right] . \qquad \text{(A.4.18)}$$

The multiplication of (A.4.17) and (A.4.18) gives

$$(1 - t^2)^{-1/2} t P_{n+1}^m(t) P_{n+1}^{m+1}(t) = \frac{1}{(2n+3)^2} \left[(n+1-m) (1 - t^2)^{-1} (P_{n+2}^{m+1}(t))^2 \right.$$

$$-(n+m+2)(1 - t^2)^{-1} (P_n^{m+1}(t))^2 + (2m+1) (1 - t^2)^{-1} P_{n+2}^{m+1}(t) P_n^{m+1}(t) \Big] ,$$

and it follows that

$$\int_{-1}^{1} (1-t^2)^{-1/2} \, t \, P_{n+1}^m(t) \, P_{n+1}^{m+1}(t) \, dt = \frac{n+1-m}{(2n+3)^2} M_{n+2}^{m+1} - \frac{n+m+2}{(2n+3)^2} M_n^{m+1}$$

$$+ \frac{2m+1}{(2n+3)^2} \int_{-1}^{1} (1-t^2)^{-1} \, P_{n+2}^{m+1}(t) \, P_n^{m+1}(t) \, dt \,. \tag{A.4.19}$$

The formula (A.4.16) now means

$$\int_{-1}^{1} (1-t^2) \, [(P_{n+1}^m(t))']^2 \, dt = N_{n+1}^{m+1} - m^2 \, N_{n+1}^m + m^2 \, M_{n+1}^m$$

$$- \frac{2m(n+1-m)}{(2n+3)^2} M_{n+2}^{m+1} + \frac{2m(n+m+2)}{(2n+3)^2} M_n^{m+1}$$

$$- \frac{2m(2m+1)}{(2n+3)^2} \int_{-1}^{1} (1-t^2)^{-1} \, P_{n+2}^{m+1}(t) \, P_n^{m+1}(t) \, dt \,. \tag{A.4.20}$$

From the recursion formula (A.4.4) it follows that

$$P_{n+1}^m(t) = \frac{1}{2n+3} \left[(1-t^2)^{-1/2} \, P_{n+2}^{m+1}(t) - (1-t^2)^{-1/2} \, P_n^{m+1}(t) \right],$$

and by squaring

$$(P_{n+1}^m(t))^2 = \frac{1}{(2n+3)^2} \left[(1-t^2)^{-1} \, (P_{n+2}^{m+1}(t))^2 + (1-t^2)^{-1} \, (P_n^{m+1}(t))^2 \right.$$

$$\left. - 2(1-t^2)^{-1} \, P_{n+2}^{m+1}(t) \, P_n^{m+1}(t) \right] \,.$$

We get

$$\int_{-1}^{1} (1-t^2)^{-1} \, P_{n+2}^{m+1}(t) \, P_n^{m+1}(t) \, dt = -\frac{(2n+3)^2}{2} \, N_{n+1}^m + \frac{1}{2} M_{n+2}^{m+1} + \frac{1}{2} M_n^{m+1} \,. \tag{A.4.21}$$

This result has to be substituted into (A.4.20),

$$\int_{-1}^{1} (1-t^2) \, [(P_{n+1}^m(t))']^2 \, dt = N_{n+1}^{m+1} - m^2 \, N_{n+1}^m + m \, (2m+1) \, N_{n+1}^m$$

$$+ m^2 \, M_{n+1}^m - \frac{m}{2n+3} \, M_{n+2}^{m+1} + \frac{m}{2n+3} \, M_n^{m+1} \,,$$

and we substitute this further into (A.4.13) getting

$$\int_0^\pi \left((A^{m,n})^2 + (B^{m,n})^2 \right) \sin\theta d\theta = \frac{1}{4} N_{n+1}^{m+1} + \frac{1}{4} \left[(n+1)^2 - m^2 \right] N_{n+1}^m$$

$$+ \frac{1}{4} m \, (2m+1) \, N_{n+1}^m + \frac{1}{4} m^2 \, M_{n+1}^m - \frac{1}{4} \frac{m}{2n+3} \, M_{n+2}^{m+1} + \frac{1}{4} \frac{m}{2n+3} \, M_n^{m+1} \,.$$

$$\tag{A.4.22}$$

If we use now (A.4.10) and (A.4.11) we have proved the following relations:

$$N_{n+1}^{m+1} = (n+m+2) \, (n+1-m) \, N_{n+1}^m \,,$$

$$M_{n+1}^m = \frac{2n+3}{2m} \, N_{n+1}^m \,, \tag{A.4.23}$$

$$M_{n+2}^{m+1} = \frac{2n+3}{2(m+1)} \, (n+m+3) \, (n+m+2) \, N_{n+1}^m \,, \tag{A.4.24}$$

and

$$M_n^{m+1} = \frac{2n+3}{2(m+1)} (n+1-m)(n-m) N_{n+1}^m .$$ (A.4.25)

Together with (A.4.22) we have obtained

$$\int_0^\pi ((A^{m,n})^2 + (B^{m,n})^2) \sin\theta d\theta = \frac{1}{4} \left[(n+1)(2n+3) - \frac{m}{2}(2n+3) \right] N_{n+1}^m .$$

(A.4.26)

The remaining integral in (A.4.12) is easy to compute:

$$\int_0^\pi (C^{m,n})^2 \sin\theta d\theta = \int_{-1}^1 (C^{m,n})^2 \, dt$$

$$= \frac{1}{4} m^2 \int_{-1}^1 (1-t^2)^{-1} (P_{n+1}^m(t))^2 \, dt$$

$$= \frac{1}{4} m^2 M_{n+1}^m$$

$$= \frac{1}{8} m (2n+3) N_{n+1}^m .$$ (A.4.27)

Substituting (A.4.26) and (A.4.27) into (A.4.12) we get finally

$$\|X_{n,0}^m\|_{0,L^2(S^2)}^2 = \frac{\pi}{4}(n+1)(2n+3) N_{n+1}^m$$

$$= \frac{\pi}{2}(n+1) \frac{(n+m+1)!}{(n+1-m)!},$$

and this is the desired result.   □

A.4.3 Scalar products of holomorphic spherical functions

Theorem A.4.2. *For all $n \in \mathbb{N}_0$ and $m = 1, \ldots, n+1$; $l = 1, \ldots, n$ we have*

$$(X_{n,0}^0, X_{n,3}^0)_{0,L^2(S^2)} = (X_{n,0}^0, X_{n,3}^l)_{0,L^2(S^2)} = (X_{n,0}^0, Y_{n,3}^l)_{0,L^2(S^2)}$$
$$= (X_{n,0}^m, X_{n,3}^0)_{0,L^2(S^2)} = (X_{n,0}^m, X_{n,3}^l)_{0,L^2(S^2)} = (Y_{n,0}^m, X_{n,3}^0)_{0,L^2(S^2)}$$
$$= (Y_{n,0}^m, Y_{n,3}^l)_{0,L^2(S^2)} = 0$$

and

$$(X_{n,0}^m, Y_{n,3}^l)_{0,L^2(S^2)} = -(Y_{n,0}^m, X_{n,3}^l)_{0,L^2(S^2)} = \begin{cases} 0, & m \neq l, \\ \frac{\pi}{2} m \frac{(n+m+1)!}{(n-m+1)!}, & m = l. \end{cases}$$

Proof. For $n \in \mathbb{N}_0$ and $m = 1, \ldots, n+1$, $l = 1, \ldots, n$ we use (10.12), (10.13) as well as the definition of $Y_{n,3}^l$ and get

$$(X_{n,0}^m, Y_{n,3}^l)_{0,L^2(S^2)} = -\int_{S^2} C^{m,n} B^{l,n} \sin m\varphi \sin l\varphi \, |do|$$

$$- \int_{S^2} B^{m,n} C^{l,n} \cos m\varphi \cos l\varphi \, |do| .$$ (A.4.28)

Because of

$$\int_0^{2\pi} \sin m\varphi \sin l\varphi \, d\varphi = \int_0^{2\pi} \cos m\varphi \cos l\varphi \, d\varphi = 0 \,, \quad m \neq l \,,$$

it follows from (A.4.28) that

$$(X_{n,0}^m, Y_{n,3}^l)_{0,L^2(S^2)} = 0 \,, \quad m \neq l \,.$$

For $m = l$ we get from (10.15), (10.16) using the notation in (A.4.10)

$$
\begin{aligned}
(X_{n,0}^m, Y_{n,3}^m)_{0,L^2(S^2)} &= -\int_{S^2} B^{m,n} \, C^{m,n} \, |do| \\
&= -2\pi \int_0^\pi B^{m,n} \, C^{m,n} \, \sin\theta \, d\theta \\
&= -2\pi \int_{-1}^1 B^{m,n} \, C^{m,n} \, dt \qquad\qquad\qquad (A.4.29)\\
&= -\frac{\pi}{2} m \int_{-1}^1 t \, (P_{n+1}^m(t))' \, P_{n+1}^m(t) \, dt + \frac{\pi}{2} m(n+1) N_{n+1}^m.
\end{aligned}
$$

To compute the integral on the right-hand side of (A.4.29) we use again the recursion formula (A.4.3):

$$(P_{n+1}^m(t))' = (1-t^2)^{-1/2} \, P_{n+1}^{m+1}(t) - m \, (1-t^2)^{-1} \, t \, P_{n+1}^m(t) \,.$$

We multiply this expression by $t \, P_{n+1}^m(t)$ and get

$$t \, (P_{n+1}^m(t))' \, P_{n+1}^m(t) = (1-t^2)^{-1/2} \, t \, P_{n+1}^{m+1}(t) \, P_{n+1}^m(t) - m \, (1-t^2)^{-1} \, t^2 \, (P_{n+1}^m(t))^2,$$

with (A.4.15) now

$$
\begin{aligned}
\int_{-1}^1 t \, (P_{n+1}^m(t))' \, P_{n+1}^m(t) \, dt &= \int_{-1}^1 (1-t^2)^{-1/2} \, t \, P_{n+1}^{m+1}(t) \, P_{n+1}^m(t) \, dt \\
&\quad -m \int_{-1}^1 (1-t^2)^{-1} \, t^2 \, (P_{n+1}^m(t))^2 \, dt \\
&= \int_{-1}^1 (1-t^2)^{-1/2} \, t \, P_{n+1}^{m+1}(t) \, P_{n+1}^m(t) \, dt \\
&\quad +m \, N_{n+1}^m - m \, M_{n+1}^m \qquad\qquad (A.4.30)
\end{aligned}
$$

follows. We calculate from (A.4.19) and (A.4.21)

$$
\begin{aligned}
\int_{-1}^1 (1-t^2)^{-1/2} \, t \, P_{n+1}^m(t) \, P_{n+1}^{m+1}(t) \, dt &= \frac{1}{2(2n+3)} M_{n+2}^{m+1} - \frac{1}{2(2n+3)} M_n^{m+1} \\
&\quad - \frac{2m+1}{2} N_{n+1}^m \,.
\end{aligned}
$$

The result will be substituted into (A.4.30), using the equations (A.4.24) and (A.4.25) we get at last

$$\int_{-1}^1 t \, (P_{n+1}^m(t))' \, P_{n+1}^m(t) \, dt = -\frac{1}{2} N_{n+1}^m \,.$$

Now we substitute (A.4.31) into (A.4.29) and use again (A.4.10):

$$(X_{n,0}^m, Y_{n,3}^m)_{0,L^2(S^2)} = \frac{\pi}{4} m \, (2n+3) \, N_{n+1}^m$$

$$= \frac{\pi}{2} m \, \frac{(n+m+1)!}{(n+1-m)!} \, ,$$

i.e., the assertion.   □

A.4.4   Complete orthonormal systems in $\mathcal{H}_{n,\mathbb{H}}^+$

Theorem A.4.3. *For all $n \in \mathbb{N}_0$ the $n+1$ holomorphic homogeneous polynomials*

$$r^n \, X_{n,0}^{0,*}, \; r^n \, X_{n,0}^{2k_1,*}, \; r^n \, Y_{n,3}^{2k_2,*}, \; k_1 = 1, \ldots, \left[\frac{n+1}{2}\right] \, , \; k_2 = 1, \ldots, \left[\frac{n}{2}\right] \quad (A.4.31)$$

form an orthogonal basis in $\mathcal{H}_{n,\mathbb{H}}^+$.

Proof. For all $n \in \mathbb{N}_0$ the set (A.4.31) contains $n+1$ functions. Thus it is enough to prove their orthogonality. We write down the scalar product of two functions $f, g \in L^2(S^2)_{\mathbb{H}}$ in detail:

$$(f,g)_{L^2(S^2)} = \int_{S^2} [\overline{f} \, g]_0 \, |do| + \int_{S^2} [\overline{f} \, g]_1 \, |do| \, e_1 \qquad (A.4.32)$$

$$+ \int_{S^2} [\overline{f} \, g]_2 \, |do| \, e_2 + \int_{S^2} [\overline{f} \, g]_3 \, |do| \, e_3 \, . \qquad (A.4.33)$$

The scalar part of (A.4.32) is the same as the real inner product (10.2), moreover the $\mathbb{H}$-holomorphic spherical functions $X_{n,0}^{0,*}, X_{n,0}^{2k_1,*}, Y_{n,3}^{2k_2,*}$ ($k_1 = 1, \ldots, \left[\frac{n+1}{2}\right]$, $k_2 = 1, \ldots, \left[\frac{n}{2}\right]$) are orthogonal relative to the real scalar product, so we have at once

$$\int_{S^2} [\overline{X_{n,0}^{2k_1,*}} \, X_{n,0}^{2k_1',*}]_0 \, |do| = 0, \; k_1 \neq k_1' \, ,$$

$$\int_{S^2} [\overline{Y_{n,3}^{2k_2,*}} \, Y_{n,3}^{2k_2',*}]_0 \, |do| = 0, \; k_2 \neq k_2' \, ,$$

$$\int_{S^2} [\overline{X_{n,0}^{0,*}} \, X_{n,0}^{2k_1,*}]_0 \, |do| = \int_{S^2} [\overline{X_{n,0}^{0,*}} \, Y_{n,3}^{2k_2,*}]_0 \, |do|$$

$$= \int_{S^2} [\overline{X_{n,0}^{2k_1,*}} \, Y_{n,3}^{2k_2,*}]_0 \, |do| = 0 \, ,$$

where $k_1, k_1' = 1, \ldots, \left[\frac{n+1}{2}\right]$, and $k_2, k_2' = 1, \ldots, \left[\frac{n}{2}\right]$.
It remains to prove that the other components of (A.4.32) are also zero. From the definition we conclude

$$X_{n,0}^{0,*} = \frac{1}{\|X_n^0\|_{L^2(S^2)}} X_n^0 \, ,$$

$$X_{n,0}^{m,*} = \frac{1}{\|X_n^m\|_{L^2(S^2)}} X_n^m \, , \; m = 2k_1, \; k_1 = 1, \ldots, \left[\frac{n+1}{2}\right] \, ,$$

$$Y_{n,3}^{l,*} = \frac{\sqrt{s_{n,l}}}{\|X_n^l\|_{L^2(S^2)}} [(n+1)Y_n^l \, e_3 - l \, X_n^l] \, , \; l = 2k_2, \; k_2 = 1, \ldots, \left[\frac{n}{2}\right] \, .$$

To show the orthogonality we work only with the not normalized polynomials X_n^0, X_n^m, given by (10.9) resp. (10.12), and we denote the functions $(n+1)Y_n^l e_3 - l X_n^l$ by Z_n^l. If we use the expressions of X_n^m and Y_n^m according to (10.12), resp. (10.13), then the functions Z_n^l have the form

$$
\begin{aligned}
Z_n^l \quad := \quad & (n+1)Y_n^l e_3 - l X_n^l \\
= \quad & -l\, A^{l,n} \cos l\varphi \\
& + \left(F^{l,n} \sin \varphi \sin l\varphi - G^{l,n} \cos \varphi \cos l\varphi \right) e_1 \\
& - \left(F^{l,n} \cos \varphi \sin l\varphi + G^{l,n} \sin \varphi \cos l\varphi \right) e_2 \\
& + (n+1) A^{l,n} \sin l\varphi\, e_3,
\end{aligned}
$$

where

$$
F^{l,n} := (n+1)B^{l,n} + l\, C^{l,n} \quad \text{and} \quad G^{l,n} := (n+1)C^{l,n} + l\, B^{l,n}
$$

with $l = 2k_2, k_2 = 1, \ldots, \left[\frac{n}{2}\right]$. We start calculating the e_1-coordinate of (A.4.32) for the functions X_n^0, X_n^m, and Z_n^l for all m and all l getting

$$
\int_{S^2} [\overline{X_n^0} X_n^m]_1 |do|
$$

$$
= \int_{S^2} \left[(A^{0,n} B^{m,n} - B^{0,n} A^{m,n}) \cos \varphi \cos m\varphi - A^{0,n} C^{m,n} \sin \varphi \sin m\varphi \right] |do|
$$

$$
= \int_0^\pi (A^{0,n} B^{m,n} - B^{0,n} A^{m,n}) \sin \theta\, d\theta \int_0^{2\pi} \cos \varphi \, \cos m\varphi \, d\varphi
$$

$$
- \int_0^\pi A^{0,n} C^{m,n} \sin \theta\, d\theta \int_0^{2\pi} \sin \varphi \, \sin m\varphi \, d\varphi.
$$

Together with

$$
\int_0^{2\pi} \cos \varphi \, \cos m\varphi \, d\varphi = \int_0^{2\pi} \sin \varphi \, \sin m\varphi \, d\varphi = 0 , \; m = 2, 3, \ldots, \tag{A.4.34}
$$

we have

$$
\int_{S^2} [\, \overline{X_n^0} \, X_n^m\,]_1 \, |do| = 0 , \; m = 2k_1, \; k_1 = 1, \ldots, \left[\frac{n+1}{2}\right].
$$

Again from (A.4.34) we find

$$
\begin{aligned}
\int_{S^2} [\, \overline{X_n^0}\, Z_n^l\,]_1 \, |do| \quad = \quad & \int_{S^2} (B^{0,n}\, A^{l,n} - A^{0,n}\, G^{l,n}) \cos \varphi \, \cos l\varphi |do| \\
& + \int_{S^2} A^{0,n}\, F^{l,n} \sin \varphi \, \sin l\varphi |do| \\
= \quad & 0, \; l = 2k_2, \; k_2 = 1, \ldots, \left[\frac{n}{2}\right].
\end{aligned}
$$

Within the set of functions X_n^m the e_1-coordinates on the right-hand side of (A.4.32) follow from

$$
\begin{aligned}
\int_{S^2} [\overline{X_n^{m_1}} X_n^{m_2}]_1 |do| \quad = \quad & \int_{S^2} \left[(A^{m_1,n} B^{m_2,n} - B^{m_1,n} A^{m_2,n}) \cos \varphi \, \cos m_1 \varphi \, \cos m_2 \varphi \right. \\
& - A^{m_1,n} C^{m_2,n} \sin \varphi \, \cos m_1 \varphi \, \sin m_2 \varphi \\
& \left. + C^{m_1,n} A^{m_2,n} \sin \varphi \, \sin m_1 \varphi \, \cos m_2 \varphi \right] |do|.
\end{aligned}
$$

Since

$$\int_0^{2\pi} \cos\varphi \cos m\varphi \cos l\varphi \, d\varphi \;=\; \int_0^{2\pi} \sin\varphi \cos m\varphi \sin l\varphi \, d\varphi = 0,$$
$$m, l = 1, 2, \ldots, \quad m \;\neq\; l+1, \; m \neq l-1,$$

we get

$$\int_{S^2} [\overline{X_n^{m_1}} X_n^{m_2}]_1 \, |do| = 0, \quad m_1 = 2k_1, \; m_2 = 2k_1', \; k_1, k_1' = 1, \ldots, \left[\frac{n+1}{2}\right].$$

For the e_1-coordinates of X_n^m and Z_n^l it follows from (A.4.32) that

$$\int_{S^2} [\overline{X_n^m} Z_n^l]_1 \, |do| = \int_{S^2} \left[\left(A^{m,n} F^{l,n} - (n+1) B^{m,n} A^{l,n}\right) \sin\varphi \cos m\varphi \sin l\varphi \right.$$
$$+ \left(-A^{m,n} G^{l,n} + l\, B^{m,n} A^{l,n}\right) \cos\varphi \cos m\varphi \cos l\varphi$$
$$- l\, C^{m,n} A^{l,n} \sin\varphi \sin m\varphi \cos l\varphi$$
$$\left. - (n+1) C^{m,n} A^{l,n} \cos\varphi \sin m\varphi \sin l\varphi\right] |do|.$$

Together with

$$\int_0^{2\pi} \cos\varphi \sin m\varphi \sin l\varphi \, d\varphi \;=\; 0, \; m, l = 1, 2, \ldots, \quad m \neq l+1, \; m \neq l-1,$$

we obtain

$$\int_{S^2} [\overline{X_n^m} Z_n^l]_1 \, |do| = 0, \; m = 2k_1, \; l = 2k_2, \; k_1 = 1, \ldots, \left[\frac{n+1}{2}\right], \; k_2 = 1, \ldots, \left[\frac{n}{2}\right].$$

Finally working with (A.4.32) and with the functions Z_n^l for the e_1-coordinates we get

$$\int_{S^2} [\overline{Z_n^{l_1}} Z_n^{l_2}]_1 |do| = \int_{S^2} \left[-\left(l_1 A^{l_1,n} F^{l_2,n} + (n+1) G^{l_1,n} A^{l_2,n}\right) \sin\varphi \cos l_1\varphi \sin l_2\varphi\right.$$
$$+ \left(l_1 A^{l_1,n} G^{l_2,n} - l_2 G^{l_1,n} A^{l_2,n}\right) \cos\varphi \cos l_1\varphi \cos l_2\varphi$$
$$+ \left(l_2 F^{l_1,n} A^{l_2,n} - (n+1) A^{l_1,n} G^{l_2,n}\right) \sin\varphi \sin l_1\varphi \cos l_2\varphi$$
$$\left. - (n+1) \left(A^{l_1,n} F^{l_2,n} + F^{l_1,n} A^{l_2,n}\right) \cos\varphi \sin l_1\varphi \sin l_2\varphi\right] |do|$$
$$= 0, \quad l_1 = 2k_2, \; l_2 = 2k_2', \; k_2, k_2' = 1, \ldots, \left[\frac{n}{2}\right].$$

An analogous consideration gives the desired result for the e_2-coordinates. For the e_3-coordinates we have initially

$$\int_{S^2} [\overline{X_n^0} X_n^m]_3 \, |do| = -\int_{S^2} B^{0,n} C^{m,n} \sin m\varphi \, |do|$$
$$= 0, \quad m = 2k_1, \; k_1 = 1, \ldots, \left[\frac{n+1}{2}\right],$$

in view of

$$\int_0^{2\pi} \sin m\varphi \, d\varphi \;=\; 0, \; m = 1, 2, \ldots.$$

Analogously we deduce

$$\int_{S^2} [\,\overline{X_n^0} \, Z_n^l\,]_3 \, |do| \;=\; \int_{S^2} \Big[(n+1) A^{0,n} A^{l,n} + B^{0,n} F^{l,n}\Big] \sin l\varphi \, |do| = 0,$$

$$l = 2k_2, \; k_2 = 1, \ldots, \left[\frac{n}{2}\right].$$

In case of our functions X_n^m the e_3-coordinates in (A.4.32) are given by

$$\int_{S^2} [\,\overline{X_n^{m_1}} \, X_n^{m_2}\,]_3 \, |do| \;=\; \int_{S^2} (-B^{m_1,n} C^{m_2,n} \cos m_1\varphi \sin m_2\varphi$$
$$+ C^{m_1,n} B^{m_2,n} \sin m_1\varphi \cos m_2\varphi) \, |do|.$$

From

$$\int_0^{2\pi} \sin m\varphi \cos l\varphi \, d\varphi = 0, \quad m, l = 1, 2, \ldots \tag{A.4.35}$$

we can conclude

$$\int_{S^2} [\,\overline{X_n^{m_1}} \, X_n^{m_2}\,]_3 \, |do| = 0, \; m_1 = 2k_1, \; m_2 = 2k_1', \; k_1, k_1' = 1, \ldots, \left[\frac{n+1}{2}\right].$$

The e_3-coordinates of Z_n^l in (A.4.32) are

$$\int_{S^2} [\,\overline{Z_n^{l_1}} \, Z_n^{l_2}\,]_3 \, |do| = \int_{S^2} \Big\{ \Big[-l_1(n+1) A^{l_1,n} A^{l_2,n} - G^{l_1,n} F^{l_2,n} \Big] \cos l_1\varphi \sin l_2\varphi$$
$$+ \Big[l_2(n+1) A^{l_1,n} A^{l_2,n} + F^{l_1,n} G^{l_2,n} \Big] \sin l_1\varphi \cos l_2\varphi \Big\} \, |do| \overset{(A.4.35)}{=} 0,$$

$$l_1 = 2k_2, \; l_2 = 2k_2', \; k_2, k_2' = 1, \ldots, \left[\frac{n}{2}\right].$$

Finally we need the e_3-coordinates in (A.4.32) for X_n^m and Z_n^l:

$$\int_{S^2} [\,\overline{X_n^m} \, Z_n^l\,]_3 |do| = \int_{S^2} \Big\{ \Big[(n+1) A^{m,n} A^{l,n} + B^{m,n} F^{l,n} \Big] \cos m\varphi \sin l\varphi$$
$$- C^{m,n} G^{l,n} \sin m\varphi \cos l\varphi \Big\} \, |do| \overset{(A.4.35)}{=} 0,$$

$$m = 2k_1, \; l = 2k_2, \; k_1 = 1, \ldots, \left[\frac{n+1}{2}\right], \; k_2 = 1, \ldots, \left[\frac{n}{2}\right].$$

Summarizing the above calculated partial results we obtain

$$(X_{n,0}^{0,*}, X_{n,0}^{m,*})_{L^2(S^2)} = (X_{n,0}^{0,*}, Y_{n,0}^{l,*})_{L^2(S^2)}$$
$$= (X_{n,0}^{m_1,*}, Y_{n,3}^{m_2,*})_{L^2(S^2)} = 0,$$

$$m = 2k_1, \; l = 2k_2, \; k_1 = 1, \ldots, \left[\frac{n+1}{2}\right], \; k_2 = 1, \ldots, \left[\frac{n}{2}\right],$$

and

$$(X_{n,0}^{m_1,*}, X_{n,0}^{m_2,*})_{L^2(S^2)} = (Y_{n,3}^{m_1,*}, Y_{n,3}^{m_2,*})_{L^2(S^2)} = 0,$$
$$m_1 \neq m_2, \ m_1 - m_2 \text{ even }.$$

Thus we have proved that for all $n \in \mathbb{N}_0$ the $n + 1$ polynomials (A.4.31) are orthogonal relative to the scalar product in $L^2(\mathbb{B}_3)_{\mathbb{H}}$, consequently they form a basis in $\mathcal{H}_{n,\mathbb{H}}^+$. $\quad\square$

A.4.5   Derivatives of holomorphic spherical functions

Theorem A.4.4. *For $n \geq 1$ we have*

$$\partial X_n^m = (n+m+1)X_{n-1}^m, \quad m = 0,\dots,n,$$
$$\partial Y_n^m = (n+m+1)Y_{n-1}^m, \quad m = 1,\dots,n.$$

Proof. As the polynomials and their derivatives have a similar form it is sufficient to show

$$\text{(i) } \overset{(1)}{A^{m,n}} = (n+m+1)A^{m,n-1}, \quad m = 0,\dots,n,$$

$$\text{(ii) } \overset{(1)}{B^{m,n}} = (n+m+1)B^{m,n-1}, \quad m = 0,\dots,n,$$

$$\text{(iii) } \overset{(1)}{C^{m,n}} = (n+m+1)C^{m,n-1}, \quad m = 1,\dots,n .$$

If we use (10.33), (10.34), (10.35), Legendre's differential equation (A.4.1), and the usual substitution $t = \cos\theta$ we are able to describe the coefficients $\overset{(1)}{A^{m,n}}$, $\overset{(1)}{B^{m,n}}$, and $\overset{(1)}{C^{m,n}}$ explicitly:

$$\overset{(1)}{A^{m,n}} = \frac{1}{2}\Big[(2n+1)\sin^2\theta\cos\theta \frac{d}{dt}[P_{n+1}^m(t)]_{t=\cos\theta} + (n+1)(2n+1)\cos^2\theta P_{n+1}^m(\cos\theta)$$
$$- ((n+1)^2 - m^2)P_{n+1}^m(\cos\theta)\Big],$$

$$\overset{(1)}{B^{m,n}} = \frac{1}{2}\Big[(2n+1)\sin\theta\cos^2\theta \frac{d}{dt}[P_{n+1}^m(t)]_{t=\cos\theta} - n\sin\theta \frac{d}{dt}[P_{n+1}^m(t)]_{t=\cos\theta}$$
$$- (n+1)(2n+1)\sin\theta\cos\theta P_{n+1}^m(\cos\theta) + m^2 \frac{\cos\theta}{\sin\theta} P_{n+1}^m(\cos\theta)\Big],$$

$$\overset{(1)}{C^{m,n}} = \frac{1}{2}\Big[m(n+1)\frac{\cos\theta}{\sin\theta} P_{n+1}^m(\cos\theta) + m\sin\theta \frac{d}{dt}[P_{n+1}^m(t)]_{t=\cos\theta}\Big] .$$

At first we check (i). We use the recursion formula (A.4.2) in $\overset{(1)}{A^{m,n}}$ with $t = \cos\theta$ and get

$$\overset{(1)}{A^{m,n}} = \frac{1}{2}(n+m+1)\left[(2n+1)\cos\theta P_n^m(\cos\theta) - (n-m+1)P_{n+1}^m(\cos\theta)\right] .$$

Application of the three-terms recurrence relations (A.4.5) gives

$$\overset{(1)}{A^{m,n}} = \frac{1}{2}(n+m+1)(n+m)P_{n-1}^m(\cos\theta) .$$

On the other hand from the recursion formula (A.4.2) it follows also that

$$A^{m,n-1} = \frac{1}{2}\left[\sin^2\theta\,\frac{d}{dt}[P_n^m(t)]_{t=\cos\theta} + n\cos\theta P_n^m(\cos\theta)\right] = \frac{1}{2}(n+m)P_{n-1}^m(\cos\theta)\,,$$

(A.4.36)

thus we get

$$\overset{(1)}{A^{m,n}} = (n+m+1)A^{m,n-1}\,.$$

To prove (ii) we use

$$2\cos\theta\,A^{m,n-1} - 2\sin\theta B^{m,n-1} = n\,P_n^m(\cos\theta)$$

(A.4.37)

and

$$2\cos\theta\,\overset{(1)}{A^{m,n}} - 2\sin\theta\,\overset{(1)}{B^{m,n}} = n\left[(n+1)\cos\theta\,P_{n+1}^m(\cos\theta) + \sin^2\theta\,\frac{d}{dt}[P_{n+1}^m(t)]_{t=\cos\theta}\right]\,.$$

Application of the recursion formula (A.4.2) together with the substitution $t = \cos\theta$ leads to

$$2\cos\theta\,\overset{(1)}{A^{m,n}} - 2\sin\theta\,\overset{(1)}{B^{m,n}} = n(n+m+1)P_n^m(\cos\theta)\,.$$

We replace $n\,P_n^m(\cos\theta)$ in the right-hand side by the left-hand side of (A.4.37) and get

$$2\cos\theta\,\overset{(1)}{A^{m,n}} - 2\sin\theta\,\overset{(1)}{B^{m,n}} = (n+m+1)\left(2\cos\theta\,A^{m,n-1} - 2\sin\theta B^{m,n-1}\right)\,.$$

We use then the already proved property (i) which gives

$$\overset{(1)}{B^{m,n}} = (n+m+1)\,B^{m,n-1}\,.$$

It remains to prove (iii). We apply (A.4.2) in $\overset{(1)}{C^m}$ together with $t = \cos\theta$ and obtain directly

$$\overset{(1)}{C^{m,n}} = (n+m+1)C^{m,n-1}\,. \qquad\qquad \square$$

A.4.6 Exercises

1. Prove that the Legendre polynomials and the Legendre functions are solutions of an ordinary differential equation of second order, for $m = 0,\dots,n+1$:

$$(1-t^2)(P_{n+1}^m(t))'' - 2t(P_{n+1}^m(t))' + \left((n+1)(n+2) - m^2\frac{1}{1-t^2}\right)P_{n+1}^m(t) = 0.$$

2. Show that the Legendre polynomials satisfy the recursion formulas

$$
\begin{aligned}
(1-t^2)(P_{n+1}^m(t))' &= (n+m+1)P_n^m(t) - (n+1)\,t\,P_{n+1}^m(t)\,,\\
(1-t^2)^{1/2}(P_{n+1}^m(t))' &= P_{n+1}^{m+1}(t) - m\,(1-t^2)^{-1/2}\,t\,P_{n+1}^m(t)\,,\\
(1-t^2)^{1/2}P_{n+1}^m(t) &= \frac{1}{2n+3}\left(P_{n+2}^{m+1}(t) - P_n^{m+1}(t)\right)\,,
\end{aligned}
$$

and the three-terms recurrence relations for $m = 0, \ldots, n+1$,

$$(n+1-m)P_{n+1}^m(t) - (2n+1)\,t\,P_n^m(t) + (n+m)P_{n-1}^m(t) = 0 .$$

Show also for $m = n \geq 1$,

$$P_m^m(t) = (2m-1)!!\,(1-t^2)^{m/2} .$$

3. Show that the Legendre polynomials and the Legendre functions are orthogonal in pairs in $L^2([-1,1])$,

$$\int_{-1}^1 P_{n+1}^m(t)P_{n+1}^l(t)\,dt = 0, \ m \neq l,$$

and that the norms are given by

$$\int_{-1}^1 \left(P_{n+1}^m(t)\right)^2 dt = \frac{2}{2n+3}\frac{(n+1+m)!}{(n+1-m)!}, \ m = 0, \ldots, n+1.$$

4. Show that the Legendre functions (not the Legendre polynomials) are also orthogonal in a weighted L^2-space:

$$\int_{-1}^1 P_{n+1}^m(t)\,P_{n+1}^k(t)\,(1-t^2)^{-1}\,dt = \begin{cases} 0, & m \neq k, \\[2mm] \dfrac{(n+1+m)!}{m\,(n+1-m)!}, & m = k, \end{cases}$$

$m, k = 1, \ldots, n+1$.

5. Using the last assertions prove that holomorphic polynomials of different orders are orthogonal on S^2.

Bibliography

[1] Ahlfors, L.V. (1966) *Lectures on quasiconformal mappings.* Van Nostrand: Princeton.

[2] Ahlfors, L.V. (1984) Old and new in Möbius groups. *Ann. Acad. Sci. Fennicae, Ser. A 1, Math.* 9: 93–105.

[3] Ahlfors, L.V. (1985) Möbius transformations and Clifford numbers. In: Chavel, I.; Farkas, H.M. (eds.): *Differential geometry and complex analysis.* H.E. Rauch Memorial Volume, Springer: Berlin etc.

[4] Ahlfors, L.V. (1986) Möbius transformations in $\mathbb{R}^n$ expressed through 2×2 matrices of Clifford numbers. *Complex Variables* 5: 215-224.

[5] Ahlfors, L.V.; Lounesto, P. (1989) Some remarks on Clifford algebras. *Complex Variables* 12: 201-209.

[6] Altmann, S.L. (1986) *Rotations, quaternions and double groups.* Clarendon Press: Oxford.

[7] Andrews, L.C. (1992) *Special functions of mathematics for engineers.* SPIE Optical Engineering Press: Bellingham; Oxford University Press: Oxford, 2nd ed., reprint 1998.

[8] Amann, H.; Escher, J. (1998 ff.) *Analysis I, II, III.* Grundstudium Mathematik. Birkhäuser: Basel.

[9] Begehr, H.; Jeffrey, A. (eds.) (1992) *Partial differential equations with complex analysis.* Pitman Research Notes Math. Ser. 262. John Wiley: New York.

[10] Begehr, H.; Xu, Z. (1992) Non-linear half-Dirichlet problems for first order elliptic equations in the unit ball of $\mathbb{R}^n$. *Applicable Anal.* 45: 3–18.

[11] Bitsadze, A.W. (1963) *Grundlagen der Theorie analytischer Funktionen einer komplexen Veränderlichen* (Russisch). Nauka: Moskau; Deutsch 1973, Akademie Verlag: Berlin.

[12] Blaschke, W. (1958) Anwendung dualer Quaternionen auf Kinematik. *Ann. Acad. Sci. Fennicae, Ser. A I, Math.* 250/3, 13 p.

[13] Borel, É. (1913) Définition et domaine d'existence des fonctions monogènes uniformes. *Proc. Fifth Internat. Congress Mathematicians, Cambridge* 1: 133–144.

[14] Brackx, F.; Delanghe, R.; Sommen, F. (1982) *Clifford analysis.* Pitman Research Notes Math. Ser. 76. Pitman: London etc.

[15] Brackx, F.; Sommen, F. (2000): Clifford-Hermite wavelets in Euclidean space. *J. Fourier Anal. Appl.* 6: 299-310.

[16] Buff, J.J. (1973) Characterization of analytic functions of a quaternion variable. *Pi Mu Epsilon J.* 5: 387-392.

[17] Bühler, F. (2000) *Die symplektische Struktur für orthogonale Gruppen und Thetareihen als Modulformen.* Dissertation, RWTH Aachen.

[18] Cação, I. (2004) *Constructive approximation by monogenic polynomials.* Dissertation, Universidade de Aveiro/Portugal.

[19] Cação, I.; Gürlebeck, K.; Malonek, H. (2001) Special monogenic polynomials and L_2-approximation. *Adv. Appl. Clifford Alg.* 11 (S2): 47–60.

[20] Calderbank, D. (1996) A function theoretic approach to the analysis of Dirac operators on manifolds with boundary. In: Sprößig, W.; Gürlebeck, K. (eds.): *Proc. of the Symposium "Analytical and numerical methods in quaternionic and Clifford analysis",* June 1996 Seiffen/Germany. TU Freiberg: 15-25.

[21] Cartan, H. (1961) *Théorie élémentaire des fonctions analytiques d'une ou plusieurs variables complexes.* Hermann: Paris.

[22] Cartan, H. (1974) *Differentialformen.* B.I.-Wissenschaftsverlag: Mannheim etc.

[23] Cauchy, A. L. (1825) Sur les intégrales définies prises entre des limites imaginaires. *Bull. Sci. Math. Astr. Phys. Chim. (Bull. Férussac)* 3: 214-221 = *Œuvres, Ser. 2,* 2: 57-65.

[24] Cauchy, A. L. (1841) Exercices d'analyse et de physique mathématique. *Œuvres, Ser. 2,* 12: 58-112.

[25] Constales, D.; Kraußhar, R.S. (2002) Szegö and polymonogenic Bergman kernels for half-space and strip domains, and single-periodic functions in Clifford analysis. *Complex Variables* 47: 349-360.

[26] Constales, D.; Kraußhar, R.S. (2002) Bergman kernels for rectangular domains and multiperiodic functions in Clifford analysis. *Math. Meth. Appl. Sci.* 25: 1509-1526.

[27] Constales, D.; Kraußhar, R.S. (2002) Representation formulas for the general derivatives of the fundamental solution of the Cauchy-Riemann operator in Clifford analysis and applications. *Z. Anal. Anwend.* 21: 579-597.

[28] Constales, D.; Kraußhar, R.S. (2003) Closed formulas for singly-periodic monogenic cotangent, cosecant and cosecant-squared functions in Clifford analysis. *J. London Math. Soc.* 67: 401-416.

[29] Crowe, M.C. (1967) *A history of vector analysis.* Dover Public.: New York, 3. Aufl.

[30] Davenport, C.M. (1991) *A commutative hypercomplex calculus with applications to special relativity.* Privately Published, Knoxville (Tennessee).

[31] Delanghe, R. (1970) On regular-analytic functions with values in a Clifford algebra. *Math. Ann.* 185: 91–111.

[32] Delanghe, R. (2001) Clifford analysis: History and perspective. *Computat. Meth. Funct. Theory* 1: 107–153.

[33] Delanghe, R.; Sommen, F., Souček, V. (1992) *Clifford algebra and spinor-valued functions.* Kluwer: Dordrecht.

[34] Delanghe, R.; Sommen, F.; Souček, V. (1992) Residues in Clifford analysis. In: Begehr, H.; Jeffrey, A. (eds): *Partial differential equations with complex analysis.* Pitman Res. Notes Math. Ser. 262: 61–92.

[35] Delanghe, R.; Souček, V. (1992) On the structure of spinor-valued differential forms. *Complex Variables* 18: 223–236.

[36] Dixon, A.C. (1904) On the Newtonian potential. *Quart. J. Math.* 35: 283-296.

[37] Dzhuraev, A.D. (1982) On the Moisil-Teodorescu system. In: Begehr, H.; Jeffrey, A. (eds): *Partial differential equations with complex analysis.* Pitman Res. Notes Math. Ser. 262: 186-203.

[38] Eisenstein, G. (1847) Genaue Untersuchung der unendlichen Doppelproducte, aus welchen die elliptischen Functionen als Quotienten zusammengesetzt sind, und der mit ihnen zusammenhängenden Doppelreihen (als eine neue Begründung der Theorie der elliptischen Functionen, mit besonderer Berücksichtigung ihrer Analogie zu den Kreisfunctionen). *J. Reine Angew. Math. (Crelles J.)* 35: 153-274.

[39] Elstrodt, J.; Grunewald, F.; Mennicke, J. (1985) Eisenstein series on three-dimensional hyperbolic space and imaginary quadratic number fields. *J. Reine Angew. Math.* 360 : 160-213.

[40] Elstrodt, J.; Grunewald, F.; Mennicke, J. (1990) Kloosterman sums for Clifford algebras and a lower bound for the positive eigenvalues of the Laplacian for congruence subgroups acting on hyperbolic spaces. *Invent. Math.* 101: 641-668.

[41] Freitag, E.; Hermann, C.F. (2000) Some modular varieties of low dimension. *Adv. Math.* 152: 203-287.

[42] Fueter, R. (1932) Analytische Theorie einer Quaternionenvariablen. *Comment. Math. Helv.* 4: 9-20.

[43] Fueter, R. (1935) Die Funktionentheorie der Differentialgleichungen $\Delta u = 0$ und $\Delta\Delta u = 0$ mit vier reellen Variablen. *Comment. Math. Helv.* 7: 307–330.

[44] Fueter, R. (1935-1936) Über die analytische Darstellung der regulären Funktionen einer Quaternionenvariablen. *Comment. Math. Helv.* 8: 371-378.

[45] Fueter, R. (1936-37) Die Singularitäten der eindeutigen regulären Funktionen einer Quaternionenvariablen. *Comment. Math. Helv.* 9: 320-334.

[46] Fueter, R. (1939) Über vierfachperiodische Funktionen. *Monatsh. Math. Phys.* 48: 161-169.

[47] Fueter, R. (1940) *Reguläre Funktionen einer Quaternionenvariablen.* Vorlesungsausarbeitung Math. Inst. Univ. Zürich.

[48] Fueter, R. (1945) Über die Quaternionenmultiplikation der vierfachperiodischen regulären Funktionen. *Experientia* 1: 57.

[49] Fueter, R. (1949) *Funktionentheorie im Hyperkomplexen.* Lecture notes written and supplemented by E. Bareiss, Math. Inst. Univ. Zürich, Herbstsemester 1948/49.

[50] Fueter, R. (1949) Über Abelsche Funktionen von zwei komplexen Variablen. *Ann. Mat. Pura Appl., Ser. IV,* 28: 211-215.

[51] Gritsenko, V. (1987) Arithmetic of quaternions and Eisenstein series. *J. Sov. Math.* 52: 3056-3063 (1990); translation from *Zap. Nauchn. Semin. Leningr. Otd. Mat. Inst. Steklova* 160: 82-90.

[52] Gürlebeck, K. (1984) *Über die optimale Interpolation verallgemeinert analytischer quaternionenwertiger Funktionen und ihre Anwendung zur näherungsweisen Lösung wichtiger räumlicher Randwertaufgaben der mathematischen Physik.* Dissertation TH Karl-Marx-Stadt.

[53] Gürlebeck, K.; Malonek, H. (1999) A hypercomplex derivative of monogenic functions in $\mathbb{R}^{n+1}$ and its applications. *Complex Variables* 39: 199-228.

[54] Gürlebeck, K.; Sprößig, W. (1990) *Quaternionic analysis and elliptic boundary value problems.* Birkhäuser: Basel.

[55] Gürlebeck, K.; Sprößig, W. (1997) *Quaternionic and Clifford calculus for physicists and engineers.* Mathematical Methods in Practice. Wiley: Chichester.

[56] Habetha, K. (1976) Eine Bemerkung zur Funktionentheorie in Algebren. In: Meister, E.; Weck, N.; Wendland, W. (eds.): *Function theoretic methods of partial differential equations.* Proc. Internat. Sympos. Darmstadt 1976. Lect. Notes Math. 561. Springer: Berlin etc.: 502-509.

[57] Habetha, K. (1986) Eine Definition des Kroneckerindex in $\mathbb{R}^{n+1}$ mit Hilfe der Cliffordanalysis. *Z. Anal. Anwend.* 5: 133-137.

[58] Haefeli, H.G. (1947) Hyperkomplexe Differentiale. *Comment. Math. Helv.* 20: 382-420.

[59] Hamilton, W.R. (1866) *Elements of Quaternions.* Longmans Green: London, reprinted by Chelsea: New York 1969.

[60] Harvey, F.R.: (1990) *Spinors and calibrations.* Perspectives in Mathematics, Boston, Academic Press, (1990).

[61] Hempfling, T. (1996) Aspects of modified Clifford analysis. In: Sprößig,W.; Gürlebeck, K. (eds): *Analytical and Numerical Methods in Quaternionic and Clifford Analysis.* Proc. Conf. Seiffen/Germany 1996. TU Freiberg : 49-59.

[62] Hempfling, T.; Kraußhar, R.S. (2003) Order theory for isolated points of monogenic functions. *Archiv d. Math.* 80: 406-423.

[63] Hestenes, D. (1968) Multivector calculus. *J. Math. Anal. Appl.* 24: 313-325.

[64] Hodge, W.V.D.; Pedoe D. (1952) *Methods of algebraic geometry.* Cambridge Univ. Press: Cambridge/UK.

[65] Holmann, H.; Rummler, H. (1972) *Alternierende Differentialformen.* B.I.-Wissenschaftsverlag: Mannheim etc.

[66] Hurwitz, A. (1922) Über die Komposition der quadratischen Formen. *Math. Ann.* 88: 1-25.

[67] John, F. (1955) *Plane waves and spherical means applied to partial differential equations.* Interscience Publ.: New York; Nachdruck, Springer: Berlin etc. 1981.

[68] Kähler, E. (1985) Die Poincaré-Gruppe. *Mathematica, Festschrift Ernst Mohr,* TU Berlin: Berlin, 117-144.

[69] Klotzek, B. (1971) *Geometrie* . Deutscher Verlag Wiss.: Berlin.

[70] Knott, C.G. (1911) *Life and scientific work of Peter Guthrie Tait.* Cambridge.

[71] Klein, F.; Fricke, R. (1890-1892) *Vorlesungen über die Theorie der elliptischen Modulfunktionen I,II.* Teubner: Leipzig.

[72] Kochendörffer, R. (1974) *Einführung in die Algebra.* Hochschulbücher Math. 18. Deutscher Verlag Wiss.: Berlin, 4. Aufl.

[73] Kraußhar, R.S. (2000) *Eisenstein series in Clifford analysis.* Dissertation RWTH Aachen. Aachener Beiträge zur Mathematik 28. Wissenschaftsverlag Mainz: Aachen.

[74] Kraußhar, R.S. (2001) Monogenic multiperiodic functions in Clifford analysis. *Complex Variables* 46: 337-368.

[75] Kraußhar, R.S. (2001) On a new type of Eisenstein series in Clifford analysis. *Z. Anal. Anwend.* 20: 1007-1029.

[76] Kraußhar, R.S. (2002) Automorphic forms in Clifford analysis. *Complex Variables* 47: 417-440.

[77] Kraußhar, R.S. (2002) Eisenstein series in complexified Clifford analysis. *Computat. Meth. Funct. Theory* 2: 29-65.

[78] Kraußhar, R.S. (2002) Monogenic modular forms in two and several real and complex vector variables. *Computat. Meth. Funct. Theory* 2: 299-318.

[79] Kraußhar, R.S. (2003) The multiplication of the Clifford-analytic Eisenstein series. *J. Number Theory* 102: 353-382.

[80] Kraußhar, R.S. (2004) A theory of modular forms in Clifford analysis, their applications and perspectives. In: Qian, T., et al. (eds.): *Advances in analysis and geometry. New developments using Clifford algebras*. Trends in Math., Birkhäuser: Basel: 311-343.

[81] Kraußhar, R.S. (2004) *Automorphic forms and functions in Clifford analysis and their applications*, Frontiers in Mathematics. Birkhäuser: Basel.

[82] Kraußhar, R.S.; Ryan, J. (2005) Clifford and harmonic analysis on cylinders and tori. *Rev. Mat. Iberoamericana* (to appear).

[83] Kraußhar, R.S.; Ryan, J. (2005) Some conformally flat spin manifolds, Dirac operators and automorphic forms. *Preprint*.

[84] Krieg, A. (1985) *Modular forms on half-spaces of quaternions*. Springer: Berlin-Heidelberg,.

[85] Krieg, A. (1988) Eisenstein series on real, complex and quaternionic half-spaces. *Pac. J. Math.* 133: 315-354.

[86] Krieg, A. (1988) Eisenstein-series on the four-dimensional hyperbolic space. *J. Number Theory* 30: 177-197.

[87] Kryloff, N.M. (1947) Sur les quaternions de W.R. Hamilton et la notion de la monogenéité. *Dokl. Akad. Nauk SSSR* 55: 787-788.

[88] Lawrentjew, M.A.; Schabat, B.W. (1967) *Methoden der komplexen Funktionentheorie*. Deutscher Verlag Wiss.: Berlin.

[89] Leray, J. (1959) Le calcule différentiel et intégral sur une variété analytique complexe (Problème de Cauchy III). *Bull. Soc. Math. France* 87: 81–180.

[90] Leutwiler, H. (2001) Quaternionic analysis in $\mathbb{R}^3$ versus its hyperbolic modification. In: Brackx, F. et al. (eds.): *Clifford Analysis and Its Applications*. Kluwer: Dordrecht, 193-211.

[91] Leutwiler H. (1996) Rudiments of a function theory in $\mathbb{R}^3$. *Exposit. Math.* 14: 97-123.

[92] Li, C.; McIntosh, A.; Qian, T. (1994) Clifford algebras, Fourier transforms and singular convolution operators on Lipschitz surfaces. *Rev. Mat. Iberoamericana* 10: 665-721.

[93] Lipschitz, R. (1886) *Untersuchungen über die Summe von Quadraten*. Max Cohen und Sohn: Bonn.

[94] Lounesto, P. (1995) Möbius transformations. Vahlen matrices and their factorization. In: Ryan J. (ed.). *Clifford algebras in analysis and related topics*. CRC Press: Boca Raton etc., 355-359.

[95] Lounesto, P. (2002) Introduction to Clifford algebras. In: Ablamovicz R.; Sobzyk, G. (eds.): *Lectures on Clifford geometric algebras*. TTU Press: Cookeville, TN/USA, 1-32.

[96] Lounesto, P.; Bergh, P. (1983) Axially symmetric vector fields and their complex potentials. *Complex Variables* 2: 139-150.

[97] Lugojan, S.(1991) Quaternionic derivability. *An. Univ. Timisoara, Seria Stiinte Mat.* 29, No. 2-3: 175-190.

[98] Lugojan S. (1992) Quaternionic derivability. *Semin. Geom. Sitopol/Univ. Timisoara* No. 105: 1-22, und in: G. Gentili et al. (eds.): *Quaternionic structures in mathematics and physics*. Proc. Meeting Trieste/Italy, Sept. 1994. SISSA: Trieste 1994.

[99] Maaß H. (1949) Automorphe Funktionen von mehreren Veränderlichen und Dirichletsche Reihen. *Abh. Math. Sem. Univ. Hamburg* 16: 72-100.

[100] Malonek, H.R. (1987) *Zum Holomorphiebegriff in höheren Dimensionen.* Habilitationsschrift. Pädagogische Hochschule Halle.

[101] Malonek, H.R. (1990) A new hypercomplex structure of the Euclidean space $\mathbb{R}^{m+1}$ and a concept of hypercomplex differentiability. *Complex Variables* 14: 25-33.

[102] Malonek, H.R. (1993) Hypercomplex differentiability and its applications. In: Brackx, F., et al. (eds.) *Clifford algebras and applications in mathematical physics.* Kluwer: Dordrecht, 141-150.

[103] Malonek, H.R.; Müller, B. (1992) Definition and properties of a hypercomplex singular integral operator. *Results Math.* 22: 713-724.

[104] Malonek, H.R.; Wirthgen, B. (1990) Zur Übertragung des Goursatschen Beweises des Cauchyschen Integralsatzes auf hyperkomplex differenzierbare Funktionen im $\mathbb{R}^{m+1}$. *Wiss. Z. Pädagogische Hochschule Halle-Köthen* 28: 34-38.

[105] Marinov, M.S. (1966) Meilikhson type theorems I. *An. Univ. Timisoara, Ser. Mat. Inform.* 34, No. 1: 95-110.

[106] Marinov, M.S. (1995) On the S-regular functions. *J. Natural Geom.* 7: 21-44.

[107] Mejlikhzhon, A.S. (1948) On the notion of monogeous quaternions (Russisch). *Dokl. Akad. Nauk SSSR* 59: 431-434.

[108] Michlin, S.G.; Prößdorf, S. (1986) *Singular integral operators.* Akademie-Verlag: Berlin.

[109] Mitrea, M.; Sabac, F. (1998) Pompeiu's integral representation formula. History and mathematics. *Revue Roumaine Math. Pures Appl.* 43: 211-226.

[110] Moisil, Gr.C. (1930) Sur les systèmes d'équations de M. Dirac, du type elliptique. *C. R. Acad. Sci. Paris* 191: 1292-1293.

[111] Nöbeling, G. (1978) *Integralsätze der Analysis.* De Gruyter Lehrbuch. Walter de Gruyter: Berlin etc.

[112] Norguet, F. (1959) Sur la théorie des résidues. *C. R. Acad. Sci. Paris* 248: 2057-2059.

[113] Peirce, B. (1881) Linear associative algebras. *Amer. J. Math.* 4: 97-215.

[114] Perwass, C.: (2000) Applications of Geometric Algebra in Computer Vision, Dissertation, Cambridge University

[115] Plemelj, J. (1908) Ein Ergänzungssatz zur Cauchyschen Integraldarstellung analytischer Funktionen. Randwerte betreffend. *Monatsh. Math. Phys.* 76: 205-210.

[116] Poincaré, H. (1954) *Oeuvres II.* Gauthier-Villars: Paris 1916/1954.

[117] Pompeiu, D. (1906) Sur la continuité des fonctions de deux variables complexes. *Ann. Fac. Sci. Toulouse* 7: 264-315.

[118] Pompeiu, D. (1909) Sur la représentation des fonctions analytiques par des intégrales définies. *C. R. Acad. Sci. Paris* 149: 1355-1357.

[119] Porteous I. (1969) *Topological geometry.* Van Nostrand-Reinhold: London.

[120] Qian, T. (1996) *Singular integrals on star-shaped Lipschitz surfaces in the quaternionic space.* Research and Technical Reports, The University of New-England, No. 120, 1-30.

[121] Qian, T. (1997) Generalization of Fueter's result to $\mathbb{R}^{n+1}$. *Atti Accad. Naz. Lincei, Rend., Cl. Fis. Mat. Nat., Ser. 8,* 9: 111-117.

[122] Remmert, R. (1991) *Funktionentheorie 1.* Springer Lehrbuch, Ser. Grundwissen Math., Springer: Berlin etc. 4. Aufl.

[123] Richter, O. (2002) Theta functions with harmonic coefficients over number fields. *J. Number Theory* 95: 101-121.

[124] Richter, O.; Skogman, H. (2004) Jacobi theta functions over number fields. *Monatsh. Math.* 141: 219-235.

[125] Ryan, J. (1982) Clifford analysis with generalized elliptic and quasi elliptic functions. *Applicable Anal.* 13: 151-171.

[126] Ryan, J. (1982) Complexified Clifford analysis. *Complex Variables* 1: 119-149.

[127] Saak, E.M. (1975) On the theory of multidimensional elliptic systems of first order. *Dokl. Akad. Nauk SSSR* 222: 47-51 = *Sov. Math., Dokl.* 16: 591-595.

[128] Sansone, G. (1959) *Orthogonal Functions*. Pure and Applied Mathematics 9. Interscience Publishers: New York.

[129] Sce, M. (1957) Osservazioni sulle serie di potenze nei moduli quadratici. *Atti. Accad. Naz. Lincei, Rend., Cl. Fis. Mat. Nat., Ser. 8*, 23: 220-225.

[130] Shapiro M.V. und N.I. Vasilevski, (1995) Quaternionic ψ-hyperholomorphic functions, singular integral operators and boundary value problems. I. ψ-hyperholomorphic function theory, *Complex Variables* 27: 17-46.

[131] Siegel, C.L. (1988) *Topics in complex function theory, Vol. III*. Wiley: New York-Chichester.

[132] Smirnov, W.I. (1967) *Lehrgang der Höheren Mathematik, Teil 5*. Deutscher Verlag Wiss.: Berlin.

[133] Snyder, H.H. (1982) An introduction to theories of regular functions on linear associative algebras. In: Draper, R.N. (ed.): *Commutative algebra: analytic methods*. Lect. Notes Pure Appl. Math. 68: 75-93.

[134] Sokhotski, J.W. (1873) *Über bestimmte Integrale und Funktionen. die für Reihenentwicklung benötigt werden* (Russisch). St. Petersburg.

[135] Sommen, F. (1981) A product and an exponential function in hypercomplex function theory. *Applicable Anal.* 12: 13-26.

[136] Sommen, F. (1982) Some connections between Clifford and complex analysis. *Complex Variables* 1: 97-118.

[137] Sommen, F. (1982) Spherical monogenic functions. *Tokyo J. Math.* 4: 427-456.

[138] Sommen, F. (1984) Monogenic differential forms and homology theory. *Proc. Royal Irish Acad., Sect A*, 84: 87-109.

[139] Sommen, F. (1988) Special functions in Clifford analysis and axial symmetry. *J. Math. Anal. Appl.* 130: 100-133.

[140] Sommen, F.; Souček, V. (1992) Monogenic differential forms. *Complex Variables* 19: 81-90.

[141] Sommen, F.; Xu Z. (1992) Fundamental solutions for operators which are polynomials in the Dirac operator. In: Micali, A.; Boudet, R.; Helmstetter, J. (eds.): *Clifford algebras and their applications in mathematical physics*. Kluwer: Dordrecht, 313-326.

[142] Souček, V. (1980) *Regularni funkce quaternionve promenne*. Thesis Charles University Prague.

[143] Souček, V. (1983) Quaternion valued differential forms in $\mathbb{R}^4$. *Suppl. Rend. Circ. Mat. Palermo, Ser. 2*, 33: 293-300.

[144] Sprößig, W. (1978) Analoga zu funktionentheoretischen Sätzen im $\mathbb{R}^n$. *Beiträge zur Analysis* 12: 113-126.

[145] Stein, E.M.; Weiss, G. (1968) Generalization of the Cauchy–Riemann equations and representations of the rotation group. *Amer. J. Math.* 90: 163-196.

[146] Stern, I. (1989) *Randwertaufgaben für verallgemeinerte Cauchy–Riemann-Systeme im Raum.* Dissertation A, Martin-Luther-Universität Halle-Wittenberg.

[147] Stern, I. (1991) Boundary value problems for generalized Cauchy–Riemann systems in the space. In: Kühnau R.; Tutschke, W. (eds.): *Boundary value and initial value problems in complex analysis.* Pitman Res. Notes Math. 256: 159-183.

[148] Stern, I. (1993) Direct methods for generalized Cauchy–Riemann systems in the space. *Complex Variables* 23: 73-100.

[149] Study, E. (1889) Über Systeme von complexen Zahlen. *Nachr. Ges. Wiss. Göttingen* 1889: 237-268.

[150] Stummel, F. (1967) Elliptische Differenzenoperatoren unter Dirichletrandbedingungen. *Math. Z.* 97: 169-211.

[151] Suchomlinov, G.A. (1938) On the extension of linear functionals in linear normed spaces and linear quaternionic spaces (Russisch). *Mat.Sbornik, Ser.* 2: 353-358.

[152] Sudbery, A. (1979) Quaternionic analysis. *Math. Proc. Cambridge Phil. Soc.* 85: 199-225.

[153] Tait, P.G. (1867) *An elementary treatise on quaternions.* Clarendon Press: Oxford; enlarged editions, Cambridge University Press: Cambridge 1873, 1890.

[154] Théodoresco, N. (1936) *La derivée aréolaire.* Ann. Roum. Math., Cahier 3, Bucharest.

[155] Titchmarsh, E.C. (1939) *The theory of functions.* Oxford University Press: London, 2nd ed.

[156] Vahlen, K.Th. (1902) Über Bewegungen und komplexe Zahlen. *Math. Ann.* 55: 585-593.

[157] van Lancker, P. (1996) *Clifford analysis on the unit sphere.* Thesis, University of Gent.

[158] Vekua, I.N. (1959) *Generalized analytic functions* (Russisch). Nauka: Moskau; Deutsch 1963, Deutscher Verlag Wiss.: Berlin; Englisch 1967, Wiley: New York.

[159] Whittaker, E.T.; Watson, G.N. (1958) *A course of modern analysis.* Cambridge University Press: London, 4th ed.

[160] Wloka, J. (1982) *Partielle Differentialgleichungen.* Teubner: Stuttgart.

[161] Zöll, G. (1987) *Ein Residuenkalkül in der Clifford-Analysis und die Möbiustransformationen der Euklidischen Räume.* Dissertation RWTH Aachen.

Index